Atmospheric Chemical Compounds

Sources, Occurrence, and Bioassay

ACADEMIC PRESS RAPID MANUSCRIPT REPRODUCTION

Atmospheric Chemical Compounds

Sources, Occurrence, and Bioassay

T. E. Graedel

AT&T Bell Laboratories
Murray Hill, New Jersey

Donald T. Hawkins

AT&T Bell Laboratories
Murray Hill, New Jersey

Larry D. Claxton

United States Environmental Protection Agency
Research Triangle Park, North Carolina

1986

ACADEMIC PRESS, INC

Harcourt Brace Jovanovich, Publishers

Orlando San Diego New York Austin
London Montreal Sydney Tokyo Toronto

The information described in this book has been reviewed by the Health Effects Research Laboratory, U.S. Environmental Protection Agency, and approved for publication. Approval does not signify that the contents necessarily reflect the views and policies of the Agency nor does mention of trade names or commercial products constitute endorsement or recommendation for use.

ACADEMIC PRESS, INC.
Orlando, Florida 32887

United Kingdom Euition published by
ACADEMIC PRESS INC. (LONDON) LTD.
24–28 Oval Road, London NW1 7DX

Library of Congress Cataloging in Publication Data

Graedel, T. E.
Atmospheric chemical compounds.

Bibliography: p.
Includes index.
1. Environmental chemistry–Technique.
2. Atmospheric chemistry–Environmental aspects.
I. Hawkins, Donald T. II. Claxton, Larry D.
III. Title.
TD193.G73 1986 628.5'3 86-70505
ISBN 0–12–294485–2 (alk. paper)

PRINTED IN THE UNITED STATES OF AMERICA

86 87 88 89 9 8 7 6 5 4 3 2 1

Contents

Preface

Atmospheric compounds are numerous and chemically diverse, and information concerning them is scattered very widely throughout the scientific literature. This dissemination of information reflects the many scientific disciplines involved in measurements of atmospheric compounds, determination of their sources, or assessments of their effects on the earth's flora and fauna, materials, radiation budget, etc. As studies of these matters become increasingly detailed and chemically sophisticated, a compendium of the scattered information becomes increasingly valuable. In addition, since specialists from different fields find need for the specific information, they often need at least an introduction to related specialties as well. The present work is an effort to fill many of these needs.

This book bears some similarities to T. E. Graedel's *Chemical Compounds in the Atmosphere* (Academic Press, 1978). There are a number of important differences which should be noted, however. The first is that the earlier work was restricted to compounds present in the troposphere as gases or as aerosol particle constituents. The diligent research of many scientists since 1978 has permitted us to broaden the scope considerably in this book: we include listings of compounds in clouds, fog, rain, snow, and ice, a listing of compounds detected in the stratosphere, and a compendium of compounds present in indoor air. (The latter regime specifically excludes air within industrial buildings, however, as being related more to manufacturing processes than to more general emission, transport, and occurrence considerations.) It is beyond the scope of this book to use the source and occurrence data to deduce atmospheric budgets of elements or compounds, but the information provided here can be used to form the basis for such studies.

A second major innovation in this book is the listing of bioassay information for the compounds. The atmospheric and biological communities are becoming increasingly aware of the interrelationships between the two fields; we hope this compendium will contribute to this closer contact.

Third, and partly because we expect the data in this book may be useful to scientists from many different specialities, we have tried within space limitations to provide some perspective on our topics rather than to content ourselves with tables and figures. We thus discuss the structure and properties of the atmosphere, review the sizes and occurrence of clouds and other forms of atmospheric water, survey the indoor environment and its interaction with outdoor air, and present an introduction to carcinogenicity and the bioassay of atmospheric compounds.

Fourth, we have provided extensive cross-indexing, since the total number of species listed herein exceeds 2800. Compounds may be located by their chemical type, their name, their Chemical Abstracts Service Registry Number, or their sources. In presenting these cross-indexes, we intend that no one seeking data on a compound included in this book should be unable to find it. Having found it, he or she can then conveniently update the data in this book or obtain more detailed information by using the Registry Number as the key to computerized literature searches.

An index to the authors cited in this book is also included.

The uses mentioned above are primarily related to individual compounds or to individual sources. The totality of the information, however, has relevance to more comprehensive atmospheric studies. In particular, the occurrence of large numbers of chemically-similar compounds, prolific sources of compounds, the implications of the tabular data for the chemistry of different atmospheric regimes, interactions with surfaces, and biological impact implications can be addressed by statistical and schematic analyses of the data. These topics are discussed in the final chapter.

Many terms used in this book are common within a specialty but uncommon outside it. A meteorologist is likely to be familiar with the term *hydrometeor*, for example, but the chemist or epidemiologist is not. Conversely, the meteorologist may not know exactly what is meant by the terms *arene* or *mutagen*. Our approach is to define such terms the first time they arise, either within the text or in a footnote, and to include in the index a reference to that definition.

Any compendium of information must deal with a decision to terminate the data acquisition and proceed with presentation and publication. For this volume, the detailed literature search extended through September,

1984 (*Chemical Abstracts* Vol. 101, No. 14, and supplementary material). A few important studies appearing from September, 1984 through March, 1985 were also included. The bioassay information includes primary data reviewed by the U. S. Environmental Protection Agency Gene-Tox Program through 1982.

Our basic approach is taxonomic rather than pedagogical. As a result, some readers may find the need for additional information on topics covered here only in passing. Among the many useful chemistry references available are the treatise on photochemistry by Calvert and Pitts[1213] and those on chemical kinetics by Benson.[1214,1215] Heicklen[1216] describes the interaction of light with molecules in the atmosphere and presents more complete discussions of many topics in atmospheric chemistry than are appropriate in this work. An overview of air quality and its analysis and control is provided by Seinfeld.[1217] Graedel and Weschler[1323] have reviewed information on chemical reactions in atmospheric water droplets. Thorough reviews of atmospheric gas phase kinetics and mechanisms have been published by Atkinson and coworkers.[1324,1325] In a field evolving as rapidly as is atmospheric chemistry, frequent reference to the current literature is also of great value. Introductory reviews of genetic toxicology can be found in the texts by Brusick,[B51] Frei and Brinkman,[B56] McElheny and Abrahamson,[B52] Hsie et al.,[B53] and Heddle.[B57]

We are grateful to many people for their encouragement and assistance in connection with this book. Preprints and useful discussions came from many of our colleagues near and far, the contributions of R. Atkinson, D. Schuetzle, and B. Simoneit being particularly helpful. T. Hughes, T. E. Kleindienst, P. B. Shepson, C. J. Weschler, and L. J. Zaragosa reviewed the manuscript and made many useful comments. We would like to thank Dr. Gordon Hueter, Dr. Michael Waters, and Dr. Joellen Lewtas who encouraged and supported the inclusion of bioassay data in this volume, Ms. Carol Evans, who helped to develop and manage the U.S. EPA software that was used to collect, store, and manipulate these data, and Ms. Mary Beth Miller, who entered most of the data into the system. The text and tables were processed and phototypeset at AT&T Bell Laboratories by Patty McCrea, Wendy Ross, Lisa Sparrow and Evelyn Wilson, whose ability and cheerfulness in the face of such a difficult redactive challenge deserves much praise. Most of the Registry Numbers were obtained through the use of online databases derived from Chemical Abstracts Service's Registry Nomenclature and Structure Service. The indexes were prepared using the facilities of the UNIX* operating system.

* UNIX is a trademark of AT&T Bell Laboratories.

On a more personal level, each of us wishes to thank people close to us for their assistance and tolerance during the completion of this project. T.E.G. is grateful for the support of his daughters Laura and Martha and his wife Susannah (who helped proofread every table in the book!). D.T.H. thanks his wife Patricia and son Michael for their support. L.D.C. thanks his wife Betty and his children Meredith and Matthew for their understanding attitude during the many extra hours spent in the preparation of this book.

T. E. Graedel
D. T. Hawkins
L. D. Claxton
February 28, 1986
Murray Hill, NJ
Research Triangle Park, NC

CHAPTER 1

Introduction

1.0 The Purpose and Plan of This Book

The primary purpose of this book is to serve as a reference to information about the chemical compounds found in the earth's atmosphere. This information comprises thirteen chapters, grouped by the chemical structures and properties of the compounds. The data are presented in a series of tables, organized in a consistent manner within and between chapters.

A secondary purpose of this book is to provide a perspective on these extensive data. The effort is inaugurated in this introductory chapter, in which we give an overview of the properties of the earth's atmosphere, the indoor atmospheric environment, and of genetic toxicology. Following these introductory overviews, the first chapter concludes with detailed discussions of the tables and of the chapter texts in the remainder of the book. These discussions explain the format, the ordering of the atmospheric species, and the selection and entry of data included in the tables, as well as the contents of the text which accompanies each chapter.

1.1 The Structure and Properties of the Earth's Atmosphere

1.1.1 Introduction

In the laboratory, in the chemical manufacturing industry, and in many other situations, chemical reactions occur under relatively uniform conditions of temperature, pressure, and irradiation. In the atmosphere these conditions vary substantially, producing noticeable effects on composition and chemical reactions. Before discussing atmospheric chemistry, therefore, it is worthwhile to present some information on the gaseous medium within which this chemistry occurs.

The atmosphere is densest at the earth's surface. The density decreases rapidly with increasing altitude, as shown (in pressure units) in Figure 1.1-1. Since many chemical reactions are pressure-dependent, the density structure alone implies differences in composition. Most reactions are temperature-dependent as well and thus reflect to some extent the atmospheric temperature structure. As seen in the figure, that structure is quite complex. Its inflection points are used by scientists to divide the atmosphere into different regions for study and reference. Beginning at the earth's surface, these regions are designated the *troposphere*, the *stratosphere*, the *mesosphere*, and the *thermosphere*. These divisions are particularly useful since mass transport across the inflection points is inhibited and the regions are therefore relatively isolated from one another. In this book we restrict our discussion to the troposphere, which is influenced in a direct way by anthropogenic and natural emissions at the earth's surface, and to the stratosphere, which is influenced in significant but often indirect ways by the same sources. The higher levels of the atmosphere are relatively independent of ground level sources, and are not addressed in the present work.

Reactions involving solar photons are central to atmospheric chemistry. One therefore needs to know the spectrum of solar radiation in some detail. As shown in Figure 1.1-2, it is approximately that of a blackbody at 5900°K, crossed by numerous absorption lines. Several gases modify the solar spectrum as the radiation penetrates further into the atmosphere. O_3, O_2, H_2O, and CO_2 are the most important modifiers, but contributions are also made by CH_4, N_2O, and other natural and anthropogenic gases.

Of special importance to atmospheric chemistry are the photons energetic enough to dissociate photosensitive molecules; for most molecules such photons have wavelengths shorter than about 400 nm. In Figure 1.1-3, we show the altitude dependence of the solar flux at two wavelengths: 300 nm, the shortest wavelength radiation that reaches the

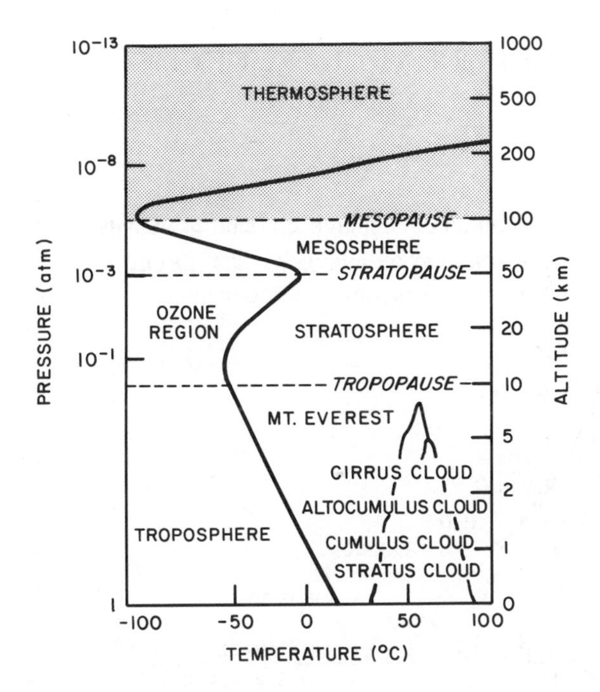

Fig. 1.1-1 The variation of atmospheric density and temperature with altitude above the earth's surface. (Reproduced with permission.[1277])

earth's surface, and 400 nm, the longest wavelength radiation generally active in atmospheric photochemistry. The fluxes increase significantly as one moves upward from the ground toward the tropopause (the troposphere-stratosphere boundary). Similarly dramatic variations are seen at constant altitudes as one moves to different latitudes or as the seasons change. Figure 1.1-4 illustrates these variations. Any reactions dependent upon solar photons obviously vary in proportion to these changes.

Another important facet of atmospheric chemistry not always present in other chemical systems is atmospheric mixing. On a small scale, these motions are turbulent and serve to disperse emissions from point sources. On a larger scale, the motions are ordered and have the ability to transport emitted species far downwind and to high altitudes. The large scale mixing is not uniform over the earth, as shown in Figure 1.1-5. Transport along a fixed latitude band is such that emittants can circumnavigate the earth in about two weeks, as shown by volcanic

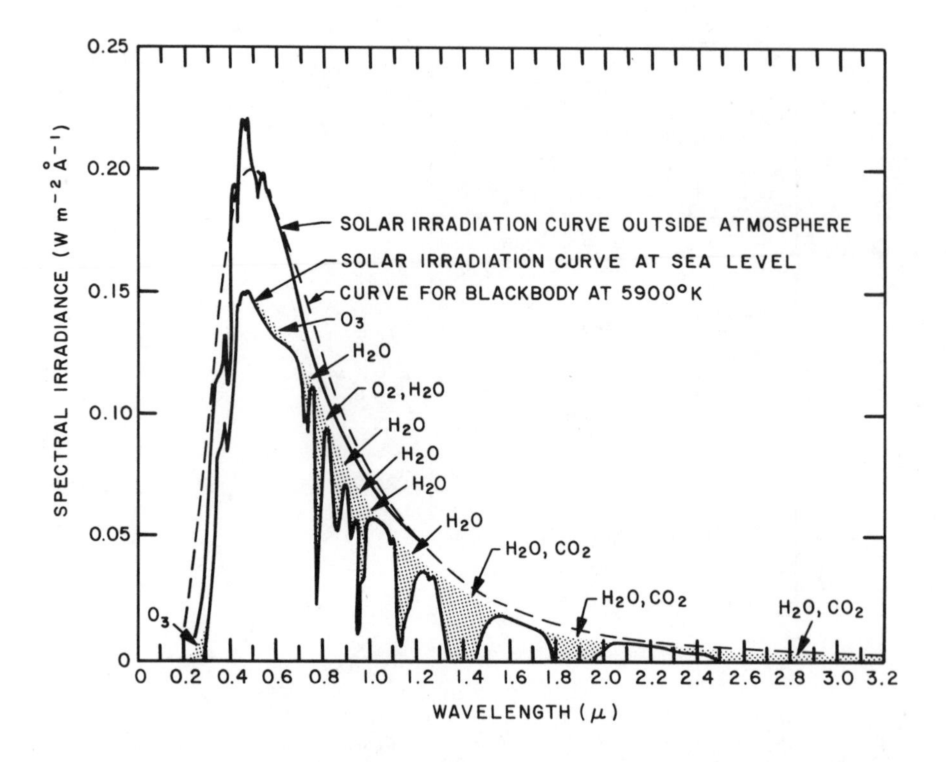

Fig. 1.1-2 Spectral distribution curves related to the sun; shaded areas indicate absorption, at sea level, due to the atmospheric constituents shown. (Reproduced with permission.[1315])

eruption clouds. In the lower troposphere, typical advection velocities are of order 25 km/hr, but variations due to local weather patterns are substantial. Within a single hemisphere, north-south transport occurs over a time scale of a few weeks. Transport across the equator is strongly inhibited by the convergence of opposed air flows, and requires six months to a year.

Vertical mixing is a strong function of local weather patterns also. A simplified conceptual approach often taken to describe this mixing is to assume that the flux of a species in a chosen direction is proportional to its mean gradient. This "eddy flux," due to the random motions of the atmosphere, is given by

$$\phi = -K[dC/dz]$$

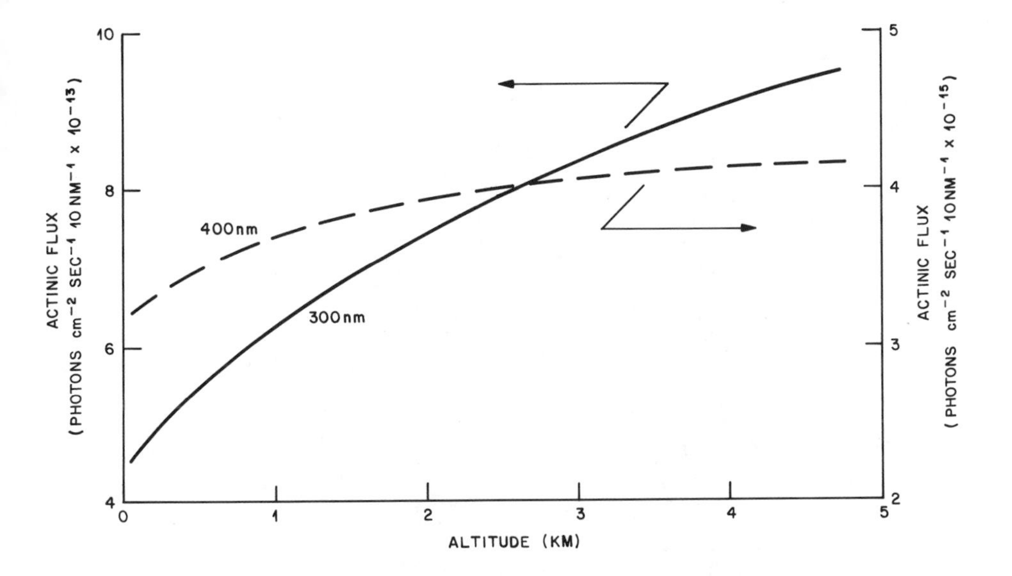

Fig. 1.1-3 The altitude dependence of the solar photon flux (the "actinic flux") in the troposphere over 10 nm intervals centered at 300 nm and 400 nm. The data are from Peterson.[1204]

where C is the concentration, z is the altitude, and K is the eddy diffusion coefficient. In Figure 1.1-6, averaged eddy coefficients are presented. The coefficient is approximately constant throughout the troposphere, indicating relatively uniform mixing. The eddy coefficients at and above the tropopause are quite low, then increase as mixing again becomes reasonably efficient in the middle and upper stratosphere.

The residence times of atmospheric trace species are closely related to the mixing reflected by these eddy coefficients. Some months are required to vertically mix species emitted at the ground throughout the troposphere. Transport across the tropopause is quite slow, since mixing is inhibited by the atmospheric temperature structure. Several years are required for an average chemically inert molecule to move from the ground to the lower stratosphere.

Our interest in this book is primarily with molecules that are reactive rather than chemically inert. It is thus of interest to see how chemical lifetimes affect the scale of influence. In Figure 1.1-7, we plot that relationship using the lifetime range imposed on atmospheric species by their reactivity and the concentrations of species with which they react. Clearly, some species have only a local influence, while those which react

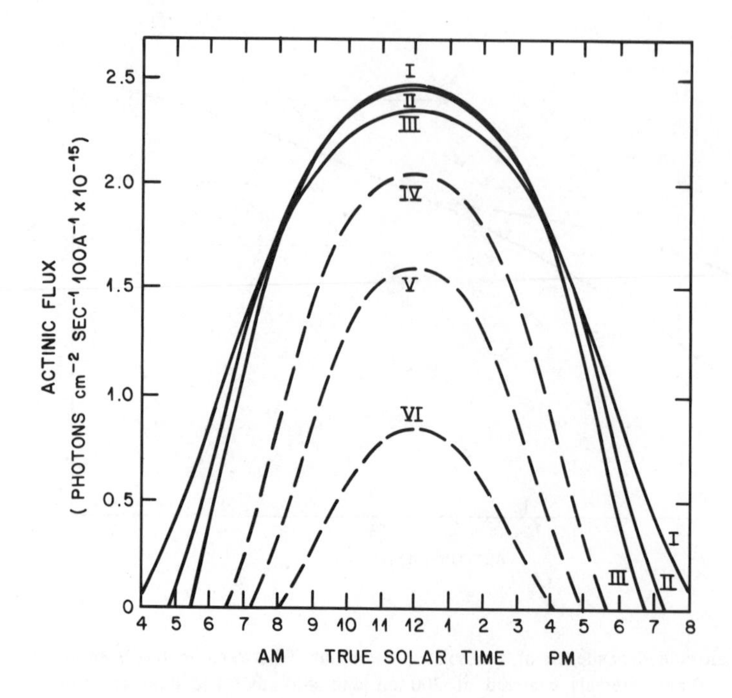

Fig. 1.1-4 Diurnal variations in actinic flux. Values illustrated are for the 100-A interval of the solar spectrum centered at 3700 A. The designations of the individual curves are: I, 20° N lat, summer solstice; II, 35° N lat, summer solstice; III, 50° N lat, summer solstice; IV, 20° N lat, winter solstice; V, 35° N lat, winter solstice; and VI, 50° N lat, winter solstice. (Reproduced with permission.[1205])

more slowly affect regional and global atmospheric composition. The same sort of lifetime dependence is shown in the vertical dimension in Figure 1.1-8. The vertical concentration profile of CF_2Cl_2, one of the chlorofluoromethane propellants, is representative of a compound with no appreciable loss mechanisms in the troposphere; its profile is essentially a reflection of atmospheric mixing. $CH_2{=}CH_2$ is seen only near its source, since it is highly reactive. The less reactive C_3H_8 represents the intermediate case.

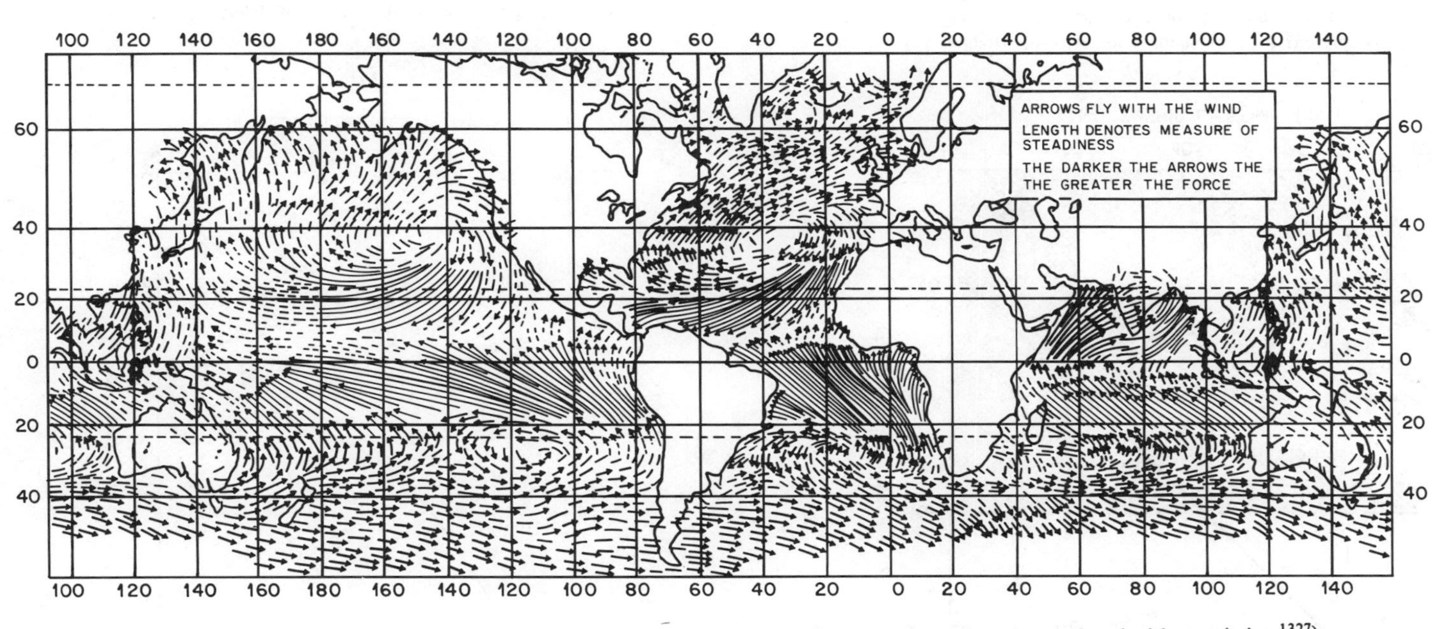

Fig. 1.1-5 Average oceanic wind flow for July and August. (Attributed to Köppen, reproduced with permission.[1327])

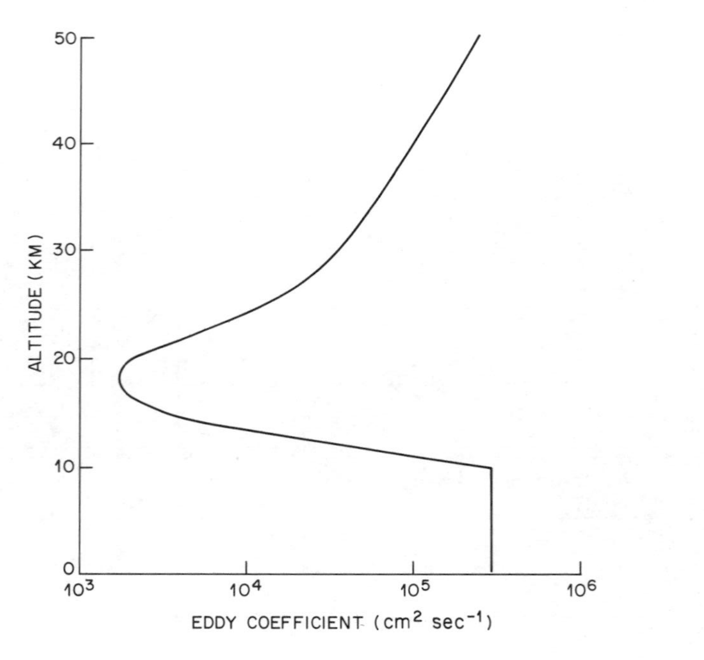

Fig. 1.1-6 Vertical eddy diffusion coefficients for the troposphere and stratosphere. The data are from Wofsy and McElroy.[1220]

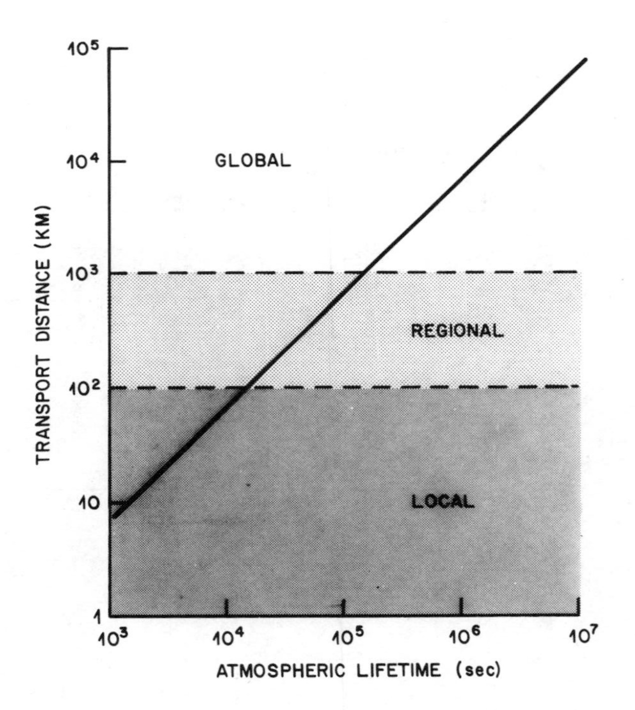

Fig. 1.1-7 Transport distance for a molecule as a function of its average atmospheric lifetime. A fixed horizontal transport velocity of 25 km/h is assumed. The limit to the lifetime for most species is the rate of reaction with either OH· or O_3; the majority of atmospheric species have lifetimes of a few hours ($\sim 10^4$ sec) to a few days ($\sim 10^6$ sec). The regimes are defined as local (<100 km), regional (100-1000 km), and global (>1000 km).

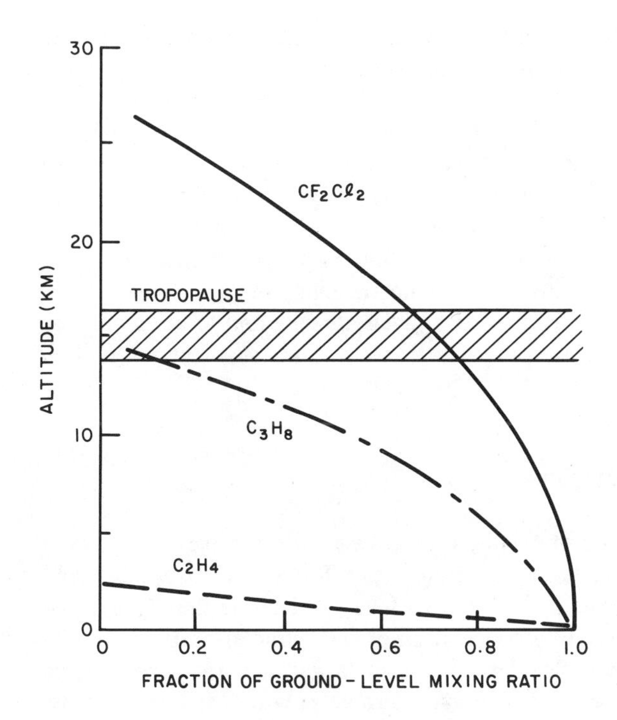

Fig. 1.1-8 Vertical distribution of three atmospheric trace gases with very different lifetimes: CF_2Cl_2,[1206] C_3H_8,[712] and $CH_2{=}CH_2$.[1207]

1.1.2 The Troposphere

1.1.2.1 Gaseous Composition

More than 99.9% of the molecules comprising the Earth's atmosphere are nitrogen (N_2), oxygen (O_2), or one of the rare gases. In addition, many trace molecules are found in the atmosphere. They are capable of influencing or controlling certain atmospheric processes, even though their total concentration is very low. Carbon dioxide, which is an important factor in the Earth's radiation balance but is chemically unreactive in the troposphere, has an average (but increasing) concentration of about 330 parts per million (ppm). The most abundant of the reactive gases is methane, which comprises less than two parts per million of the tropospheric gas. Other reactive species are still less prevalent; the combined concentration of all of the reactive trace gases in the atmosphere seldom totals 10 ppm.

A most important molecule for atmospheric chemistry, especially droplet chemistry, is water vapor. The mixing ratio of water vapor in the troposphere varies by some five orders of magnitude, from a few parts per hundred in the tropics near the surface to less than one part per thousand over the poles at the surface and to a few parts per million near the tropopause. As a consequence of the mixing of the atmosphere, instantaneous water concentrations at a given altitude and location vary by about a factor of ten depending on the instantaneous vertical motion patterns.[1201]

1.1.2.2 Aerosol Particles

Aerosol particles in the atmosphere are chemically complex and have a broad range of sizes and transport properties. The size distributions tend to be bimodal, as shown in Figure 1.1-9. The particles at the smaller end of the spectra are produced by coagulation of even smaller particles and by conversion of gas phase molecules. Those of larger size are generated by mechanical rather than chemical means: bubble bursting, soil erosion, etc. Representative estimates of global particle production from various major natural and anthropogenic sources are given in Table 1.1-1. In many cases data are scarce or missing.

Table 1.1-1. Estimates of Global Particle Production (10^6 tons/yr)*

Anthropogenic	
Direct sources	50
Formed from gases	250
Natural	
Sea salt	1000
Windblown dust	100
Forest fire	100
Formed from gases	1300
Grand total	2800

* Adapted from Bach.[1202]

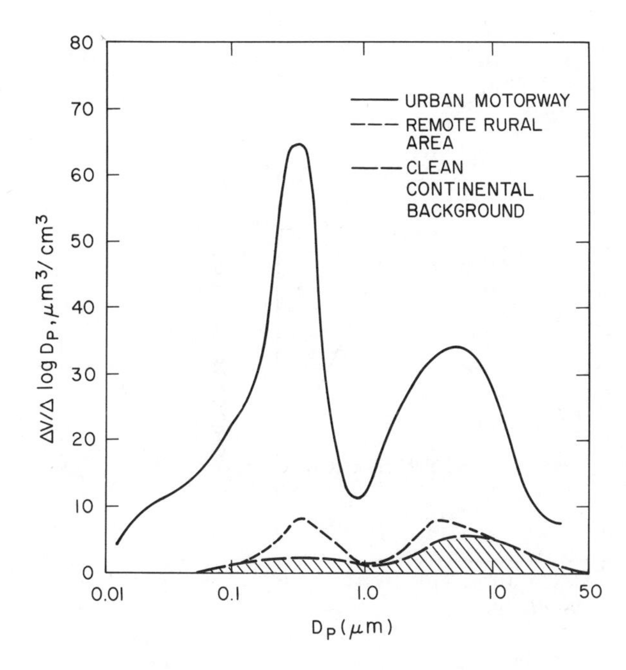

Fig. 1.1-9 Atmospheric aerosol size spectra in different atmospheric regimes. V is the total aerosol volume, D_p the aerosol particle diameter. (Adapted from Whitby.[1200])

The physical processes limiting the lifetimes of aerosol particles are directly related to particle size. If they are very small, the lifetime is limited by coagulation. If they are very large, it is limited by gravitational settling. At intermediate sizes (a few tenths of a micrometer in diameter), the lifetime is much longer, perhaps a hundred days or so. Figure 1.1-10 illustrates this dependence.

It is of interest to compare the mass of atmospheric aerosol particles with that of the reactive trace gases. The aerosol mass is rarely lower than 10 $\mu g/m^3$ near the earth's surface and rarely higher than 100 $\mu g/m^3$. The trace gases fluctuate between about 2 and 10 ppm. If we assume an average molecular weight of 30 for the gases, this gives a mass range for the trace gases of about 10^3 to 10^4 $\mu g/m^3$. On average, therefore, the mass of the reactive gaseous trace constituents in the atmosphere exceeds that of the aerosol particles by a factor of about 100.

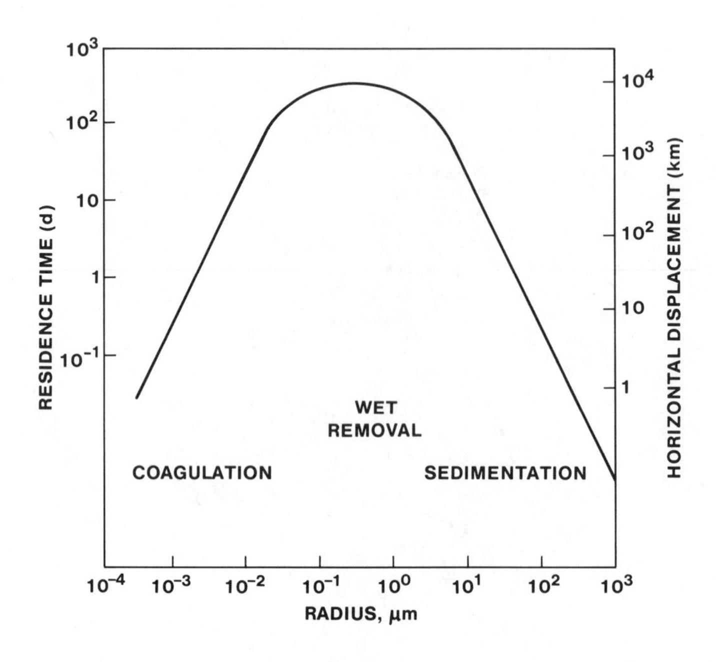

Fig. 1.1-10 The lifetimes (residence times) and horizontal displacement of atmospheric aerosol particles as a function of particle radius. The principal mechanisms responsible for particle removal in each size range are indicated. The displacement calculation assumes a horizontal transport velocity of 8 m/s and a vertical uplift velocity of 2 cm/s. (Adapted from Jaenicke.[1209])

1.1.2.3 Clouds and Precipitation

As with aerosol particles, atmospheric droplets cover a very wide range of sizes. Typical size spectra for clouds, fog, rain, and snow are illustrated in Figure 1.1-11. Each has a spread of more than an order of magnitude in size; the total size variation of these hydrometeors* is four orders of magnitude. Fog droplets are very numerous but very small, with liquid water content generally less than 0.1 gm^{-3}. Cloud droplets are fewer but larger; clouds have liquid water contents of 0.5-1.0 gm^{-3}. Raindrops and snowflakes are much larger than cloud and fog droplets.

* *Hydrometeor*: A body of solid or liquid water falling through or suspended in the air.

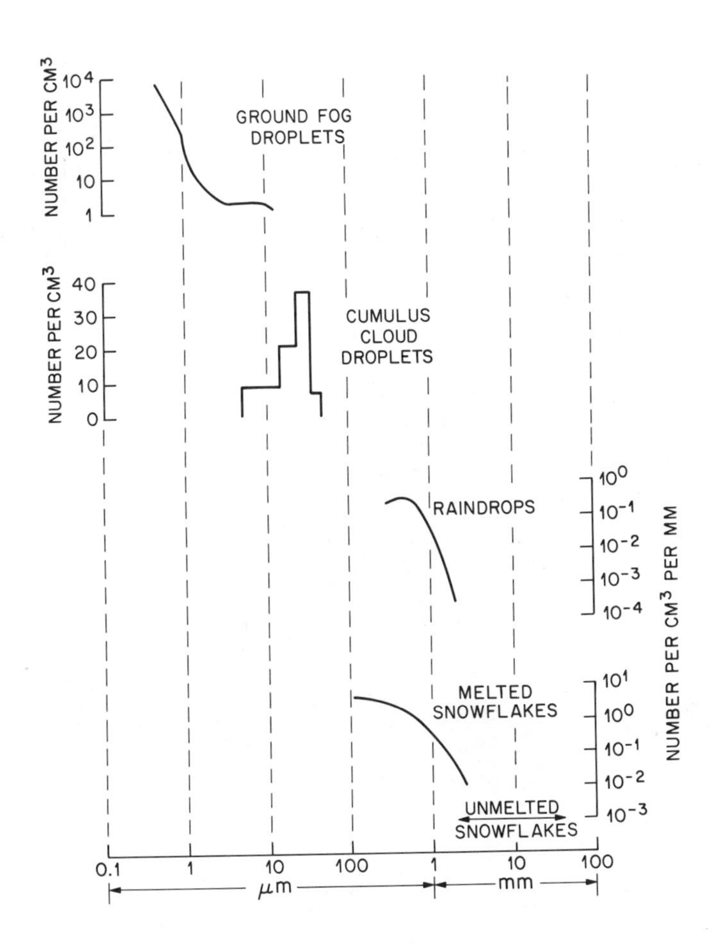

Fig. 1.1-11 Size distributions for aqueous droplets in the atmosphere. Distributions vary widely under different conditions and those presented here are intended merely to be illustrative of typical observations. Note that the ordinates are logarithmic, except for cloud droplets. The figure was constructed from data in Pruppacher and Klett.[1203]

The lifetime ranges of the hydrometeors are quite variable: from a few minutes for raindrops to a few hours for cloud droplets. If we recall that aerosol particles have lifetimes as long as a few days[1210] and are often coated with water films, the aqueous atmospheric constituents cover some seven orders of magnitude in size and four orders of magnitude in lifetime. It would be very surprising indeed if their chemical makeup and processes were not strikingly different.

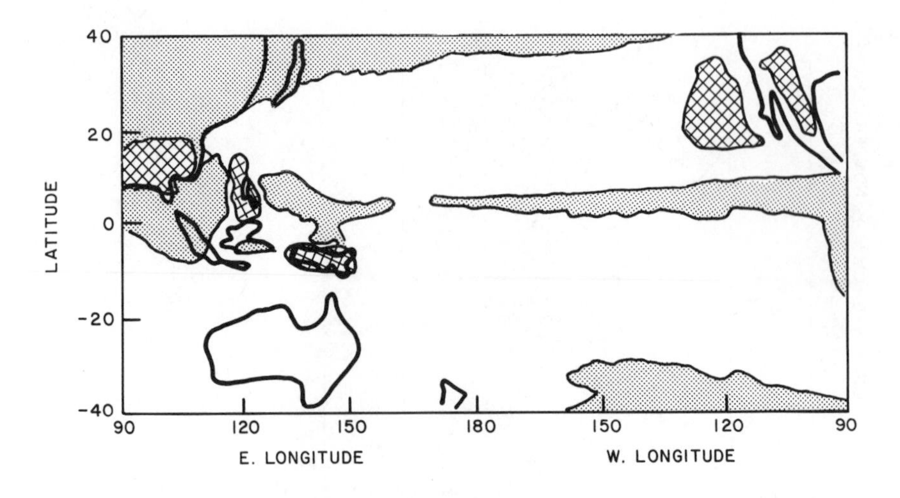

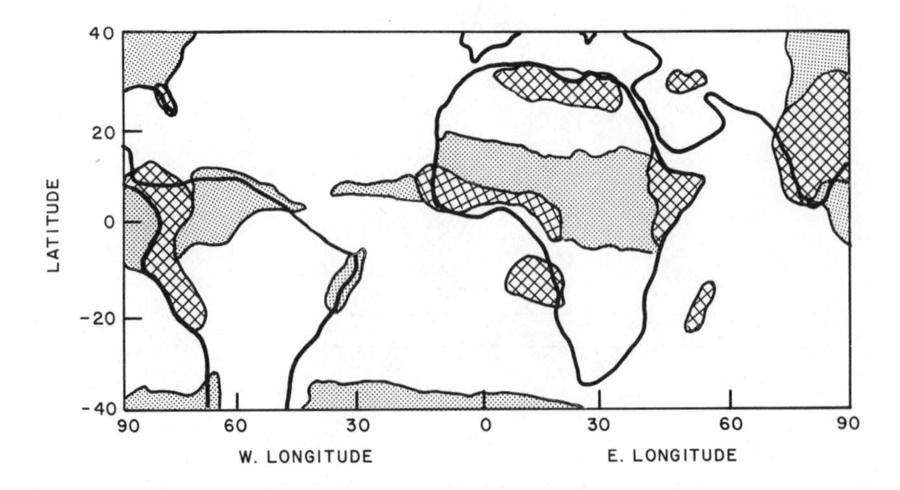

Fig. 1.1-12 The global distribution of the occurrence of persistent (hatching) and moderately persistent (stippling) cloud cover. This figure is adapted from illustrations of average reflectivity determined from satellites for the month of July, 1967 to 1970.[1313]

The impact of water droplets on atmospheric chemistry depends not only on droplet sizes and lifetimes, but on the frequency with which droplets are present. Measures of this are the occurrence of clouds (Figure 1.1-12) and fog (Figure 1.1-13), which show dramatic geographical variation. Another is the volume and variability of precipitation. As seen in Figure 1.1-14, the annual precipitation at different locations on the globe

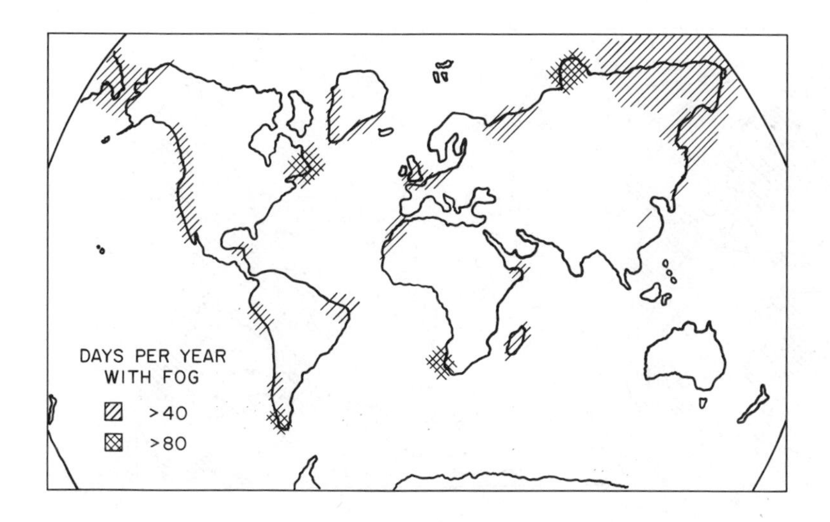

Fig. 1.1-13 The global distribution of the occurrence of frequent fog. In polar regions fog predominates in summer because of the constant cooling of the air near the ground as it is warmed by radiation or transported from lower latitudes over ice covered land or cold seas. In the tropics fog happens throughout the year when air humidity is high because of the relative long duration of nocturnal cooling or constant cooling near the sea surface in areas of cold upwelling water. (Adapted from Rudloff.[1319])

varies from almost nothing to more than 300 cm y^{-1}. Different geographical areas also show wide fluctuations about the mean rate. A measure of these fluctuations can be obtained by expressing the average precipitation amount P of the wettest month as a percentage of the total annual precipitation:[1218]

$$C = \frac{100\ P_{max}}{\sum P}$$

The results are given in Figure 1.1-15. One would anticipate that the chemical composition of frequent light rainfalls would be quite different from that of sporadic heavy ones.[1219] Since a large fraction of the trace species in the atmosphere is returned to the surface by precipitation, the deposition rate of trace substances varies widely over the globe.

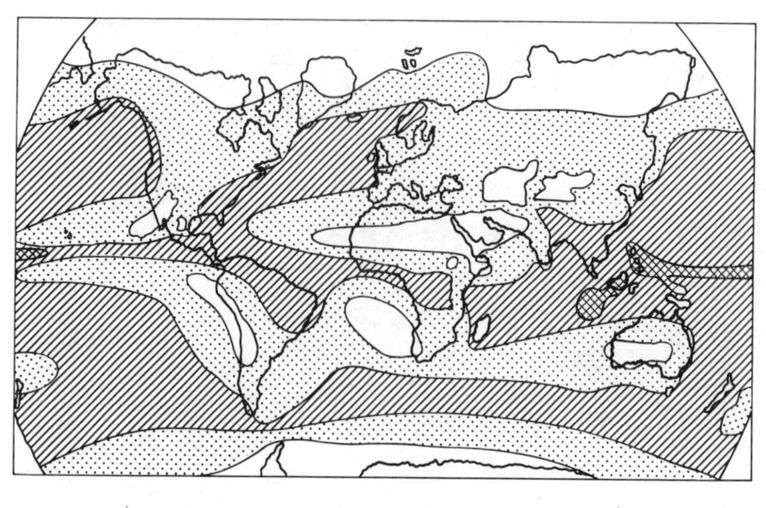

Fig. 1.1-14 The global distribution of the annual depth of rainfall in centimeters. (Adapted from Rudloff.[1319])

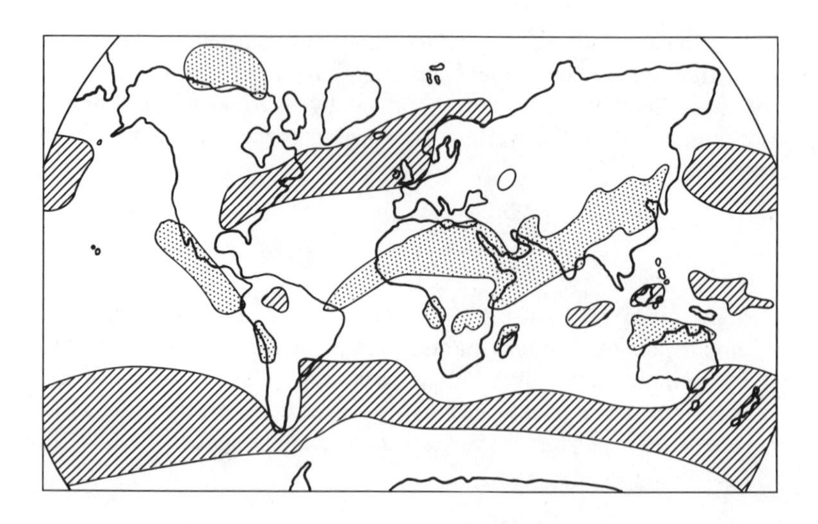

Fig. 1.1-15 Areas of strong (C > 25%, stippling) and weak (C < 12.5%, hatching) fluctuations in precipitation amount. (Adapted from Péczeley.[1218])

1.2 The Indoor Atmosphere

Most people spend most of their time indoors. As a result, the constituents in indoor air, their concentrations, and their bioassay are of considerable interest. After decades of neglect except in the industrial workplace, measurements of the chemical characteristics of indoor air are being made with increasing frequency. The results of these measurements are included within the tables of this book.

The occurrence and concentrations of chemical compounds in indoor air are largely controlled by three factors: whether the sources of the compounds are indoor or outdoor sources, the diurnal emission patterns of those sources, and the degree of air exchange between indoors and outdoors. To demonstrate how differences in these factors may dominate indoor air quality, we reproduce two data plots. The first, in Figure 1.2-1,

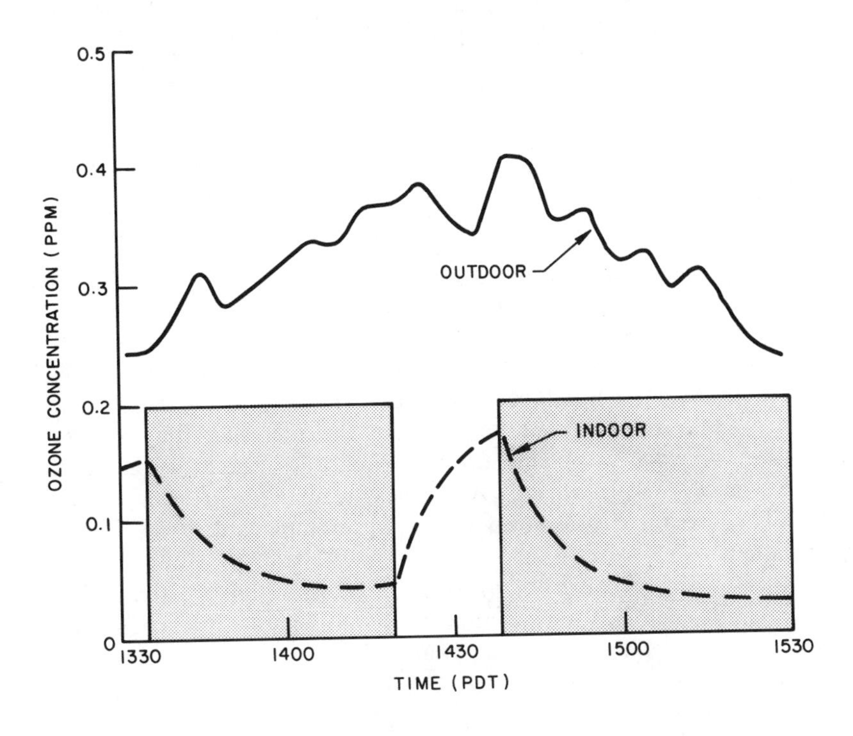

Fig. 1.2-1 Relative indoor and outdoor concentrations of ozone in the Spalding Laboratory, California Institute of Technology, July 22, 1975. The shaded areas indicate times when the building's charcoal filtering system for incoming air was in operation. (Adapted from Committee on Indoor Pollutants.[1290])

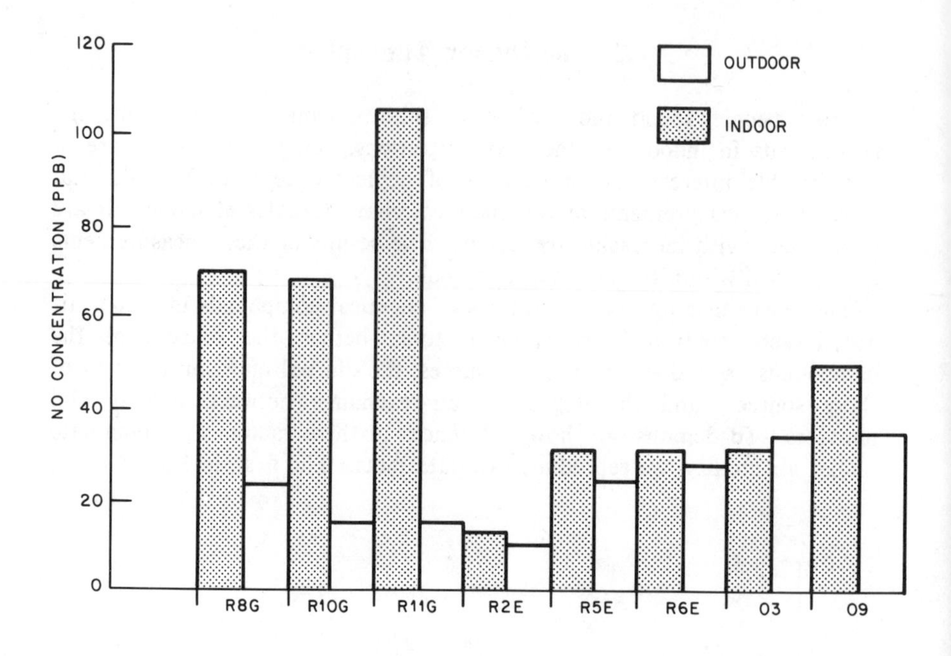

Fig. 1.2-2 Indoor and outdoor nitric oxide concentrations: means of hourly concentrations for buildings in the Boston metropolitan area. R = residence, O = office, G = gas heat, E = electric heat. The office heating systems are diverse, but are well isolated from the points of measurement. (Adapted from Moschandreas et al.[1288])

is for ozone. In the case shown, ozone was present within the building because of its presence in outdoor air (as a result of smog chemistry) and its injection with that air into the building by the ventilating system. Figure 1.2-2 provides an example for nitric oxide, which is generated within residences by gas stoves. As can be seen, air in residences and office buildings without gas stoves differs little from outdoor air in NO concentrations.

The rates of air exchange for different buildings differ greatly. Office and commercial buildings with ventilating systems normally exchange air within a room two or three times an hour. The "fresh" air is a composite of outdoor air and recirculated, filtered, indoor air. In the case of most residences, however, the exchange of air occurs by infiltration, a much slower process. Figure 1.2-3 shows air exchange rates for a sample of houses in North Dakota. The mean is 0.8 exchange per hour, with individual differences of as much as a factor of seven. Substantial air quality differences can thus result from differences in air infiltration rates.

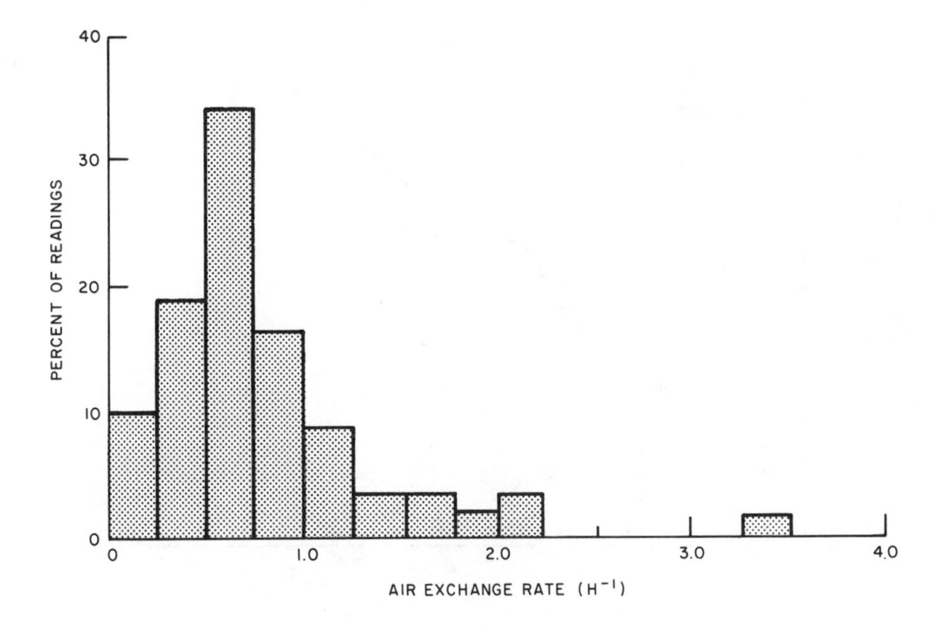

Fig. 1.2-3 A histogram of measured natural air infiltration rates for seventeen houses in Fargo, North Dakota. (Adapted from Grot and Clark.[1289])

In general, air exchange rates are sufficiently low so that indoor air quality related to organic species is determined primarily by the types and magnitudes of indoor sources. Most water-soluble inorganic species have few indoor sources, however, and thus tend to originate outdoors.

Another significant difference between the indoor and outdoor environments is the photon spectrum. The outdoor spectrum is determined by the sun and was shown in Fig. 1.1-2. Indoors, three possible sources of light may be present. One is solar radiation that enters through windows. The spectrum of this radiation has almost no flux in the photochemically active ultraviolet region ($\lambda < 380$ nm) because of absorption and reflection by the window glass.[1314] A second source is incandescent lamps, which are strong in the yellow portion of the spectrum but quite weak in the ultraviolet. The third source is fluorescent lighting, which has both line and continuum components and is reasonably strong in the ultraviolet. Typical spectra for the latter two sources are shown in Fig. 1.2-4.

A final difference between the indoor and outdoor environments is the much higher surface-to-volume ratio indoors, which increases the probability for deposition of indoor gases and particles to surfaces.

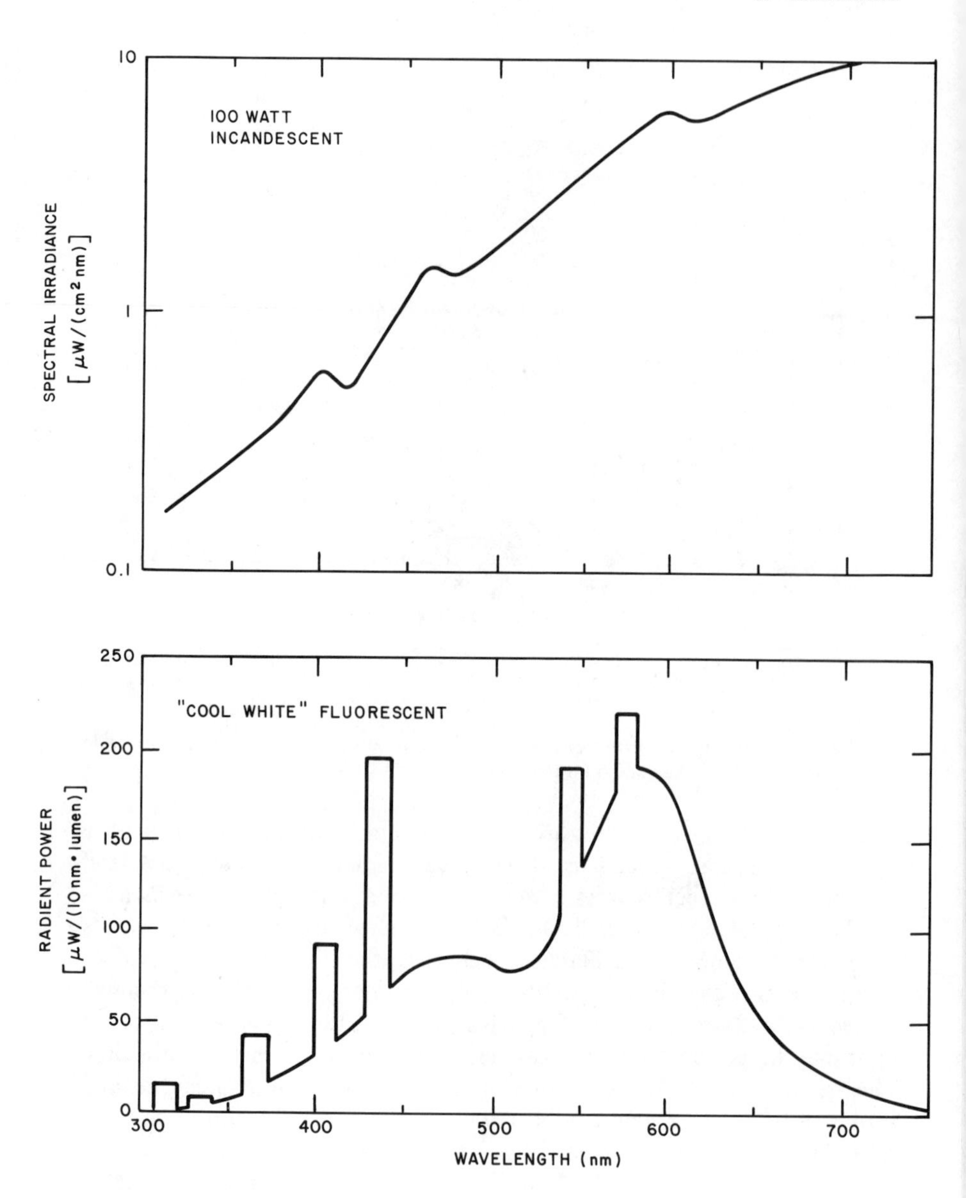

Fig. 1.2-4 Spectra of artificial sources of photons in the indoor atmosphere: top, 100 watt incandescent lamp measured at 50 cm distance [R. C. Peterson, AT&T Bell Laboratories, private communication, 1984]; bottom, "cool white" fluorescent light. (Adapted from Kaufman.[1308])

Deposition aids in cleaning the air, but accelerates the degradation rate of sensitive surfaces (paintings, corrodible metals, the human lung). A further complication is presented by Camuffo,[1291] who has shown that solar radiation incident on interior walls can cause cycles of release and absorption of water vapor. Since a number of indoor gases are water soluble, this finding implies cycles for corrosive gases as well.

1.3 Genetic Toxicology and the Bioassay of Chemical Compounds

1.3.1 Introduction

Society, in recent years, has come to realize that chemicals present a "two-edged sword." On one hand, chemicals enhance, protect, and prolong our lives. Medicines have alleviated diseases, reduced pain, and saved lives. Synthetic chemicals, such as plastics, have reduced costs, made new products available, and made our lives more comfortable. On the other hand, it has become clear that there are risks associated with the production and use of many chemicals. One of the major efforts in the field of genetic toxicology is the effort to identify and evaluate those chemicals that are likely to produce the long-term health effects of cancer and mutation. Since man is exposed to thousands of chemicals and the lag time between exposure and expression of a disease state can be many generations, clinical and epidemiological approaches alone have been unable to identify most of the presently known carcinogens and mutagens.[B58] Toxicological tests using species other than man, however, have shown that a number of chemicals within our environment can initiate cancer-producing and genetic disease-producing processes. Most of these tests monitor for evidence demonstrating that the chemical(s) under investigation interacts with the basic hereditary material (DNA) of the cell. Except for some viruses, DNA (deoxyribonucleic acid) is the substance that provides the genetic code for all species of plants and animals. When alterations are made within the DNA code, a mutation is said to have occurred. Because human DNA has basically the same chemical structure as some non-human DNA, it is surmised that chemicals mutagenic within these lower species are mutagenic to the human population.

The purpose of this section is to provide an overview to the field of environmental genetic toxicology, and to provide basic information concerning the tests referenced within this volume. The information should assist those unfamiliar with bioassay data in understanding its importance and help the reader associate bioassay information with the other types of information presented in the various tables. By associating bioassay data with information on the chemicals in the atmosphere, one also begins to arrive at an understanding of the types and amounts of airborne toxicants to which individuals are exposed. The task of testing all compounds for all the various genotoxicity endpoints would be enormous. Approximately 70,000 commercial chemicals are in use in the United States[B59] and many more are introduced each year.[B60] In addition, there are many environmental chemicals, some of which are degradation products of these commercial chemicals, to which the population is exposed. It is important, therefore, to set priorities for testing and research. One way that this priority setting can be enhanced is to understand to which chemicals the population is exposed and what information is or is not available concerning the sources and effects of those chemicals.

1.3.2 Mutation, genetic disease, and cancer

In very simple terms, a mutation is any suddenly occurring alteration of genetic material that is later inherited by subsequent generations of a cell or organism. Before describing more precisely what a mutation is, one needs to decide upon the importance of mutations to man. Most mutations are known to be deleterious, producing effects that range from very minor perturbations all the way to lethality.

How common is mutation in humans? If genetic disease is rare or if genetic alterations within the cells of an individual have no detrimental effect, mutation is unimportant to human health. On the other hand, mutations are important if disease associated with aberrant hereditary material is common. In addition to many non-genetic congenital* variants, there are now more than 2,000 types of human genetic variants known.[B33] Many of these are seen as morphological variants with an abnormality that develops during gestation. Although most of the morphological variants appear as dominants (that is, they are detectable in first generation progeny), the biochemical alterations responsible are generally not known.

The most commonly recognized congenital malformations involve the heart, the nervous system, the face, and the skeletal system. The

* *Congenital*: Associated with or observed at birth.

estimated prevalence of congenital malformations varies according to the type and location of the study. By pooling the same types of information from a large number of epidemiological studies, Kennedy[B34] showed that government records and retrospective questionnaires gave an average incidence of 0.83% anatomical malformations per live birth, whereas hospital records showed an average of 1.25%, and extensive examination procedures give 4.05%. These estimations are still underestimates for two reasons: (1) molecular diseases such as phenylketonuria generally were not included, and (2) few of the studies reviewed included follow-up to detect abnormalities that appear at later stages of development.

Although the major anatomical malformations are revealed quite successfully by diagnosis at birth, the biochemical and nervous system disorders that precipitate other handicapping conditions are not as easily diagnosed. A good example of this has been phenylketonuria, commonly called PKU. At birth the child may show no outward manifestions of disease; however, if the condition goes untreated the child shows signs of mental retardation starting at six months of age. Although no chromosomal abnormality can be seen in these children, they cannot manufacture a needed enzyme (phenylalanine hydroxylase). This is a recessive Mendelian trait that can be carried for many generations before it is expressed. It is fortunate that this condition can be diagnosed at birth and the mental retardation prevented through proper diet control. Over 15 million newborns have now been screened for PKU and the incidence of PKU alone is 1 per 11,000.[B35] The term *molecular disease* can be applied to this type of disease because an alteration in a biochemical pathway produces the clinical symptoms seen. Although metabolic "blocks" may affect protein, nucleic acid, lipid, pigment, or carbohydrate metabolism, the precipitating abnormality invariably lies within the genetic material that specifies the synthesis of a protein. There are a host of these metabolic diseases that are caused by single gene products.

In contrast to those monogenic traits just described, traits determined by the collaboration of several genes at different loci (gene sites) are called polygenic traits. In polygenic inheritance, each of many genes contribute a minor effect. Polygenic traits, such as height, therefore, show a continuous variation and a definite familial tendency. This type of inheritance is used to explain the familial incidence of conditions such as diabetes mellitus, anencephaly, spina bifida and congenital dislocation of the hip.

Another indication of the prevalence of genetic abnormalities can be seen in the study of spontaneous abortions. Although only 0.5% of all liveborn infants have chromosomal abnormalities, approximately 24% of all conceptuses show chromosomal abnormalities.[B39,B40] These figures demonstrate that more than 90% of all chromosomally abnormal fetuses are lost through spontaneous abortion. Since most parents do not carry the associated chromosomal abnormality, it seems obvious that lethal chromosomal mutations are responsible for the majority of these fetal deaths. One should remember, however, that recognizable chromosomal abnormalities account for only 20-25% of all abortions.

Some individuals with chromosomal abnormalities survive beyond fetal development. Many, if not most, of these individuals suffer from an associated handicapping disease. Approximately one in 200 newborns demonstrate a structural and/or numerical chromosomal abnormality.[B41] Errors in chromosome number can occur with the sex chromosomes (e.g., Kleinfelter's syndrome) or with autosomes (e.g., Trisomy 21). Structural rearrangements such as deletions, insertions, translocations, etc. also produce disease states. For example, deletion of the short arm of one chromosome (chromosome number 5) produces the *cri du chat* syndrome, a condition which presents various physical abnormalities, mental retardation, and a characteristic cat-like cry in infants.

What, then, is the total incidence of genetic disease in man? Approximately 0.5% of all births show chromosomal aberrations, approximately 1.6% show polygenic birth defects, and approximately 1.5% show single gene disorders.[B42] This would give a total of 3.6% of all live births having some form of genetic disease; however, this does not include other polygenic diseases that are not detectable at birth. One out of every 1,000 infants is born deaf and approximately 50% of these are due to a genetic cause. Genetic disorders account for 40% to 50% of all childhood blindness. Other multifactorial diseases such as diabetes mellitus, schizophrenia, hypertension, and arteriosclerosis have genetic components. Benirschke et al.[B42] estimate that 11% of the population is affected by these multifactorial, polygenic diseases. If we concede that only one-half of these diseases (6%) are due to genetic factors, then nearly 10% of the human population carries some type of genetically-induced burden. Although much of genetic disease seems to be due to the transmission of mutations already in the population, it is reasonable to presume that exposure to environmental mutagens can cause an increase in disease by increasing the frequency of mutations.

When there is damage to the genetic material (DNA) of sex cells (gametes) then the genetic diseases just discussed may occur. Similarly, many believe that cancer is the result of the other cells (somatic cells) within the body receiving genetic damage and thus altering the control of the normally ordered and restrained proliferation of cells. What evidence supports this contention? Several observations show the relationship of cancer to genetic events. The evidence supporting somatic mutation as the initiation of cancer is circumstantial but strong. One prediction of all somatic mutation theories is that resultant tumors will have a clonal origin. In other words, the neoplastic (tumor causing) transformation occurs in a single cell which then undergoes uncontrolled proliferation. If a highly infectious agent (such as a virus) was the initiating factor, then a rarely seen multicellular origin would be anticipated. Several investigators[B43,B44] have shown that genetic marker studies suggest a clonal origin for most examined neoplasms. This type of evidence suggests that the initial event in most cancers is a mutation or a series of mutations within a single cell.

Another factor of note is that the chromosomal changes found in certain classes of neoplasms are not random. For example, in 34 patients with hematologic disorders, all the patients demonstrated an excess of chromosome number in an area that cytologists label 1q25-1q32. While several different chromosomal regions have been associated with neoplasia, most of the presently meager evidence, not directly related to oncogenes,* associates these areas with nucleic acid metabolism.[B44] In other words, those genes that affect DNA and RNA (ribonucleic acid) synthesis are the genes involved in chromosomal changes associated with neoplasia. Not all neoplasias, however, have observable chromosomal changes, but this may merely be due to cytological detection limits. It has been estimated that a band structure revealed by the newest chromosome cytological techniques contains 5,000,000 nucleotide base pairs, and that a deletion or duplication of up to one third of these pairs (2,000,000 base pairs) would not be detected. Therefore, undetectable chromosomal changes could be occurring in some cancerous states. All in all, these studies support the hypothesis that specific genes are probably involved in the alteration of a cell from a normal to a cancerous state.

Recently, cancer research has received a new impetus from the identification and manipulation of cellular oncogenes using recombinant DNA techniques. As would be expected, there are multiple oncogenes that have been identified, and each oncogene may be activated ("turned on") only in certain tissue compartments and may be limited to transforming cells of that tissue. This new effort also demonstrates that cellular oncogenes are activated in at least two different ways. In one, some proto-oncogenes** are activated when they become associated with a retrovirus,† or secondly, they can be activated when altered via mutational

* *Oncogene*: A gene, either viral or cellular, that initiates or promotes the transformation of a cell to a tumor-producing state.

** *Proto-oncogene*: A normal and naturally-occurring gene which can be transformed to an oncogene by any of several genetic mechanisms (e.g., mutation).

† *Retrovirus*: A single-stranded RNA virus, certain species of which are known to be capable of carrying some specific oncogenes with their RNA code.

events (without interaction with a retrovirus). Alternatively, more than one oncogene may be needed before cellular transformation (alteration of a normal cell to a cancerous cell) can proceed. The process of *in vivo* tumorigeneis (tumor development) is even more complex. For an introduction to cellular oncogenes and the processes of carcinogenesis refer to the review of Land et al.[B45]

The increased incidence of leukemia associated with many genetic disorders, including Down's syndrome, Trisomy D, Kleinfelter's syndrome, and Bloom's syndrome, suggests a genetic susceptibility to cancer.[B46] The genetic component of cancerous diseases is also suggested by variation among different ethnic groups of different types of cancer.[B47] Also, it is well known that a variety of congenital malformations precede neoplasms in man. In 371 carefully studied cases of childhood malignant disease in Tokyo,[B48] 41% of the children demonstrated congenital malformation, in contrast to 13% of the children without neoplasms. Other examples, such as aniridia (the absence of the iris) associated with Wilm's tumor, also exist. Several tumors and tumor syndromes, such as retinoblastoma and polyposis of the colon appear to be caused by dominant gene mutations.[B46] Although one disease type may promote or develop into the other, these congenital diseases and neoplasms appear to have a common origin.

Finally, known chemical carcinogens (cancer-causing agents) are known to give positive results in mutation testing systems. One of the earliest demonstrations of this phenomena was by Ames et al.[B49] Among 300 chemicals previously tested for carcinogenicity in whole animal studies, a simple bacterial assay showed that 90% of the known carcinogens are also mutagenic.[B49] Although this type of approach using bacteria, mammalian cells, or other short term assays does not detect all classes of carcinogens, toxicologists have continued to find a high degree of correlation between the results of short-term bioassays and whole animal carcinogen bioassays.

Is cancer initiated by a genetic event or events such as mutation? The evidence is debatable but tends to say that, yes, genetic alteration is the first step in the production of cancer. One must realize, however, that many other alterations must occur before a neoplasm is produced from this initial step, including alterations in cell metabolism, the body's immunologic defense system, hormonal controls, and other health factors. At present, however, multiple mutations within a single cell seem the most reasonable way of explaining the induction of cancer.

Just as the study of mutations may lead to increased understanding of tumorigenesis and cancer, the study of tumorigenesis and cancer epidemiology may provide researchers with an estimate of mutation rates within the human population. If one assumes that cancer is the demonstration of the occurrence of mutation(s) within the somatic cells of an organism and that these mutations must occur within specific genes, one could determine the average number of mutations needed to produce a transformed (cancerous) cell. If one could then estimate the fraction of transformed cells that survive to produce a tumor, an estimate of somatic cell mutation frequencies in man could be determined from the frequency of cancerous diseases. A general increase or decrease in the frequency of cancer within persons of childbearing age might indicate, therefore, a corresponding increase or decrease in heritable mutations. Also, one would suspect that any cancer causing agent (carcinogen) to which human gametes (sex cells) are exposed could produce heritable mutations, thus increasing the incidence of genetic diseases. Fortunately, there are special physiological barriers that prevent the exposure of human gametes to many types of substances.

As was noted previously, a major difficulty in identifying human carcinogens and mutagens is the long time between exposure and expression. The extended period allows for many and varied exposures to occur before the health effect is detected. With carcinogenesis the consequences may not be evident for years; with mutagenesis, the consequences may not be evident for generations. In both cases it may not be possible to associate an increase in disease incidence with its cause. A primary goal in genetic toxicology, therefore, is the identification of genotoxins and their sources in order to prevent or lessen human exposure.

1.3.3 Concepts of Mutation

A mutation is any sudden change in a cell or organism that can be inherited. Although mutations can occur in RNA viruses and the DNA of cytoplasmic organelles,* the mutations of greatest interest occur within genes in the nucleus of the cell. Most people understand that genes are needed for the development of a new individual but do not understand that genes are needed on a day-to-day basis for control of normal body functions. In order to understand the various types of mutation and how mutations produce an effect, a clear understanding of the basic components is helpful.

* *Organelle*: A membrane-bound substructure of the cell that performs a specialized function (e.g., the nucleus or the mitochondria).

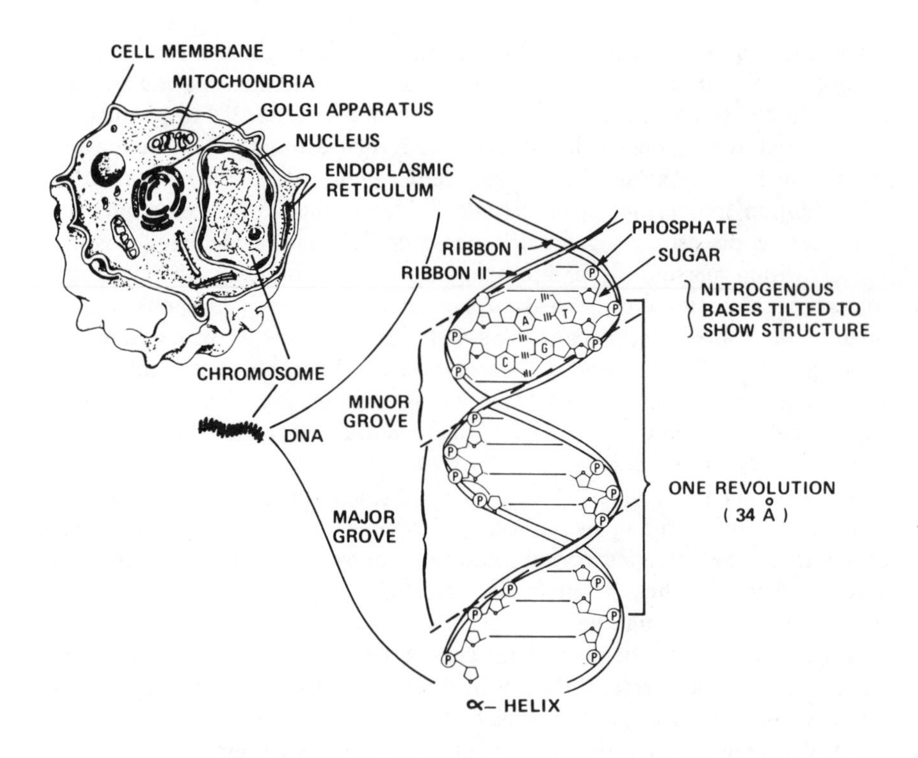

Fig. 1.3-1 An illustration of the relationship of genetic components (DNA, chromosome, and cell nucleus) to other components in a mammalian cell. The nitrogenous bases in the DNA are adenine (A), cytosine (C), guanine (G), and thymine (T).

Our bodies contain more than 10 trillion cells, and at some stage in their life cycle each cell contains a full complement of the genes needed by the entire organism. Separated from the cytoplasm by a membrane is the nucleus, which contains genes clustered together in specific units called chromosomes. Genes are composed of DNA, which is a large, complex molecule. Figure 1.3-1 illustrates the relationships and structures of a cell and its genetic components.

1.3.3.1 The Basic Structure

The basic chemical compounds involved in the formation of DNA are: (1) two purine nitrogenous bases, (2) two pyrimidine nitrogenous bases, (3) a sugar, and (4) phosphoric acid. Nitrogenous bases plus a sugar and

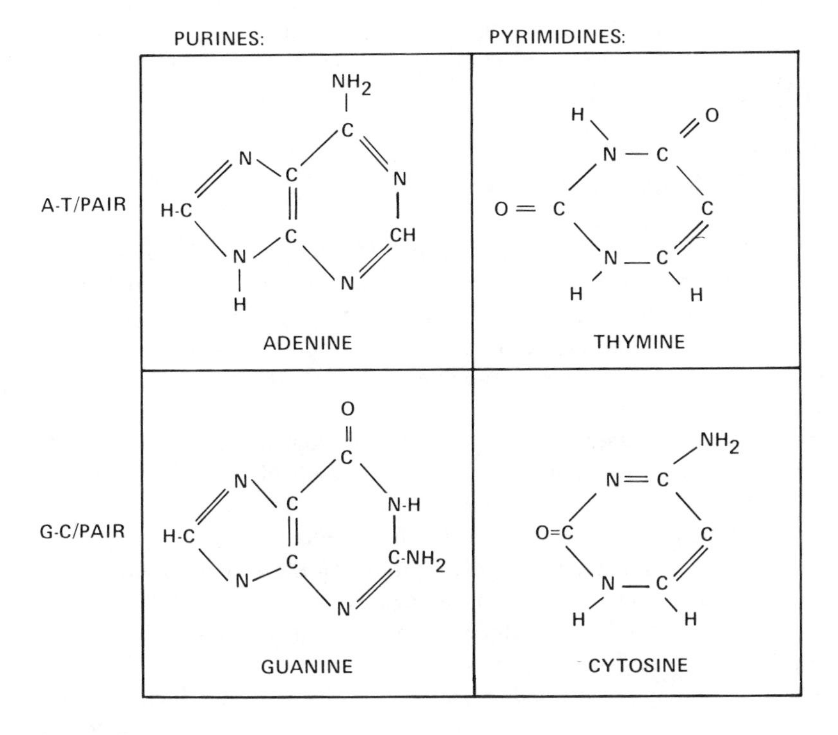

SUGAR:

PHOSPHORIC ACID:

H – O – P – OH

Fig. 1.3-2 The chemical structure of nucleotide components.

phosphoric acid give a complex molecule called a nucleotide. As seen in Figure 1.3-1, the phosphoric acid and sugar form two strands to which the nitrogenous bases are attached. The two strands usually create an alpha helix and the bases form a stairstep-like array between the two strands. Figure 1.3-2 shows the chemical structures of these basic components. Within the DNA helix adenine pairs with thymine (the A-T pair), and the guanine pairs with cytosine (the G-C pair). Because these bases are held together by relatively loose bonding forces, the two strands of the DNA helix can separate. This separation allows for the production of two daughter strands.

1.3.3.2 The Functions of DNA

When the two strands serve as templates for the production of two new strands, their separation allows for the production of two new daughter cells. Since the bases pair only with their respective partners, two duplicates of the original DNA are produced. Upon cell division, one copy of the original DNA goes to one daughter cell while the other copy goes to the second daughter cell.

The nitrogenous bases not only allow for replication of the DNA molecule but also provide the means for coding all of the cell's proteins. These proteins in turn provide structure for the cell and control of all other metabolic activities. Genes that code for specific proteins are called structural genes, and a structural gene contains on the average about 1,000 nucleotides arranged in a specific linear sequence. The code used by DNA is called a triplet code because three consecutive nitrogenous bases specify a particular amino acid within a protein.

On a day-to-day basis, DNA is responsible for producing three components needed in the production of protein. Figure 1.3-3 illustrates this process. First, DNA produces messenger RNA (mRNA), which codes for the proper sequence of amino acids (the building blocks of protein). Each group of three nucleotides (a codon) codes for one amino acid. Secondly, DNA codes for the transfer RNAs (tRNA). These tRNAs carry specific amino acids to the site of protein construction. Each tRNA has a anti-codon site consisting of three nucleotides that attaches at a specific codon of the mRNA. Next, the ribosomal RNA (rRNA) is produced by the DNA. Ribosomal-RNA brings together the mRNA and tRNA-amino acid complex in an orderly manner so that the protein is produced. These three components (mRNA, rRNA, and tRNA) when complexed together are called a polyribosome. If any of these components has been altered, a proper protein may not be produced. With an altered protein, cellular functions may be compromised or halted. Therefore an alteration (mutation) in DNA causes an alteration in RNA, which in turn produces defective proteins or lack of protein function.

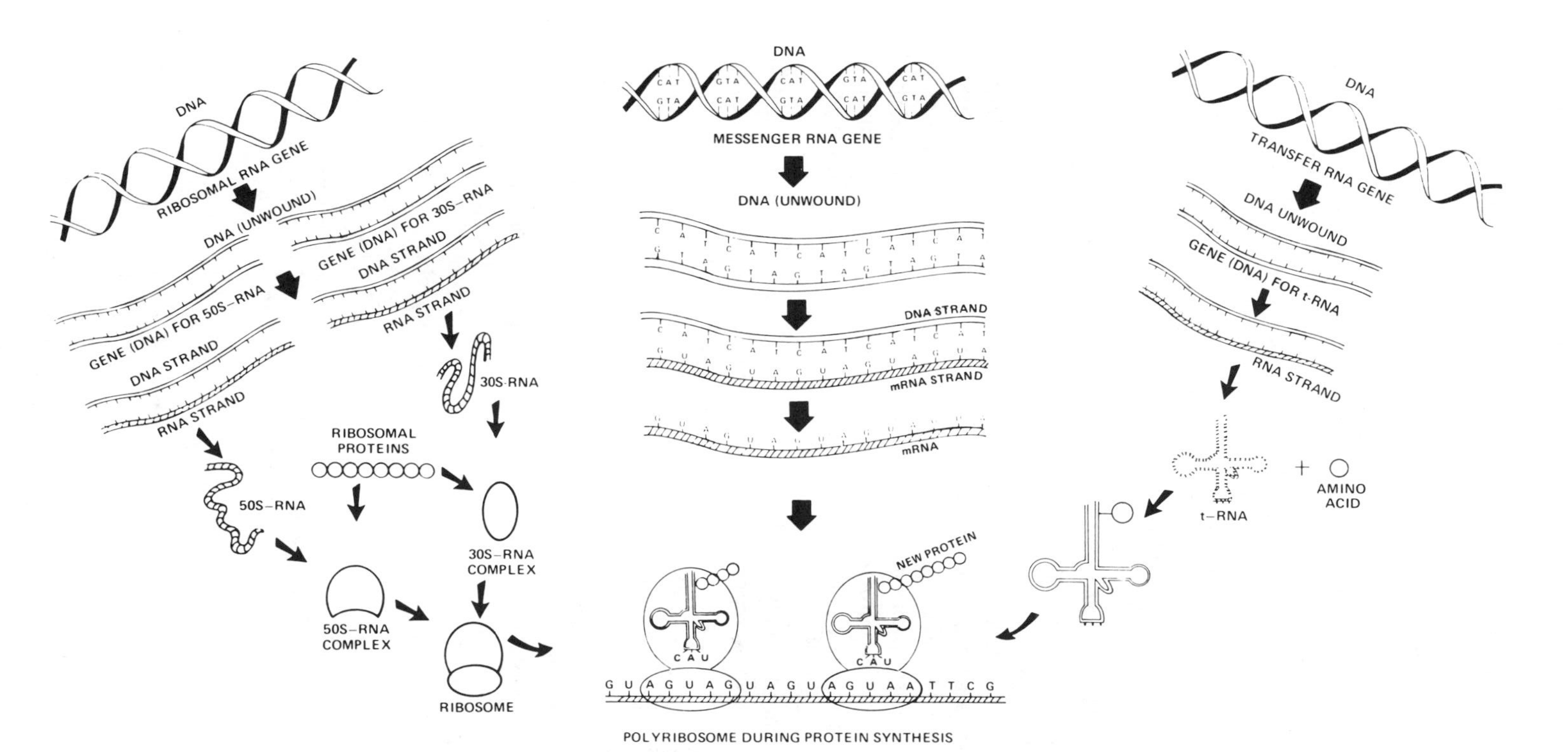

Fig. 1.3-3 The role of DNA in coding for the three components required for protein synthesis. *Center*: DNA codes for the messenger RNA (mRNA) that codes for the proper sequence of amino acids. *Right*: DNA codes for transfer RNA (tRNA) which carries specific amino acids to the site of protein construction. *Left*: DNA codes for ribosomal RNA (rRNA) which complexes mRNA and tRNA for protein production.

1.3.3.3 Types of Mutation

Geneticists generally divide mutations into two broad categories: (1) mutations that are cytologically visible mutations (chromosomal aberrations), and (2) cytologically "invisible" mutations that occur at the submicroscopic level (gene mutations).

Chromosomal aberrations: A human cell normally has 23 pairs of autosomal* chromosomes and a pair of sex chromosomes. (A normal male will have an X and a Y sex chromosome, and a normal female will have two X chromosomes.) This gives a total complement of 46 chromosomes per cell. Modern staining techniques that cause chromosomes to have a banded appearance allow for the identification of each chromosome and sometimes segments of a chromosome.

When seen cytologically (i.e., with the aid of a microscope), mutation produces either a change in the number of chromosomes or a change in the structure of individual chromosomes. The mutational alterations easiest to observe are changes in chromosome number (Figure 1.3-4). Changes that involve entire sets of chromosomes are called euploidy. For example, a human cell with 46 pairs of chromosomes would be euploid. This type of change is not seen in viable mammals, including man. Variations in number that involve only single chromosomes within a set are called aneuploidy. The most common type of aneuploidy in man is trisomy, examples of which are Klinefelter's syndrome and Down's syndrome.

The karyotype* of an individual may be altered not only by changes in entire chromosomes or chromosome sets, but also by structural changes within a chromosome (Figure 1.3-5). This alteration occurs when the chromosomes fracture and the broken ends rejoin in new combinations. Included among the structural changes are deletions, duplications, inversions, and translocations. A deletion involves the loss of a portion of a chromosome while a duplication involves the insertion of an additional copy of a portion of the chromosome. Deletions and duplications upset the metabolic balance of an organism by altering the amount of gene products produced. Sometimes the order of genes on a chromosome is reversed in one area. This is called an inversion. When a broken portion of a chromosome attaches to a second chromosome it is termed a translocation. Since the position of a gene affects its regulation and activity, translocations and inversions may be detrimental to an organism.

* *Autosomal*: Pertaining to any chromosomes other than sex chromosomes.

* *Karyotype*: The arrangement of chromosomes in a systematic pattern in order to aid observation and comparison.

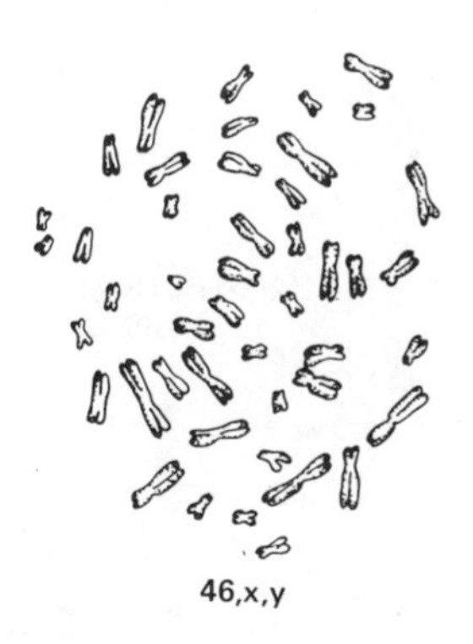

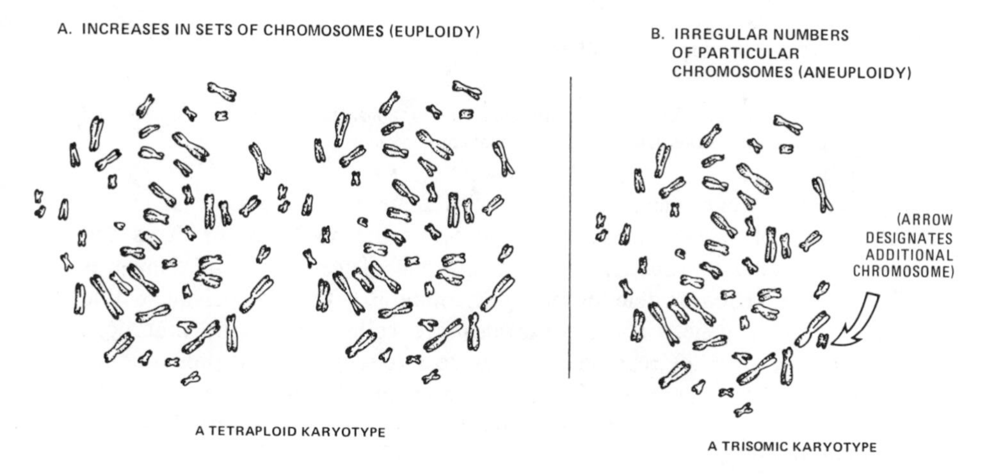

Fig. 1.3-4 The normal chromosome karyotype (chromose complement) and the first class of chromosomal aberrations, in which there is a change in the number of chromosomes.

Gene mutations: When the alteration is in the nucleotide sequence of a gene and cannot be observed microscopically, it is referred to as a gene mutation. Two classes of gene mutations (Figure 1.3-6) are recognized: point mutations and intragenic deletions. Within point mutations there are two types. The first type, termed a base pair substitution, involves the

2. CHANGES IN CHROMOSOMAL STRUCTURE

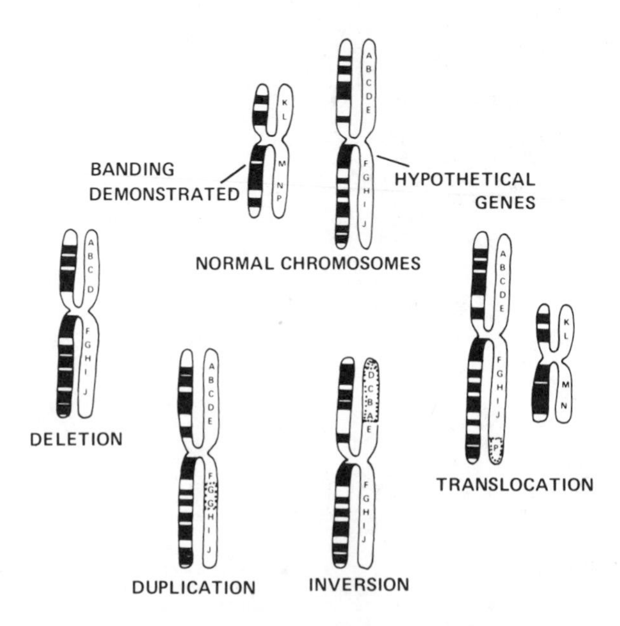

Fig. 1.3-5 The second class of chromosomal aberrations, in which the chromosomal rearrangements involve the structure of individual chromosomes.

replacement of one nucleic acid base with another. This may result in the substitution of one amino acid for another in the final gene products and thus alter cellular function. A second type of gene mutation involves either the insertion or deletion of a nucleotide(s) within the polynucleotide sequence of a gene. These mutations are called frameshift mutations because they alter the reading of the nucleotide sequence and thus produce an incorrect gene product. When a more extensive deletion occurs within a gene so that the informational material of that gene is essentially lost, it is called an intragenic deletion.

Other classifications: Mutations may also be classified in other ways. For example, a mutation may be lethal; that is, produce death in the cell or organism. A mutation also may be termed recessive, dominant, or X-linked. If a mutation is dominant, only one such mutation need occur in a diploid* organism such as man for it to be expressed in the first generation

* *Diploid*: Having corresponding (homologous) pairs of chromosomes for each genetically-controlled characteristic except sex. The total number of chromosomes in a diploid organism's somatic cells is twice that of a gametic cell.

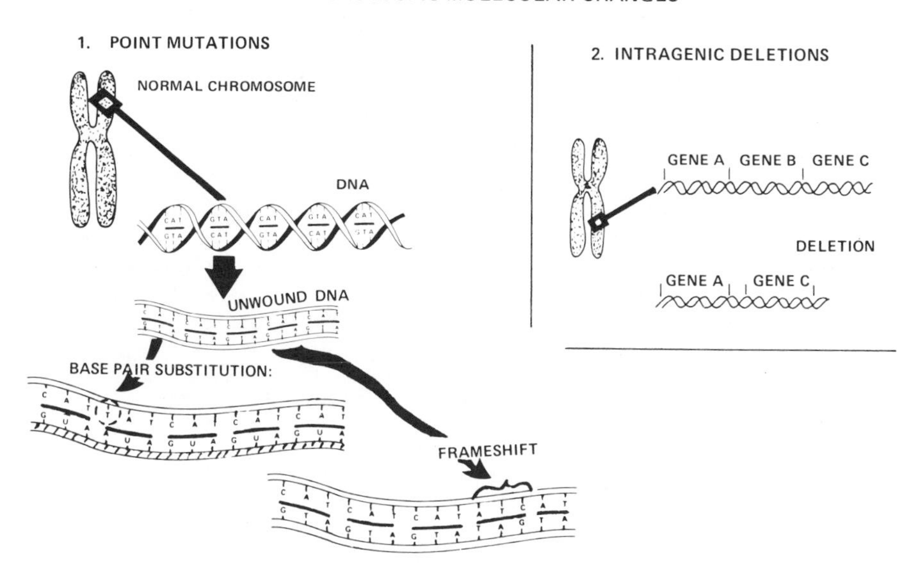

Fig. 1.3-6 Two classes of submicroscopic gene mutation.

in which it occurs. If a mutation is recessive, it can only be completely expressed if it encounters the same mutation in the homologous chromosome. Recessive mutations can, therefore, be carried by many successive generations of an organism or cell before they are expressed. X-linked mutations can act as either dominant or recessive mutations in females. However, since the male normally has only one X chromosome, X-linked mutations act as dominants in the male. Some examples are known for each of these modes of inheritance. Brachydactyly (short-fingeredness) is a dominant trait, albinism is a recessive trait, and color blindness is X-linked.

1.3.3.4 Induction of a Mutation

Scientists have found that there are a variety of chemical processes that can alter DNA and cause a mutation. Malling and Wassom[B50] used seven major categories when classifying chemicals according to their action on

DNA. These seven categories are: (1) alkylation; (2) arylation; (3) intercalation; (4) base analog incorporation; (5) metaphase poisons; (6) deamination; and (7) enzyme inhibition. Table 1.3-1 provides a summary of the mechanisms in each of these seven categories.

Table 1.3-1. Mechanisms of Mutagenic Agents*

Chemical Action	Mechanism of Action	Result
Alkylation	Adding of an alkyl group ($-CH_3-CH_2CH_3$, etc.) to a nucleotide	BPS, CLI, IGD
Arylation	Covalent bonding of an aryl group	FS
Intercalation	"Wedging" of the compound into the DNA helix	FS
Base Pair Analog Incorporation	Base-pairing errors due to mispairing	BPS
Metaphase Poisons	Interfere with spindle formation and disrupts the migration and segregation of chromosomes	Aneuploidy
Deamination	Removal of an amine group (NH_2) from cytosine, guanine or adenine	BPS
Enzyme Inhibition	Interferes with biosynthesis of purines or pyrimidines and interfere with repair	Enhances all types of mutation

* Summarized from Malling and Wassom.[B50]

Abbreviations BPS, base-pair substitution; CLI, cross-linkage inhibiting replication; IGD, intragenic deletion; FS, frameshift mutation.

Note: Continuing work has refined and added to this information.

1.3.3.5 Induction of Disease by Mutation

As has been seen, a mutation can alter the sequence of bases within a codon. This process can produce disease in the following way:

One of the most important molecules within man is the hemoglobin of red blood cells, since it carries oxygen to other cells. One of the protein chains that is a part of the hemoglobin molecule contains 146 amino acids; therefore, the portion of DNA that codes for these 146 amino acids must contain at least 438 bases (3 bases for each amino acid coded). The sixth codon in this sequence normally codes for glutamine. Assume for the purpose of illustration that this codon is CTC. If a mutagenic agent causes a base pair substitution at the thymine (T) location of this codon, then one could obtain a codon which reads CAC. CAC would give a mRNA molecule with a GUG codon which would pair with a CAC anticodon of a tRNA which carries valine and not glutamine. Therefore, the sixth amino acid in the resulting hemoglobin would have valine in place of glutamine. This hemoglobin is called sickle cell hemoglobin because the red blood cell assumes an irregular shape due to presence of the variant hemoglobin. If this occurs in a somatic cell, then the individual will contain a sickle cell erythrocyte (red blood cell) among millions of normal cells and this alteration of a single cell would be of no consequence. When this occurs in the germinal cell DNA that produces a new individual, the new individual either is a sickle cell carrier or has sickle cell anemia. He, therefore, will be able to pass the potential for this disease on to future generations. Normally, this mutant gene is transmitted from one generation to the next through normal hereditary processes; however, mutations such as that outlined above can maintain and/or increase the burden of genetic disease within the human population. This is only one illustration of how one type of damage can cause disease. Geneticists now understand a variety of other possible mechanisms and results of mutation.

In summary, genetic information is recorded in our cells and this genetic material can be altered to produce a mutation. We also see that a small molecular change in DNA has the potential to cause an increase in major hereditary diseases.

1.3.4 The Use of Bioassay Screening Systems for Detecting Carcinogens and Mutagens

The present efforts of genetic toxicology are centered around the defining of genetic mechanisms involved in disease processes and the screening of chemicals and environmental substances for the ability to produce a genotoxic effect, e.g., carcinogenesis or mutagenesis. Since there are advantages to preventing disease rather than using treatment schemes, toxicologists are active in developing and evaluating bioassays for identifying carcinogens and mutagens. It is helpful to understand some of the types of test systems that are presently available.

1.3.4.1 Screening with *In Vivo* Mammalian Systems

It has been recognized for many years that the physiology and genetics of most mammals are similar to man's. Subhuman mammals are thus used to screen chemicals for both toxicity and beneficial health effects. Since the basic genetic material for all mammals has, biochemically, the same structural elements and similar mechanisms, mammalian systems also are used to test the mutagenic/carcinogenic potential of a substance. Several of these systems have been developed and they can be used to monitor for cytological mutations, gene mutations, and tumor production. These animals also can be exposed in the same manner as man (inhalation, ingestion, etc.). The major drawbacks to these systems is the number of animals that must be used to evoke a measurable response and the difficulty of quantifying a predicted effect in man based on test data from subhuman mammals. In addition, these complex experiments are very costly. This cost factor, when resources are limited, dictates that only a few substances can be examined by these procedures each year.

1.3.4.2 Screening with *In Vivo* Non-mammalian Systems

Because DNA structure for all animals and plants is very similar, tests conducted on non-human organisms can be used to ascertain the effect of a substance on DNA. Many of these systems are especially useful because a great deal of genetic information about specific organisms (e.g., Drosophila, Habobracon, Tradescantia, etc.) is available. These organisms are less difficult to use, less costly, have relatively short generation times and can be exposed in a manner similar to humans. It is difficult, however, to extrapolate results from these systems to humans. Most of these systems do not have mammalian-like metabolism, and mammalian metabolic systems have not been incorporated into most of these test systems. Also, some of the genetic repair and regulatory mechanisms may be different from those in humans.

1.3.4.3 Screening with *In Vitro* Systems

In vitro systems are systems in which cells or organs are grown in artificial media outside their natural environment. *In vitro* systems include bacterial and fungal systems as well as tissue culture of both human and non-human cells. These systems are utilized because they are rapid, relatively inexpensive, and highly definable physiologically and genetically. One can screen several compounds in multiple *in vitro* systems for the same effort, cost, and time involved in one *in vivo* system. One must realize, however, that these systems do not completely mimic a whole mammalian organism. These systems in themselves may activate an inactive substance to an active one, or may inactivate one that is active in man. These types of actions would give both false positive and false negative results.

1.3.4.4 Description of Test Systems

Test systems are too numerous and complicated to describe fully in one summary. Instead, the terms used in describing test systems will be briefly described and a summary table which demonstrates the comparative aspects of the assay systems will be given. Several texts either provide or are centered around a discussion of these various bioassays and their use.[B31,B51,B52,B53,B54,B55,B56,B57] Test systems are described in the following ways:

1. *By the Test Organism Used*: Organisms used for screening range from bacteria (*Salmonella typhimurium* and *Escherichia coli*) to mammals (mice and rats). They include many diverse forms, including insects such as Habobracon, plants like Tradescantia, and fungi such as Neurospora. Prokaryotes (e.g., bacteria) do not have a nucleus; therefore, the DNA-chromosomal material is contained within the cytoplasm. Eukaryotes (e.g., yeast, mammals) have a cytologically distinguishable nucleus and most are diploid during some stage of their life cycle.

2. *By Type of Mutation Observed*: Systems are classified by the type of gene mutations and/or chromosomal aberrations they detect. In addition, several other terms are used to classify the mutational type. Several mutation systems detect the genetic alteration in predetermined loci; these are termed specific locus mutation systems. Mutation systems detect forward mutation or reverse mutation. A forward mutation is when a wild type loci (as would be found in nature) is mutated. Reverse mutations result by a mutation restoring mutant DNA back to its original sequence and/or function; therefore, the reverse mutation must be highly specific, in contrast to a forward mutation, which can produce its results by a variety of mechanisms.

3. *By the Type of Metabolism Included*: Every organism has biochemical pathways that will metabolize compounds to new forms (or metabolites); however, certain types of metabolism are unique to mammals. Some non-mammalian systems, therefore, cannot mimic the metabolism of mammals. Many of the non-mammalian systems can be

Table 1.3-2. Comparison of Types of Mutation Systems

Type of Organism	Treated Entity	Type of Assay(s)	Mammalian Metabolic Activation	Time/ Cost
Bacteria	Cells	Primary DNA damage, Forward and reverse gene mutation	Added	Weeks/ Low
Fungi	Conidia and mycellium	Forward and reverse gene mutation	Added	Month/ Medium
Yeasts	Cell	Mitotic crossing over, Gene conversion, Forward and reverse gene mutation	Added	Months/ Low
Insects	Egg, Larvae, Adult	Most types of damage including sexed linked recessive lethal	Not used	Months/ Medium
Plants	Cuttings and root tips	Gene mutation and chromosomal effects	Not used	Months/ Medium to Low
Mammalian Cells	Cells	Gene mutation, chromosomal effects, cell transformation	Added and endogenous	Months/ Medium
Mammals	Whole animals	Gene mutation, chromosomal effects, DNA damage, carcinogenicity	Endogenous	Months to years /Medium to High

made to mimic this type of metabolism by the addition of mammalian enzymes to the system or by the intermediate use of a mammal to produce the metabolites.

4. *Other Considerations*: When one is screening substances for the benefit of human health one must ask how sensitive, costly, and time-consuming is the test system. As the number of organisms involved and the number of genetic loci involved are increased, the sensitivity of the system increases. With microbial and tissue culture systems it is easy to survey a large number of end points. *In vivo* mammalian systems generally do not involve large numbers because of the high cost, long time periods, and technical complexities involved. Some *in vivo* systems require one to two years for completion, whereas other systems require only one to two weeks' time.

Table 1.3-2 lists some of these general characteristics for the types of test systems used today.

1.4 The Tables

The chemical compounds in the atmosphere have been divided into groups containing compounds of similar chemical structure and behavior. For each of these groups a detailed table of information is presented. The format, guidelines, and explanations of the entries in these tables are presented below.

1.4.1 Selection Requirements and Ordering of Compounds

As mentioned in the preface, we seek to include in this compendium all atmospheric compounds known to be emitted into the air and/or those found as gases or as constituents of aerosol particles, clouds, fog, ice, rain, or snow. The regimes of interest are the troposphere, the stratosphere, and that indoor air to which the public may be exposed. We exclude air within industrial facilities and air at altitudes above the stratopause.

In determining the order in which compounds are listed, we have been guided by the recommendations of the International Union of Pure and Applied Chemistry.[1211] Inorganic compounds appear first, in Chapter 2. Hydrocarbon compounds comprise Chapter 3: they are subdivided into alkanic (saturated) aliphatic compounds, olefinic (unsaturated) aliphatic compounds, cyclic compounds, and aromatic compounds of several levels of complexity. Similar subdivisions occur in the subsequent chapters. Chapters 4 to 9 follow IUPAC preferential ordering in which the relevant substituent group has increasing priority for citation as the principal group. Consequently, ethers are contained in Chapter 4, alcohols in Chapter 5, ketones in Chapter 6, aldehydes in Chapter 7, derivatives of organic acids in Chapter 8, and organic acids in Chapter 9. Heterocyclic oxygen compounds (compounds in which an oxygen atom replaces a carbon atom as a ring constituent) comprise Chapter 10.

Studies of the cycles of the elements in the atmosphere are the focus of much current research and it has seemed useful to present separately those organic compounds containing atoms other than carbon, hydrogen, and oxygen. Thus, Chapter 11 lists organic compounds containing nitrogen, Chapter 12 those containing sulfur, Chapter 13 those containing halogens, and Chapter 14 those containing any other element.

Within a table we follow rigorous ordering policies as well. The basic policy is to proceed from the smallest (the simplest) compound within the tabular group to the largest (the most complex). The listing for a parent compound precedes those of its derivatives. The latter appear in the following order; alkyl derivatives, olefinic derivatives, cyclic or arene derivatives (where attached to the parent by a single bond [e.g., phenylnaphthalene]), and cyclic or arene derivatives incorporated into the ring structure (e.g., benzo[*a*]pyrene). Oxygenated derivatives of the parent compound then follow, in the same order as given above. In each case the derivatives are given in order of multiplicity and then of group complexity.

Each compound appears only once in these tables, even in the case where its structural or functional properties would make it eligible for multiple inclusions. For such compounds, the preferred location is the latest possible (i.e., an hydroxyacid appears in Chapter 9 (Carboxylic Acids) rather than Chapter 5 (Alcohols), a chlorinated aldehyde in Chapter 13 (Halogen-Containing Organic Compounds) rather than Chapter 7 (Aldehydes)).

Two groups of compounds that are included in these tables require special mention. The first is the emittants from tobacco smoke. The literature on such compounds is extensive, and its incorporation here reflects the widespread incidence of vegetation combustion, not only in cigarette and cigar smoking but also in refuse combustion, slash burning of agricultural lands, etc. The tobacco literature thus provides insight into a wide variety of processes with potential atmospheric impact. We have intentionally been illustrative rather than comprehensive in including data from this voluminous literature. Interested readers may wish to consult Johnstone and Plimmer,[396] Holzer et al.,[421] and Stedman[634] for more extensive information.

Also deserving comment are the natural vegetative emissions. It is clear that some of these compounds are important factors in atmospheric chemistry, isoprene from trees, for example. It seems less useful to clutter the compendium with detailed listings of the volatile aroma compounds of apples, oranges etc., particularly since the fluxes seem very likely to be small and since a single aroma can be the composite of several hundred compounds (e.g., Willaert et al.[1212]). We have thus attempted to include vegetative compounds that appear to have significant fluxes to the atmosphere, but to exclude minor aroma volatiles.

1.4.2 Numbering

Each chemical compound recorded in this book has associated with it two unique numbers. The first is an identification number assigned in the present work. The number consists of the number of the table within which the compound appears followed by the sequential number of that compound within the table, e.g., 5.2-12. It is this number that appears in the cross-reference indexes at the end of the book. The second number is the Registry Number assigned to the compound by Chemical Abstracts Service (CAS). This number is useful for computerized literature searching and for making certain of the identity of a compound. In some cases, "several" appears in this column, indicating that more than one number may apply. This occurs in cases where an analytical technique does not specify the position of a ligand. For example, "hydroxyindanone" allows the OH ligand to be at any one of several positions, each associated with a different registry number. In one case, this volume lists alcohols for which CAS assigned Registry Numbers only to the methylated derivatives which were analyzed by the experimenters; for these compounds the derivative registry number followed by M (i.e., 70561-57-8M) is given. In a few cases CAS has not assigned a number to a recently reported compound, a mixed salt or ion, or a mineral without a fixed molecular formula.

1.4.3 Nomenclature of Compounds

Conventions in nomenclature are all but unnecessary for the simpler compounds with which one deals most of the time. For more complicated structures, however, conventions rapidly become essential. In assembling the information for this book, the most difficult task has been to make sure which compound an author was talking about. For example, in one article cited herein a compound is listed (in correct CAS form) as 1-phenanthrenecarboxylic acid, 1,2,3,4,4a,9,10,10a-octahydro-1,4a-dimethyl-7-(1-methylethyl)-. If one uses this name to draw the structure, the result is

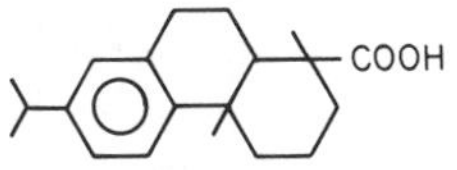

COOH

where the basic phenanthrene skeleton is obvious. Another article cited herein includes dehydroabietic acid in a list of vegetative compounds emitted during wood processing. Scientists dealing with natural plant products typically draw the structure of dehydroabietic acid as

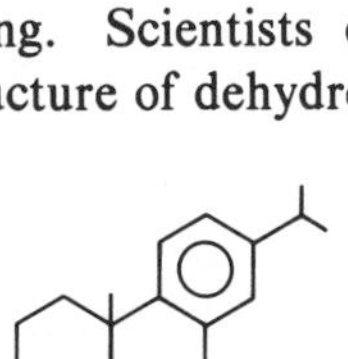

COOH

Clearly, the compounds are identical, but one knows this only if one has a reference to the structures of natural plant products, the knowledge that structures are drawn in different orientations, and the assurance that both authors used the nomenclature of their choice correctly. (It is surprising how often authors do not give sufficient care to nomenclature, and several entries in this book are labeled 'tentative' because the compound as listed was impossible and it was necessary to make a judicious guess as to what was meant).

If one knows for certain the compound with which one is dealing, it is then necessary to decide how to name it in this book. Alternate choices are often possible, given the CAS name, the IUPAC (International Union of Pure and Applied Chemistry) name, and the trivial name. To give two examples, structure A below is named 'Toluene' by IUPAC, but 'Benzene, 1-methyl' by CAS, and structure B carries the trivial name 'Crotonic acid' but is named '2-Butenoic acid' by IUPAC and CAS. Our approach to nomenclature has been threefold. First, in choosing a name we have

COOH

A  B

tried not to avoid using a familiar trivial name merely for the sake of consistency. (We thus retain, for example, α-pinene, a name familiar to many.) For the most part, however, we name compounds by the IUPAC recommendations, judging those names to be more comprehensible (if less computer-searchable) than their CAS equivalents. For those trivial names we judge to be infrequently used or inadvisable (e.g., caprilic acid), we have provided an alphabetical cross-referenced index to the name used herein (octanoic acid) as well as to all the compounds in the book. Finally, we have indexed all compounds by their Chemical Abstracts Service Registry Numbers. We urge all scientists to use Registry Numbers when reporting on any chemical compound, even the simplest.

1.4.4 Emission Sources

The aim of these columns is to tabulate what is known of the origins of the chemical species. Many of the species are known to be emitted by specific processes, or from specific sources. These processes or sources may be natural, but are more often anthropogenic. The source lists are intended to demonstrate the known diversity of origins, and the references are illustrative rather than exhaustive; no more than two references for each type of source are given. The source descriptions are brief: adding "manufacturing" (as in "charcoal manufacturing") often makes them more understandable. The abbreviation "comb." is used to indicate a combustion process. Further information is provided by the reference titles, and, of course, by the references themselves. Sources for each compound are listed in alphabetical order.

A compound may be emitted either as an aerosol (indicated by "a") or as a gas. (In the former case it is more properly regarded as "a component of the liquid or solid aerosol"). A few compounds, such as some of the higher alkanes, are apparently emitted in both phases, because of the wide range of source temperatures. The literature is sometimes unspecific as to the state of the compound upon emission, often because analyses of atmospheric samples collected on filter material are not able to determine the natural state of the detected compounds. The division of emission into gaseous and solid states is thus one that should not be regarded as rigorous. In many cases for plant volatiles, the listed compounds have been identified in the "essential" (i.e., volatile) oil of the plants rather than in the air above them; such identifications carry a letter e after the reference number (i.e., 552e).

It is beyond the scope of this work to critically evaluate each atmospheric measurement cited. The listing of two references to the detection of a specific compound is intended to serve as independent confirmation of that detection. If only one reference is given, however, the detection should be regarded as tentative. In some cases, authors themselves describe identification as tentative, in which circumstance a "t" follows the reference number (e.g., 188t).

Concentrations of compounds emitted from their several sources are not presented in these tables. Such concentrations are extremely dependent on the characteristics of the specific source being measured, on the proximity of the detector to the source, and on the techniques and methodology utilized for measurement. Inclusion of emission stream concentrations in the tables would therefore require substantial additional information to be presented as well; it seems preferable to have interested readers examine the references themselves. To aid in this examination, source references that contain concentration data in addition to species identification are denoted in the tables with an asterisk (e.g., 257*).

1.4.5 Detection of Compounds

The distinction between measurement of a given species from a source and its detection as a gas or aerosol is that in the latter case the detection is made in the absence of any known or suspected sources in the vicinity of the measurement. Authors are sometimes vague on this point; we have placed entries in the Sources columns in those cases where a choice based on inadequate evidence was required, since it is more chemically conservative not to assume a compound to be long-lived enough to be present in the ambient atmosphere. The division between gas phase and aerosol phase is often useful, but, as noted above, such a distinction is sometimes difficult and those presented here must not be regarded as rigorous. Tentative ambient detections are noted by a "t" following the reference number e.g., 121t.

The letters given in parentheses in the Detection column indicate the regime in which detection occurred. The following code is used:

No letter	gas
(a)	aerosol
(c)	cloud
(f)	fog
(i)	ice
(r)	rain
(s)	snow
(I)	indoor
(S)	stratosphere

In the case of ice, entries are made only for species found in ice pellets in the atmosphere, or in glacial ice isolated from sources of compounds other than the atmosphere.

As in the Sources entries, references marked with an asterisk (e.g., 879*(c)) indicate that the reference contains quantitative concentration data as well as qualitative identification information.

In the interest of presenting the source information in the most usable form, we have permitted substantial overlap between certain source listings. For example, many compounds are recorded as being emitted both by 'biomass combustion' and by 'wood combustion'. The latter is obviously included in the former. However, readers interested in emission from residential fireplaces will find the subdivision useful.

1.4.6 Bioassay Information

The intent of these columns is to provide a summary of reviewed and evaluated genetic toxicology bioassay data. Most of the summarized data was reviewed through the U. S. Environmental Protection Agency Gene-Tox Program. This program, described previously[B62,B63] and published as reports in *Mutation Research*,[B1-B30,B36-B38] is intended as a service to those who need a summary of bioassay data and does not reflect EPA policy. Within this program, a panel of experts for each selected bioassay was convened by the EPA and this panel of experts reviewed the data, primarily information existing in peer reviewed journal articles. Only articles containing data sufficient for evaluation were used; abstracts, short communications, and articles without supporting data were deleted from consideration. In most cases, the reviewers agreed with the evaluation of the journal articles' authors; however, if the panel disagreed with the evaluation of the authors, the panel's evaluation was recorded. The user of these columns should be aware that the Gene-Tox evaluation appears here and that this evaluation may differ from that in the open literature. For this reason, users are urged to refer to both the Gene-Tox reports and the original reports for the most complete interpretation of bioassay results. At the time of the writing of this book, two Gene-Tox articles had not been published. The authors felt that these two articles, which described the results of Salmonella bacterial mutagenicity[B36] and whole animal carcinogenicity,[B37] were of such general importance that they requested advance copies of the information. This information was supplied; however, the reader must be aware that some updating of this information may occur before these two articles are published. In addition, another article[B38] which describes the testing of 250 compounds using the Salmonella preincubation assay was included. The authors would warn that the information, summarized in a very qualitative manner, should not be used for hazard and/or risk evaluation. Instead, this summary can be used to locate published reports, to identify the type and amounts of information available, and to ascertain knowledge gaps, especially in relation to air pollutants.

Table 1.4-1 lists the codes for the various bioassays that are summarized. Table 1.4-2 provides a summary of codes used in the two results columns. One column labeled "-MA" records the results of the bioassay when no exogenous metabolic activation system is used. Any metabolism affecting these results is due to the endogenous metabolism of the test organism. The other results column is labeled "+MA," and the results of tests that used an exogenous metabolic activation system are recorded within this column. Although liver homogenates of various mammalian species are the most common form of exogenous metabolic activation, this column includes all forms of exogenous activation (e.g., cell-mediated activation). The final column provides the number for the article used as a reference. These numbers refer to the bioassay reference list.[B1ff]

Table 1.4-1. Codes for Bioassay Assessments

Abbreviation	Bioassay
ALC	Allium cytogenetics assays
ARA	*Arabidopsis thaliana* mutagen assay
ASPD	Aspergillus, diploid systems
ASPH	Aspergillus, haploid systems
CCC	Carcinogen bioassay, whole animal tests
CHOM	CHO/HGPRT mutation system
CTL	Cell transformation, established cell lines
CTP	Cell transformation, primary cells, limited lifetime strains
CTV	Cell transformation, viral enhancement
CYB	Mammalian cytogenetic bone-marrow assay
CYC	Mammalian cytogenetic in vitro cell culture assay, all cell types
CYG	Mammalian in vivo cytogenetic assay, spermatogonial stem cells treated, spermatocytes observed
CYI	Mammalian cytogenetic leukocyte or lymphocyte assay
CYO	Mammalian cytogenetic oocyte and early embryo assay
CYS	Mammalian cytogenetic spermatogonial assay
CYT	Mammalian in vivo cytogenetic assay, differentiating spermatogonia or spermatocytes treated, spermatocytes observed
HOC	*Hordeum vulgare* (barley) chromosome aberration assays

Continued

Table 1.4-1. Continued

Abbreviation	Bioassay
HOM	*Hordeum vulgare* (barley) chlorophyll-deficient mutant assay
HMACH	Host-mediated assay using Chinese Hamster Ovary Cells
HMAEA	Host-mediated assay using Ehrlich ascites tumor cells
HMAEC	Host-mediated assay using *Escherichia coli*
HMAHL	Host-mediated assay using human lymphoid cells
HMALA	Host-mediated assay using Lettre's ascites tumor cells
HMAML	Host-mediated assay using Murine leukemic cells
HMANC	Host-mediated assay using *Neurospora crassa*
HMASC	Host-mediated assay using *Saccharomyces cerevisiae*
HMASM	Host-mediated assay using *Serratia marcescens*
HMASP	Host-mediated assay using *Schizosaccharomyces pombe*
HMAST	Host-mediated assay using *Salmonella typhimurium*
HMAV7	Host-mediated assay using Chinese hamster V79 cells
HMAWC	Host-mediated assay using Walker carcinoma cells
HTM	Heritable translocations in the mouse
L5	L5178Y mouse lymphoma assay
MDR	Mammalian cell DNA repair assays
MNT	Micronucleus test, all species
MST	Mouse spot test
NEU	*Neurospera crassa*, all tests

Continued

Table 1.4-1. Continued

Abbreviation	Bioassay
REC	DNA repair-deficient bacterial assays
SCE	Sister chromatid exchange, all tests
SLF	Mouse specific locus, female, morphology
SLG	Mouse specific locus, male, spermatogonia, morphology
SLP	Mouse specific locus, male, postspermatogonia, morphology
SLV	Mouse specific locus, variable stages, morphology
SPH	Sperm, human studies
SPI	Sperm morphology test in the mouse
SPL	Sperm acrosomal abnormalities in the mouse
SRL	Sex-linked recessive lethal test using *Drosophila melanogaster*
STD00	*S. typhimurium*, Desiccator test, Strain TA100
STD35	*S. typhimurium*, Desiccator test, Strain TA1535
STD37	*S. typhimurium*, Desiccator test, Strain TA1537
STD38	*S. typhimurium*, Desiccator test, Strain TA1538
STD98	*S. typhimurium*, Desiccator test, Strain TA98
STF00	*S. typhimurium*, Fluctuation test, Strain TA100
STF98	*S. typhimurium*, Fluctuation test, Strain TA98
STI00	*S. typhimurium*, Preincubation test, Strain TA100
STI35	*S. typhimurium*, Preincubation test, Strain TA1535
STI37	*S. typhimurium*, Preincubation test, Strain TA1537
STI38	*S. typhimurium*, Preincubation test, Strain TA1538

Continued

Table 1.4-1. Continued

Abbreviation	Bioassay
STI98	*S. typhimurium*, Preincubation test, Strain TA98
STP00	*S. typhimurium*, Plate test, Strain TA100
STP35	*S. typhimurium*, Plate test, Strain TA1535
STP37	*S. typhimurium*, Plate test, Strain TA1537
STP38	*S. typhimurium*, Plate test, Strain TA1538
STP98	*S. typhimurium*, Plate test, Strain TA98
STS00	*S. typhimurium*, Spot test, Strain TA100
STS35	*S. typhimurium*, Spot test, Strain TA1535
STS37	*S. typhimurium*, Spot test, Strain TA1537
STS38	*S. typhimurium*, Spot test, Strain TA1538
STS98	*S. typhimurium*, Spot test, Strain TA98
STU00	*S. typhimurium*, Suspension test, Strain TA100
STU35	*S. typhimurium*, Suspension test, Strain TA1535
STU37	*S. typhimurium*, Suspension test, Strain TA1537
STU38	*S. typhimurium*, Suspension test, Strain TA1538
STU98	*S. typhimurium*, Suspension test, Strain TA98
TRC	Tradescantia cytogenetic tests
TRM	Tradescantia assay for gaseous mutagens
VIC	*Vicia faba* cytogenetic tests
V79A	V79 Chinese Hamster Cells, 8-Azaguanine resistance
V79O	V79 Chinese Hamster Cells, Ouabain resistance
V79T	V79 Chinese Hamster Cells, 6-Thioguanine resistance
WP2	*E. coli* reverse mutation assays
WPU	*E. coli* uvrA reverse mutation assay
YEA	Yeast (*Schizosaccharomyces pombe*) mutation tests
ZMS	*Zea mays* mutation tests

Table 1.4-2. Codes for Bioassay Results

Result	Abbreviation	Notes/Explanations
Negative	NEG	Consensus Negative, usually no comments are given.
	–	Negative with constraints, may be followed by a comment or question mark (?).
	SN*	Sufficient Negative in whole animal carcinogen bioassays (at least two species are used)
Positive	+	Used alone, this is a consensus positive with no necessary comments; otherwise, it is followed by a comment or question mark.
	SP*	Sufficient Positive in whole animal carcinogen bioassays (at least two species are used)
Inconclusive	?	Used alone, even preliminary results could not be assigned. Preliminary and/or tentative indication of results may be given with a + or –. Results may be followed with comments.
	"/"	Multiple results usually seen within a single study or publication differ and/or are inconclusive. (e.g., +/–)
	","	Separates results from different studies that are reported together in a summary publication. (e.g., +,+,–)

* Used for whole animal carcinogen bioassay results.[B37]

Table 1.4-2. Continued

Result	Abbreviation	Notes/Explanations
	NE	Not Evaluated: data or its presentation did not allow adequate evaluation, etc. Usually followed by comments.
	I*	Inadequate: data is inadequate for determination.
	I,N*	Inadequate, Negative: data is inadequate for final determination; however, it is an apparent negative.
	I,P*	Inadequate, Positive: data is inadequate for final determination; however, it is an possible positive compound.
	LN*	Limited Negative: The chemical demonstrated a clear negative response but the data are limited (e.g., by number of strains of animals used.)
	LP*	Limited Positive: The chemical demonstrated a clear positive response but the data are limited.
Comments		Letters following results (May be after a "," or enclosed in parentheses)
	A	No negative control (solvent) data
	B	No positive control data

Continued

Table 1.4-2. Continued

Result	Abbreviation	Notes/Explanations
	C	Not tested to high enough dose
	D	Non-standard metabolic activation
	E	Not enough doses
	F	Reproducibility is in question
	G	Data not adequately presented
	H	Only transformed data
	I	Other committee reasons for "flagging" data.

1.4.7 Structures

The structures of a number of the chemical compounds are included in this book. If a structure is presented for a compound, the species number is followed by an S (e.g., 5.2-10S). The structures for each table are collected and placed at the end. Virtually all of the structures were confirmed by comparison with structures provided by Chemical Abstracts Service, and all are presented in CAS format. Abbreviations for ligands used in the structures are Me: methyl group ($-CH_3$); Et: ethyl group ($-C_2H_5$); Pr-i: isopropyl group ($-CH(CH_3)_2$); and Ac: acetyl group ($-C(O)CH_3$). In cases where the identity of a ligand is known but its locant is not, the ligand and its bonding are indicated beneath the parent structure by the derivative (D_i) notation. For a methyl derivative, for example, the notation is $D_1 - Me$.

The inclusion of only a limited number of structures represents a compromise struck between maximum utility (all structures) and maximum compactness (no structures). We have tried to include a sufficient number so that the working scientist with a reasonable knowledge of chemical nomenclature will seldom be forced to search elsewhere for a structure. For example, we provide a structure for anthracene, but not for 1-methylanthracene nor benz[*a*]anthracene. We also attempt to provide structures in every case where we have chosen a trivial name in preference to a systematic one. We recommend reference 1211 for assistance in relating nomenclature to structure for compounds which are named systematically. Some of these relationships become very complicated, particularly for polynuclear aromatic hydrocarbons and their derivatives, and an unambiguous structure can always be procured by using the Registry Number for access to *Chemical Abstracts*.

1.5 The Text Information

A relatively brief discussion is provided with each of the tables. The intent of this textual material is to place the compounds of the table into proper perspective with the atmosphere as a whole, and to explore their most common sources and removal mechanisms. The first part of the text presents information on the structural similarities of the compounds in the table and on the most common sources of the compounds. Next, the relative occurrences of the compounds in different atmospheric regimes is noted. Finally, the text explores the most important or most probable atmospheric chemical reactions of the compounds.

Atmospheric chemistry has progressed rapidly in the past few years and many of the important chemical processes have now been specified. Such information forms a good base for many of the chemical reaction sequences outlined in succeeding chapters. Conversely, in a substantial number of cases relevant laboratory work remains to be performed. The structural and chemical similarity of many systems to other systems already studied often enables one to proceed by analogy, however, and such an approach is used freely throughout this volume to examine the probable fates of many of the atmospheric compounds.

CHAPTER 2

Inorganic Compounds

2.0 Introduction

Inorganic compounds consist of acids, bases, salts, and oxides of metals or nonmetals, together with the elements that comprise them and the ions[‡] and radicals[*] derived from them. In this chapter, we divide the inorganic compounds into seven groups: (1) elements, elemental ions, and elemental radicals, (2) compounds comprised solely of oxygen and/or hydrogen atoms, (3) compounds containing nitrogen (and H and O but no other

‡ *Ions* are atoms or groups of atoms that carry an electrical charge and do not normally exist in the free state. Negatively-charged ions are termed *anions*; positively charged ions are termed *cations*.

* *Radicals* are uncharged atoms or groups of atoms that do not normally exist in the free state. They possess unpaired electrons, which are indicated by a centered dot following the formula or superimposed on a chemical structure at the site of unsatisfied bonding. Radicals are produced in the atmosphere when a solar photon is absorbed by a molecule and the energy is used to break one or more chemical bonds. In this book, at least, the terms *radical* and *free radical* are synonymous.

elements), (4) compounds containing sulfur (as well as H, O, and N), (5) compounds containing any of the *halogen* (F, Cl, Br, I) atoms (as well as H, O, N, and S), (6) other compounds with fixed composition (largely hydrides, oxides, and carbonates), and (7) minerals.

2.1 Elements, Elemental Ions, Elemental Radicals

More than twenty species of elements and elemental ions are present in the atmosphere. Except for the rare gases, elements are rare except for mercury, which is emitted by a number of natural and anthropogenic processes. In contrast to the elements, elemental ions are ubiquitous. Many are found in aerosols and in all forms of precipitation. They arise from industrial processes such as coal combustion and smelting, but are probably produced in greater quantity by the ionization of soil or seawater constituents present in the atmosphere.

In the case of the particulate chloride ion, sufficient data have been acquired to allow a concentration plot to be drawn, as in Figure 2.1-1. This figure is the prototype for a number of similar diagrams, used throughout this book to show absolute and relative concentrations of abundant species in different atmospheric regimes. The figure shows chloride ion concentrations to be highest over the oceans, presumably as a result of heavy sea spray. Urban areas have high concentrations as well, a reflection of industrial activity. In more remote regions, low concentrations indicate the absence of sources and the effects of diffusion from high concentration regions. Chloride ion is present in the stratosphere as well as the troposphere, probably as a result of haloalkane chemistry. The absence of concentration flags in some regimes indicates that measurements of Cl^- have not been made there, to the authors' knowledge. The chloride ion is a common constituent of precipitation.

Most of the elements that are present in the atmosphere are chemically inert. Chlorine is important to stratospheric chemistry, however (see Chapter 15), and mercury may form alkyl derivatives in the lower atmosphere.

Elemental ion chemistry occurs on aerosol particles and in atmospheric water droplets. The elemental ions are not easily transformed in such environments; their chief impact is to establish the acidity and ionic strength of atmospheric aqueous solutions.

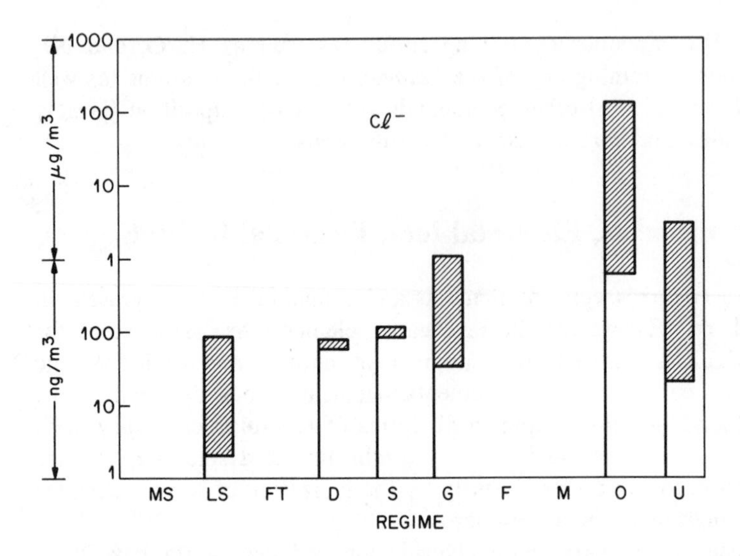

Fig. 2.1-1 Approximate concentration ranges (the upper and lower limits of the rectangles) of particulate chloride ion in different atmospheric regimes. The regime code for this figure and others like it is as follows: U=urban; O=oceanic; M=marshland; F=forest; G=grassland; S = steppes and mountains; D = desert (all of the previous measured within the boundary layer); FT = free troposphere (~5 km altitude); LS = lower stratosphere (15-20 km altitude); MS = middle stratosphere (~25 km altitude). (Adapted from Graedel.[1257])

2.2 Compounds of Hydrogen and Oxygen

The eight compounds in this group are all airborne or dissolved gases which occur throughout the troposphere and stratosphere. Molecular hydrogen has many biogenic and combustion-related sources, while several of the compounds are emitted in geothermal steam and volcano venting.

Two of the most important atmospheric compounds are ozone and the hydroxyl radical, which between them and together with photons initiate virtually all of the oxidation chains in the atmosphere. The concentration diagram for ozone, shown in Figure 2.2-1, shows wide variations in nearly every atmospheric regime. This occurs because ozone is so responsive to the presence of other molecules and to the flux of solar radiation. In the gas phase in the troposphere it is formed by the photolysis of nitrogen

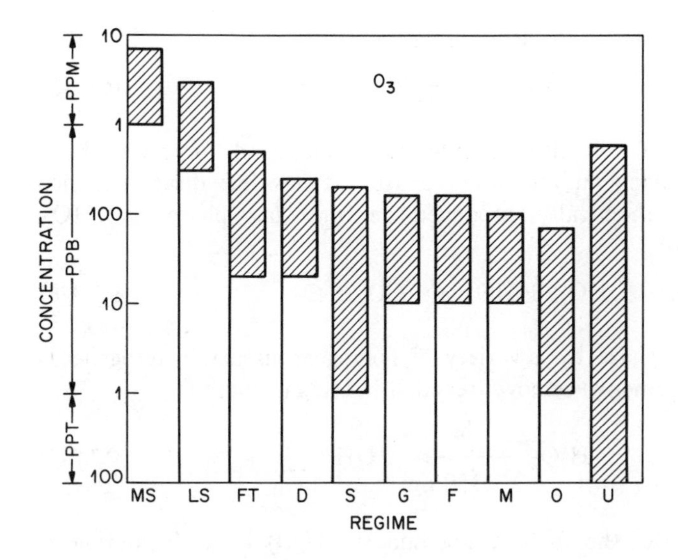

Fig. 2.2-1 Approximate concentration ranges of O_3 in different atmospheric regimes. The symbols are explained in the caption to Fig. 2.1-1.

$$NO_2 \xrightarrow[\lambda<420\ \text{nm}]{h\nu} NO + O \tag{R2.2-1}$$

$$O + O_2 \xrightarrow{M} O_3 \tag{R2.2-2}$$

dioxide and in the stratosphere by the photolysis of molecular oxygen

$$O_2 \xrightarrow[\lambda<234\ \text{nm}]{h\nu} O + O \tag{R2.2-3}$$

followed by (R2.2-2). Removal reactions are many, but among the most important are reactions with NO in the troposphere

$$O_3 + NO \rightarrow O_2 + NO_2 \tag{R2.2-4}$$

and with oxygen atoms in the stratosphere:

$$O_3 + O \rightarrow 2O_2 \qquad \text{(R2.2-5)}$$

Hydrogen peroxide is also an important compound for atmospheric chemistry, particularly in the liquid phase. It has few direct emission sources, but is chemically produced in the gas phase by $HO_2\cdot$ disproportionation*

$$HO_2\cdot + HO_2\cdot \rightarrow H_2O_2 + O_2 \qquad \text{(R2.2-6)}$$

and in the liquid phase by a variety of homogeneous and heterogeneous processes.[1267] Its principal removal reaction in the gas phase is

$$H_2O_2 \xrightarrow[\lambda<350\text{ nm}]{h\nu} 2OH\cdot . \qquad \text{(R2.2-7)}$$

In the liquid phase, the favored reaction for H_2O_2 is the oxidation of bisulfite ion:

$$\left\{H_2O_2 + HSO_3^- \rightarrow H_2O + HSO_4^-\right\}_{aq} \qquad \text{(R2.2-8)}$$

Proton hydrates appear to be common in the stratosphere, where they form by the clustering of water molecules around protons:

$$H^+(H_2O)_m + H_2O \rightarrow H^+(H_2O)_{m+1} \qquad \text{(R2.2-9)}$$

The details of stratospheric ion chemistry remains to be established, but the mechanisms and rates of the most significant processes seem reasonably well understood.[1322]

The gas phase chemistry of the $HO_x\cdot$ ($OH\cdot$, $HO_2\cdot$) radicals is crucial to most atmospheric oxidation processes. Although (R2.2-7) contributes to hydroxyl radical production, $OH\cdot$ is produced primarily by the high-energy photodissociation of ozone, followed by reaction of the excited oxygen atom with water vapor:

* *Disproportionation*: A chemical reaction in which a single compound serves as both oxidizing and reducing agent.

$$O_3 \xrightarrow[\lambda<320\ nm]{h\nu} O_2 + O(^1D) \qquad (R2.2\text{-}10)$$

$$O(^1D) + H_2O \rightarrow 2OH\cdot\ . \qquad (R2.2\text{-}11)$$

The gas phase $HO_2\cdot$ radical is generated largely through hydrocarbon reaction chains which will be discussed in Chapter 3. The concentrations of $OH\cdot$ and $HO_2\cdot$ are very low and their measurement is among the most difficult analytical problems of gas phase atmospheric chemistry.

There is uncertainty concerning the presence of $HO_x\cdot$ radicals in aqueous atmospheric droplets. It is thought that they may be incorporated from the gas phase,[1277,1323] although no satisfactory measurements of this process have yet been made. An alternative proposal[1326] is that iron ions in solution can complex hydroxyl ions and undergo photoreduction to produce $OH\cdot$:

$$\left\{Fe^{3+} + OH^- \rightarrow [Fe^{3+}OH^-]^{2+}\right\}_{aq} \qquad (R2.2\text{-}12)$$

$$\left\{[Fe^{3+}OH^-]^{2+} \xrightarrow{h\nu} Fe^{2+} + OH\cdot\right\}_{aq} \qquad (R2.2\text{-}13)$$

If present from either source, the radicals will promote extensive chemical reactions in atmospheric droplets.

2.3 Inorganic Nitrogen Compounds

Twenty species comprise Table 2.3. About half commonly occur in the gas phase, about half in condensed phases. Several are found in the stratosphere. The sources of these species are diverse. In the case of N_2O and NH_3, the molecules are emitted by many natural and anthropogenic processes. NO and NO_2, in contrast, are almost entirely products of any type of high temperature combustion and are thus emitted by power plants, vehicles, forest fires, refuse burning, etc. No significant direct sources of nitrates are known.

The location and strength of sources and the chemical reactivity of the compounds establish the atmospheric concentrations of the inorganic nitrogen species. Several have been subject to extensive analytical investigation, permitting concentration diagrams to be constructed for them. That for ammonia is shown in Figure 2.3-1. The concentrations are highest in urban areas and in agricultural regions where fertilizer and

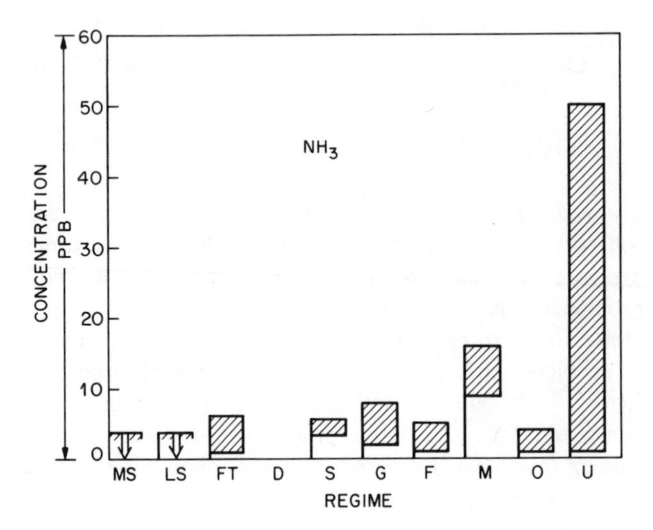

Fig. 2.3-1 Approximate concentration ranges of ammonia in different atmospheric regimes. Most of the symbols are explained in the caption to Fig. 2.1-1. The symbols for the stratosphere indicate that only upper limit concentrations have been established.

cattle feedlot sources exist. Elsewhere, ammonia levels are much lower. The urban-rural dichotomy is even plainer for NO (Figure 2.3-2, logarithmic scale), with its dominant combustion sources. HNO_3 is largely a product of NO chemistry and thus its concentration diagram (Figure 2.3-3) is similar to that for NO. We do not show a diagram for N_2O, although it has been monitored extensively, since its long lifetime has permitted a uniform concentration of about 330 ppb to become established throughout the troposphere.

Nitrogen compounds are common in atmospheric aerosol particles as well as within the gas phase. For both ammonium and nitrate ions, sufficient data exist to permit concentration plots to be drawn. In the case of ammonium ion, as shown in Figure 2.3-4, the pattern reflects the sources of ammonia: fertilizer, animal urine, and industrial processes. Some ammonium ion is apparently present in the stratosphere, although measurements are not yet definitive.

The concentration plot for nitrate ion is presented in Figure 2.3-5. Urban concentrations are highest, but considerably more uniformity exists for NO_3^- than for NH_4^+. This reflects the production of NO_3^- by a sequence involving gas phase reactions of NO and NO_2; because of the multistep process, atmospheric mixing has time to disperse the nitrate precursors more uniformly than is possible for the NH_3/NH_4^+ couple.

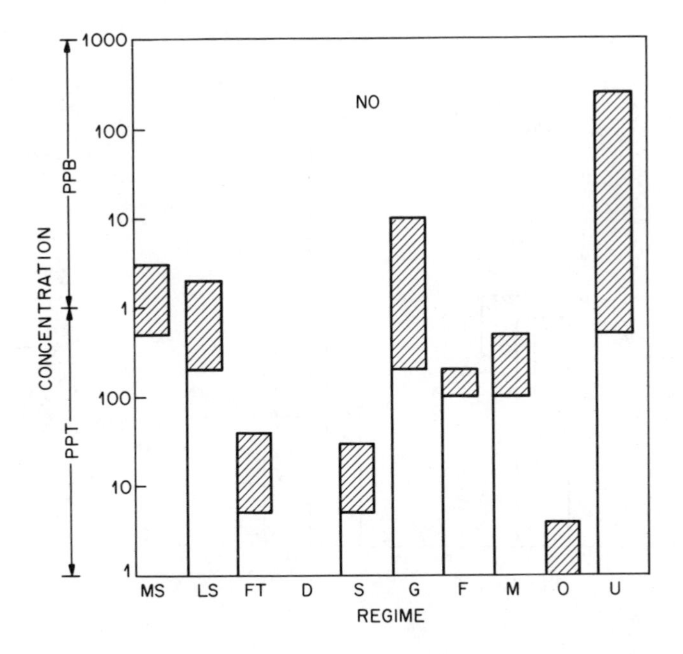

Fig. 2.3-2 Approximate concentration ranges of nitric oxide in different atmospheric regimes. The symbols are explained in the caption to Fig. 2.1-1.

NH_4^+ and NO_3^- are also invariably found in precipitation.

Gaseous ammonia reacts rather slowly with OH· to produce a nitrogen-containing radical:

$$NH_3 + OH\cdot \rightarrow NH_2\cdot + H_2O \qquad \text{(R2.3-1)}$$

The fate of this radical is uncertain, but it seems most likely to react with ozone[1256] in a chain that eventually forms oxides of nitrogen. The aqueous solubility of ammonia is high, however, and most NH_3 is lost through surface deposition rather than through gas-phase atmospheric chemistry.

In contrast to the situation with ammonia, the oxides of nitrogen are vital components of tropospheric gas phase chemistry. The species initially emitted is generally NO, which is oxidized within concentrated exhaust plumes by

$$2NO + O_2 \rightarrow 2NO_2 \qquad \text{(R2.3-2)}$$

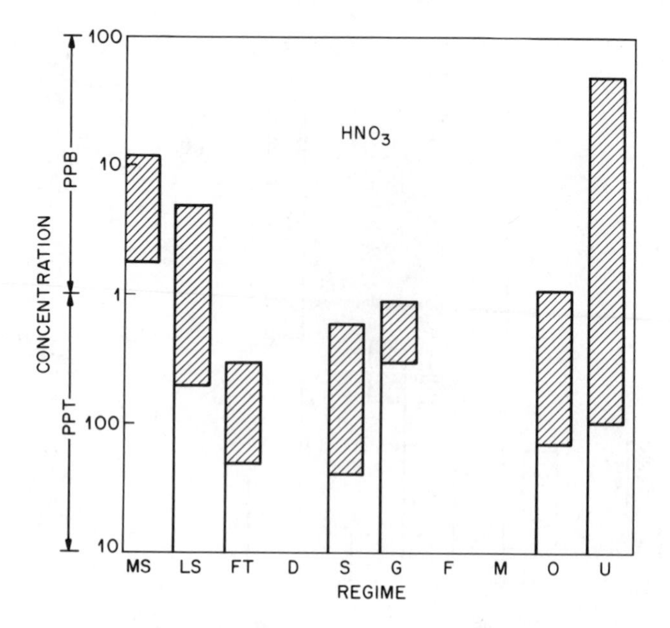

Fig. 2.3-3 Approximate concentration ranges of gaseous nitric acid in different atmospheric regimes. The symbols are explained in the caption to Fig. 2.1-1.

and in the free troposphere by ozone:

$$NO + O_3 \rightarrow NO_2 + O_2 \,. \qquad \text{(R2.3-3)}$$

As seen previously (Section 2.2), NO_2 can be photoreduced to produce ozone. It can also be oxidized to either the nitrate radical or to nitric acid

$$NO_2 + O_3 \rightarrow NO_3 + O_2 \qquad \text{(R2.3-4)}$$

$$NO_2 + OH\cdot \xrightarrow{M} HNO_3 \qquad \text{(R2.3-5)}$$

Ozone and the nitrate radical are two of the more vigorous atmospheric oxidizing compounds and are responsible for limiting the lifetimes of a number of atmospheric trace gases in and near urban areas.

The gas phase chemistry of inorganic nitrogen compounds in the stratosphere is no less important. Much of that chemistry is initiated by the photolysis of O_3 followed by reaction with N_2O

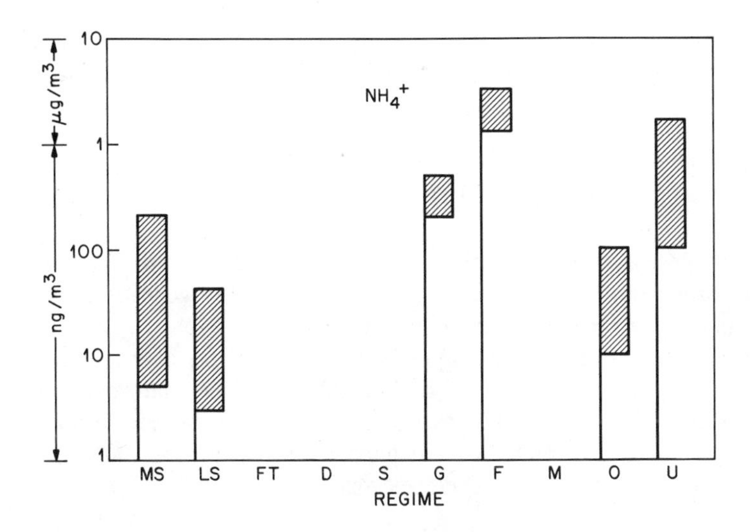

Fig. 2.3-4 Approximate concentration ranges of particulate ammonium ion in different atmospheric regimes. The symbols are explained in the caption to Fig. 2.1-1.

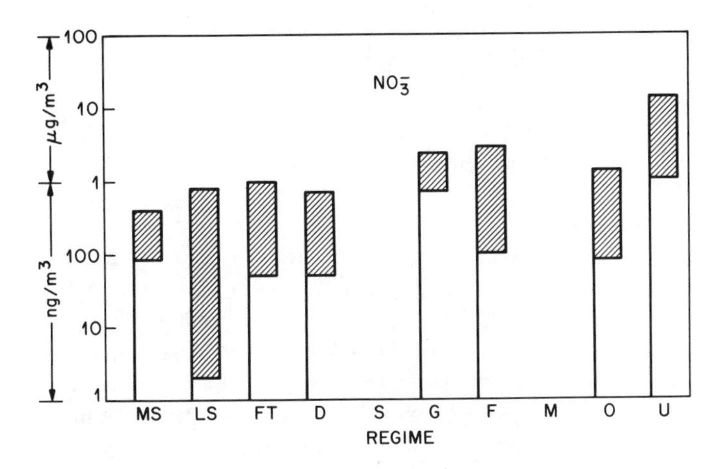

Fig. 2.3-5 Approximate concentration ranges of particulate nitrate ion in different atmospheric regimes. The symbols are explained in the caption to Fig. 2.1-1.

$$O_3 \xrightarrow[\lambda<320\ \text{nm}]{h\nu} O_2 + O(^1D) \tag{R2.3-6}$$

$$N_2O + O(^1D) \rightarrow 2NO \tag{R2.3-7}$$

The oxides of nitrogen react with "odd oxygen" (O, O_3) to balance stratospheric ozone production and thus help to define the concentration profiles of important stratospheric trace compounds.

In contrast to the gas phase, the aqueous phase chemistry of inorganic nitrogen compounds appears to be of limited interest. The main species are NH_4^+ and NO_3^-. NH_4^+ does not participate in any known reactions in atmospheric droplets. The NO_3^- ion participates in one process of some importance: a weak photolysis to generate ozone by

$$\left\{NO_3^- \xrightarrow[\lambda<330\ \text{nm}]{h\nu} NO_2^- + O\right\}_{aq} \tag{R2.3-8}$$

$$\{O + O_2 \rightarrow O_3\}_{aq} \,. \tag{R2.3-9}$$

2.4 Inorganic Sulfur Compounds

More than sixty inorganic sulfur compounds are known to be emitted into or detected in the atmosphere. Most of these occur in aerosol particles, but several gas phase species are also commonly present in the troposphere. The stratosphere contains several sulfur compounds.

A number of gases containing sulfur in a reduced valence state are generated by natural biological processes. Industrial processes are important sources for such compounds as well. Condensed phase sulfates are primarily produced as byproducts of fossil fuel combustion.

The concentrations of several gas phase inorganic sulfur compounds are well established. For hydrogen sulfide, the concentration diagram is shown in Figure 2.4-1; it clearly suggests the principal sources of H_2S to be in urban areas and in marshland. This diagram forms an interesting contrast with that for COS (Figure 2.4-2). This latter gas has a much lower flux to the atmosphere than does H_2S, but it is so unreactive that its lifetime is very long and it is evenly mixed throughout the troposphere.

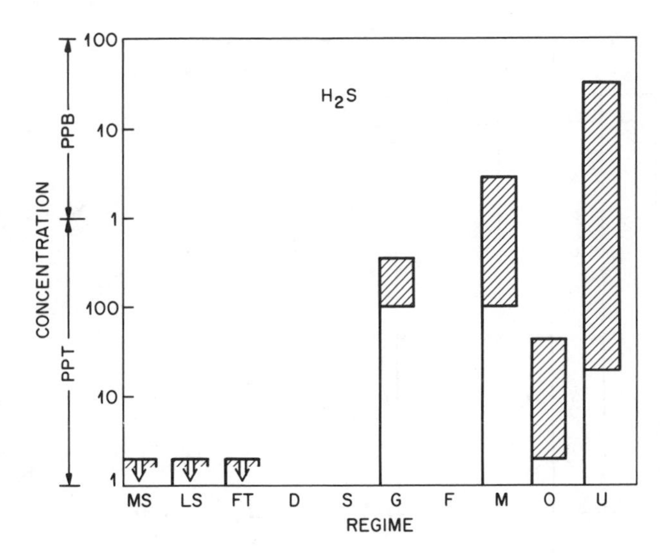

Fig. 2.4-1 Approximate concentration ranges of hydrogen sulfide in different atmospheric regimes. The values for the free troposphere and stratosphere are upper limits. The symbols are explained in the caption to Fig. 2.1-1.

The concentration diagram for SO_2 is shown in Figure 2.4-3. SO_2 has several very strong anthropogenic sources and a few weak natural ones. As a result, its concentrations are much higher in urban areas than elsewhere. In remote areas, SO_2 is thought to be produced chemically from naturally emitted CS_2.

The sulfate ion is a major component of most atmospheric particulate matter. Its concentration diagram, shown in Figure 2.4-4, indicates that the urban levels are the highest in the atmosphere. Sulfate concentrations are nearly as high over a few forests which have been studied, but these levels are attributed to transport from urban areas. Concentrations are much lower elsewhere. They are comparable in the free troposphere and in the stratosphere, a circumstance that reflects the active sulfur chemistry that occurs in both those regimes.

The gas phase atmospheric chemistry of hydrogen sulfide is initiated by $OH\cdot$ radical reaction and proceeds to sulfur dioxide formation:

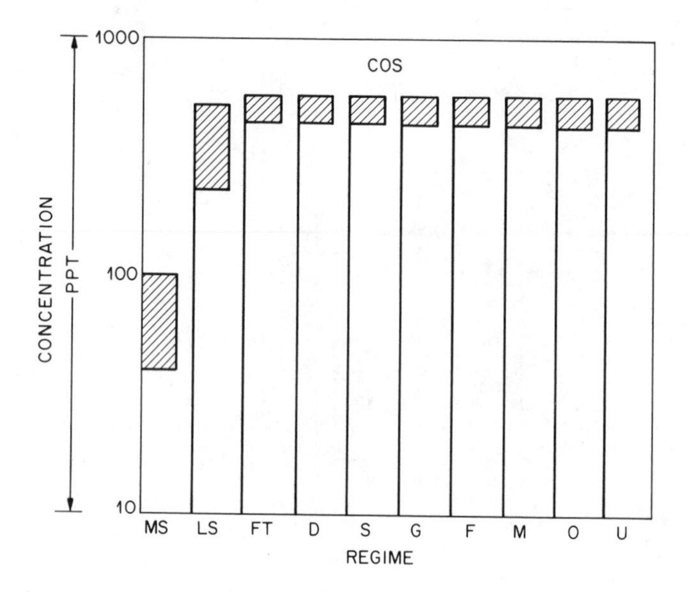

Fig. 2.4-2 Approximate concentration ranges of carbonyl sulfide in different atmospheric regimes. The symbols are explained in the caption to Fig. 2.1-1.

$$H_2S + OH\cdot \rightarrow HS\cdot + H_2O \qquad \text{(R2.4-1)}$$

$$HS\cdot + 2O_2 \rightarrow HO_2\cdot + SO_2 \qquad \text{(R2.4-2)}$$

Sulfur dioxide also reacts exclusively in the atmosphere with the hydroxyl radical, the eventual product being sulfuric acid:

$$SO_2 + OH\cdot \xrightarrow{M} HSO_3\cdot \qquad \text{(R2.4-3)}$$

$$HSO_3\cdot + O_2 \rightarrow SO_3 + HO_2\cdot \qquad \text{(R2.4-4)}$$

$$SO_3 + H_2O \rightarrow H_2SO_4 . \qquad \text{(R2.4-5)}$$

Sulfuric acid is an efficient condensation center for atmospheric water vapor, creating the sulfate aerosol particles seen throughout the troposphere.

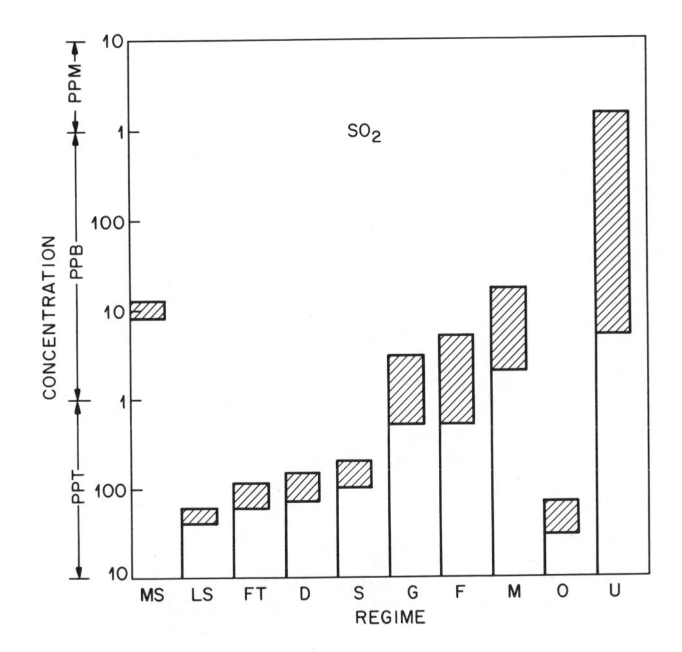

Fig. 2.4-3 Approximate concentration ranges for sulfur dioxide in different atmospheric regimes. The symbols are explained in the caption to Fig. 2.1-1.

An alternative fate for SO_2 is dissolution in atmospheric water droplets. At the moderately acidic conditions normally present, dissolved SO_2 ionizes to bisulfite. This ion is then transformed to sulfate by any of several oxidizers, H_2O_2 being perhaps the most important.

$$\left\{SO_2 \cdot H_2O \rightleftarrows H^+ + HSO_3^-\right\}_{aq} \qquad \text{(R2.4-6)}$$

$$\left\{HSO_3^- \xrightarrow{H_2O_2} SO_4^{2-}\right\}_{aq} \qquad \text{(R2.4-7)}$$

COS is the dominant sulfur gas in the stratosphere, where it photolyzes to form SO_2:

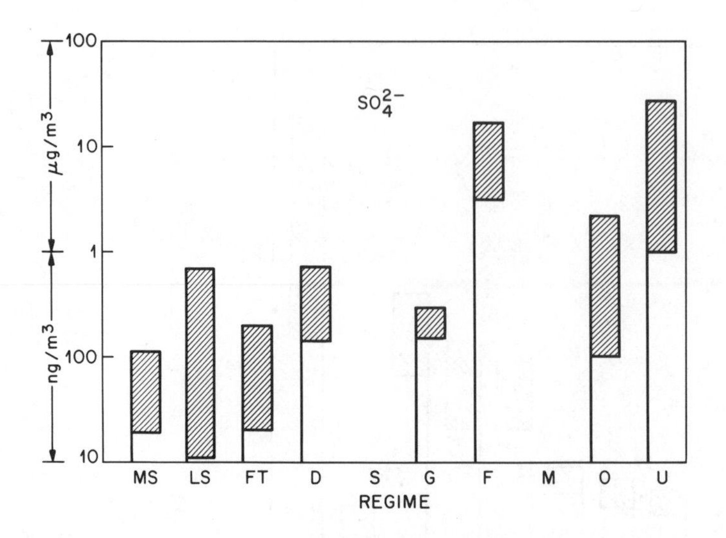

Fig. 2.4-4 Approximate concentration ranges for particulate sulfate ion in different atmospheric regimes. The symbols are explained in the caption to Fig. 2.1-1.

$$COS \xrightarrow[\lambda<255\ nm]{h\nu, O_2} CO + SO_2 \qquad (R2.4\text{-}8)$$

As in the troposphere, the SO_2 is oxidized to sulfuric acid and transformed to the aerosol phase, in which form it is eventually removed by gravitational setting.

2.5 Inorganic Halogenated Compounds

Nearly fifty inorganic halogenated compounds have been found in the atmosphere. Most of those in the lower atmosphere occur in the condensed phase. A few have been detected in the stratosphere. The simplest of the compounds, especially the halogen acids, are emitted from diverse industrial processes. Metal forming and refining generate most of the aerosol chlorides. The chloride and bromide salts of lead are emitted from internal combustion engines.

Small amounts of the dihalogen molecules enter the atmosphere from industrial processes. Their lifetimes are short, as they are photosensitive. For chlorine, for example,

$$Cl_2 \xrightarrow[\lambda<430\ \text{nm}]{h\nu} Cl\cdot + Cl\cdot \qquad \text{(R2.5-1)}$$

followed by reaction of the chlorine radical with virtually any organic molecule (RH) to produce hydrochloric acid

$$RH + Cl\cdot \rightarrow R\cdot + HCl \qquad \text{(R2.5-2)}$$

The halogen acids are highly water soluble, and are thus promptly incorporated into water droplets or onto a variety of ground surfaces.

In the stratosphere, the halogen atoms and small molecules play major chemical roles. The most important is a catalytic cycle with ozone and oxygen atoms:

$$Cl\cdot + O_3 \rightarrow ClO\cdot + O_2 \qquad \text{(R2.5-3)}$$

$$ClO\cdot + O \rightarrow Cl\cdot + O_2 \,. \qquad \text{(R2.5-4)}$$

The stratospheric halogen atoms arise from photolysis of chlorofluoromethanes, and the two reactions above play an important role in determining the concentration of stratospheric ozone.

2.6 Hydrides, Oxides, Carbonates

The compounds in this group have great chemical diversity. Carbon appears here rather than in Table 2.1 because it exists as an element in the atmosphere only in chain and layer structures as shown in Figure 2.6-1. It is common in atmospheric aerosols and in precipitation, being emitted from fossil fuel combustion sources and several industrial processes. It is a major factor in visibility degradation.[1265] Because of the propensity of carbon to adsorb gaseous and dissolved species, it is suspected of playing catalytic roles in condensed phase chemistry, including acceleration of the oxidation of S(IV) to S(VI).[1266] (The roman numerals indicate the valence state of sulfur.)

Arsine and phosphine are hydrides used in electronics and light industry. They oxidize readily upon exposure to the atmosphere and their toxicity is high. They are monitored closely enough so that the atmospheric release rates are very small.

Most of the oxides in Table 2.6 are derived from windblown dust or from industries processing ore or rock (smelting, cement manufacture, etc.). They are quite unreactive and, with the possible exception of transition metal oxides, have no chemical role in the atmosphere. Two

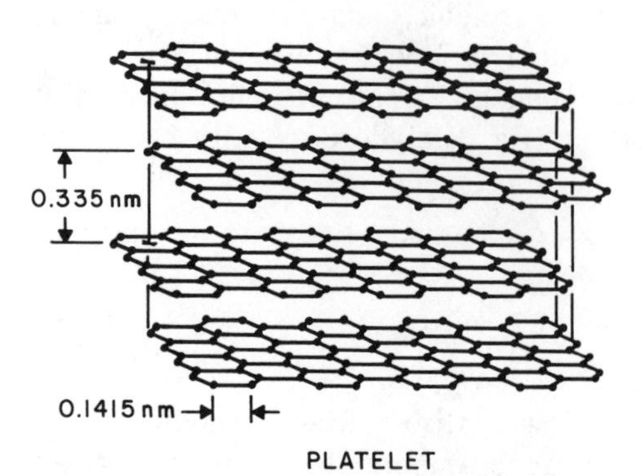

Fig. 2.6-1 Structural morphology of elemental carbon in the atmosphere. The elemental carbon found in the atmosphere shows only microcrystalline structure and does not reflect the macroscopic crystalline properties of graphite. (Reproduced with permission.[1264])

oxides, however, those of carbon, are very important. CO and CO_2 are both produced at high rates by all combustion processes involving carbon compounds. CO concentrations near combustion sources can be very high, as seen in Figure 2.6-2. Away from those sources the concentration is rather constant, both because CO is not very reactive and because it is produced at a moderate rate throughout the troposphere by the photolysis of formaldehyde and by the reaction of formaldehyde with OH·. The pattern of CO_2 concentrations (not shown) is very uniform. CO_2 is completely unreactive in the atmosphere and, in addition to its generation by combustion sources, is produced in urban regions and throughout the troposphere by the reaction

$$CO + OH\cdot \rightarrow CO_2 + H\cdot \qquad (R2.6\text{-}1)$$

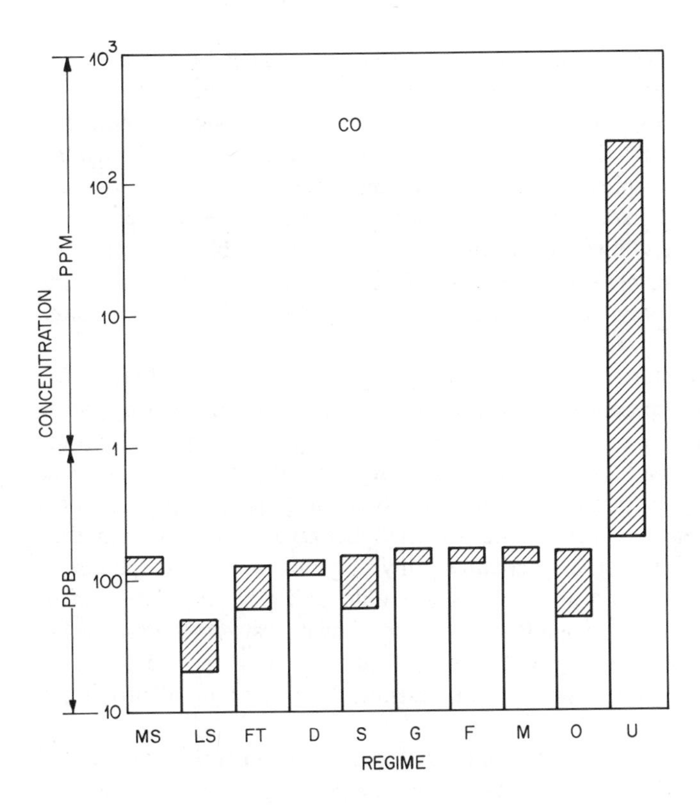

Fig. 2.6-2 Approximate concentration ranges of CO in different atmospheric regimes. The regime code is given in the caption of Figure 2.2-1. (Adapted from Graedel.[1257])

The carbonates and other compounds in Table 2.6 are primarily derived from windblown dust or industrial processes. They are not known to participate in atmospheric chemistry to any significant degree.

2.7 Minerals

Minerals are crystalline materials of definite chemical composition. A distinguishing feature between minerals and the other compounds in this book is that any of several ions may substitute for each other within many mineral formulas, so that the chemical formula of a mineral is stoichiometric only in a general sense. One thus has formulas such as that of hypersthene, (Mg, Fe) SiO_3, in which either a magnesium atom or an iron atom, but not both, is present for each occurrence of SiO_3.

Some mineral formulas fall into the inorganic divisions of Tables 2.3 to 2.6 and are sufficiently specific to appear there. These compounds may be present in an atmospheric particle as a result of anthropogenic industrial activity as well as by the mechanical injection of a portion of the earth's crust into the atmosphere by windblown soil processes, cyclones, volcanic emissions, and the like. The compounds in Table 2.7, however, enter the atmosphere entirely by natural processes (or presumably by industrial processes merely involving rock or soil dispersal). They are present in the condensed phase only, and in the form of crystals rather than isolated molecules.

The first characteristic used to identify minerals is the crystal structure, the second is the composition. The crystal structure is indicated by the x-ray diffraction pattern, which serves to identify the mineral group. Further analysis may then be used to identify the particular mineral. This procedure may be thought of as roughly analogous to the analytical chemistry sequence of identifying an alkyl benzene compound and then going on to identify the particular alkyl ligands and their points of attachment. Just as some scientists are able to report the detection of specific gas phase compounds while some must restrict themselves to identifying a compound group, some scientists report the detection of specific minerals while others must restrict themselves to identifying a mineral group.

Table 2.7 contains 31 entries and includes both minerals and mineral groups as reported in the atmospheric literature. We do not differentiate here between them, nor do we describe their interrelationships. To do so would involve presenting a short course on mineralogy, which is obviously inappropriate. For those who wish more information, however, we note that we have found Frye[1285] particularly helpful.

Table 2.1. Elements and Elemental Ions

Species Number	Registry Number	Name	Emission Source	Emission Ref.	Detection Ref.	Bioassay Code	Bioassay −MA	Bioassay +MA	Bioassay Ref.
2.1-1	12586-59-3	**Hydrogen ion**			889*(a),911*(a) 750*(c),751*(c) 746*(f),796*(f) 776*(i),779*(i) 760*(r),764*(r) 752*(s),756*(s)				
2.1-2	12768-75-1	**Helium atom**	natural gas volcano	456* 73	294*,558* 294*(S),558*(S)				
2.1-3	17778-80-2	**Oxygen atom**			294*(S),704*(S)				
2.1-4	16984-48-8	**Fluoride ion**	coal comb. smelting volcano	889(a) 889(a) 1038(a)	882*(a),889*(a) 746*(f) 777*(i) 915*(r) 939(s)	CYI	NEG		B8
2.1-5	7440-01-9	**Neon atom**			294*,558*				

Table 2.1. *(Continued)*

Species Number	Registry Number	Name	Emission Source	Emission Ref.	Detection Ref.	Bioassay Code	−MA	+MA	Ref.
2.1-6	17341-25-2	**Sodium ion**	coal comb.	889(a)	889*(a),911*(a)				
			smelting	889(a)	768*(c),769*(c)				
					746*(f)				
					760*(r),765*(r)				
					752*(s),756*(s)				
					806*(I)				
2.1-7	22537-22-0	**Magnesium ion**			911*(a),1014*(a)				
					769*(c)				
					746*(f)				
					760*(r),765*(r)				
2.1-8	22537-23-1	**Aluminum ion**			764(r)				
2.1-9	22537-15-1	**Chlorine atom**			701*(S)				
2.1-10	16887-00-6	**Chloride ion**	coal comb.	889(a)	889*(a),909*(a)				
			smelting	889(a)	768*(c),769*(c)				
			volcano	1038(a)	746*(f),796*(f)				
					777*(i)				
					760*(r),765*(r)				
					752*(s),756*(s)				
					1033*(S)				
					806*(I)				
2.1-11	7440-37-1	**Argon atom**	geothermal steam	984,998*	294*,558*				
			natural gas	456*					
			volcano	106*,234					

Species Number	Registry Number	Name	Emission Source	Ref.	Detection Ref.	Bioassay Code	−MA	+MA	Ref.
2.1-12	24203-36-9	**Potassium ion**	coal comb.	889(a)	889*(a),911*(a)				
			smelting	889(a)	769*(c),912*(c)				
					746*(f)				
					760*(r),765*(r)				
					756*(s)				
2.1-13	14127-61-8	**Calcium ion**	coal comb.	889(a)	911*(a)				
					768*(c),769*(c)				
					746*(f)				
					760*(r),765*(r)				
					756*(s),939(s)				
2.1-14	20074-52-6	**Iron (III) ion**	coal comb.	889(a)					
2.1-15	14701-22-5	**Nickel (II) ion**	coal comb.	889(a)					
2.1-16	23713-49-7	**Zinc (II) ion**	coal comb.	889(a)					
2.1-17	7782-49-2	**Selenium atom**			825(a)				
2.1-18	24959-67-9	**Bromide ion**			909*(a),934(a)				
2.1-19	7439-90-9	**Krypton atom**			294*,558*				
2.1-20	7440-63-3	**Xenon atom**	nuclear power	391*	294*,558*				

Table 2.1. *(Continued)*

Species Number	Registry Number	Name	Emission Source	Emission Ref.	Detection Ref.	Bioassay Code	−MA	+MA	Ref.
2.1-21	7439-97-6	**Mercury atom**	cement mfr.	907	286,529*				
			chlorine mfr.	27,58*					
		coal comb. 490*,907; geothermal steam 489,998*; mercury mining 286; paint 66,222; refuse comb. 403*,1044; sewage tmt. 529*; vegetation 1043*; volcano 61,914*							
2.1-22	14280-50-3	**Lead (II) ion**	coal comb.	889(a)					
2.1-23	10043-92-2	**Radon atom**	geothermal steam	998*	494*(a)				
			soil	467,483	1035(I)				
			volcano	430*					

Table 2.2. Compounds of Hydrogen and Oxygen

Species Number	Registry Number	Name	Emission Source	Emission Ref.	Detection Ref.	Code	Bioassay −MA	Bioassay +MA	Bioassay Ref.
2.2-1	1333-74-0	**Dihydrogen**	auto	401*	291*,294*				
			biomass comb.	354,655*	291*(S),718*(S)				
		geothermal steam 372,402*; HCl mfr. 622; insects 785*; microbes 210; natural gas 423*,456*; oceans 291; rocket 536,1102; sewage tmt. 1013*; turbine 836*; vegetation 461; volcano 106*,246							
2.2-2	7782-44-7	**Dioxygen**	geothermal steam	984,998*	294*,558*	V79A	+		B7
			volcano	73,604	294*(S),558*(S)				
2.2-3	10028-15-6	**Ozone**	lightning	318	283*,410*	TRM	+		B21
			power trans.	318	700*(S),734*(S)	SCE	NEG		B6
					965*(I),979*(I)	CYB	NEG		B8
						CYG	NEG		B8
						CYI	+/−		B8
						VIC	+		B18
2.2-4	7732-18-5	**Water vapor**	geothermal steam	998*	294,1201*				
			volcano	106*,604	738*(S),739*(S)				
2.2-5S	65058-03-9	**Proton hydrates**			723(S),730(S)				
2.2-6S	3352-57-6	**Hydroxyl radical**	turbine	835*	259*,287*				
					705*(S),742*(S)				

Table 2.2. *(Continued)*

Species Number	Registry Number	Name	Emission Source	Emission Ref.	Detection Ref.	Bioassay Code	Bioassay −MA	Bioassay +MA	Bioassay Ref.
2.2-7S	3170-83-0	**Hydroperoxyl radical**			1041*,1137*				
					743*(S),995*(S)				
2.2-8S	7722-84-1	**Hydrogen peroxide**	auto	910	641*,667*	STI98	−(BE)		B36
					750*(c),767*(c)	STI00	−(B)		B36
					1115*(f)	REC		+	B9
					1114*(i)	VIC	NEG		B18
					749*(r),766*(r)	YEA	+		B27
						NEU	+		B32

$H^{+}\cdot(H_2O)_n$ 2.2-5

$OH\cdot$ 2.2-6

$HO_2\cdot$ 2.2-7

HOOH 2.2-8

Table 2.3. Inorganic Nitrogen Compounds

Species Number	Registry Number	Name	Emission Source	Emission Ref.	Detection Ref.	Code	Bioassay −MA	Bioassay +MA	Bioassay Ref.
2.3-1	7727-37-9	**Dinitrogen**	geothermal steam	372,402*	294*,558*				
			propellant	559,563	294*(S),558*(S)				
			volcano	73,106*					
2.3-2	7664-41-7	**Ammonia**	acrylonitrile mfr.	626	23*,342*	SRL	?		B30
			adhesives	11					
		ammonia mfr. 58*; animal waste 141,160; auto 565,581; cement mfr. 1094; chemical mfr. 1147*; coal comb. 895,1094; coke mfr. 58*,1094; diesel 795*,1094; fertilizer mfr. 58*,256*; fish processing 110,255*; foundry 279*,1094; geothermal steam 372,402*; lacquer mfr. 427,428; leather mfr. 1151*,1152; microbes 302,1094; Na_2CO_3 mfr. 58*; petroleum mfr. 58*,221; plastics comb. 46,354; polymer comb. 611; refrigeration 559; refuse comb. 395,1094; rendering 637,1078; rocket 536; sewage tmt. 174*,1013*; starch mfr. 695,1071; tobacco smoke 396; vegetation 1068*; volcano 73,913							
2.3-3	14798-03-9	**Ammonium ion**	coal comb.	889(a)	889*(a),911*(a)				
			smelting	889(a)	750*(c),751*(c)				
			volcano	1038(a)	746*(f),796*(f)				
					760*(r),764*(r)				
					752*(s),756*(s)				
					1033*(S)				
					806*(I)				

Table 2.3. *(Continued)*

Species Number	Registry Number	Name	Emission Source	Emission Ref.	Detection Ref.	Bioassay Code	−MA	+MA	Bioassay Ref.
2.3-4		**Ammonium ion-water (1/2)**			1081t				
2.3-5	151-50-8	**Potassium cyanide**	steel mfr.	225(a)		VIC	+		B18
						ARA	NEG		B17
2.3-6	1336-21-6	**Ammonium hydroxide**	petroleum stor.	58*					
2.3-7	10102-43-9	**Nitric oxide**	acrylonitrile mfr.	626	282*,1181*				
			auto	282,444*	399(a)				
		biomass comb. 655*; coal comb. 282,895; diesel 136*,895; HNO_3 mfr. 669*; lightning 644; microbes 210,302; refuse comb. 1093; rocket 622,922*; sewage tmt. 1013*; tobacco smoke 634; turbine 835*,836*; vegetation 649; volcano 1149			700*(S),709*(S) 963*(I),964*(I)				
2.3-8	10102-44-0	**Nitrogen dioxide**	acrylonitrile mfr.	626	282*,1182*				
			auto	282,444*	399(a)				
		biomass comb. 662*,1155; coal comb. 282,895; diesel 136*,895; HNO_3 mfr. 669*; lightning 644,1037; microbes 975*,976*; refuse comb. 661,1093; rocket 922*,935; tobacco smoke 4*; turbine 836*,895; vegetation 649; volcano 234; wood comb. 830*,890			709*(S),735*(S) 893*(I),963*(I)				
2.3-9	14797-65-0	**Nitrite ion**	coal comb.	889(a)	909*(a),934(a) 933(s),941(s)				
2.3-10	12033-49-7	**Nitrate radical**			683*,996* 708*(S)				

Species Number	Registry Number	Name	Emission Source	Emission Ref.	Detection Ref.	Bioassay Code	Bioassay −MA	Bioassay +MA	Bioassay Ref.
2.3-11	14797-55-8	**Nitrate ion**	smelting	889(a)	1081t				
			volcano	1038(a)	889*(a),909*(a)				
					750*(c),751*(c)				
					746*(f),796*(f)				
					777*(i),779*(i)				
					760*(r),765*(r)				
					753*(s),756*(s)				
					798(S)t,1168(S)t				
					806*(I),1035(I)				
2.3-12	10024-97-2	**Nitrous oxide**	auto	807*,895	296*,299*	TRM	+		B21
			biomass comb.	655*	399(a)	V79A	NEG		B7
		coal comb. 353*,656; lightning 644,740*; ocean 297*; polymer comb. 611; propellant 562,563; tobacco smoke 634; turbine 836*; vegetation 547; volcano 783*,1091			983*(r)	CCC	I,P		B37
					983*(s)	SPI	NEG		B25
					713*(S),718*(S)				
2.3-13	10102-03-1	**Dinitrogen pentoxide**	fertilizer mfr.	278*					
2.3-14	7782-77-6	**Nitrous acid**	auto	885*	320*,696*	REC	+		B9
						YEA	+		B27
						ASPD	+		B10
						ASPH	+		B11
						NEU	+		B32
2.3-15	7697-37-2	**Nitric acid**	coal comb.	961	226*,608*				
			diesel	774	151(a),213(a)				
		explosives mfr. 58*; HNO_3 mfr. 35,58*; rocket 536; volcano 1149			753*(s)				
					702*(S),710*(S)				

Table 2.3. *(Continued)*

Species Number	Registry Number	Name	Emission Source	Emission Ref.	Detection Ref.	Code	Bioassay −MA	Bioassay +MA	Bioassay Ref.
2.3-16S		**Hydrated nitrate ion**			798(S)t				
2.3-17S		**Nitrate-nitric acid ion**			1081t 719(S)t,745(S)t				
2.3-18	7631-99-4	**Sodium nitrate**			213(a),292(a)				
2.3-19	6484-52-2	**Ammonium nitrate**	fertilizer mfr. rocket	669* 536	81(a),214*(a)				
2.3-20	7757-79-1	**Potassium nitrate**			292(a)				

$NO_3^- \cdot (H_2O)_n$
2.3-16

$NO_3^- \cdot (HNO_3)_n$
2.3-17

Table 2.4. Inorganic Sulfur Compounds

Species Number	Registry Number	Name	Emission Source	Emission Ref.	Detection Ref.	Code	Bioassay −MA	Bioassay +MA	Bioassay Ref.
2.4-1	7783-06-4	**Hydrogen sulfide**	animal waste	157,522	23*,83*				
			auto	418*,640					
		coffee mfr. 598; coke oven 623; fish processing 58*,255*; food decay 1074; geothermal steam 372,402*; leather mfr. 1151*,1152; marshland 1049*,1056*; microbes 210,670*; natural gas 423*,456*; onion odor 636; paint 695,1071; petroleum mfr. 127*,987; plastics comb. 354; plastics mfr. 695; polymer comb. 987; refuse comb. 787,1093; rendering 637,1078; rubber mfr. 119*; sewage tmt. 174,204; SO_2 scrubbing 1104; soil 1049*; starch mfr. 523*,695; synthetic fiber mfr. 58*,191; tobacco smoke 158,396; vegetation 1067*; volcano 106*,246; water treatment 185; wood pulping 19,58*							
2.4-2S	463-58-1	**Carbonyl sulfide**	animal waste	522,651	89*,100*				
			auto	418*,771*	399(a)				
		biomass comb. 354,655*; diesel 772*; fish processing 385*; geothermal steam 673*,1054; marshland 1056*,1057*; microbes 646,670*; minerals 1090t; natural gas 423*,454*; oceans 1055*; petroleum mfr. 252*,987; plastics comb. 354; polymer comb. 611,987; refuse comb. 787, 26(a); rubber abrasion 606; SO_2 scrubbing 1104; soil 1049*; starch mfr. 523*; synthetic fiber mfr. 191,253*; tobacco smoke 396; volcano 234,604; wood pulping 1053*			715*(S),716*(S) 1185(I)				

Table 2.4. *(Continued)*

Species Number	Registry Number	Name	Emission Source	Emission Ref.	Detection Ref.	Bioassay Code	−MA	+MA	Ref.
2.4-3	75-15-0	**Carbon disulfide**	animal waste	522	100*,258	STI00	NEG	NEG	B38
			biomass comb.	597	399(a)	ST135	NEG	NEG	B38
		chemical mfr. 679; coke oven 623; fish processing 385*; geothermal steam 673*,1054; marshland 1049*,1057*; microbes 646; minerals 1090t; natural gas 423*,454*; petroleum mfr. 127*,252*; plastics comb. 354; polymer comb. 987; refuse comb. 439,787,26(a); rubber abrasion 331,606; rubber mfr. 119*; SO_2 scrubbing 1104; soil 1049*; starch mfr. 523*,695; synthetic fiber mfr. 58*,191; turbine 414; volcano 234,783*; wood pulping 695,1071			820(I)	ST137	NEG	NEG	B38
						STI98	NEG	NEG	B38
2.4-4	1313-82-2	**Sodium sulfide**	wood pulping	53(a),256*(a)					
2.4-5S	1309-36-0	**Pyrite**			1069(a)				
2.4-6	1314-98-3	**Zinc sulfide**	zinc mfr.	256*(a)					
2.4-7	1314-87-0	**Lead sulfide**	industrial	101(a)					
			smelting	256*(a)					
2.4-8S	1303-18-0	**Arsenopyrite**	foundry	328(a)					
2.4-9	10544-50-0	*cyclo*-**Octasulfur**	coal comb.	828(a)t	213(a),589(a)				
			diesel	698(a)					
			foundry	279(a)					
		refuse comb. 840(a); sulfur mfr. 58*; titanium mfr. 412*(a); volcano 538,1045(a)							

Species Number	Registry Number	Name	Emission Source	Emission Ref.	Detection Ref.	Bioassay Code	−MA	+MA	Ref.
2.4-10	7446-09-5	**Sulfur dioxide**	acrylonitrile mfr.	626	289,1183*	TRM	+		B21
			animal waste	1071	399(a)	TRC	+		B20
		auto 541,885*; biomass comb. 597t; coal comb. 316,630; coke mfr. 316; diesel 58*,772*; H_2SO_4 mfr. 316; petroleum mfr. 316,987; plastics comb. 46; polymer comb. 611,987; refuse comb. 316,787; rocket 922*; rubber abrasion 606; sewage tmt. 932,1161; smelting 316; starch mfr. 523*,1071; tire pyrolysis 609; turbine 71; volcano 28*,235; whiskey mfr. 1158; wood comb. 830*; wood pulping 623			711*(S),715*(S) 917*(I),966*(I)				
2.4-11	7782-99-2	**Sulfurous acid**	coal comb.	203	862(a)				
			smelting	150					
2.4-12S		**Hydrogensulfite radical**			733*(S)				
2.4-13S	15181-46-1	**Hydrogensulfite ion**			767*(c) 1085*(r),1101*(r)				
2.4-14S		**Hydrogensulfite-nitric acid ion**			719(S)t,745(S)t				
2.4-15S		**Hydrogensulfite-nitric acid-water ion**			798(S)t,1168(S)t				
2.4-16	10196-04-0	**Ammonium sulfite**			213(a)				
2.4-17	50820-24-1	**Iron (III) sulfite**			839(a),937(a)				
2.4-18	35788-00-2	**Copper (II) sulfite**			937(a)				

Table 2.4. *(Continued)*

Species Number	Registry Number	Name	Emission Source	Emission Ref.	Detection Ref.	Bioassay Code	−MA	+MA	Ref.
2.4-19	13597-44-9	**Zinc (II) sulfite**			937(a)				
2.4-20	7446-10-8	**Lead (II) sulfite**			937(a)t				
2.4-21	7446-11-9	**Sulfur trioxide**	auto	541					
			battery mfr.	379					
		brick mfr. 268; cement mfr. 539; oil comb. 317*,356; H_2SO_4 mfr. 93; turbine 883							
2.4-22	7664-93-9	**Sulfuric acid**	auto	58*,520*	22(a),257(a)				
			coal comb.	580*,1050*,630(a)	721*(S),744(S)				
		explosives mfr. 58*; furnace soot 51*(a); H_2SO_4 mfr. 35,58*; oil comb. 1050*; refuse comb. 1093; steel mfr. 142*(a); volcano 28*,1045(a)							
2.4-23	14808-79-8	**Sulfate ion**	coal comb.	889*(a)	889*(a),909*(a)				
			smelting	889(a)	750*(c),751*(c)				
			volcano	1038(a)	746*(f),796*(f)				
					777*(i),779*(i)				
					760*(r),765*(r)				
					753*(s),756*(s)				
					1033*(S)				
					806*(I),1035(I)				
2.4-24S		**Hydrogensulfite-sulfuric acid ion**			719(S)t,745(S)t				
2.4-25S		**Hydrogensulfite-nitric acid-sulfuric acid ion**			719(S)t,745(S)t				

Species Number	Registry Number	Name	Emission Source	Emission Ref.	Detection Ref.	Code	Bioassay −MA	Bioassay +MA	Bioassay Ref.
2.4-26S		**Hydrogensulfite-nitric acid-sulfuric acid-water ion**			1168(S)t				
2.4-27S	7782-78-7	**Nitrosylhydrogen sulfate**			741(S)t				
2.4-28S	13826-67-0	**Dinitryl peroxodisulfurate**			741(S)t				
2.4-29	7803-63-6	**Ammonium hydrogen sulfate**	oil comb.	1051(a)	257(a),292(a)				
2.4-30	7783-20-2	**Ammonium sulfate**			22(a),292(a) 741(S)				
2.4-31	7727-54-0	**Diammonium peroxodisulfurate**			741(S)				
2.4-32	12398-89-9	**Ammonium sulfate-ammonium nitrate (1/2)**			1116(a)				
2.4-33	15668-97-0	**Ammonium sulfate-calcium sulfate (1/2)**			668(a),1116(a)				
2.4-34	14871-68-2	**Ammonium sulfate-lead sulfate (1/1)**			605(a),668(a)				
2.4-35	13775-30-9	**Triammonium hydrogen bis (sulfate)**			36(a),680(a)				
2.4-36	7681-38-1	**Sodium hydrogen sulfate**			213(a),589(a)				

Table 2.4. *(Continued)*

Species Number	Registry Number	Name	Emission Source	Emission Ref.	Detection Ref.	Code	Bioassay −MA	Bioassay +MA	Bioassay Ref.
2.4-37	7757-82-6	**Sodium sulfate**	oil comb. wood pulping	1051(a),1061(a) 45(a),53(a)	292(a),668(a)				
2.4-38	7727-73-3	**Sodium sulfate decahydrate**	wood pulping	10(a)					
2.4-39	7783-10-0	**Ammonium sulfate-sodium sulfate-water (1/1/4)**			1116(a)				
2.4-40	7487-88-9	**Magnesium sulfate**	oil comb.	1061(a)	859(a)				
2.4-41	14168-73-1	**Magnesium sulfate monohydrate**	oil comb.	1051(a)					
2.4-42	10043-01-3	**Aluminum sulfate**	coal comb. oil comb.	1061(a) 1051(a)	859(a)				
2.4-43	7778-18-9	**Calcium sulfate** oil comb. 1051(a),1061(a); wood pulping 45(a)	coal comb. lime mfr.	1061(a) 256*(a)	280(a),617(a)				
2.4-44S	13397-24-5	**Gypsum**		668(a),680(a)					
2.4-45S	24189-51-3	**Koktaite**			1117(a)				
2.4-46	63311-56-8	**Vanadium (IV) oxide sulfate**	oil comb.	1061(a)					
2.4-47	12440-32-3	**Vanadium (IV) oxide sulfate trihydrate**	oil comb.	1051(a)					

Species Number	Registry Number	Name	Emission Source	Emission Ref.	Detection Ref.	Code	Bioassay −MA	Bioassay +MA	Bioassay Ref.
2.4-48	7778-80-5	**Potassium sulfate**	cement mfr. oil comb.	468(a) 1051(a)	859(a),1062(a)				
2.4-49	18432-25-2	**Chromium (II) sulfate**			859(a)				
2.4-50	7785-87-7	**Manganese (II) sulfate**			859(a)				
2.4-51	7720-78-7	**Iron (II) sulfate**	coal comb.	1061(a)	859(a)	SRL	NEG		B30
2.4-52	7782-63-0	**Iron (II) sulfate heptahydrate**	oil comb.	1051(a)					
2.4-53	10138-04-2	**Iron (III) sulfate-ammonium sulfate (1/3)**			680(a)				
2.4-54S	24389-93-3	**Mohrite**			1117(a)				
2.4-55	7786-81-4	**Nickel (II) sulfate**	oil comb.	1051(a),1061(a)	280(a)	CCC	I		B37
2.4-56	7758-98-7	**Copper (II) sulfate**			839(a),859(a)				
2.4-57	7733-02-0	**Zinc (II) sulfate**			107(a),859(a)				
2.4-58	7446-19-7	**Zinc (II) sulfate monohydrate**	oil comb.	1051(a)	680(a)				
2.4-59	13814-87-4	**Diammonium zinc (II) bis (sulfate)**			107(a),680(a)				
2.4-60	13494-91-2	**Gallium sulfate**			859(a)				
2.4-61	7759-02-6	**Strontium sulfate**			839(a),1069(a)				

Table 2.4. *(Continued)*

Species Number	Registry Number	Name	Emission Source	Emission Ref.	Detection Ref.	Bioassay Code	−MA	+MA	Bioassay Ref.
2.4-62	7727-43-7	**Barium sulfate**			859(a),1069(a)				
2.4-63	7446-14-2	**Lead sulfate**	auto	1132(a)	280(a),605(a)				
			industrial	101(a)					
		oil comb. 1051(a); refuse comb. 196(a); smelting 648(a)							
2.4-64	12202-17-4	**Lead oxide-lead sulfate (1/1)**	smelting	648(a)	680(a),1132(a)				

OCS 2.4-2 — FeS_2 2.4-5 — $FeAsS$ 2.4-8 — $HSO_3\cdot$ 2.4-12 — $HOS(O)O^-$ 2.4-13 — $HSO_4^-\cdot(HNO_3)_n$ 2.4-14 — $HSO_4^-\cdot(HNO_3)_m\cdot(H_2O)_n$ 2.4-15

$HSO_4^-\cdot(H_2SO_4)_n$ 2.4-24 — $HSO_4^-\cdot(HNO_3)_m\cdot(H_2SO_4)_n$ 2.4-25 — $HSO_4^-\cdot(HNO_3)_m\cdot(H_2SO_4)_n\cdot(H_2O)_p$ 2.4-26 — $NOOSO_3H$ 2.4-27

$O_2NOSO_2ONO_2$ 2.4-28 — $CaSO_4\cdot 2H_2O$ 2.4-44 — $(NH_4)_2\cdot[Ca(SO_4)]_2\cdot H_2O$ 2.4-45 — $(NH_4)_2[Fe(SO_4)_2]\cdot H_2O$ 2.4-54

Table 2.5. Inorganic Halogenated Compounds

Species Number	Registry Number	Name	Emission Source	Emission Ref.	Detection Ref.	Code	Bioassay −MA	Bioassay +MA	Ref.
2.5-1	14989-30-1	**Chlorine monoxide radical**			701*(S),720*(S)				
2.5-2	7782-50-5	**Dichlorine**	aluminum mfr.	58*,457					
			chlorine mfr.	58*,76					
		HCl mfr. 622; refuse comb. 395; sewage tmt. 1161; titanium mfr. 115; wood pulping 76; zinc mfr. 256*							
2.5-3	7726-95-6	**Dibromine**	industrial	1150		ALC	+		B19
2.5-4	7553-56-2	**Diiodine**	iodine mfr.	398	213(a),589(a)				
2.5-5	7664-39-3	**Hydrogen fluoride**	aluminum mfr.	115,256*	44,271	SRL	+		B30
			brick mfr.	58*	725*(S),732*(S)				
		ceramics mfr. 417*; electronics mfr. 616; fertilizer mfr. 58*,93; HF mfr. 58*; lacquer mfr. 427; phosph. acid mfr. 58*; rocket 536; steel mfr. 58*; volcano 30*, 234							
2.5-6	7647-01-0	**Hydrogen chloride**	auto	386*	129*,782*				
			biomass comb.	354	725*(S),727*(S)				
		cement mfr. 1050; ceramics mfr. 642; coal comb. 76,1050; HCl mfr. 58*,622; lacquer mfr. 427; polymer comb. 304,354; refuse comb. 31,76; rocket 536,622; sea salt 470,1105; sewage tmt. 1013*; titanium mfr. 115; volcano 30*,234							
2.5-7S		**Nitrate-hydrochloric acid ion**			1168(S)t				

Table 2.5. *(Continued)*

Species Number	Registry Number	Name	Emission Source	Emission Ref.	Detection Ref.	Code	Bioassay −MA	Bioassay +MA	Bioassay Ref.
2.5-8S		**Nitrate-hypochlorous acid ion**			1168(S)t				
2.5-9	10035-10-6	**Hydrogen bromide**	auto	386*	728*(S)				
			volcano	152					
2.5-10	10034-85-2	**Hydrogen iodide**	volcano	152					
2.5-11	16961-83-4	**Silicon dihydrogen hexafluoride**	chemical mfr.	93					
			fertilizer mfr.	256*(a)					
2.5-12	7637-07-2	**Boron trifluoride**			114				
2.5-13	12125-02-9	**Ammonium chloride**	auto	599	213*(a),214*(a)				
			fertilizer mfr.	58*,93,256*(a)					
			zinc mfr.	256*(a)					
2.5-14	7681-49-4	**Sodium Fluoride**	aluminum mfr.	256*(a)		STP00	NEG	NEG	B36
			oceans	585(a)		STP35	−(C)	−(C)	B36
			wood pulping	256*(a)		STP37	−(C)	−(C)	B36
						STP38	−(C)	−(C)	B36
						STP98	−(C)	−(C)	B36
						SCE	NEG		B6
						STI00	NEG	NEG	B38
						ST135	NEG	NEG	B38
						STI37	NEG	NEG	B38
						STI98	NEG	NEG	B38
						ALC	+		B19
						SRL	?		B30

Species Number	Registry Number	Name	Emission Source	Emission Ref.	Detection Ref.	Code	Bioassay −MA	Bioassay +MA	Bioassay Ref.
						NEU	NEG		B32
2.5-15	7647-14-5	**Sodium chloride**	ceramics mfr.	642	280(a),292(a)	CYC	NEG		B8
			ocean	585(a)		MNT	?/−		B29
			wood pulping	256*(a)		SPI	NEG		B25
2.5-16	7786-30-3	**Magnesium chloride**	titanium mfr.	115	911(a)t	REC	NEG		B9
2.5-17	7784-18-1	**Aluminum fluoride**	aluminum mfr.	256*(a),492(a)					
			lead mfr.	115(a)					
2.5-18	1327-41-9	**Aluminum chloride**	aluminum mfr.	256*(a)	460(a)t				
			rocket	536					
2.5-19	13775-53-6	**Aluminum trisodium hexafluoride**	aluminum mfr.	256*(a),492(a)					
2.5-20	2551-62-4	**Sulfur hexafluoride**	dispersion tracer	590,593	37*,135* 731*(S),737*(S)				
2.5-21	2699-79-8	**Sulfur difluoride dioxide**	industrial	400					
2.5-22	7783-61-1	**Silicon tetrafluoride**	aluminum mfr.	256*	12*(a)				
			electronics mfr.	616					
		fertilizer mfr. 58*,93; phosph. acid mfr. 58*							
2.5-23	7447-40-7	**Potassium chloride**			1070(a)				
2.5-24	7789-75-5	**Calcium fluoride**	aluminum mfr.	256*(a),492(a)					

Table 2.5. *(Continued)*

Species Number	Registry Number	Name	Emission Source	Emission Ref.	Detection Ref.	Bioassay Code	Bioassay −MA	Bioassay +MA	Bioassay Ref.
2.5-25S	1306-05-4	**Apatite**			839(a),1069(a)				
2.5-26	7550-45-0	**Titanium tetrachloride**	titanium mfr.	115					
2.5-27	7789-28-8	**Iron (II) fluoride**			859(a)				
2.5-28	7758-94-3	**Iron (II) chloride**	rocket	1102(a)	409(a)	REC	NEG		B9
2.5-29	7646-85-7	**Zinc chloride**	foundry	1069(a)		HMAST		+	B13
			zinc mfr.	425(a)		STI35	?(H)		B36
						STI37	+(H)		B36
						STU37		+ABD	B36
						CYI	NEG		B8
						REC	NEG		B9
						SPI	NEG		B25
2.5-30	7487-94-7	**Mercuric chloride**			388	NEU	NEG		B32
2.5-31	7758-95-4	**Lead chloride**	auto	1132(a)	1132(a)	REC	NEG		B9
			foundry	1069(a)					
			industrial	101(a)					
2.5-32S	15887-88-4	**Lead chloride hydroxide**	auto	1132(a)	1132(a)				
2.5-33	10031-22-8	**Lead bromide**	auto	1132(a)	1132(a)				
2.5-34	16651-91-5	**Lead bromide hydroxide**	auto	1132(a)	1132(a)				

Species Number	Registry Number	Name	Emission Source	Emission Ref.	Detection Ref.	Code	Bioassay −MA	Bioassay +MA	Bioassay Ref.
2.5-35	13778-36-4	**Lead bromide chloride**	auto	96(a),311(a)	668(a),1069(a)t				
2.5-36	12205-70-8	**Lead chloride-lead oxide (1/2)**	auto	1132(a)	1132(a)				
2.5-37	12301-73-4	**Lead bromide-lead oxide (1/2)**	auto	1132(a)	1132(a)				
2.5-38		**Lead bromide chloride-lead oxide (1/2)**	auto	1132(a),1133(a)	1132(a)				
2.5-39		**Ammonium chloride-lead bromide chloride (1/1)**			1069(a)t				
2.5-40		**Ammonium chloride-lead bromide chloride (1/2)**	auto	96(a),1133(a)	668(a),1130(a)				
2.5-41		**Lead bromide chloride-ammonium chloride (1/2)**	auto	1130(a),1134(a)	1130(a)				
2.5-42		**Ammonium bromide chloride-lead bromide chloride (1/1)**			1130(a)				
2.5-43	7790-98-9	**Ammonium perchlorate**	rocket	536					
2.5-44S		**Chlorate-nitric acid ion**			1168(S)t				

$NO_3^- \cdot HCl$ 2.5-7 | $NO_3^- \cdot HOCl$ 2.5-8 | $CaF_2 \cdot 3Ca_3P_2O_8$ 2.5-25 | $ClPbOH$ 2.5-32 | $ClO_3^- \cdot HNO_3$ 2.5-44

Table 2.6. Inorganic Hydrides, Oxides, Carbonates, etc.

Species Number	Registry Number	Name	Emission Source	Emission Ref.	Detection Ref.	Bioassay Code	Bioassay −MA	Bioassay +MA	Bioassay Ref.
2.6-1	7440-44-0	**Carbon (soot)**	aluminum mfr.	256*(a)	947*(a), 948*(a)				
			auto	1064*(a)	949*(r), 950*(r)				
		coal comb. 256*(a); diesel 1064*(a); foundry 279(a); natural gas comb. 1064*(a); titanium mfr. 412*(a); wood comb. 1064*(a)							
2.6-2	7803-51-2	**Phosphine**	industrial	618, 1150	147				
2.6-3	7784-42-1	**Arsine**	aluminum mfr.	116					
			smelting	904					
2.6-4	1304-56-9	**Beryllium oxide**	rocket	536		CCC	SP		B37
2.6-5	1303-86-2	**Boron oxide**	rocket	536					

Species Number	Registry Number	Name	Emission Source	Emission Ref.	Detection Ref.	Bioassay Code	Bioassay −MA	Bioassay +MA	Bioassay Ref.
2.6-6	630-08-0	**Carbon monoxide**	acrylonitrile mfr.	626	290*,293*				
			auto	281,444*	399(a)				
		biomass comb. 655*, 662*; combustion 281,395; diesel 136*; electronics mfr. 616; geothermal steam 998*; HCl mfr. 622; lightning 644; microbes 210,302; natural gas 423*; ocean 1175*; plastics comb. 46; polymer comb. 611,612; refuse comb. 661,787; rocket 536,622; sewage tmt. 932,1013*; soil 1176*; tobacco smoke 4*, 168; turbine 71,594; vegetation 199,1174*; volcano 106*, 604; wood comb. 830*, 890; wood pulping 19			763*(s) 718*(S), 724*(S) 868*(I), 893*(I)				
2.6-7	124-38-9	**Carbon dioxide**	auto	281,592*	295*, 305*				
			biomass comb.	655,662*	399(a)				
		combustion 281; diesel 136*; foaming agent 562; geothermal steam 372,402*; HCl mfr. 622; insects 785*; microbes 646; natural gas 456*; plastics comb. 46; polymer comb. 611,612; propellant 559,563; refuse comb. 661,1093; rocket 536; sewage tmt. 932,1013*; tobacco smoke 396,445; turbine 836*; volcano 73,106*; wood comb. 890			780*(i), 1178(i) 715*(S), 736*(S) 929(I), 1035(I)				
2.6-8	463-79-6	**Carbonic acid**			764*(r)				
2.6-9	71-52-3	**Hydrogen carbonate ion**			1195(r) 939(s),941(s)				
2.6-10	3812-32-6	**Carbonate ion**			1168(S)t				

Table 2.6. *(Continued)*

Species Number	Registry Number	Name	Emission Source	Emission Ref.	Detection Ref.	Bioassay Code	−MA	+MA	Ref.
2.6-11	1313-59-3	**Sodium oxide**	aluminum mfr.	256*(a)	1096(a)t				
			cement mfr.	256*(a)					
			coal comb. 256*(a), 1106(a); refuse comb. 256*(a); rock dust 473(a); zinc mfr. 256*(a)						
2.6-12	1309-48-4	**Magnesium oxide**	cement mfr.	256*(a), 539(a)	1096(a)t				
			coal comb.	256*(a), 1106(a)					
			fertilizer mfr. 256*(a); foundry 279(a); lime mfr. 256*(a); magnesium mfr. 58*(a); oil comb. 1051(a); refuse comb. 256*(a); steel mfr. 161(a), 256*(a); titanium mfr. 412*(a); zinc mfr. 256*(a)						
2.6-13	1344-28-1	**Aluminum oxide**	aluminum mfr.	256*(a)	280(a), 460(a)t				
			cement mfr.	256*(a)	703(S)				
			coal comb. 256*(a), 1106(a); fertilizer mfr. 256*(a); foundry 279(a); lead mfr. 115(a); lime mfr. 256*(a); oil comb. 1051(a); refuse comb. 256*(a); rock dust 473(a); rocket 536(a), 922*(a); steel mfr. 256*(a); titanium mfr. 412*(a); wood pulping 256*(a); zinc mfr. 256*(a)						
2.6-14	21645-51-2	**Aluminum oxide trihydrate**			460(a)t				
2.6-15	24623-77-6	**Aluminum hydroxide oxide**			1096(a)t				
2.6-16	7631-86-9	**Silicon dioxide**	aluminum mfr.	256*(a)	280(a), 468(a)				
			cement mfr.	256*(a), 539(a)	939(s)				
			coal comb. 256*(a), 1106(a); fertilizer mfr. 256*(a); foundry 279(a); oil comb. 1051(a); refuse comb. 256*(a); rock dust 473(a); steel mfr. 256*(a); tire pyrolysis 607(a); titanium mfr. 412*(a); wood pulping 256*(a); zinc mfr. 256*(a)						

Species Number	Registry Number	Name	Emission Source	Emission Ref.	Detection Ref.	Bioassay Code	Bioassay −MA	Bioassay +MA	Bioassay Ref.
2.6-17	14265-44-2	**Phosphate ion**			934(a)				
					1195(r)				
2.6-18	1314-56-3	**Diphosphorus pentaoxide**	coal comb.	256*(a), 1106(a)					
			fertilizer mfr.	187*, 256*(a)					
		foundry 279(a); phosph. acid mfr. 58*, 93; rock dust 473(a); steel mfr. 256*(a)							
2.6-19	12136-45-7	**Potassium oxide**	cement mfr.	256*(a)	280(a), 1096(a)t				
			coal comb.	256*(a), 1106(a)					
		refuse comb. 256*(a); rock dust 473(a)							
2.6-20	1305-78-8	**Calcium oxide**	cement mfr.	256*(a), 539(a)	1096(a)t				
			coal comb.	256*(a), 1106(a)	933(s)				
		fertilizer mfr. 256*(a); foundry 279(a); lime mfr. 256*(a); oil comb. 1051(a); refuse comb. 256*(a); rock dust 473(a); steel mfr. 161(a), 256*(a); tire pyrolysis 607(a); titanium mfr. 412*(a); wood pulping 45(a), 256*(a); zinc mfr. 256*(a)							
2.6-21	13463-67-7	**Titanium dioxide**	coal comb.	256*(a)	1062(a)	CCC	SN		B37
			foundry	279(a)					
		refuse comb. 256*(a); rock dust 473(a); tire pyrolysis 607(a); titanium mfr. 115(a), 412*(a)							
2.6-22	18252-79-4	**Vanadium (V) dioxide ion**	coal comb.	889(a)					

Table 2.6. *(Continued)*

Species Number	Registry Number	Name	Emission Source	Emission Ref.	Detection Ref.	Bioassay Code	−MA	+MA	Ref.
2.6-23	1314-62-1	**Divanadium (V) pentaoxide**	foundry	279(a)					
			oil comb.	1051(a)					
		petroleum mfr. 1135; titanium mfr. 412*(a)							
2.6-24	1308-38-9	**Chromium (III) dioxide**	foundry	279(a)					
			steel mfr.	256*(a)					
			titanium mfr.	412*(a)					
2.6-25	1344-43-0	**Manganese (II) oxide**	coal comb.	1106(a)					
			foundry	279(a)					
		rock dust 473(a); steel mfr. 65(a), 256*(a); turbine 957(a)							
2.6-26	1313-13-9	**Manganese (IV) oxide**	titanium mfr.	412*(a)					
2.6-27	1317-35-7	**Trimanganese tetraoxide**	auto	631(a), 1087(a)					
2.6-28	1345-25-1	**Iron (II) oxide**	coal comb.	256*(a), 1106(a)	280(a), 1096(a)t				
			foundry	279(a)					
		rock dust 473(a); steel mfr. 161(a), 256*(a)							
2.6-29	1309-37-1	**Iron (III) oxide**	aluminum mfr.	256*(a)	280(a), 336(a)				
			auto	1133(a)	1121(S)t				
		cement mfr. 256*(a), 539(a); coal comb. 256*; fertilizer mfr. 256*(a); foundry 279(a), 1069(a); lime mfr. 256*(a); oil comb. 1051(a); refuse comb. 256*(a); rock dust 473(a); smelting 648(a); steel mfr. 161(a), 256*(a); titanium mfr. 412*(a); wood pulping 256*(a)							

Species Number	Registry Number	Name	Emission Source	Emission Ref.	Detection Ref.	Code	Bioassay −MA	Bioassay +MA	Bioassay Ref.
2.6-31	1317-61-9	**Triiron tetraoxide**	smelting	648(a)	280(a), 336(a) 1126(S), 1127(S)				
2.6-32	1308-04-9	**Tricobalt tetraoxide**	foundry	279(a)					
2.6-33	1313-99-1	**Nickel (II) oxide**	foundry	279(a)					
			oil comb.	1051(a)					
		steel mfr. 256*(a); zinc mfr. 256*(a)							
2.6-34	1317-38-0	**Copper (II) oxide**	steel mfr.	256*(a)					
2.6-35	1314-13-2	**Zinc oxide**	copper mfr.	256*(a), 1059(a)	1069(a)				
			foundry	1069(a)					
		lead mfr. 256*(a); smelting 648(a); steel mfr. 256*(a); tire pyrolysis 607(a), 609(a); zinc mfr. 256*(a), 350(a)							
2.6-36	15502-74-6	**Arsenite ion**			775(r)				
2.6-37	15584-04-0	**Arsenate ion**	coal comb.	889(a)t	775(r)				
2.6-38	1327-53-3	**Arsenic oxide**	copper mfr.	302,256*(a)	112*, 672	REC	+		B9
			foundry	1069(a)	213(a), 589(a)				
		gold mfr. 328(a), 930(a); lead mfr. 115,256*(a); smelting 657,648(a)							
2.6-39	7446-08-4	**Selenium dioxide**	lead mfr.	256*(a)	213(a), 589(a)	REC	+		B9
			refuse comb.	220*, 530					
2.6-40	1313-27-5	**Molybdenum (III) oxide**	foundry	279(a)					
2.6-41	1306-19-0	**Cadmium (II) oxide**	lead mfr.	256*(a)		CCC	SP		B37

Table 2.6. *(Continued)*

Species Number	Registry Number	Name	Emission Source	Emission Ref.	Detection Ref.	Bioassay Code	Bioassay −MA	Bioassay +MA	Bioassay Ref.
2.6-42	21651-19-4	**Tin (II) oxide**	lead mfr.	256*(a)					
2.6-43	18282-10-5	**Tin (IV) oxide**	zinc mfr.	256*(a)					
2.6-44	1309-64-4	**Antimony trioxide**	copper mfr. lead mfr.	256*(a) 115		REC	+		B9
2.6-45	7446-07-3	**Tellurium dioxide**	lead mfr.	256*(a)					
2.6-46	1304-29-6	**Barium oxide**	coal comb.	1106(a)					
2.6-47	12680-02-3	**Lanthanum oxide**	petroleum mfr.	1135					
2.6-48	11129-18-3	**Cerium oxide**	petroleum mfr.	1135					
2.6-49	11113-81-8	**Praseodymium oxide**	petroleum mfr.	1135					
2.6-50	12648-30-5	**Neodymium oxide**	petroleum mfr.	1135					
2.6-51	1317-36-8	**Lead (II) oxide**	auto copper mfr.	1132(a) 256*(a)	280(a), 1132(a)				
		foundry 1069(a); lead mfr. 256*(a); steel mfr. 101(a), 256*(a)							
2.6-52	1309-60-0	**Lead (IV) oxide**	industrial	101(a)					
2.6-53	1314-41-6	**Trilead tetraoxide**	lead mfr.	256*(a)					
2.6-54	1304-76-3	**Bismuth trioxide**			1119(a)				

Species Number	Registry Number	Name	Emission Source	Emission Ref.	Detection Ref.	Bioassay Code	Bioassay −MA	Bioassay +MA	Bioassay Ref.
2.6-55	506-87-6	**Ammonium carbonate**	chemical mfr.	337(a)					
2.6-56	497-19-8	**Sodium carbonate**	foundry	1069(a)t					
			lead mfr.	115(a)					
		lime mfr. 256*(a); wood pulping 45(a), 256*(a)							
2.6-57	546-93-0	**Magnesium carbonate**	lime mfr.	256*(a)					
			wood pulping	256*(a)					
2.6-58	471-34-1	**Calcium carbonate**	lime mfr.	256*(a)	280(a), 468(a)				
			wood pulping	10(a), 256*(a)	933(s)				
2.6-59	598-62-9	**Manganese carbonate**	auto	1087(a)					
2.6-60	1633-05-2	**Strontium carbonate**			1069(a)t				
2.6-61	598-63-0	**Lead (II) carbonate**	auto	1132(a)	1132(a)				
2.6-62	12326-84-0	**Lead (II) carbonate-lead (II) oxide (1/2)**	auto	1132(a)	1132(a)				
2.6-63	15281-98-8	**Iron carbonyl**	tobacco smoke	493(a)					
2.6-64	58207-38-8	**Cobalt carbonyl**	tobacco smoke	493(a)					
2.6-65	13463-39-3	**Nickel carbonyl**	tobacco smoke	493(a)					

Table 2.6. *(Continued)*

Species Number	Registry Number	Name	Emission Source	Emission Ref.	Detection Ref.	Bioassay Code	−MA	+MA	Ref.
2.6-66	10043-35-3	**Boric acid**	fibreglass mfr.	348	366,818t	STI00	?	?	B38
			geothermal steam	372		STI35	NEG	NEG	B38
		ocean 586; volcano 349,913				STI37	NEG	NEG	B38
						STI98	NEG	NEG	B38
2.6-67	7664-38-2	**Phosphoric acid**	lacquer mfr.	428					
2.6-68	27875-33-8	**Iron (III) phosphate**			859(a)				
2.6-69	7446-27-7	**Lead phosphate**	auto	1132(a)	839(a), 1132(a)	CCC	SP		B37
2.6-70	1308-14-1	**Chromic acid**	industrial	58*, 108					

Table 2.7. Minerals

Species Number	Registry Number	Name	Emission Source	Emission Ref.	Detection Ref.	Code	Bioassay −MA	Bioassay +MA	Bioassay Ref.
2.7-1S	12244-10-9	**Albite**			1052(a), 1069(a)				
2.7-2S		**Aluminosilicates**	petroleum mfr.	1135					
2.7-3S	12244-31-4	**Austenite**			1127(S),1128(S)				
2.7-4S	1302-27-8	**Biotite**			1069(a), 1096(a)				
2.7-5S	1318-59-8	**Chlorite**			1069(a), 1070(a)				
2.7-6S	12001-29-5	**Chrysotile**			1069(a),1197(a)	WPU	NEG		B1
					1129(r)	WP2	NEG		B1
						CCC	SP		B37
2.7-7S	12173-11-4	**Cohenite**			1127(S), 1128(S)				
2.7-8S	16389-88-1	**Dolomite**			1069(a)t, 1096(a)				
2.7-9S	1310-14-1	**Goethite**			1062(a)				
2.7-10S	12178-42-6	**Hornblende**			1069(a), 1070(a)				
2.7-11S	17068-62-1	**Hypersthene**			1070(a)				
2.7-12S	12173-60-3	**Illite**			1062(a), 1069(a)				

Table 2.7. *(Continued)*

Species Number	Registry Number	Name	Emission Source	Emission Ref.	Detection Ref.	Code	Bioassay −MA	Bioassay +MA	Bioassay Ref.
2.7-13S	12168-52-4	**Ilmenite**			839(a)				
2.7-14S		**ϵ - Iron-nickel carbide**			1126(S), 1127(S)				
2.7-15S	12173-68-1	**Kamacite**			1127(S)				
2.7-16S	1318-74-7	**Kaolinite**			1062(a), 1069(a)				
2.7-17S	12173-78-3	**Labradorite**			1052(a)				
2.7-18S	1317-63-1	**Limonite**			839(a)				
2.7-19S	68563-18-8	**Mica**			839(a)				
2.7-20S		**Microcline**			1052(a)				
2.7-21S	1306-41-8	**Monazite**			839(a), 1069(a)				
2.7-22S	1318-93-0	**Montmorillonite**			839(a), 1069(a)				
2.7-23S	1318-94-1	**Muscovite**			1052(a), 1069(a)				
2.7-24S	1317-71-1	**Olivine**			1069(a)				
2.7-25S	12251-44-4	**Orthoclase**			839(a), 1069(a)				
2.7-26S	12174-11-7	**Palygorskite**			1070(a)				

Species Number	Registry Number	Name	Emission Source	Emission Ref.	Detection Ref.	Code	Bioassay −MA	Bioassay +MA	Bioassay Ref.
2.7-27S		**Plagioclase felspar**			1070(a), 1096(a)				
2.7-28S	12174-37-7	**Pyroxene**			839(a), 1069(a)				
2.7-29S	14807-96-6	**Talc**			1069(a)				
2.7-30S	12135-61-4	**Titanite**			1069(a)				
2.7-31S	13983-17-0	**Wollastinite**			459(a)				

$Na_2O \cdot Al_2O_3 \cdot 6SiO_2$
2.7-1

$Al_xSi_yO_z$
2.7-2

$(Fe,Ni)_3C$
2.7-3

$(K,H)_2(Mg,Fe)_2(Al,Fe)_2(SiO_4)_2$
2.7-4

$Mg_3(Si_4O_{10})(OH)_2 \cdot Mg_3(OH)_6$
2.7-5

$Mg_3Si_2O_5(OH)_4$
2.7-6

$(Fe,Ni)_3C$
2.7-7

$(Ca,Mg)CO_3$
2.7-8

$Fe(O)OH$
2.7-9

$Ca_2(Fe^{2+},Mg)_4Al(Si,Al)O_{22}(OH,F)_2$
2.7-10

$(Mg,Fe^{2+})_2Si_2O_6$
2.7-11

$(K,H_3O)(Al,Mg,Fe)_2(Si,Al)_4O_{10}[(OH)_2H_2O]$
2.7-12

$FeO \cdot TiO_2$
2.7-13

$(Fe,Ni)_3C$
2.7-14

Fe_xNi_y
2.7-15

Table 2.7. *(Continued)*

$Al_2O_3 \cdot 2SiO_2 \cdot 2H_2O$ 2.7-16	$(NaAlSi_3O_8)_x \cdot (CaAl_2Si_2O_8)_y$ 2.7-17	$2Fe_2O_3 \cdot 3H_2O$ 2.7-18	$K_2O \cdot 2Al_2O_3 \cdot Fe_2O_3 \cdot 6SiO_2 \cdot 2H_2O$ 2.7-19
$K_2O \cdot Al_2O_3 \cdot 6SiO_2$ 2.7-20	$(Ca,Nd,Pr,La)PO_4(+Th_3[PO_4]_4$ 2.7-21	$Al_6Mg_2(Si_4O_{10})_3(OH)_{10} \cdot 12H_2O$ 2.7-22	$K_2O \cdot Al_2O_3 \cdot 6SiO_2 \cdot 2H_2O$ 2.7-23
$(Mg,Fe)_2SiO_4$ 2.7-24	$K_2O \cdot Al_2O_3 \cdot 6SiO_2$ 2.7-25	$(Mg,Al)_2Si_4O_{10}(OH) \cdot 4H_2O$ 2.7-26	$(Na,Ca)Al(Al,Si)Si_2O_8$ 2.7-27
ABZ_2O_6 $A=Ca,Fe^{2+},Li,Mg,Na$ $B=Al,Cr^{3+},Fe^{2+},Fe^{3+},Mg,Mn^{2+}$ $Z=Al,Si$ 2.7-28	$3MgO \cdot 4SiO_2 \cdot H_2O$ 2.7-29	$CaO \cdot TiO_2 \cdot SiO_2$ 2.7-30	$CaSiO_3$ 2.7-31

CHAPTER 3

Hydrocarbons

3.0 Introduction

Organic chemistry is the chemistry of the compounds of carbon; its compounds comprise the remainder of the tabulated data in this book. The hydrocarbons (organic compounds containing only C and H atoms) are the basic compounds from which the more complex organic structures are derived.

We will not provide a detailed description of organic chemical nomenclature here, but a few terms are so widely used (here and elsewhere) that it is useful to define them precisely. *Aliphatic* compounds (also called *acyclic*) are those which contain no ring structures of the atoms. This group includes *alkanes*, in which only single bonds occur (i.e., the maximum number of hydrogen atoms is present), *olefins* (also called *alkenes*), in which double bonds occur, and *alkynes*, in which triple bonds occur. *Carbocyclic* compounds (generally simply called *cyclic*) are those which contain rings made up solely of carbon atoms.

The hydrocarbon compounds in this chapter are divided into eight groups. The first three are the alkanes, the unsaturated aliphatics (olefins and alkynes), and the monocyclic hydrocarbons. The fourth group is comprised of terpenes* and other polycyclic hydrocarbons. The *arenes* or *aromatic* compounds (cyclic hydrocarbons containing one or more rings and possessing "conjugated" or alternating unsaturation) are divided into four groups on the basis of the number of contiguous rings in the compound. This division is occasionally disregarded if a multiring structure is named as a derivative of a smaller parent (e.g., anthracene has three contiguous rings and benz[*a*]anthracene has four, but it has been deemed more convenient to the reader to include both in the same table). The nomenclature of the more complex arenes becomes particularly involved and it is in tables such as these where registry numbers and structural drawings are particularly useful.

3.1 Alkanes

More than 130 alkanes have been identified as atmospheric species. Of those detected in ambient air, about two-thirds are found in the gas phase and about one-third in the aerosol phase. A large number have been found indoors, and the higher alkanes are found in rain and snow.

Methane, the smallest of the alkanes, is the most abundant reactive gas in the atmosphere. Its reactivity is not high, however, and its atmospheric concentrations tend to be reasonably uniform except in urban areas, as seen in its concentration plot (Figure 3.1-1).

Even when grouped and considered as a single species, the concentrations of the higher alkanes are small compared with that of methane. In general, the abundance patterns, seen in Figure 3.1-2, reflect diffusion from urban sources as well as slow removal by OH· reaction. Near the ground, the relative amounts of C_2-C_5 alkanes are similar to each other. The preferential removal of larger compounds is apparent in the free troposphere and stratosphere.

Alkanes are emitted in profusion from any process involving the combustion of fossil fuels or other organic material: automobile, diesel, and turbine combustion, biomass burning, refuse combustion, etc., as a consequence of their presence in the original combusted material. Some of the smaller alkanes are used as propellants and in miscellaneous industrial applications. A few are emitted by vegetation.

* *Terpenes* are compounds which contain multiples of the basic isoprene ($H_2C{:}C(CH_3)CH{:}CH_2$) structure. They are widely produced by vegetation. Those volatile enough to be present in the atmosphere generally have the molecular formula $C_{10}H_{18}$.

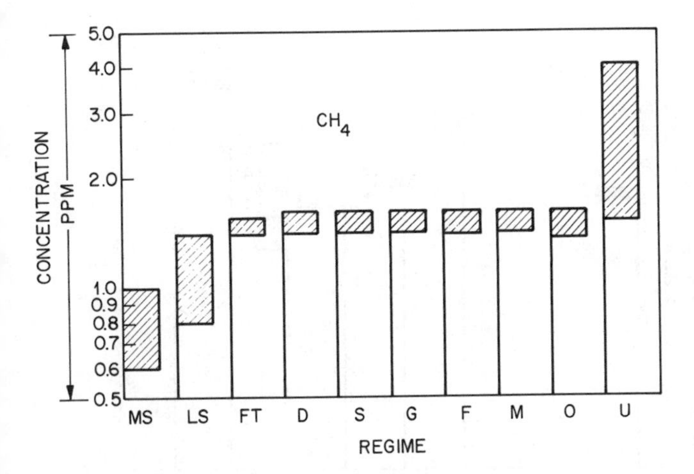

Fig. 3.1-1 Approximate concentration ranges of CH_4 in different atmospheric regimes. The regime code is given in the caption of Figure 2.1-1. (Adapted from Graedel.[1257])

The gas phase chemistry of methane contains many features common to organic chemistry in the atmosphere. Its chemical chain is initiated by reaction with the hydroxyl radical, a process in which a hydrogen atom is abstracted from methane:[1221]

$$CH_4 + OH\cdot \rightarrow CH_3\cdot + H_2O\ . \qquad \text{(R3.1-1)}$$

The $CH_3\cdot$ (methyl) radical then adds molecular oxygen

$$CH_3\cdot + O_2 \xrightarrow{M} CH_3O_2\cdot \qquad \text{(R3.1-2)}$$

(M is a third body capable of removing excess energy from the adduct.) In urban areas where high concentrations of nitric oxide are present, (R3.1-2) is followed by

$$CH_2O_2\cdot + NO \rightarrow CH_3O\cdot + NO_2 \qquad \text{(R3.1-3)}$$

$$CH_3O\cdot + O_2 \rightarrow HCHO + HO_2\cdot \qquad \text{(R3.1-4)}$$

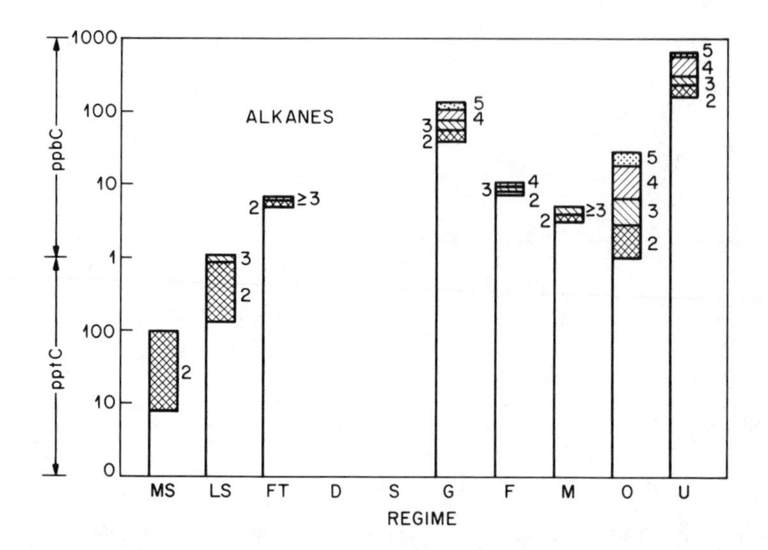

Fig. 3.1-2 Approximate concentration ranges of alkanes other than methane in different atmospheric regimes. Within the flags, the segments indicate the typical fraction of total concentration due to alkanes with the number of carbon atoms shown. For example, oceanic nonmethane alkanes are typically comprised of about 29 per cent ethane, 26 percent propane, 31 per cent butanes, and 14 per cent pentanes. The regime code is given in the caption of Figure 2.1-1.

If NO concentrations are low, the favored reaction chain is

$$CH_3O_2\cdot + HO_2\cdot \rightarrow CH_3OOH + O_2 \qquad \text{(R3.1-5)}$$

$$CH_3OOH \xrightarrow[\lambda<350\ \text{nm}]{h\nu} CH_3O\cdot + OH\cdot \qquad \text{(R3.1-6)}$$

followed by (R3.1-4).

Similar reactions obtain for the higher alkanes. Initial H abstraction is followed by O_2 addition to produce an alkylperoxyl radical, $RO_2\cdot$, (R, R$'$, and R$''$ in this and following discussions refer to unspecified alkyl [C_xH_{2x+1}] groups.) The $RO_2\cdot$ radical then reacts with NO. The favored reaction channel involves O atom transfer followed by reaction with O_2 to form either an aldehyde (R$'$CHO) or ketone (R$'$C(O)R$''$), depending on the point of initial OH$\cdot$ attack:

$$RO_2\cdot + NO \rightarrow RO\cdot + NO_2 \qquad (R3.1\text{-}7)$$

$$RO\cdot + O_2 \rightarrow R'CHO + HO_2\cdot$$

$$\rightarrow R'C(O)R'' + HO_2\cdot \qquad (R3.1\text{-}8)$$

Alternatives to (R3.1-7) and (R3.1-8) are the addition reactions

$$RO_2\cdot + NO \xrightarrow{M} RONO_2 \qquad (R3.1\text{-}9)$$

$$RO\cdot + NO_2 \xrightarrow{M} RONO_2 \qquad (R3.1\text{-}10)$$

Reaction (R3.1-9), at least, becomes increasingly efficient as the carbon number of the alkane increases.[1248] Reaction (R3.1-9) and (R3.1-10) are probably the source of a number of the organic nitrates found in the atmosphere (see Chapter 11).

Alkanes are relatively insoluble in water and appear to play no significant role in aqueous atmospheric chemistry.

3.2 Alkenes and Alkynes

Nearly 150 unsaturated aliphatic hydrocarbons have been found in the atmosphere, with detection in the gas phase being somewhat more common than in the aerosol phase. Some are found indoors. There is an almost total absence of alkene detection in precipitation (reflecting the low aqueous solubility of alkenes) or in the stratosphere (reflecting the high reactivity of alkenes in the troposphere).

The concentrations of the alkenes decrease rapidly away from their urban sources, as seen in Figure 3.2-1. This suggests that they are more reactive than the alkanes, which will be seen to be the case.

Alkenes and alkynes are emitted from sources similar to those of the alkanes. By far the largest sources are those involving combustion of fossil fuels. A few of the lower molecular weight alkenes have industrial uses, and a few are emitted by vegetation.

The presence of an unsaturated bond in these compounds provides a location for chemical attack not present in the alkanes. The hydroxyl radical is again the most common reactant,[1221] adding at the unsaturated bond.[1224] E.g., for propene,

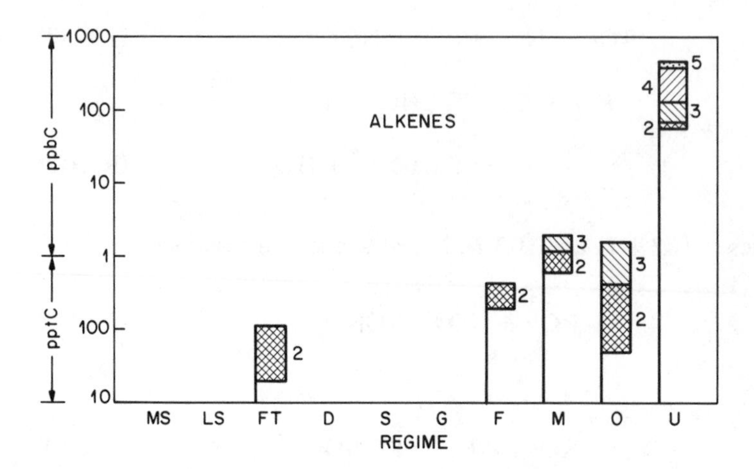

Fig. 3.2-1 Approximate concentration ranges of alkenes in different atmospheric regimes. The regime code is given in the caption of Figure 2.1-1 and the segmented division of the flags explained in the caption of Figure 3.1-2.

$$CH_3CH{=}CH_2 + OH\cdot \rightarrow CH_3\dot{C}HCH_2OH \qquad (R3.2\text{-}1)$$

followed by reactions similar to those of the alkanes;

$$CH_3\dot{C}HCH_2OH + O_2 \rightarrow CH_3CH(\dot{O}_2)CH_2OH \qquad (R3.2\text{-}2)$$

$$CH_3CH(\dot{O}_2)CH_2OH + NO \rightarrow CH_3CH(\dot{O})CH_2OH + NO_2 \qquad (R3.2\text{-}3)$$

$$CH_3CH(\dot{O})CH_2OH + O_2 \rightarrow CH_3CHO + HCHO + HO_2\cdot \qquad (R3.2\text{-}4)$$

to produce hydroxycarbonyl compounds. The initial OH· reaction is much more rapid in the case of alkenes (but not alkynes) than for alkanes; as a result, the atmospheric lifetimes of the alkenes are short.

Atmospheric alkenes also react with ozone molecules. The initial step is the formation of a molozonide which rapidly decomposes to a carbonyl and an energy-rich biradical, as shown in Figure 3.2-2. The degradation processes for the biradical are not yet well understood, but the formation of aldehyde or organic acid products seems likely.[1324]

Another alternative reaction chain of interest is with the nitrate radical. As with OH·, $NO_3\cdot$ adds to alkenes at the double bond and preferentially at the terminal carbon,[1223] e.g., for propene,

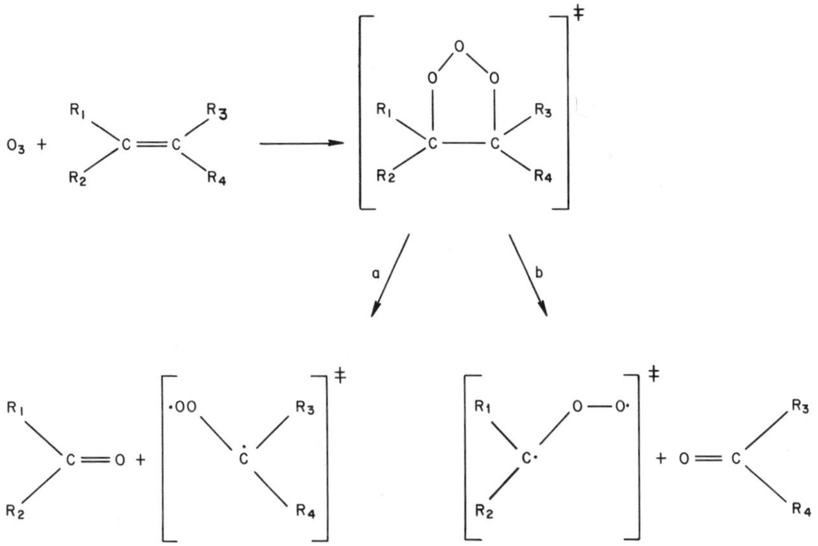

Fig. 3.2-2 Sequences for ozone-alkene reactions. (Reproduced with permission.[1324])

$$CH_3CH{=}CH_2 + NO_3\cdot \rightarrow CH_3\dot{C}HCH_2NO_3 \qquad (R3.2\text{-}5)$$

The subsequent products are known to be alkylnitrates.[1223] Since few direct sources of the alkylnitrates are known but a number of them have been detected (see Chapter 11), they probably originate from alkene-$NO_3\cdot$ reactions.

Under most atmospheric conditions, the primary loss process for the lower alkenes is expected to be reaction with OH·. For the higher alkenes, reaction with ozone can be dominant if $NO_3\cdot$ is low. If $NO_3\cdot$ concentrations are high, $NO_3\cdot$ reaction will dominate the loss processes of the higher alkenes.[1324]

Alkynes react slowly with OH· in the atmosphere; the details of the chemical sequences have not yet been determined.

Alkenes and alkynes possess very low water solubilities and are not expected to participate in aqueous phase atmospheric chemistry.

3.3 Monocyclic Hydrocarbons

Nearly a hundred monocyclic hydrocarbons have been detected in the atmosphere, about two-thirds of them in the gas phase. Many are found in indoor air. These compounds are almost entirely produced by fossil fuel or biomass combustion, although cyclohexene is widely used as a solvent.

These compounds have low solubility and have not been found in precipitation. Their concentrations are moderate in urban air and low elsewhere.

The principal fate of cycloalkanes is H abstraction by OH·,[1225,1226] e.g., for cyclohexane,

$$c\text{-}C_6H_{12} + OH\cdot \longrightarrow c\text{-}C_6H_{11}\cdot + H_2O \qquad \text{(R3.3-1)}$$

This reaction is followed by O_2 addition, reaction with NO, and ring cleavage to form an aliphatic aldehyde[1228]

$$c\text{-}C_6H_{11}\cdot + O_2 \longrightarrow c\text{-}C_6H_{11}O_2\cdot \qquad \text{(R3.3-2)}$$

$$c\text{-}C_6H_{11}O_2\cdot + NO \longrightarrow c\text{-}C_6H_{11}O\cdot + NO_2 \qquad \text{(R3.3-3)}$$

$$c\text{-}C_6H_{11}O\cdot \longrightarrow OHC(CH_2)_4CH_2\cdot \qquad \text{(R3.3-4)}$$

An alternative to (R3.3-3) is an addition reaction to produce cyclonitrates, a process which appears to occur 5-10% of the time:

$$c\text{-}C_6H_{11}O_2\cdot + NO \longrightarrow c\text{-}C_6H_{11}ONO_2 \qquad \text{(R3.3-5)}$$

In the case of cycloalkenes, the OH· and $NO_3\cdot$ reactants add efficiently at the double bond,[1227,1229] the products of the reactions being well defined. The cycloalkenes react very rapidly with ozone as well, probably to produce difunctional aliphatic compounds through ring cleavage of the ozonide, followed by fragmentation to give a variety of small molecules and radicals:[1324]

O–O
O_3 → O → CHO $\dot{O}_2$ → CO, CO_2, OTHER PRODUCTS

3.4 Terpenes and Polycyclic Hydrocarbons

Some sixty of these compounds are found in the atmosphere, about two-thirds of them in the aerosol phase. The single ring compounds and some of the bicyclo compounds are emitted profusely by vegetation; the larger molecules are emitted during biomass and fossil fuel combustion. A number of these compounds are present indoors as a result of their emission from wood or wood-based building materials. Within forests, the concentrations of terpenes are moderately high; elsewhere, these compounds are not abundant.

The major fate of the terpenes is reaction with ozone.[1259] Various reaction sequences have been suggested,[1259,1260] but much remains to be determined. As with other hydrocarbons, a variety of oxygenated carbonyl products are anticipated. The terpenes react with $NO_3\cdot$ [1229] and $OH\cdot$ [1230] as well as with ozone, and their atmospheric lifetimes are only a few hours.

3.5 Monocyclic Arenes

Benzene and more than seventy of its derivatives are found in the atmosphere, most often in the gas phase. Many occur in indoor air. They are produced by fossil fuel combustion and are also emitted into the atmosphere by a very wide spectrum of industrial processes. Their concentrations decrease relatively slowly away from their principal anthropogenic sources, as seen in Figure 3.5-1.

The atmospheric reaction chains of the monocyclic arenes are quite complex, but are relatively well defined as a result of detailed studies in several laboratories. The principal chain initiator is $OH\cdot$, which can add to the aromatic ring or abstract a hydrogen atom from a side group. A variety of aromatic aldehydes, alcohols, and nitrates are produced, as well as some products of ring cleavage. An example is a proposed reaction mechanism for toluene that is consistent with experiments and atmospheric observations is given in Figure 3.5-2. The products of arene oxidation have moderately high molecular weight and moderate solubility in water and thus can readily be deposited on aerosol particle surfaces.

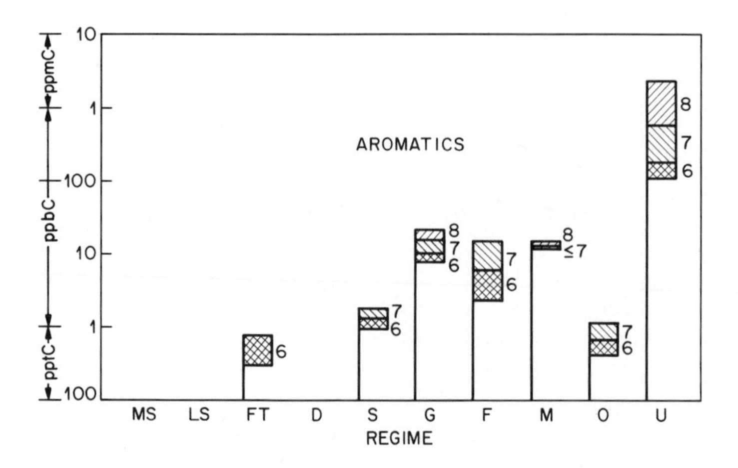

Fig. 3.5-1 Approximate concentration ranges of monocyclic arenes in different atmospheric regimes. The regime code is given in the caption of Figure 2.1-1 and the segmented division of the flags explained in the caption of Figure 3.1-2. 6 = Benzene, 7 = toluene, 8 = xylenes and ethylbenzene.

3.6-3.8 Higher Polycyclic Arenes

More than 200 arenes with two or more rings have been detected in the atmosphere. Some of the smaller, more volatile, molecules are found in both gas and aerosol phases, but most occur only in the aerosol phase. A few of these compounds are found in indoor air. Of the compounds in precipitation, more polycyclic arenes have been found than is the case for any other chemical group except the aliphatic acids.

The higher arenes enter the atmosphere almost entirely as a result of the combustion or industrial use of coal, coal products, or petroleum with a high aromatic content. Biomass combustion produces small amounts of the bicyclic arenes.

Atmospheric chemists have only a rudimentary understanding of the chemistry of these compounds. If a polycyclic arene is completely hydrogenated, its gas phase reaction with OH· will proceed by H atom abstraction.[1232] In the more usual case of an unsaturated system, both OH·[1233] and NO_3·[1223] will add readily at an olefinic bond. The products of these reactions are not known, but hydroxy, carbonyl, and nitrate derivates of the parent arenes seem likely. From the perspective of

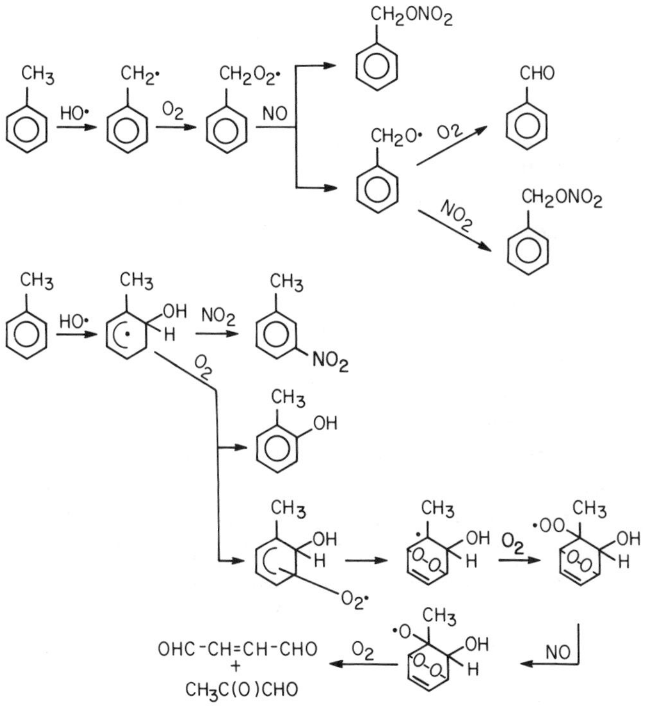

Fig. 3.5-2 Reaction pathways for the hydroxyl radical-initiated oxidation of toluene in the troposphere: (a) H abstraction from the side chain; (b) addition to the aromatic ring.

laboratory kinetic studies of OH· reactions with monocyclic and bicyclic arenes and of limited studies on polycyclic arenes, the gas phase reactions of OH· with higher polycyclic arenes are predicted to be quite rapid.[1233,1334]

Most of the atmospheric chemical reactions involving the higher arenes are likely to occur in the condensed phase. Studies of these processes are limited, but the arenes are known to be moderately reactive in the environment.[1261] Their photooxidation on wet aerosols is expected to lead to the formation of endoperoxides.[424,1262] Thus, for anthracene,

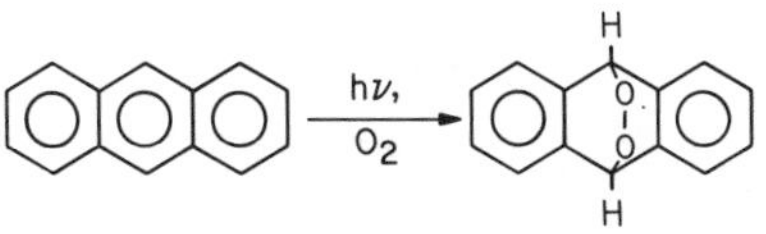

(R3.6-1)

If the bridgehead carbon atoms are also bonded to hydrogen atoms, conversion to quinones occurs rapidly:[424,1263]

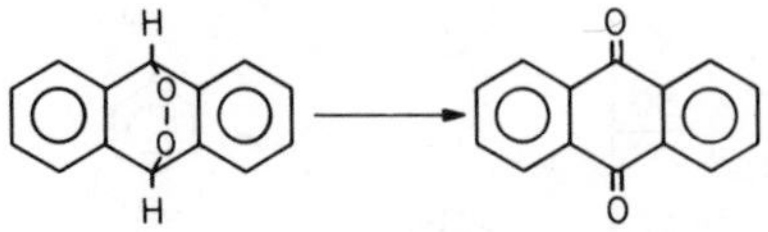

(R3.6-2)

Oxidation to ketones can also take place by oxygen addition to a saturated C-H bond.[424] For fluorene,

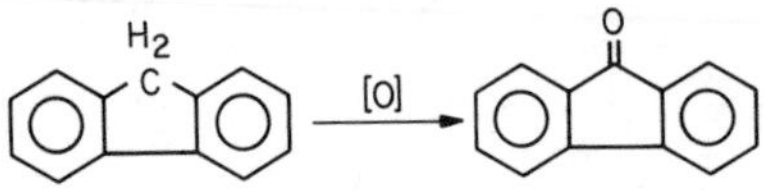

(R3.6-3)

Finally, limited studies have been conducted regarding the reactivity of polycyclic arenes on particles to ozone and nitrogen dioxide. In these studies, aerosol particles are doped with polynuclear aromatic hydrocarbons and exposed to the sunlit atmosphere or to a laboratory environment containing photons, ozone, and/or nitrogen dioxide. Transformation processes have been observed to be relatively efficient, with ketones, quinones, and nitro compounds being produced.[1301–1303]

Table 3.1. Alkanes

Species Number	Registry Number	Name	Emission Source	Emission Ref.	Detection Ref.	Code	Bioassay −MA	Bioassay +MA	Bioassay Ref.
3.1-1	74-82-8	**Methane**	acrylonitrile mfr.	626	6*,232*				
			animal waste	651	778*(i)				
			auto	159*,526	706*(S),707*(S)				
			biomass comb. 183,655*; coal outgassing 619; diesel 435*,812; foundry 279*; geothermal steam 372,402*; HCl mfr. 622; insects 471,784*; microbes 301,302; natural gas 232,423*; petroleum mfr. 232,543*;polymer comb. 118,354; refuse comb. 439,1093; sewage tmt. 932,1013*; tobacco smoke 396,448; turbine 359*,414; veneer drying 678; volcano 106*,234; water tmt. 185; wood pulping 19						
3.1-2	74-84-0	**Ethane**	acrylonitrile mfr.	626	123*,178				
			auto	159*,465*	399(a)				
			biomass comb.	232,439	712*(S),717*(S)				
			diesel 434*; gasoline vapor 465*; geothermal steam 402*,984; insects 471; microbes 302; natural gas 232,423*; petroleum mfr. 232,543*; polymer comb. 118,304; refuse comb. 1093; sewage tmt. 1013*; soil 1097; tobacco smoke 396,453; turbine 359*,414; veneer drying 678,791*; volcano 234,783*; water treatment 185; wood pulping 19						

Table 3.1. *(Continued)*

Species Number	Registry Number	Name	Emission Source	Emission Ref.	Detection Ref.	Code	Bioassay −MA	Bioassay +MA	Bioassay Ref.
3.1-3	74-98-6	**Propane**	acrylonitrile mfr.	626	170*,472*				
			auto	57,170*	399(a)				
		biomass comb. 232,439; diesel 136*,433*; gasoline vapor 232,465*;			712*(S)				
		geothermal steam 984,998*; insects 471; microbes 302;			820(I)				
		natural gas 232,423*; petroleum mfr. 232,543*; polymer comb. 118,304;							
		propellant 559,563; soil 1097; tobacco smoke 396,453;							
		turbine 414; veneer drying 678,791*; volcano 234;							
		water tmt. 185; wood pulping 19							
3.1-4	106-97-8	**Butane**	acrylonitrile mfr.	626	6*,170*				
			auto	159*,465*	399(a)				
		biomass comb. 232,524; building mtls. 82; diesel 136*,435*;			820(I)				
		gasoline vapor 232,465*; geothermal steam 998*; microbes 302;							
		natural gas 423*,465*; petroleum mfr. 221,232; polymer comb. 118,304;							
		printing 620; propellant 559; tobacco smoke 396, 634;							
		turbine 414; veneer drying 678,791*; water tmt. 185							
3.1-5	75-28-5	**Isobutane**	auto	47,170*	6*,170*				
			biomass comb.	232,524					
		diesel 434*,435*;gasoline vapor 232,465*; microbes 302;							
		natural gas 423*,465*; petroleum mfr. 232,543*; propellant 559,563;							
		tobacco smoke 396,634; turbine 414; veneer drying 678,791*;							
		volcano 234							
3.1-6	75-83-2	**2,2-Dimethylbutane**	auto	47,153*	164,263				
			biomass comb.	232,524					
		gasoline vapor 232; petroleum mfr. 232							

Species Number	Registry Number	Name	Emission Source	Emission Ref.	Detection Ref.	Bioassay Code	Bioassay −MA	Bioassay +MA	Bioassay Ref.
3.1-7	79-29-8	**2,3-Dimethylbutane**	auto	153*,159*	232*,524*				
			biomass comb.	232					
			gasoline vapor	232,486*					
3.1-8	464-06-2	**2,2,3-Trimethylbutane**	auto	47,526	597				
			biomass comb.	597					
			turbine	414					
3.1-9	594-82-1	**2,2,3,3-Tetramethyl-butane**	turbine	414					
3.1-10	109-66-0	**Pentane**	acrylonitrile mfr.	626	6*,170*				
			auto	159*,465*	364(a)				
		biomass comb. 232,524; building mtls. 82; diesel 136*,435*; gasoline vapor 232,465*; natural gas 423*,465*; petroleum mfr. 58*,232; polymer comb. 304,612; syn. rubber mfr. 58*; tobacco smoke 634; turbine 414; veneer drying 678,791*; volcano 234; water tmt. 185			820(I),1108(I)				
3.1-11	78-78-4	**Isopentane**	auto	159*,465*	6*,170*				
			biomass comb.	232,524	399(a)				
		building mtls. 82; diesel 136*,434*; gasoline vapor 232,465*; natural gas 423*,465*; petroleum mfr. 58*,232; polymer comb. 612; tobacco smoke 634; turbine 414; vegetation 552e; veneer drying 791*							
3.1-12S	463-82-1	**Neopentane**	biomass comb.	821	886*				
			turbine	414					

Table 3.1. *(Continued)*

Species Number	Registry Number	Name	Emission Source	Emission Ref.	Detection Ref.	Code	Bioassay −MA	Bioassay +MA	Bioassay Ref.
3.1-13	107-83-5	**2-Methylpentane**	auto	153*,159*	178,232*				
			biomass comb.	232,597	399(a)				
		gasoline vapor 232,465*; natural gas 465*; paint 1017*; petroleum mfr. 232,1000*; tobacco smoke 634; turbine 414; vegetation 552e			820(I),1124(I)				
3.1-14	96-14-0	**3-Methylpentane**	auto	153*,159*	164,232*				
			biomass comb.	232,597	820(I),1124(I)				
		gasoline vapor 232,465*; natural gas 465*; paint 1017*; petroleum mfr. 232,1000*; vegetation 552e							
3.1-15	590-35-2	**2,2-Dimethylpentane**	auto	284	232*,582				
			biomass comb.	597					
3.1-16	565-59-3	**2,3-Dimethylpentane**	auto	284,465*	164,232*				
			gasoline vapor	465*,621					
		natural gas 465*; paint 1017*; turbine 414							
3.1-17	108-08-7	**2,4-Dimethylpentane**	auto	159*,170*	170*,263				
			biomass comb.	597	820(I)				
		gasoline vapor 486,621; paint 1017*; petroleum mfr. 1000*							
3.1-18	562-49-2	**3,3-Dimethylpentane**	auto	284	582,601				
3.1-19	564-02-3	**2,2,3-Trimethylpentane**	auto	47,526	374,1099				
			biomass comb.	597	820(I)				
		petroleum mfr. 1000*; turbine 414							

Species Number	Registry Number	Name	Emission Source	Emission Ref.	Detection Ref.	Code	Bioassay −MA	Bioassay +MA	Ref.
3.1-20	540-84-1	**2,2,4-Trimethylpentane**	auto	159*,170*	170*,263				
			biomass comb.	597	1108(I)t				
		gasoline vapor 465*,621; natural gas 465*; turbine 414							
3.1-21	560-21-4	**2,3,3-Trimethylpentane**	auto	526,810	374,525				
			gasoline vapor	486					
			turbine	414					
3.1-22	565-75-3	**2,3,4-Trimethylpentane**	auto	159*,526	374,582				
			gasoline vapor	486					
3.1-23	7154-79-2	**2,2,3,3-Tetramethyl-pentane**	turbine	414					
3.1-24	1186-53-4	**2,2,3,4-Tetramethyl-pentane**	turbine	414					
3.1-25	617-78-7	**3-Ethylpentane**	biomass comb.	597	582				
3.1-26	609-26-7	**3-Ethyl-2-methylpentane**	auto	810	582,597 820(I)				
3.1-27	1067-08-9	**3-Ethyl-3-methylpentane**			582,597				
3.1-28	16747-33-4	**3-Ethyl-2,3-dimethyl-pentane**	turbine	414					
3.1-29	1067-20-5	**3,3-Diethylpentane**	auto	682t	682				

Table 3.1. *(Continued)*

Species Number	Registry Number	Name	Emission Source	Emission Ref.	Detection Ref.	Bioassay Code	Bioassay −MA	Bioassay +MA	Bioassay Ref.
3.1-30	110-54-3	**Hexane**	acrylonitrile mfr.	626	164,170*				
			auto	159*,465*	399(a),458(a)				
		biomass comb. 232,524; building mtls. 82,856; chemical mfr. 566*; diesel 435*,812; gasoline vapor 232,465*; natural gas 465*; paint 1017*; petroleum mfr. 48*,232; polymer comb. 304,612; tobacco smoke 634; turbine 414; vegetation 552e			820(I),1108(I)				
3.1-31	591-76-4	**2-Methylhexane**	auto	159*,170*	86*,170*				
			turbine	414	399(a)				
					820(I),1108(I)				
3.1-32	589-34-4	**3-Methylhexane**	auto	47,153*	57,123*				
			biomass comb.	597	820(I),1108(I)				
		gasoline vapor 465*,486; natural gas 465*; paint 1017*; petroleum mfr. 1000*; rendering 837t; turbine 414							
3.1-33	590-73-8	**2,2-Dimethylhexane**	auto	284,465*	582				
			gasoline vapor	465*					
		natural gas 465*; turbine 414							
3.1-34	584-94-1	**2,3-Dimethylhexane**	auto	284,682t	597,1017*				
			paint	1017*					
3.1-35	589-43-5	**2,4-Dimethylhexane**	auto	159*	582,597				
			biomass comb.	597					
			turbine	414					
3.1-36	592-13-2	**2,5-Dimethylhexane**			582,591				

Species Number	Registry Number	Name	Emission Source	Emission Ref.	Detection Ref.	Bioassay Code	Bioassay −MA	Bioassay +MA	Bioassay Ref.
3.1-37	563-16-6	**3,3-Dimethylhexane**	turbine	414	582				
3.1-38	583-48-2	**3,4-Dimethylhexane**	biomass comb.	597	374				
3.1-39	16747-26-5	**2,2,4-Trimethylhexane**			1108(I)				
3.1-40	3522-94-9	**2,2,5-Trimethylhexane**	auto gasoline vapor turbine	47,486* 486 414	374,597				
3.1-41	921-47-1	**2,3,4-Trimethylhexane**	turbine	414					
3.1-42	1069-53-0	**2,3,5-Trimethylhexane**	auto gasoline vapor	486*,682t 486	682				
3.1-43	1071-81-4	**2,2,5,5-Tetramethylhexane**			820(I)				
3.1-44	619-99-8	**3-Ethylhexane**	petroleum mfr.	1000*					
3.1-45	3074-77-9	**3-Ethyl-4-methylhexane**	turbine	414					
3.1-46	3074-75-7	**3-Ethyl-5-methylhexane**	turbine	414					
3.1-47	52896-99-8	**3-Ethyl-5,5-dimethylhexane**	turbine	414					

Table 3.1. *(Continued)*

Species Number	Registry Number	Name	Emission Source	Emission Ref.	Detection Ref.	Bioassay Code	Bioassay −MA	Bioassay +MA	Bioassay Ref.
3.1-48	142-82-5	**Heptane**	auto	159*,465*	86*,591*				
			biomass comb.	597,969*	399(a),458(a)				
		building mtls. 82,856; diesel 812; gasoline vapor 58*,465*; natural gas 465*; paint 1017*; petroleum mfr. 543*,1000*; rendering 837,879; solvent 78,598; tobacco smoke 634; turbine 414			820(I),1108(I)				
3.1-49	592-27-8	**2-Methylheptane**	auto	284,465*	86*,582				
			building mtls.	856	399(a)				
		gasoline vapor 58*,465*; natural gas 465*; paint 1017*; petroleum mfr. 1000*; turbine 414t			820(I),1108(I)				
3.1-50	589-81-1	**3-Methylheptane**	auto	284,486*	263,582				
			building mtls.	856	820(I)				
		gasoline vapor 486; paint 1017*; tobacco smoke 421							
3.1-51	589-53-7	**4-Methylheptane**	auto	284,810	582,674				
					820(I)				
3.1-52	1071-26-7	**2,2-Dimethylheptane**	turbine	414					
3.1-53	3074-71-3	**2,3-Dimethylheptane**	auto	682t	682				
3.1-54	2213-23-2	**2,4-Dimethylheptane**	auto	486*,810	591,682				
			gasoline vapor	486					
		syn. rubber mfr. 58*; turbine 414							
3.1-55	2216-30-0	**2,5-Dimethylheptane**	auto	682t	582,682				
			turbine	414	1124(I)				

Species Number	Registry Number	Name	Emission Source	Emission Ref.	Detection Ref.	Bioassay Code	Bioassay −MA	Bioassay +MA	Bioassay Ref.
3.1-56	1072-05-5	**2,6-Dimethylheptane**	auto	682t	597,682				
3.1-57	4032-86-4	**3,3-Dimethylheptane**	turbine	414					
3.1-58	922-28-1	**3,4-Dimethylheptane**	turbine	414					
3.1-59	79004-86-7	**Trimethylheptane**	auto	682t	682 820(I)				
3.1-60	30586-18-6	**Pentamethylheptane**			1124(I)				
3.1-61	15869-80-4	**3-Ethylheptane**	tobacco smoke turbine	421 414					
3.1-62	52896-91-0	**3-Ethyl-4-methyl-heptane**	turbine	414					
3.1-63	13475-78-0	**3-Ethyl-6-methyl-heptane**	turbine	414					
3.1-64	2216-32-2	**4-Ethylheptane**	auto turbine	682t 414	682				
3.1-65	3178-29-8	**4-Propylheptane**			597				

Table 3.1. *(Continued)*

Species Number	Registry Number	Name	Emission Source	Emission Ref.	Detection Ref.	Code	Bioassay −MA	Bioassay +MA	Bioassay Ref.
3.1-66	111-65-9	**Octane**	adhesives	1111	86*,200				
			auto	47,465*	364(a),458(a)				
		biomass comb. 597; brewing 144; building mtls. 82,856; diesel 812; gasoline vapor 465*,486; landfill 365; natural gas 465*; paint 1017*; polymer comb. 853; rendering 837,879; solvent 134,598; tobacco smoke 421,634; turbine 414			820(I),1108(I)				
3.1-67	3221-61-2	**2-Methyloctane**	auto	465*,810	86,525				
			gasoline vapor	465*	399(a)				
		natural gas 465*; rendering 837t			1108(I)				
3.1-68	2216-33-3	**3-Methyloctane**	auto	682t	582,674				
			tobacco smoke	421					
			turbine	414					
3.1-69	2216-34-4	**4-Methyloctane**	auto	682t	582,597				
			turbine	414					
3.1-70	7146-60-3	**2,3-Dimethyloctane**	turbine	414	597				
3.1-71	15869-89-3	**2,5-Dimethyloctane**	auto	682t	597,682t				
3.1-72	2051-30-1	**2,6-Dimethyloctane**	auto	682t	682				
			turbine	414					
3.1-73	62016-31-3	**2,3,4-Trimethyloctane**	turbine	414					

Species Number	Registry Number	Name	Emission Source	Emission Ref.	Detection Ref.	Code	Bioassay −MA	Bioassay +MA	Bioassay Ref.
3.1-74	62016-41-5	**3,3,5-Trimethyloctane**	turbine	414					
3.1-75	111-84-2	**Nonane**	adhesives auto	1111 465*,486*	525*,591* 364(a),458(a) 820(I),1108(I)				
		biomass comb. 597; brewing 144; building mtls. 856; coal comb. 1159; diesel 812; gasoline vapor 486; petroleum mfr. 543; rendering 837,879; tobacco smoke 421,634; turbine 414; volcano 1007							
3.1-76	871-83-0	**2-Methylnonane**	turbine	414	86,200 399(a) 1108(I)				
3.1-77	5911-04-6	**3-Methylnonane**	auto	682t	597,674				
3.1-78	17301-94-9	**4-Methylnonane**			597				
3.1-79	15869-85-9	**5-Methylnonane**			597				
3.1-80	17302-27-1	**2,5-Dimethylnonane**	turbine	414					
3.1-81	17302-23-7	**4,5-Dimethylnonane**	turbine	414					
3.1-82	124-18-5	**Decane**	adhesives auto	1111 47,526	200*,525* 364(a),458(a) 820(I),1108(I)				
		biomass comb. 597; brewing 144; building mtls. 856; coal comb. 1159; diesel 176,812; rendering 837,879; tobacco smoke 421; turbine 359*,414; volcano 1007							

Table 3.1. *(Continued)*

Species Number	Registry Number	Name	Emission Source	Emission Ref.	Detection Ref.	Bioassay Code	Bioassay −MA	Bioassay +MA	Bioassay Ref.
3.1-83	6975-98-0	**2-Methyldecane**	rendering	837t	86,200				
			tobacco smoke	421	399(a)				
3.1-84	13151-34-3	**3-Methyldecane**	diesel	309(a)					
			tobacco smoke	421					
			turbine	414					
3.1-85	2847-72-5	**4-Methyldecane**	refuse comb.	840(a)	628				
			turbine	414	820(I),1108(I)				
3.1-86	13151-35-4	**5-Methyldecane**	tobacco smoke	421	597				
					1108(I)				
3.1-87	1120-21-4	**Undecane**	adhesives	1111	86*,200*				
			biomass comb.	597	364(a),458(a)				
		brewing 144; building mtls. 856; diesel 812; rendering 837,879; tobacco smoke 421; turbine 414; volcano 1007			820(I),1108(I)				
3.1-88	7045-71-8	**2-Methylundecane**	rendering	837t	86,200				
					399(a)				
					1108(I)				
3.1-89	1002-43-3	**3-Methylundecane**	diesel	309(a)					

Species Number	Registry Number	Name	Emission Source	Emission Ref.	Detection Ref.	Code	Bioassay −MA	Bioassay +MA	Bioassay Ref.
3.1-90	112-40-3	**Dodecane**	auto	972	86*,200*				
			biomass comb.	597	364(a),458(a)				
		brewing 144; building mtls. 856; rendering 837,879; tobacco smoke 421,634(a); volcano 1007							
3.1-91	1560-97-0	**2-Methyldodecane**	rendering	837t	86,200				
					399(a)				
3.1-92	629-50-5	**Tridecane**	biomass comb.	597	86,525				
			brewing	144	364(a),458(a)				
		refuse comb. 790(a); rendering 837,879; tobacco smoke 421,634(a); turbine 414			820(I), 1108(I)				
3.1-93	1560-96-9	**2-Methyltridecane**	rendering	837t	86,200				
3.1-94	26730-14-3	**7-Methyltridecane**			597				
3.1-95	629-59-4	**Tetradecane**	biomass comb.	597	86,525				
			brewing	144	310(a),363(a)				
		diesel 309(a),685*(a); refuse comb. 790(a)t; rendering 837,879 tobacco smoke 634(a)			820(I),1110*(I)				
3.1-96	1560-95-8	**2-Methyltetradecane**	rendering	837t	399(a)				
3.1-97	18435-22-8	**3-Methyltetradecane**	diesel	309(a)					

Table 3.1. *(Continued)*

Species Number	Registry Number	Name	Emission Source	Emission Ref.	Detection Ref.	Bioassay Code	−MA	+MA	Ref.
3.1-98	629-62-9	**Pentadecane**	biomass comb.	1089(a)	86*,633*				
			brewing	144	310(a),363(a)				
		diesel 309(a); refuse comb. 788(a),790(a)t; tobacco smoke 634(a)			570(r)				
					820(I),1110*(I)				
3.1-99	1560-93-6	**2-Methylpentadecane**	diesel	309(a)	399(a)				
3.1-100S	1921-70-6	**Pristane**	auto	956(a)	677*,834				
			diesel	956(a)	882(a),906(a)				
3.1-101	544-76-3	**Hexadecane**	auto	311(a)	200*,633*				
			biomass comb.	1089(a)	310*(a),363(a)				
		diesel 309(a),956(a); refuse comb. 788(a),790(a)t;			570(r)				
		tobacco smoke 634(a)			820(I),1110*(I)				
3.1-102	1560-42-5	**2-Methylhexadecane**			399(a)				
3.1-103S	638-36-8	**Phytane**	auto	956(a)	906(a)				
			diesel	956(a)					
3.1-104	629-78-7	**Heptadecane**	auto	310(a)	200*,633*				
			biomass comb.	1089(a)	310*(a),363(a)				
		diesel 309(a),956(a); refuse comb. 788(a),790(a)t; tobacco smoke 634(a)			570(r)				
3.1-105	1560-89-0	**2-Methylheptadecane**			399(a)				

Species Number	Registry Number	Name	Emission Source	Emission Ref.	Detection Ref.	Bioassay Code	Bioassay −MA	Bioassay +MA	Bioassay Ref.
3.1-106	593-45-3	**Octadecane**	auto	310(a)	200*,633*				
			biomass comb.	1089(a)	310*(a),363(a)				
		diesel 309(a),956(a); refuse comb. 788(a),878(a); tobacco smoke 634(a)			570(s)				
3.1-107	1560-88-9	**2-Methyloctadecane**			399(a)				
3.1-108	629-92-5	**Nonadecane**	auto	310(a),682t	86*,633*				
			biomass comb.	1089(a)	310*(a),363(a)				
		diesel 309(a),956(a); refuse comb. 788(a),790(a)t; tobacco smoke 634(a)			570(r)				
					570(s)				
3.1-109	112-95-8	**Eicosane**	auto	310(a),628t	86*,633*				
			biomass comb.	1089(a)	310*(a),363*(a)				
		diesel 309(a); refuse comb. 788(a),790(a)t; tobacco smoke 634(a)			570(s)				
3.1-110	629-94-7	**Heneicosane**	auto	310(a),311(a),682t	633*,677*				
			biomass comb.	1089(a)	310*(a),633*(a)				
		diesel 310(a),956(a); refuse comb. 788(a),790(a)t; tobacco smoke 634(a)			570(r)				
					570(s)				
3.1-111	629-97-0	**Docosane**	auto	310(a),311(a),682t	633*,677*				
			biomass comb.	1089(a)	633*(a),660*(a)				
		diesel 310(a),956(a); refuse comb. 788(a),790(a)t; tobacco smoke 634(a)			570(r)				
					570(s)				

Table 3.1. *(Continued)*

Species Number	Registry Number	Name	Emission Source	Emission Ref.	Detection Ref.	Bioassay Code	−MA	+MA	Ref.
3.1-112	638-67-5	**Tricosane**	aluminum mfr.	1058*(a)	633*,677*				
			auto	310(a),311(a),682t	310*(a),633*(a)				
		biomass comb. 1089(a); diesel 310(a),956(a); refuse comb. 788(a),878(a);			570(r)				
		tobacco smoke 634(a)			570(s)				
					1109*(I)				
3.1-113	646-31-1	**Tetracosane**	aluminum mfr.	1058*(a)	633*,677*				
			auto	310(a),311(a),682t	310*(a),633(a)				
		biomass comb. 1089(a); diesel 310(a),956(a); refuse comb. 788(a), 790(a)t;			570(r),762*(r)				
		tobacco smoke 634(a)			570(s)				
					1109*(I)				
3.1-114	629-99-2	**Pentacosane**	aluminum mfr.	1058*(a)	633*,677*				
			auto	310(a),311(a),682t	633*(a),660*(a)				
		biomass comb. 1089(a); diesel 310(a),956(a); refuse comb. 790(a)t,878(a);			570(r),762*(r)				
		tobacco smoke 634(a)			570(s),762*(s)				
					1109*(I)				
3.1-115	630-01-3	**Hexacosane**	aluminum mfr.	1058*(a)	633*,677*				
			auto	310(a),311(a),682t	633*(a),660*(a)				
		biomass comb. 1089(a); diesel 310(a),956(a); refuse comb. 790(a)t,878(a);			570(r),762*(r)				
		tobacco smoke 634(a)			570(s)				
					1109*(I)				
3.1-116	1561-02-0	**2-Methylhexacosane**	tobacco smoke	396(a),634(a)					

Species Number	Registry Number	Name	Emission Source	Emission Ref.	Detection Ref.	Code	Bioassay −MA	Bioassay +MA	Bioassay Ref.
3.1-117	593-49-7	**Heptacosane**	aluminum mfr.	1058*(a)	633*,677*				
			auto	310(a),311(a)	633*(a),660*(a)				
		biomass comb. 1089(a); diesel 310(a); refuse comb. 790(a),878(a); tobacco smoke 396(a),634(a)			570(r),762*(r)				
					570(s),762*(s)				
					1109*(I)				
3.1-118	1561-00-8	**2-Methylheptacosane**	tobacco smoke	634(a)					
3.1-119	630-02-4	**Octacosane**	aluminum mfr.	1058*(a)	633*,677*				
			auto	310(a),311(a)	633*(a),660*(a)				
		biomass comb. 1089(a); diesel 310(a); refuse comb. 878(a); tobacco smoke 396(a),634(a)			570(r),762*(r)				
					570(s)				
					1109*(I)				
3.1-120	1560-98-1	**2-Methyloctacosane**	tobacco smoke	396(a),634(a)					
3.1-121	630-03-5	**Nonacosane**	aluminum mfr.	1058*(a)	633*				
			auto	310(a),311(a)	633*(a),660*(a)				
		biomass comb. 1089(a); diesel 310(a); refuse comb. 878(a); tobacco smoke 396(a),634(a)			570(r),762*(r)				
					570(s),762*(s)				
					1109*(I)				
3.1-122	1560-75-4	**2-Methylnonacosane**	tobacco smoke	396(a),634(a)					

Table 3.1. *(Continued)*

Species Number	Registry Number	Name	Emission Source	Emission Ref.	Detection Ref.	Bioassay Code	−MA	+MA	Ref.
3.1-123	638-68-6	**Triacontane**	aluminum mfr.	1058*(a)	633*				
			auto	311(a)	633*(a),660*(a)				
		diesel 310(a); refuse comb. 790(a)t,878(a); tobacco smoke 396(a),634(a)			570(r)				
					570(s)				
					1109*(I)				
3.1-124	1560-72-1	**2-Methyltriacontane**	tobacco smoke	396(a),634(a)					
3.1-125	630-04-6	**Hentriacontane**	aluminum mfr.	1058*(a)	633*				
			auto	311(a)	633*(a),660*(a)				
		diesel 310(a); refuse comb. 878(a); tobacco smoke 396(a),634(a)			570(r),762*(r)				
					570(s),762*(s)				
					1109*(I)				
3.1-126	1720-12-3	**2-Methylhentriacontane**	tobacco smoke	396(a)					
3.1-127	544-85-4	**Dotriacontane**	aluminum mfr.	1058*(a)	633				
			auto	311(a)	310*(a),660*(a)				
		diesel 310(a); refuse comb. 878(a); tobacco smoke 396(a),634(a)			570(r)				
3.1-128	1720-11-2	**2-Methyldotriacontane**	tobacco smoke	396(a)					
3.1-129	630-05-7	**Tritriacontane**	aluminum mfr.	1058*(a)	633				
			auto	311(a)	310*(a),363(a)				
		diesel 310(a); refuse comb. 878(a); tobacco smoke 396(a),634(a)			570(r)				
3.1-130	14167-59-0	**Tetratriacontane**	refuse comb.	878(a)	310*(a),363(a)				
			tobacco smoke	634(a)					

Species Number	Registry Number	Name	Emission Source	Emission Ref.	Detection Ref.	Code	Bioassay −MA	Bioassay +MA	Bioassay Ref.
3.1-131	630-07-9	**Pentatriacontane**	refuse comb.	878(a)	310*(a),363(a)				
			tobacco smoke	634(a)					
3.1-132	630-06-8	**Hexatriacontane**	auto	311(a)	310*(a),1023(a)				
			refuse comb.	878(a)					
			tobacco smoke	634(a)					
3.1-133	7194-84-5	**Heptatriacontane**	refuse comb.	878(a)	310*(a),1023(a)				
3.1-134	7194-85-6	**Octatriacontane**	refuse comb.	878(a)	1023(a)				
3.1-135	7194-86-7	**Nonatriacontane**	refuse comb.	878(a)	1023(a)				
3.1-136	4181-95-7	**Tetracontane**			1023(a)				

CMe_4

3.1-12

$Me_2CH(CH_2)_3CHMe(CH_2)_3CHMe(CH_2)_3CHMe_2$

3.1-100

$Me_2CH(CH_2)_3CHMe(CH_2)_3CHMe(CH_2)_3CHMeCH_2Me$

3.1-103

Table 3.2. Alkenes and Alkynes

Species Number	Registry Number	Name	Emission Source	Emission Ref.	Detection Ref.	Bioassay Code	Bioassay −MA	Bioassay +MA	Bioassay Ref.
3.2-1	74-85-1	**Ethene**	acrylonitrile mfr.	626	3*,123*				
			auto	159*,465*					
		biomass comb. 183,662*; chemical mfr. 1157; diesel 136*,434*; foundry 279*; fruit ripening 146; microbes 172,210; petroleum mfr. 232,1000*; polymer comb. 118,304; refuse comb. 419*,439; sewage tmt. 1013*; soil 1097,1098; solvent 134; tobacco smoke 396,453; turbine 359*,414; vegetation 461; veneer drying 678,791*; volcano 234; wood comb. 890; wood pulping 19							
3.2-2	74-86-2	**Acetylene**	animal waste	931	90*,232*				
			auto	159*,465*	717*(S)				
		biomass comb. 207,232; calc. carbide mfr. 58*; diesel 136*,433*; foundry 279*; petroleum mfr. 232,524; polymer comb. 611,853; refuse comb. 439,543; tobacco smoke 396,634; turbine 359*,414; vegetation 461; veneer drying 791*; volcano 783*							
3.2-3	115-07-1	**Propene**	acrylonitrile mfr.	626	6*,170				
			auto	159*,465*					
		biomass comb. 183,232; chemical mfr. 1157; diesel 136*,433*; microbes 302; natural gas 465*; petroleum mfr. 232,1000*; polymer comb. 118,304; refuse comb. 439,1093; tobacco smoke 396,453; turbine 359*,414; veneer drying 678,791*; volcano 234; wood pulping 19							

Species Number	Registry Number	Name	Emission Source	Emission Ref.	Detection Ref.	Code	Bioassay −MA	Bioassay +MA	Bioassay Ref.
3.2-4	463-49-0	**Propadiene**	auto	49*,486*	57,170*				
			diesel	433*					
		tobacco smoke 634; turbine 359*,414							
3.2-5	74-99-7	**Propyne**	auto	159*,465*	170*,524*				
			biomass comb.	232,524					
		petroleum mfr. 232; polymer comb. 853; tobacco smoke 396,634; turbine 359*,414							
3.2-6	460-12-8	**Propadiyne**			399(a)				
3.2-7	106-98-9	**1-Butene**	auto	284,972	232*,260				
			biomass comb.	232,524	399(a)				
		chemical mfr. 1157; diesel 136*,223*; gasoline vapor 232; microbes 302; petroleum mfr. 232; polymer comb. 118,304; tobacco smoke 634; turbine 414; veneer drying 791*; volcano 234			820(I)				
3.2-8	563-46-2	**2-Methyl-1-butene**	auto	47,170*	123*,207				
			biomass comb.	232,524					
		gasoline vapor 232,621; petroleum mfr. 1000*; tobacco smoke 634; turbine 359*,414							
3.2-9	563-45-1	**3-Methyl-1-butene**	auto	153*,284	582				
			diesel	433*,434*					
		tobacco smoke 634; turbine 359*; volcano 234							
3.2-10	563-78-0	**2,3-Dimethyl-1-butene**	auto	284	582				
			tobacco smoke	634	399(a)				

Table 3.2. *(Continued)*

Species Number	Registry Number	Name	Emission Source	Emission Ref.	Detection Ref.	Bioassay Code	Bioassay −MA	Bioassay +MA	Bioassay Ref.
3.2-11	558-37-2	**3,3-Dimethyl-1-butene**	tobacco smoke	634	582				
3.2-12	594-56-9	**2,3,3-Trimethyl-1-butene**	auto	284	582				
			turbine	414					
3.2-13	760-21-4	**2-Ethyl-1-butene**	auto	284,810	582				
			turbine	359*					
3.2-14	107-00-6	**1-Butyne**	petroleum mfr.	371					
			syn. rubber mfr.	58*					
			tobacco smoke	634					
3.2-15	687-97-4	**1-Butyn-3-ene**	petroleum mfr.	371					
			tobacco smoke	634					
3.2-16	115-11-7	**Isobutene**	auto	284,527	162,232*				
			biomass comb.	232,524	399(a)				
		diesel 136*; gasoline vapor 232,621; petroleum mfr. 232; syn. rubber mfr. 58*; tobacco smoke 396,634; turbine 414; veneer drying 791*; volcano 234							
3.2-17	590-19-2	**1,2-Butadiene**	petroleum mfr.	371					
			tobacco smoke	634					
3.2-18	106-99-0	**1,3-Butadiene**	auto	159,810	170*,232*				
			biomass comb.	232,524	399(a)				
		chemical mfr. 1157; diesel 136,433; petroleum mfr. 111,371; plastics mfr. 104; syn. rubber mfr. 58; tobacco smoke 396,634							

Species Number	Registry Number	Name	Emission Source	Emission Ref.	Detection Ref.	Code	Bioassay −MA	Bioassay +MA	Bioassay Ref.
3.2-19S	78-79-5	**Isoprene**	auto	159,465	198*,472*				
			biomass comb.	597,821	399(a)				
		gasoline vapor 465; rubber abrasion 331; tobacco smoke 396,446; turbine 359,414; vegetation 198,207; wood pulping 695,1071			820(I)				
3.2-20	3404-63-5	**2-Ethyl-1,3-butadiene**			364(a)				
3.2-21	590-18-1	*cis*-**2-Butene**	auto	49*,170*	123*,170*				
			biomass comb.	232,524					
		diesel 433*,812; gasoline vapor 232							
3.2-22	624-64-6	*trans*-**2-Butene**	acrylonitrile mfr.	626	170*,207				
			auto	49,486					
		biomass comb. 232,524; diesel 136,433; gasoline vapor 232,486; refuse comb. 1093; solvent 134; tobacco smoke 634; turbine 414; veneer drying 678							
3.2-23	513-35-9	**2-Methyl-2-butene**	auto	159,170	123*,232				
			biomass comb.	232,597					
		diesel 433; gasoline vapor 232,486; tobacco smoke 634; turbine 359,414							
3.2-24	563-79-1	**2,3-Dimethyl-2-butene**	auto	153,284	57,582				
3.2-25	503-17-3	**2-Butyne**	turbine	359*					

Table 3.2. *(Continued)*

Species Number	Registry Number	Name	Emission Source	Emission Ref.	Detection Ref.	Code	Bioassay −MA	Bioassay +MA	Bioassay Ref.
3.2-26	109-67-1	**1-Pentene**	auto	47,486*	57,170*				
			biomass comb.	597,821	364(a),458(a)				
		diesel 136*,435*; gasoline vapor 486; polymer comb. 304,612; tobacco smoke 634; turbine 359*,414			820(I)				
3.2-27	763-29-1	**2-Methyl-1-pentene**	auto	47	57				
			biomass comb.	597					
		tobacco smoke 634; turbine 414							
3.2-28	760-20-3	**3-Methyl-1-pentene**	tobacco smoke	634	820(I)				
3.2-29	691-37-2	**4-Methyl-1-pentene**	auto	170*	57,170*				
			biomass comb.	597	364(a)				
		diesel 433*; polymer comb. 853t; tobacco smoke 634; turbine 414							
3.2-30	3404-72-6	**2,3-Dimethyl-1-pentene**	microbes	302	582				
3.2-31	2213-32-3	**2,4-Dimethyl-1-pentene**	biomass comb.	597					
			turbine	414					
3.2-32	7385-78-6	**3,4-Dimethyl-1-pentene**	auto	284,810	628t				
			biomass comb.	597					
3.2-33	762-62-9	**4,4-Dimethyl-1-pentene**	biomass comb.	597					
			turbine	414					
3.2-34	107-39-1	**2,4,4-Trimethyl-1-pentene**	auto	284	374	STP00		−(C)	B36
			turbine	414					

Species Number	Registry Number	Name	Emission Source	Ref.	Detection Ref.	Code	Bioassay −MA	+MA	Ref.
3.2-35	4038-04-4	**3-Ethyl-1-pentene**	auto	526,810	374				
3.2-36	627-19-0	**1-Pentyne**	turbine	414					
3.2-37	591-95-7	**1,2-Pentadiene**	tobacco smoke	634					
3.2-38	504-60-9	**1,3-Pentadiene**	biomass comb.	597	582				
			syn. rubber mfr.	58*	399(a)				
			tobacco smoke	396,634					
3.2-39	1118-58-7	**2-Methyl-1,3-pentadiene**			399(a)				
3.2-40	1000-86-8	**2,4-Dimethyl-1,3-pentadiene**			820(I)				
3.2-41	591-93-5	**1,4-Pentadiene**	tobacco smoke	634	597				
3.2-42	627-20-3	*cis*-**2-Pentene**	auto	47,170*	57,170*				
			diesel	433*					
		gasoline vapor 486,621; turbine 414							
3.2-43	646-04-8	*trans*-**2-Pentene**	auto	47,153*	123*,232*				
			biomass comb.	232,524					
		gasoline vapor 486; tobacco smoke 634							

Table 3.2. *(Continued)*

Species Number	Registry Number	Name	Emission Source	Emission Ref.	Detection Ref.	Code	Bioassay −MA	Bioassay +MA	Ref.
3.2-44	625-27-4	**2-Methyl-2-pentene**	auto	153*,486*	582				
			gasoline vapor	486					
		tobacco smoke 634; turbine 414							
3.2-45	922-61-2	**3-Methyl-2-pentene**	auto	153*	582,597				
			biomass comb.	597					
			turbine	414					
3.2-46	4461-48-7	**4-Methyl-2-pentene**	auto	284,810	582				
			tobacco smoke	634					
			turbine	359*					
3.2-47	10574-37-5	**2,3-Dimethyl-2-pentene**	biomass comb.	597					
3.2-48	625-65-0	**2,4-Dimethyl-2-pentene**	biomass comb.	969*	170,582				
			turbine	414					
3.2-49	24910-63-2	**3,4-Dimethyl-2-pentene**	auto	284					
3.2-50	26232-98-4	**4,4-Dimethyl-2-pentene**	auto	810	597				
			turbine	414					
3.2-51	565-77-5	**2,3,4-Trimethyl-2-pentene**			597				

Species Number	Registry Number	Name	Emission Source	Emission Ref.	Detection Ref.	Code	Bioassay −MA	Bioassay +MA	Bioassay Ref.
3.2-52	107-40-4	**2,4,4-Trimethyl-2-pentene**	auto	284,526	374				
3.2-53	816-79-5	**3-Ethyl-2-pentene**	biomass comb.	597					
3.2-54	1000-87-9	**2,4-Dimethyl-2,3-pentadiene**			597t				
3.2-55	592-41-6	**1-Hexene**	auto biomass comb.	49* 597,821	123*,263 458(a) 820(I),1112(I)				
		diesel 812; polymer comb. 612,853; tobacco smoke 634; turbine 414							
3.2-56	6094-02-6	**2-Methyl-1-hexene**	turbine	414	582				
3.2-57	3769-23-1	**4-Methyl-1-hexene**	auto turbine	47,810 414	399(a)				
3.2-58	3524-73-0	**5-Methyl-1-hexene**	auto	47,526	374,582				
3.2-59	6975-92-4	**2,5-Dimethyl-1-hexene**	tobacco smoke	421					
3.2-60	4316-65-8	**3,5,5-Trimethyl-1-hexene**	turbine	359*					
3.2-61	1632-16-2	**2-Ethyl-1-hexene**	auto biomass comb.	284 597					
3.2-62	42296-74-2	**Hexadiene**			399(a),458(a)				

Table 3.2. *(Continued)*

Species Number	Registry Number	Name	Emission Source	Emission Ref.	Detection Ref.	Code	Bioassay −MA	Bioassay +MA	Bioassay Ref.
3.2-63	1119-14-8	**2-Methylhexadiene**			399(a)				
3.2-64	18669-52-8	**2,3-Dimethyl-1,4-hexadiene**	biomass comb.	597					
3.2-65	2235-12-3	**1,3,5-Hexatriene**	biomass comb.	597					
3.2-66	693-02-7	**1-Hexyne**			399(a)				
3.2-67		**Methyl-1-hexyne**	polymer comb.	853					
3.2-68	7688-21-3	*cis*-**2-Hexene**	auto	49*,170*	57,170*				
3.2-69	4050-45-7	*trans*-**2-Hexene**	auto tobacco smoke	49*,526 634	57,374				
3.2-70	17618-77-8	**3-Methyl-2-hexene**			582				
3.2-71	3404-62-4	**5-Methyl-2-hexene**	auto	284					
3.2-72	7145-20-2	**2,3-Dimethyl-2-hexene**	auto refuse comb.	284 26(a)					
3.2-73	3404-78-2	**2,5-Dimethyl-2-hexene**	auto	810					
3.2-74	592-46-1	**2,4-Hexadiene**	biomass comb.	597					
3.2-75	2809-69-0	**2,4-Hexadiyne**	turbine	414					

Species Number	Registry Number	Name	Emission Source	Emission Ref.	Detection Ref.	Code	Bioassay −MA	Bioassay +MA	Bioassay Ref.
3.2-76	7642-09-3	***cis*-3-Hexene**	auto	526,810	57,582				
3.2-77	15840-60-5	**2-Methyl-*cis*-3-hexene**			582				
3.2-78	13269-52-8	***trans*-3-Hexene**	auto	49*,526	374,582				
3.2-79	692-24-0	**2-Methyl-*trans*-3-hexene**	turbine	414					
3.2-80	692-70-6	**2,5-Dimethyl-*trans*-3-hexene**	auto	810					
3.2-81	3161-99-7	**Hexa-1,3,5-triyne**	microbes	302					
3.2-82	592-76-7	**1-Heptene**	auto biomass comb.	49*,153* 821	374,591 399(a),458(a) 820(I)				
		building mtls. 856; diesel 812; rendering 837,879; turbine 359*,414							
3.2-83	15870-10-7	**2-Methyl-1-heptene**	auto	682t	682 399(a)				
3.2-84	42441-75-8	**Heptadiene**			458(a)				
3.2-85	26856-31-5	**Heptyne**	polymer comb.	853					

Table 3.2. *(Continued)*

Species Number	Registry Number	Name	Emission Source	Emission Ref.	Detection Ref.	Bioassay Code	Bioassay −MA	Bioassay +MA	Bioassay Ref.
3.2-86	6443-92-1	***cis*-2-Heptene**	auto turbine	526,810 359*	582,682				
3.2-87	14686-13-6	***trans*-2-Heptene**	auto biomass comb. turbine	153* 597 359*					
3.2-88	592-78-9	**3-Heptene**	auto turbine	284,810 414	886*				
3.2-89	692-96-6	**2-Methyl-*trans*-3-heptene**			597				
3.2-90	2738-18-3	**2,6-Dimethyl-3-heptene**	auto turbine	284 414					
3.2-91	18804-49-4	**3,5-Dimethyl-3,4-heptadiene**			597				
3.2-92	111-66-0	**1-Octene**	auto biomass comb.	284,810 597,821	591,601 364(a),458(a)				
		brewing 144; building mtls. 856; diesel 812; rendering 837t,879t; turbine 414							
3.2-93	4588-18-5	**2-Methyl-1-octene**			582 399(a)				
3.2-94	6874-29-9	**2,6-Dimethyl-1-octene**	turbine	414					

Species Number	Registry Number	Name	Emission Source	Emission Ref.	Detection Ref.	Code	Bioassay −MA	Bioassay +MA	Bioassay Ref.
3.2-95	63597-41-1	**Octadiene**			833 399(a),458(a)				
3.2-96	32073-03-3	**Octyne**			399(a)				
3.2-97	7642-04-8	*cis*-**2-Octene**	auto	526,810	374				
3.2-98	13389-42-9	*trans*-**2-Octene**	auto biomass comb.	284,526 597	374				
		tobacco smoke 421; turbine 359*							
3.2-99	14850-23-8	*trans*-**4-Octene**	auto biomass comb.	682t 597	597,682				
3.2-100	124-11-8	**1-Nonene**	auto biomass comb.	682t 597,821	591,601 364(a),458(a)				
		brewing 144; building mtls. 856; diesel 812; rendering 837,879; tobacco smoke 421; turbine 359*							
3.2-101	2980-71-4	**2-Methyl-1-nonene**	rendering	837t	582 399(a)				
3.2-102	3452-09-3	**1-Nonyne**	tobacco smoke	421					
3.2-103	2198-23-4	**4-Nonene**			582				
3.2-104	20184-91-2	**4-Nonyne**			1113(I)				

Table 3.2. *(Continued)*

Species Number	Registry Number	Name	Emission Source	Emission Ref.	Detection Ref.	Code	Bioassay −MA	Bioassay +MA	Bioassay Ref.
3.2-105	872-05-9	**1-Decene**	biomass comb.	597	591,601				
			brewing	144	364(a),458(a)				
		building mtls. 856; diesel 812; tobacco smoke		421,634;					
		turbine 359*							
3.2-106	13151-27-4	**2-Methyl-1-decene**			399(a)				
3.2-107	764-93-2	**1-Decyne**	biomass comb.	597					
3.2-108	2384-70-5	**2-Decyne**	biomass comb.	597					
3.2-109	821-95-4	**1-Undecene**	biomass comb.	597	601,674				
			brewing	144	364(a),458(a)				
		rendering 837t,879t; tobacco smoke 634; turbine 359*			820(I)				
3.2-110	18516-37-5	**2-Methyl-1-undecene**			399(a)				
3.2-111	112-41-4	**1-Dodecene**	auto	526,810	200,601				
			brewing	144	364(a),458(a)				
			tobacco smoke	421,634					
3.2-112	16435-49-7	**2-Methyl-1-dodecene**			399(a)				
3.2-113S	77129-48-7	**7,11-Dimethyl-3-methylene-1,6,10-dodecatriene**	vegetation	888					

Species Number	Registry Number	Name	Emission Source	Emission Ref.	Detection Ref.	Code	Bioassay −MA	Bioassay +MA	Ref.
3.2-114	2437-56-1	**1-Tridecene**	biomass comb.	597	200,591				
			brewing	144	364(a),458(a)				
		rendering 837t,879t; tobacco smoke 634; volcano 1007							
3.2-115	18094-01-4	**2-Methyl-1-tridecene**	volcano	1007					
3.2-116	1120-36-1	**1-Tetradecene**	biomass comb.	597	591,597				
			brewing	144	364(a),458(a)				
		diesel 309(a); tobacco smoke 634; volcano 1007							
3.2-117	52554-38-3	**2-Methyl-1-tetradecene**			399(a)				
3.2-118	27251-68-9	**Pentadecene**	brewing	144	833				
			refuse comb.	878(a)t	364(a),458(a)				
			tobacco smoke	634					
3.2-119	629-73-2	**1-Hexadecene**	refuse comb.	878(a)t	364(a),458(a)				
			tobacco smoke	634					
3.2-120	61868-19-7	**2-Methyl-1-hexadecene**			399(a)				
3.2-121	6765-39-5	**1-Heptadecene**	diesel	309(a)	364(a),458(a)				
			refuse comb.	878(a)t					
			tobacco smoke	634					
3.2-122	42764-74-9	**2-Methyl-1-heptadecene**			399(a)				
3.2-123	112-88-9	**1-Octadecene**	refuse comb.	878(a)t	364(a),458(a)				
			tobacco smoke	634					

Table 3.2. *(Continued)*

Species Number	Registry Number	Name	Emission Source	Emission Ref.	Detection Ref.	Code	Bioassay −MA	Bioassay +MA	Bioassay Ref.
3.2-124	61868-20-0	**2-Methyl-1-octadecene**			399(a)				
3.2-125	18435-45-5	**1-Nonadecene**	refuse comb. tobacco smoke	878(a)t 634	364(a)				
3.2-126	3452-07-1	**1-Eicosene**	refuse comb. tobacco smoke	878(a)t 634					
3.2-127	1599-68-4	**1-Heneicosene**	refuse comb. tobacco smoke	878(a)t 634					
3.2-128	1599-67-3	**1-Docosene**	refuse comb. tobacco smoke	878(a)t 634					
3.2-129	18835-32-0	**1-Tricosene**	refuse comb. tobacco smoke	878(a)t 634					
3.2-130	10192-32-2	**1-Tetracosene**	refuse comb. tobacco smoke	878(a)t 634					
3.2-131	16980-85-1	**1-Pentacosene**	refuse comb. tobacco smoke	878(a)t 634					
3.2-132	18835-33-1	**1-Hexacosene**	tobacco smoke	634					
3.2-133	15306-27-1	**1-Heptacosene**	tobacco smoke	634					
3.2-134	18835-34-2	**1-Octacosene**	tobacco smoke	634					

Species Number	Registry Number	Name	Emission Source	Emission Ref.	Detection Ref.	Code	Bioassay −MA	Bioassay +MA	Bioassay Ref.
3.2-135	18835-35-3	**1-Nonacosene**	tobacco smoke	634					
3.2-136	18435-53-5	**1-Triacontene**	tobacco smoke	634					
3.2-137	18435-54-6	**1-Hentriacontene**	tobacco smoke	634					
3.2-138	18435-55-7	**1-Dotriacontene**	tobacco smoke	634					

$H_2C{=}CHCMe{=}CH_2$

3.2-19

$Me_2C{=}CHCH_2CH_2CMe{=}CHCH_2CH_2C({=}CH_2)CH{=}CH_2$

3.2-113

Table 3.3. Monocyclic Hydrocarbons

Species Number	Registry Number	Name	Emission Source	Emission Ref.	Detection Ref.	Code	Bioassay −MA	Bioassay +MA	Bioassay Ref.
3.3-1S	75-19-4	**Cyclopropane**	auto	810	582,597				
3.3-2	2402-06-4	*trans*-**1,2-Dimethylcyclopropane**	turbine	414					
3.3-3	53778-43-1	**1-Ethyl-1-methylcyclopropane**	turbine	414					
3.3-4	3638-35-5	**Isopropylcyclopropane**	turbine	414					
3.3-5S	287-23-0	**Cyclobutane**	turbine	414					
3.3-6	598-61-8	**Methylcyclobutane**			597 1113(I)				
3.3-7	4806-61-5	**Ethylcyclobutane**	turbine	414					
3.3-8	27538-13-2	**3-Methylenecyclobutene**			820(I)				
3.3-9S	287-92-3	**Cyclopentane**	auto biomass comb.	47,153* 597	170*,232* 820(I)				
		gasoline vapor 232,621; natural gas 456*; petroleum mfr. 58*,232; polymer comb. 612; tobacco smoke 634; turbine 414							

Species Number	Registry Number	Name	Emission Source	Emission Ref.	Detection Ref.	Code	Bioassay −MA	Bioassay +MA	Bioassay Ref.
3.3-10	96-37-7	**Methylcyclopentane**	adhesives	1111	170*,263				
			auto	159*,465*	820*(I),1124(I)				
		biomass comb. 597; gasoline vapor 465*,621; natural gas 465*; paint 1017*; petroleum mfr. 122,1000*; polymer comb. 612; tobacco smoke 634; turbine 359*; vegetation 552e							
3.3-11	1638-26-2	**1,1-Dimethylcyclopentane**	turbine	414	601,628				
3.3-12	1192-18-3	*cis*-**1,2-Dimethylcyclopentane**	biomass comb.	597	597,674				
3.3-13	822-50-4	*trans*-**1,2-Dimethylcyclo pentane**	auto	682t	674,682t				
3.3-14	2453-00-1	*cis*-**1,3-Dimethylcyclo-pentane**	biomass comb.	597	582,597				
			turbine	414					
3.3-15	1759-58-6	*trans*-**1,3-Dimethyl-cyclopentane**	auto	682t	674,682t				
3.3-16	2815-57-8	**1,2,3-Trimethylcyclo-pentane**	biomass comb.	597	582,601				
			tobacco smoke	421					
			turbine	414					
3.3-17	2815-58-9	**1,2,4-Trimethylcyclo-pentane**	auto	682t	674,682t 399(a)				
3.3-18	1640-89-7	**Ethylcyclopentane**	auto	682t	597,682				

Table 3.3. *(Continued)*

Species Number	Registry Number	Name	Emission Source	Emission Ref.	Detection Ref.	Code	Bioassay −MA	Bioassay +MA	Bioassay Ref.
3.3-19	3726-46-3	**1-Ethyl-2-methylcyclopentane**			597				
3.3-20	3726-47-4	**1-Ethyl-3-methylcyclopentane**	auto	682t	582,597 820(I)t				
3.3-21	2040-96-2	**Propylcyclopentane**	auto	682t	597,682t				
3.3-22S	142-29-0	**Cyclopentene**	auto biomass comb.	153*,170* 524,597	170*,524 458(a) 820(I)				
		polymer comb. 612; tobacco smoke 634							
3.3-23	693-89-0	**1-Methylcyclopentene**	gasoline vapor tobacco smoke	486 634	364(a),458(a)				
3.3-24	1120-62-3	**3-Methylcyclopentene**	tobacco smoke	634					
3.3-25	1759-81-5	**4-Methylcyclopentene**	tobacco smoke	634					
3.3-26	542-92-7	**Cyclopentadiene**	biomass comb. polymer comb.	597 853	820(I)t				
		polymer mfr. 251; tobacco smoke 634							
3.3-27S	2175-91-9	**6,6-Dimethylfulvene**	turbine	414					
3.3-28	7338-50-3	**6-Phenylfulvene**			628				

Species Number	Registry Number	Name	Emission Source	Emission Ref.	Detection Ref.	Bioassay Code	Bioassay −MA	Bioassay +MA	Bioassay Ref.
3.3-29S	110-82-7	**Cyclohexane**	adhesives	1111	123*,170*	STP00	NEG	NEG	B36
			auto	47,526	820(I),1108(I)	STP35	NEG	NEG	B36
		biomass comb. 597; petroleum mfr. 58*,543*; polymer comb. 853;				STP37	NEG	NEG	B36
		refuse comb. 840(a); solvent 78,134; tobacco smoke 634;				STP98	NEG	NEG	B36
		turbine 414; vegetation 552e; volcano 234							
3.3-30	108-87-2	**Methylcyclohexane**	auto	47,465*	123*,566*				
			biomass comb.	597	820(I),1108(I)				
		diesel 812; gasoline vapor 465*,486; natural gas 465*; paint 1017*; petroleum mfr. 1000*; refuse comb. 840(a); turbine 414; volcano 234							
3.3-31	590-66-9	**1,1-Dimethylcyclohexane**	biomass comb.	597	582				
			tobacco smoke	421	820(I)t				
3.3-32	2207-01-4	*cis*-**1,2-Dimethylcyclohexane**	auto	526,810	374,566*				
3.3-33	6876-23-9	*trans*-**1,2-Dimethylcyclohexane**	auto	526,810	374,582				
			biomass comb.	597					
3.3-34	591-21-9	**1,3-Dimethylcyclohexane**	tobacco smoke	421					
3.3-35	2207-04-7	*trans*-**1,4-Dimethylcyclohexane**			674				
3.3-36	3073-66-3	**1,1,3-Trimethylcyclohexane**	auto	682t	601,682				
			tobacco smoke	421					

Table 3.3. *(Continued)*

Species Number	Registry Number	Name	Emission Source	Emission Ref.	Detection Ref.	Code	Bioassay −MA	Bioassay +MA	Bioassay Ref.
3.3-37	1678-97-3	**1,2,3-Trimethylcyclo-hexane**	auto	682t	597,682t				
3.3-38	2234-75-5	**1,2,4-Trimethylcyclo-hexane**			1108(I)				
3.3-39	1839-63-0	**1,3,5-Trimethylcyclo-hexane**			628,674 399(a),458(a) 820(I),1108(I)				
3.3-40	1678-91-7	**Ethylcyclohexane**	auto	284,526	597,601 820(I),1124(I)				
3.3-41	30677-34-0	**Ethylmethylcyclohexane**	building mtls.	856	674 820(I)				
3.3-42	1331-43-7	**Diethylcyclohexane**			591 1108(I)				
3.3-43	695-12-5	**Vinylcyclohexane**			597				
3.3-44	1678-92-8	**Propylcyclohexane**	auto	682t	597,628 1108(I)				
3.3-45	696-29-7	**Isopropylcyclohexane**	auto	682t	682t 820(I)				
3.3-46	99-82-1	**1-Isopropyl-4-methylcyclo-hexane**			597				

Species Number	Registry Number	Name	Emission		Detection		Bioassay		
			Source	Ref.	Ref.	Code	−MA	+MA	Ref.
3.3-47	1678-93-9	**Butylcyclohexane**	auto turbine	682t 414	597,674				
3.3-48	1678-98-4	**Isobutylcyclohexane**			820(I)				
3.3-49	7058-01-7	*sec*-**Butylcyclohexane**			597				
3.3-50	3178-22-1	*tert*-**Butylcyclohexane**	auto	682t	582,628				
3.3-51	4292-92-6	**Pentylcyclohexane**	auto	682t	682				
3.3-52	4292-75-5	**Hexylcyclohexane**	auto	682t	682				
3.3-53	5617-41-4	**Heptylcyclohexane**	auto	682t	682				
3.3-54	1795-15-9	**Octylcyclohexane**	auto	682t	682				
3.3-55	2883-02-5	**Nonylcyclohexane**	auto	682t	682				
3.3-56	1795-16-0	**Decylcyclohexane**	auto	682t	682				
3.3-57	54105-66-7	**Undecylcyclohexane**	auto	682t	682				
3.3-58	1795-17-1	**Dodecylcyclohexane**	auto	682t,311(a)	682				
3.3-59	6006-33-3	**Tridecylcyclohexane**	auto	682t	682				
3.3-60	1795-18-2	**Tetradecylcyclohexane**	auto	682t	682				

Table 3.3. *(Continued)*

Species Number	Registry Number	Name	Emission Source	Emission Ref.	Detection Ref.	Bioassay Code	Bioassay −MA	Bioassay +MA	Bioassay Ref.
3.3-61	6006-95-7	**Pentadecylcyclohexane**	auto	682t	682				
3.3-62	6812-38-0	**Hexadecylcyclohexane**	auto	682t	682				
3.3-63	19781-73-8	**Heptadecylcyclohexane**	auto	682t	682,956(a)				
3.3-64	22349-03-7	**Nonadecylcyclohexane**			956(a)				
3.3-65	4443-55-4	**Eicosylcyclohexane**	auto	956(a)	956(a)				
3.3-66	6703-99-7	**Heneicosylcyclohexane**	auto	956(a)	956(a)				
3.3-67	61828-07-7	**Docosylcyclohexane**	auto	956(a)	956(a)				
3.3-68	61828-08-8	**Tricosylcyclohexane**	auto	956(a)	956(a)				
3.3-69	61828-09-9	**Tetracosylcyclohexane**	auto	956(a)	956(a)				
3.3-70	61828-10-2	**Pentacosylcyclohexane**	auto	956(a)	956(a)				
3.3-71	61828-11-3	**Hexacosylcyclohexane**	auto	956(a)	956(a)				
3.3-72	61828-12-4	**Heptacosylcyclohexane**	auto	956(a)	956(a)				
3.3-73	61828-13-5	**Octacosylcyclohexane**	auto	956(a)	956(a)				
3.3-74	61828-14-6	**Nonacosylcyclohexane**			956(a)				

Species Number	Registry Number	Name	Emission Source	Emission Ref.	Detection Ref.	Code	Bioassay −MA	Bioassay +MA	Bioassay Ref.
3.3-75	61828-15-7	**Triacontylcyclohexane**			956(a)				
3.3-76	827-52-1	**Phenylcyclohexane**			364(a)				
3.3-77	29188-43-0	**Ethylphenylcyclo-hexane**			628t 399(a),458(a)				
3.3-78S	110-83-8	**Cyclohexene**	auto biomass comb.	153*,810 821	597 364(a),458(a)				
		diesel 812; rubber abrasion 331; tobacco smoke 634							
3.3-79	591-49-1	**1-Methylcyclohexene**	auto rubber abrasion	284 331					
3.3-80	591-47-9	**4-Methylcyclohexene**	auto diesel	527,810 812					
3.3-81	100-40-3	**4-Vinylcyclohexene**	rubber abrasion vulcanization	606 308	597				
3.3-82	1611-21-8	**2,4-Dimethyl-4-vinylcyclohexene**	tobacco smoke	634					
3.3-83	5502-88-5	**4-Isopropyl-1-methylcyclohexene**	tobacco smoke	634					
3.3-84	31017-40-0	**Phenylcyclohexene**			364(a)				
3.3-85	628-41-1	**1,4-Cyclohexadiene**	refuse comb.	26(a)	886*				

Table 3.3. *(Continued)*

Species Number	Registry Number	Name	Emission Source	Emission Ref.	Detection Ref.	Bioassay Code	Bioassay −MA	Bioassay +MA	Bioassay Ref.
					364(a)				
3.3-86	4318-57-9	**1-Methylcyclohexadiene**	biomass comb.	597					
3.3-87	291-64-5	**Cycloheptane**	biomass comb. petroleum mfr.	597 543*	597				
3.3-88	292-64-8	**Cyclooctane**	petroleum mfr.	543*					
3.3-89	111-78-4	**1,5-Cyclooctadiene**	vulcanization	308					
3.3-90	629-20-9	**Cyclooctatetraene**			597,628				
3.3-91	293-96-9	**Cyclodecane**			597t				
3.3-92	294-62-2	**Cyclododecane**	biomass comb.	597					
3.3-93	4904-61-4	**1,5,9-Cyclododecatriene**	vulcanization	308					

CMe2

3.3-1 3.3-5  3.3-9  3.3-22 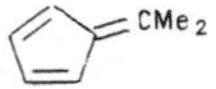 3.3-27 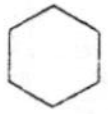 3.3-29  3.3-78

Table 3.4. Terpenes and Polycyclic Hydrocarbons

Species Number	Registry Number	Name	Emission Source	Emission Ref.	Detection Ref.	Bioassay Code	Bioassay −MA	Bioassay +MA	Bioassay Ref.
3.4-1S	80-56-8	**α-Pinene**	building mtls.	856	14*,198				
			microbes	302	399(a)				
		solvent 439; tobacco smoke 421;			820(I),1112*(I)				
		vegetation 74,198; veneer drying 197,678; wood pulping 19,267							
3.4-2S	127-91-3	**β-Pinene**	biomass comb.	821	14*,171				
			microbes	302	820(I)				
		tobacco smoke 421,634; vegetation 74,198;							
		veneer drying 678,791*; wood pulping 19,267							
3.4-3S	464-17-5	**Bornylene**	animal waste	837					
3.4-4S	123-35-3	**Myrcene**	biomass comb.	597	14*,198				
			microbes	302	820(I)				
		vegetation 461,801; veneer drying 678,791*; wood pulping 267,545							
3.4-5S	29714-87-2	**Ocimene**	vegetation	496,502	820(I)				

Table 3.4. *(Continued)*

Species Number	Registry Number	Name	Emission Source	Emission Ref.	Detection Ref.	Bioassay Code	Bioassay −MA	Bioassay +MA	Bioassay Ref.
3.4-6S	5989-27-5	**Limonene**	biomass comb.	597	14*,525*	STI00	NEG	NEG	B38
			building mtls.	856	820(I),1112*(I)	STI35	NEG	NEG	B38
		microbes 302; rendering 837; rubber abrasion 606; tobacco smoke 396,421;				STI37	NEG	NEG	B38
		vegetation 74,198; veneer drying 678,791*; wood pulping 19,267				STI98	NEG	NEG	B38
3.4-7S	138-86-3	**Dipentene**	rubber abrasion	606					
3.4-8S	99-86-5	**α-Terpinene**	vegetation	198,499	805,886*				
			wood pulping	19,267					
3.4-9S	99-85-4	**γ-Terpinene**	vegetation	663	886*				
			veneer drying	791*					
			wood pulping	19,267					
3.4-10S	586-62-9	**Terpinolene**	vegetation	461,675					
			veneer drying	791*					
			wood pulping	19,267					
3.4-11S	13466-78-9	**Δ^3-Carene**	biomass comb.	821	472,773				
			building mtls.	856					
		microbes 302; vegetation 461,675; veneer drying 678,791*; wood pulping 19,267							
3.4-12S	554-61-0	**Δ^4-Carene**			886*				

Species Number	Registry Number	Name	Emission Source	Emission Ref.	Detection Ref.	Code	Bioassay −MA	Bioassay +MA	Bioassay Ref.
3.4-13S	79-92-5	**Camphene**	biomass comb.	597t	674,773				
			vegetation	515e,663	820(I),1112*(I)				
		veneer drying 678,791*; wood pulping 19,545							
3.4-14S	58037-87-9	**α-Thujene**	rendering	833	844t				
			vegetation	1066					
3.4-15S	3387-41-5	**Sabinene**	vegetation	517e	833				
					820(I)				
3.4-16S	4221-98-1	**α-Phellandrene**	wood pulping	267,545	833,886*				
3.4-17S	555-10-2	**β-Phellandrene**	microbes	302	773,844t				
			vegetation	531,663					
		veneer drying 791*; wood pulping 19,267							
3.4-18	77-73-6	**3,3a,7,7a-Tetrahydro-4,7-methanoindene**	polymer mfr.	251		STP00	NEG	NEG	B36
						STP35	NEG	NEG	B36
						STP37	NEG	NEG	B36
						STP38	NEG	NEG	B36
						STP98	NEG	NEG	B36
3.4-19S	29799-19-7	**Bisabolene**	vegetation	514e,531					
3.4-20S	29350-73-0	**Cadinene**	vegetation	506e,514e					
3.4-21S	470-40-6	**Thujopsene**	vegetation	510e					
3.4-22S	489-40-7	**α-Gurjunene**	vegetation	497e					

Table 3.4. *(Continued)*

Species Number	Registry Number	Name	Emission Source	Emission Ref.	Detection Ref.	Bioassay Code	Bioassay −MA	Bioassay +MA	Bioassay Ref.
3.4-23S	275-51-4	**Azulene**	tobacco smoke	79(a),396(a)	209(a)				
3.4-24S	529-05-5	**Chamazulene**	vegetation	510e					
3.4-25S	6753-98-6	**α-Caryophyllene**	vegetation	499					
3.4-26S	87-44-5	**β-Caryophyllene**	vegetation	506e,531					
3.4-27S		**Triterpane**	auto diesel	956(a) 956(a)	956(a)				
3.4-28		**Methyltriterpane**	diesel	956(a)	956(a)				
3.4-29		**Ethyltriterpane**	diesel	956(a)	956(a)				
3.4-30		**Propyltriterpane**	diesel	956(a)	956(a)				
3.4-31		**Butyltriterpane**	auto diesel	956(a) 956(a)	956(a)				
3.4-32		**Pentyltriterpane**	diesel	956(a)	956(a)				
3.4-33		**Hexyltriterpane**	diesel	956(a)	956(a)				
3.4-34		**Heptyltriterpane**	diesel	956(a)	956(a)				
3.4-35		**Octyltriterpane**	diesel	956(a)	956(a)				

Species Number	Registry Number	Name	Emission Source	Emission Ref.	Detection Ref.	Code	Bioassay −MA	Bioassay +MA	Bioassay Ref.
3.4-36		**Nonyltriterpane**	diesel	956(a)	956(a)				
3.4-37S	35241-40-8	**Abieta-7,13-diene**			955*(a)				
3.4-38S	50-24-8	**Sterane**	auto	956(a)	956(a)				
3.4-39		**Methylsterane**	auto	956(a)	956(a)				
3.4-40		**Ethylsterane**	auto	956(a)	956(a)				
3.4-41S		**Diasterane**			956(a)				
3.4-42		**Methyldiasterane**			956(a)				
3.4-43		**Ethyldiasterane**			956(a)				
3.4-44S		**Trisnorneohopane**			956(a)				
3.4-45S	13849-96-2	**17α(*H*)-Hopane**	auto	956(a)	956(a)				
			diesel	956(a)					
3.4-46		**Ethyl-17α(*H*)-hopane**	auto	956(a)	956(a)				
			diesel	956(a)					
3.4-47		**Propyl-17α(*H*)-hopane**	auto	956(a)	956(a)				
			diesel	956(a)					
3.4-48		**Butyl-17α(*H*)-hopane**	auto	956(a)	956(a)				
			diesel	956(a)					

Table 3.4. *(Continued)*

Species Number	Registry Number	Name	Emission Source	Emission Ref.	Detection Ref.	Code	Bioassay −MA	Bioassay +MA	Bioassay Ref.
3.4-49		**Pentyl-17α(*H*)-hopane**	auto diesel	956(a) 956(a)	956(a)				
3.4-50		**Hexyl-17α(*H*)-hopane**	auto diesel	956(a) 956(a)	956(a)				
3.4-51		**Heptyl-17α(*H*)-hopane**	auto diesel	956(a) 956(a)	956(a)				
3.4-52		**Octyl-17α(*H*)-hopane**	diesel	956(a)	956(a)				
3.4-53S		**17α(*H*),8α(*H*),21β(*H*)-28,30-bisnorhopane**			956(a)				
3.4-54S		**17α(H)-22R,S-homohopane**			956(a)				
3.4-55		**Methyl-17α(*H*)-22R,S-homohopane**			956(a)				
3.4-56		**Ethyl-17α(*H*)-22R,S-homohopane**			956(a)				
3.4-57		**Propyl-17α(*H*)-22R,S-homohopane**			956(a)				
3.4-58		**Butyl-17α(*H*)-22R,S-homohopane**			956(a)				

Species Number	Registry Number	Name	Emission Source	Emission Ref.	Detection Ref.	Bioassay Code	Bioassay −MA	Bioassay +MA	Bioassay Ref.
3.4-59S	111-02-4	**Squalene**	tobacco smoke	396,634					
3.4-60S	11030-10-7	**Isosqualene**	tobacco smoke	396					

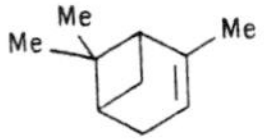

3.4-1

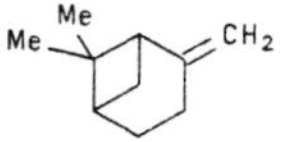

3.4-2

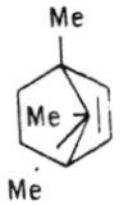

3.4-3

$Me_2C:CHCH_2CH_2C(:CH_2)CH:CH_2$

3.4-4

$Me_2CH(CH_2)_3CHMeCH_2Me$

3.4-5

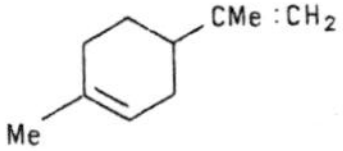

3.4-6

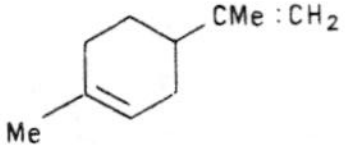

3.4-7

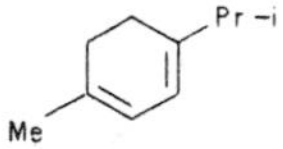

3.4-8

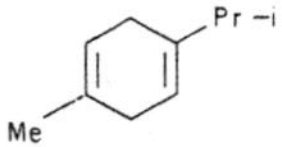

3.4-9

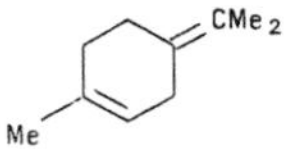

3.4-10

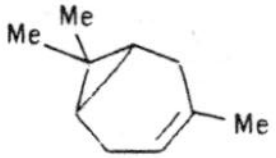

3.4-11

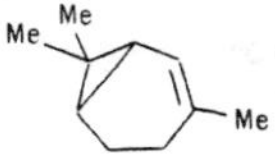

3.4-12

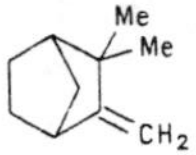

3.4-13

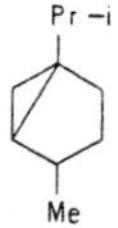

3.4-14

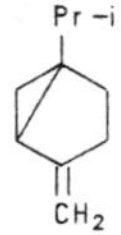

3.4-15

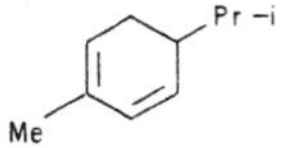

3.4-16

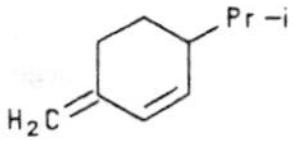

3.5-17

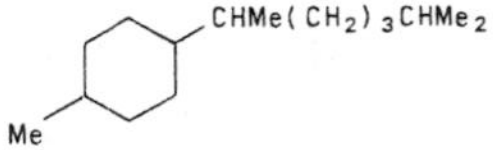

3.4-19

Table 3.4. *(Continued)*

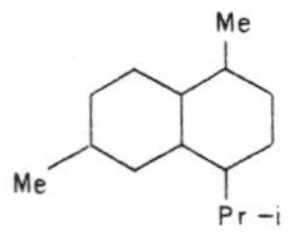

3.4-20 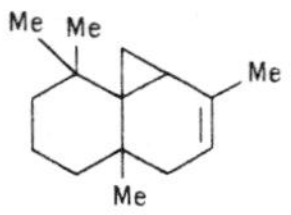 3.4-21 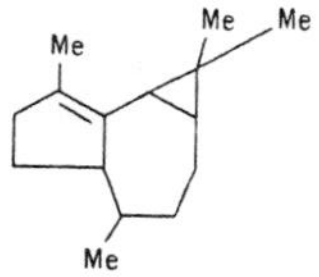 3.4-22 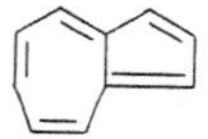 3.4-23 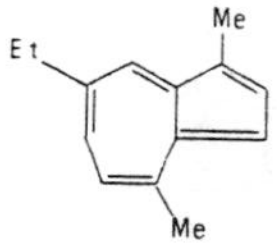 3.4-24 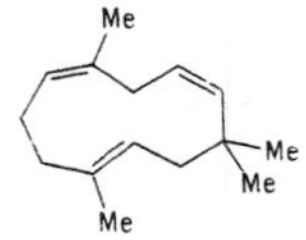 3.4-25 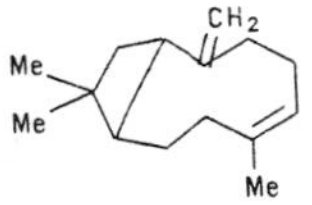 3.4-26

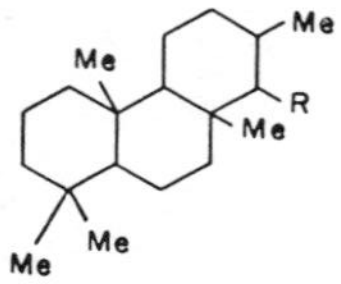

3.4-27 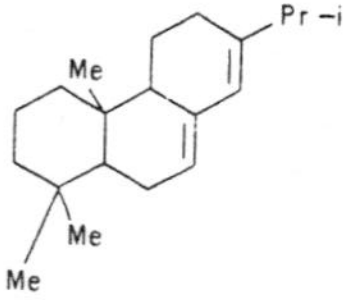 3.4-37 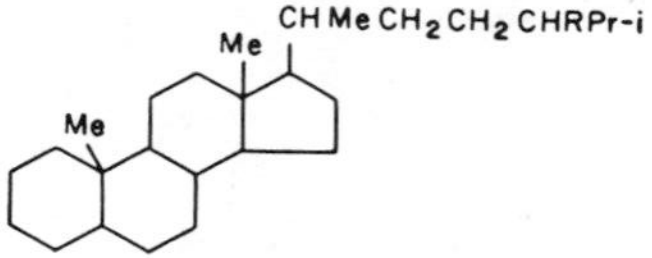 3.4-38 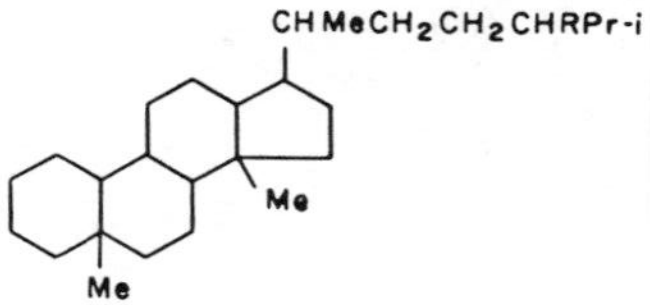 3.4-41 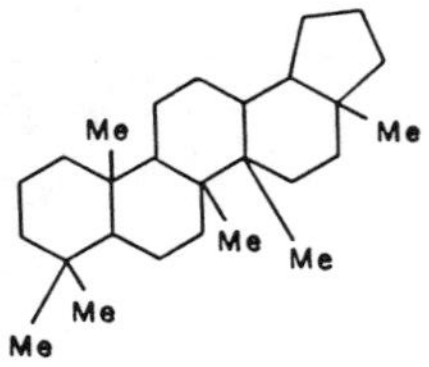 3.4-44

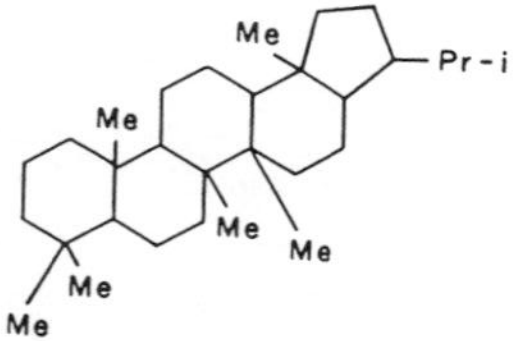

3.4-45 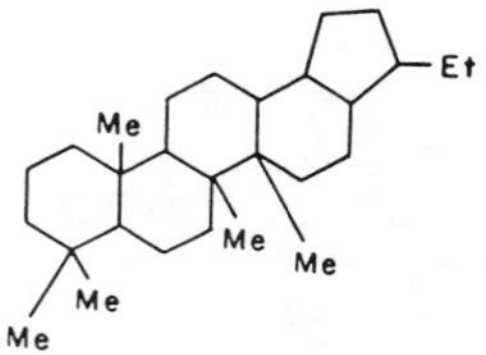 3.4-53 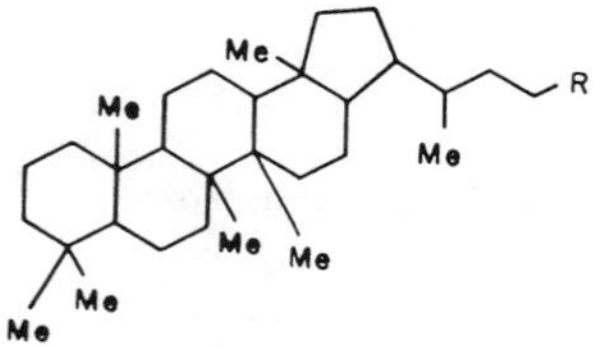 3.4-54 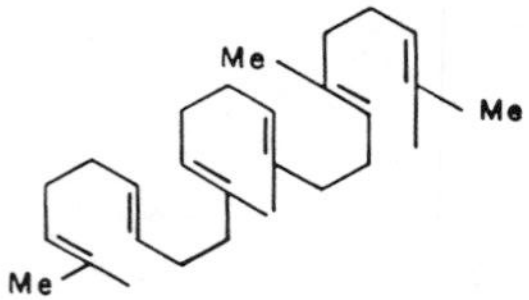 3.4-59 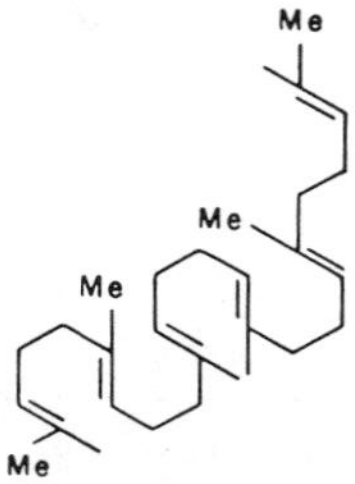 3.4-60

Table 3.5. Monocyclic Arenes

Species Number	Registry Number	Name	Emission Source	Emission Ref.	Detection Ref.	Code	Bioassay −MA	Bioassay +MA	Bioassay Ref.
3.5-1S	71-43-2	**Benzene**	acrylonitrile mfr.	626	139*,170*	TRM	+		B21
			animal waste	695,1071	364(a),458(a)	CCC	LP		B37
		auto 159*,376; biomass comb. 354,439; chemical mfr. 566*;			902(r)	SCE	NEG		B6
		coal comb. 1159; diesel 434*,812; fish processing 110*,695;			820(I),952(I)	CYI	+		B8
		gasoline vapor 58*,232;geothermal steam 998*; lacquer mfr. 428;				MNT	+		B29
		landfill 365,628; oil comb. 624; paint 695,1071; petroleum mfr. 127*,543*;				SPI	+		B25
		plastics mfr. 695,1071; polymer comb. 46,304; polymer mfr. 103;							
		refuse comb. 439,26(a); rendering 837,879; sewage tmt. 204,695;							
		solvent 60,598; tobacco smoke 396,446; turbine 359*,414;							
		vegetation 552e; veneer drying 791*; volcano 234;							
		vulcanization 331; whiskey mfr. 1158; wood comb. 890;							
		wood preserv. 1048*; wood pulping 695,1071							
3.5-2S	108-88-3	**Toluene**	acrylonitrile mfr.	626	139*,170*	SCE	NEG		B6
			adhesives	1111	364(a),458(a)	STI00	NEG	NEG	B38
		animal waste 695,1071; auto 159*,465*; biomass comb. 354,439;			755*(r),876*(r)	STI35	NEG	NEG	B38
		building mtls. 856; chemical mfr. 566*; coal comb. 1159;			632(I),820(I)	STI37	NEG	NEG	B38
		diesel 176,812; fish processing 110*,695; gasoline vapor 58*,232;				STI98	NEG	NEG	B38
		geothermal steam 998*; landfill 365,628; oil comb. 624;				SPI	NEG		B25
		paint 695,1017*; petroleum mfr. 127*,543*; plastics mfr. 695,1071;				STP38	NEG	NEG	B35
		polymer comb. 304,354; polymer mfr. 103; printing 620;				STP98	NEG	NEG	B35
		rendering 833,879; sewage tmt. 695,1071; solvent 78,134;				STP37	NEG	NEG	B35
		starch mfr. 695,1071; tobacco smoke 396,421,339(a); turbine 359*,414;				STP35	NEG	NEG	B35
		vegetation 552e; veneer drying 791*; volcano 234; vulcanization 308,331;				STP00	NEG	NEG	B35
		whiskey mfr. 1158; wood preserv. 1048*; wood pulping 19,267							

Table 3.5. *(Continued)*

Species Number	Registry Number	Name	Emission Source	Emission Ref.	Detection Ref.	Bioassay Code	−MA	+MA	Ref.
3.5-3S	95-47-6	***o*-Xylene**	auto	159*,465*	139*,566*	STI00	NEG	NEG	B38
			biomass comb.	439,597	820(I),952(I)	STI35	NEG	NEG	B38
		building mtls. 856; coal comb. 1159; diesel 812;				STI37	NEG	NEG	B38
		fish processing 110*,695; gasoline vapor 232,465*; paint 695,1017*;				STI98	NEG	NEG	B38
		petroleum mfr. 1000*; phthal. anhyd. mfr. 64*; plastics mfr. 1071;				STP35	NEG	NEG	B35
		polymer comb. 304; printing 620; rendering 833;				STP00	NEG	NEG	B35
		sewage tmt. 695,1071; tobacco smoke 421,634; vegetation 552e;				STP38	NEG	NEG	B35
		wood pulping 695,1071				STP98	NEG	NEG	B35
						STP37	NEG	NEG	B35
3.5-4	108-38-3	***m*-Xylene**	auto	159*,465*	123*,139*	STI00	NEG	NEG	B38
			biomass comb.	821	820(I),1108(I)	STI35	NEG	NEG	B38
		building mtls. 856; coal comb. 1159; diesel 176,812;				STI37	NEG	NEG	B38
		fish processing 110,695; gasoline vapor 232,465*; paint 695;				STI98	NEG	NEG	B38
		plastics mfr. 695,1071; polymer comb. 304; rendering 833;				STP00	NEG	NEG	B35
		sewage tmt. 695,1071; solvent 134,439; tobacco smoke 634;				STP38	NEG	NEG	B35
		turbine 414; vulcanization 308,331; wood pulping 695,1071				STP98	NEG	NEG	B35
						STP37	NEG	NEG	B35
						STP35	NEG	NEG	B35
3.5-5	106-42-3	***p*-Xylene**	auto	159*,465*	139*,232*	STI00	NEG	NEG	B38
			biomass comb.	597,821	458(a)	STI35	NEG	NEG	B38
		building mtls. 856; coal comb. 1159; diesel 176,812;			820(I),940(I)t	STI37	NEG	NEG	B38
		fish processing 110*,695; gasoline vapor 232,465*; lacquer mfr. 428;				STP98	NEG	NEG	B38
		paint 695,1017*t; petroleum mfr. 1000*; plastics mfr. 695,1071;				STP38	NEG	NEG	B35
		polymer comb. 304,853t; rendering 833,879t; sewage tmt. 695,1071;				STP98	NEG	NEG	B35
		tobacco smoke 634; turbine 414; veneer drying 791*;				STP37	NEG	NEG	B35
		vulcanization 308; wood pulping 695,1071				STP35	NEG	NEG	B35
						STP00	NEG	NEG	B35

Species Number	Registry Number	Name	Emission Source	Emission Ref.	Detection Ref.	Code	Bioassay −MA	Bioassay +MA	Bioassay Ref.
3.5-6	526-73-8	**1,2,3-Trimethylbenzene**	auto	465*,526	232*,566*				
			biomass comb.	597					
		gasoline vapor 232,465*; tobacco smoke 634; turbine 414; vegetation 552e							
3.5-7	95-63-6	**1,2,4-Trimethylbenzene**	auto	159*,465*	139*,232*				
			biomass comb.	821					
		building mtls. 856; gasoline vapor 232,465*; paint 1017*; printing 620; tobacco smoke 634,396(a); vegetation 552e							
3.5-8	108-67-8	**1,3,5-Trimethylbenzene**	auto	284,486*	123*,139*				
			biomass comb.	597	99(a)				
		building mtls. 856; coal comb. 1159; diesel 812,309(a); gasoline vapor 232,486; paint 1017*; rendering 837,879; solvent 134; tobacco smoke 634,396(a); turbine 359*,414; vegetation 552e			820(I)t,1108(I)				
3.5-9	488-23-3	**1,2,3,4-Tetramethylbenzene**	building mtls.	856	86,582				
			refuse comb.	878(a)t					
3.5-10	527-53-7	**1,2,3,5-Tetramethylbenzene**	diesel	309(a)	200*,566*				
3.5-11	95-93-2	**1,2,4,5-Tetramethylbenzene**	auto	526,810	86,200*				
3.5-12	87-85-4	**Hexamethylbenzene**			200				

Table 3.5. *(Continued)*

Species Number	Registry Number	Name	Emission Source	Emission Ref.	Detection Ref.	Bioassay Code−MA+MARef.
3.5-13	100-41-4	**Ethylbenzene**	auto	159*,526	139*,202	
			biomass comb.	597,821	399(a),458(a)	
		building mtls. 856; chemical mfr. 566*; coal comb. 1159; diesel 176,812; fish processing 695,1071; gasoline vapor 232; paint 1017*,1071; petroleum mfr. 1000*; plastics mfr. 695,1071; polymer mfr. 103; printing 620; rendering 833,879; sewage tmt. 695,1071; solvent 134,439; tobacco smoke 421,634 turbine 359*; vulcanization 308; wood preserv. 1048*; wood pulping 695,1071			755*(r),876*(r) 820(I),952(I)	
3.5-14	611-14-3	*o*-**Ethyltoluene**	auto	526,970*	86*,200*	
			biomass comb.	597	820(I)	
		building mtls. 856; gasoline vapor 232; paint 1017*; printing 620; tobacco smoke 634(a); vegetation 552e				
3.5-15	620-14-4	*m*-**Ethyltoluene**	auto	465*,970*	86*,232*	
			biomass comb.	597	820(I)	
		building mtls. 856; gasoline vapor 232,465*; paint 1017*; rendering 833t; tobacco smoke 634(a); turbine 414; vegetation 552e				
3.5-16	622-96-8	*p*-**Ethyltoluene**	auto	159*,972	86*,232*	
			biomass comb.	597	458(a)	
		building mtls. 856; gasoline vapor 232; paint 1017*; polymer comb. 853t; printing 620; tobacco smoke 634(a); turbine 414; vegetation 552e			820(I)	
3.5-17	933-98-2	**1-Ethyl-2,3-dimethyl-benzene**	auto	682t	682	

Species Number	Registry Number	Name	Emission Source	Ref.	Detection Ref.	Bioassay Code	−MA	+MA	Ref.
3.5-18	874-41-9	**1-Ethyl-2,4-dimethyl-benzene**	auto	682t	682				
3.5-19	1758-88-9	**1-Ethyl-2,5-dimethyl-benzene**	auto	682t	682				
3.5-20	2870-04-4	**1-Ethyl-2,6-dimethyl-benzene**	auto	465*,1036	566*t,682				
			biomass comb.	597	820(I)				
			gasoline vapor	465*t					
3.5-21	934-80-5	**1-Ethyl-3,4-dimethyl-benzene**	auto	682t	682				
3.5-22	934-74-7	**1-Ethyl-3,5-dimethyl-benzene**	auto	682t	682				
3.5-23	135-01-3	**1,2-Diethylbenzene**	auto	682t	200,374				
			biomass comb.	597					
			turbine	414					
3.5-24	141-93-5	**1,3-Diethylbenzene**	auto	526,810	582,597				
			biomass comb.	597					
		building mtls. 856; paint 1017*; rendering 837t							
3.5-25	105-05-5	**1,4-Diethylbenzene**	biomass comb.	597	597,1017*				
			paint	1017*	1108(I)				
3.5-26	25550-13-4	**Diethylmethylbenzene**			525t				

Table 3.5. *(Continued)*

Species Number	Registry Number	Name	Emission Source	Emission Ref.	Detection Ref.	Bioassay Code	−MA	+MA	Bioassay Ref.
3.5-27S	100-42-5	**Styrene**	adhesives	11,1111	123*,566*	HMASC		+	B13
			auto	1036	364(a),458(a)	CCC	LP		B37
		biomass comb. 597,1018*; building mtls. 9*,856; chemical mfr. 566*; oil comb. 624; plastics mfr. 104,117*; polymer comb. 853; polymer mfr. 251,407*; rendering 837,879; river water odor 202; rubber abrasion 606; solvent 134; tobacco smoke 421,634; vegetation 531; vulcanization 308; wood pulping 695,1071			820(I),952(I)	STP00	?	+(B)	B36
						STP37	−(B)	−(B)	B36
						STP38	−(B)	−(B)	B36
						STP98	−(B)	−(B)	B36
						STP35	−?	+	B36
						CYI	+		B8
						YEA	NEG	?	B27
						SRL	+		B30
3.5-28S	98-83-9	**α-Methylstyrene**	auto	880t	591,601				
			biomass comb.	597	364(a)				
		building mtls. 856; diesel 176; solvent 134; tobacco smoke 421							
3.5-29	611-15-4	***o*-Methylstyrene**	biomass comb.	597	820(I)				
			coal comb.	1159					
			tobacco smoke	634					
3.5-30	100-80-1	***m*-Methylstyrene**	biomass comb.	597	597				
			tobacco smoke	634	399(a),458(a)				
3.5-31	622-97-9	***p*-Methylstyrene**	biomass comb.	597	399(a),458(a)				
			polymer comb.	853					
			polymer mfr.	407					
3.5-32	1195-32-0	***p*,α-Dimethylstyrene**	biomass comb.	597	601,628				
			tobacco smoke	634	820(I)t				

Species Number	Registry Number	Name	Emission Source	Emission Ref.	Detection Ref.	Bioassay Code	Bioassay −MA	Bioassay +MA	Bioassay Ref.
3.5-33	5379-20-4	**3,5-Dimethylstyrene**	biomass comb.	1089(a)					
			polymer mfr.	407*					
3.5-34	28106-30-1	**Ethylstyrene**	polymer comb.	853	399(a)				
3.5-35	108-57-6	**1,3-Diethenylbenzene**	auto	682t	682				
3.5-36	5676-32-4	**Propylstyrene**			399(a)				
3.5-37S	536-74-3	**Phenylacetylene**	oil comb.	624	597				
			tobacco smoke	396(a),634(a)	820(I)				
3.5-38	103-65-1	**Propylbenzene**	auto	47,486*	139*,833				
			biomass comb.	597	399(a),458(a)				
		building mtls. 856; gasoline vapor 232,486; paint 1017* plastics mfr. 695,1071; polymer comb. 853; rendering 837; turbine 414; vegetation 552e; wood pulping 695,1071			820(I),1108(I)				
3.5-39	1074-17-5	**2-Propyltoluene**	auto	526,810	582,601				
					820(I)t				
3.5-40	1074-43-7	**3-Propyltoluene**	auto	682t	674,682				
3.5-41	1074-55-1	**4-Propyltoluene**	biomass comb.	597	86,525				
					399(a)				

Table 3.5. *(Continued)*

Species Number	Registry Number	Name	Emission Source	Emission Ref.	Detection Ref.	Code	Bioassay −MA	Bioassay +MA	Bioassay Ref.
3.5-42S	98-82-8	**Cumene**	auto	284,465*	123*,139*	STD00	NEG	NEG	B36
			biomass comb.	597	399(a),458(a)	STD35	NEG	NEG	B36
		building mtls. 856; coal comb. 1159; gasoline vapor 232; polymer comb. 1169; solvent 134; tobacco smoke 421,634(a); turbine 359*,414; vegetation 552e			1108(I)	STD37	NEG	NEG	B36
						STD38	NEG	NEG	B36
						STD98	NEG	NEG	B36
						STP00	NEG	NEG	B36
						STP35	NEG	NEG	B36
						STP37	NEG	NEG	B36
						STP38	NEG	NEG	B36
						STP98	NEG	NEG	B36
3.5-43S	527-84-4	***o*-Cymene**	biomass comb.	597	597,674				
3.5-44	535-77-3	***m*-Cymene**	auto	526,810	582,597				
3.5-45	99-87-6	***p*-Cymene**	auto	810	14*,566*				
			biomass comb.	597					
		solvent 134; tobacco smoke 634(a); vegetation 461,663; wood pulping 19							
3.5-46	4706-90-5	**Dimethylcumene**			597				
3.5-47	4218-48-8	**Ethylcumene**			525				
3.5-48S	300-57-2	**Allylbenzene**	vegetation	531					
3.5-49	28654-77-5	**Allylmethylbenzene**	auto	880t	628				

Species Number	Registry Number	Name	Emission Source	Emission Ref.	Detection Ref.	Code	Bioassay −MA	Bioassay +MA	Bioassay Ref.
3.5-50	104-51-8	**Butylbenzene**	auto	465*,972	139,374				
			biomass comb.	597	399(a),882(a)				
			rendering	837,879					
3.5-51	27458-20-4	**Butylmethylbenzene**			601				
3.5-52	538-93-2	**Isobutylbenzene**	auto	810	525,597				
			turbine	414	364(a)				
			vegetation	552e	820(I)				
3.5-53	135-98-8	***sec*-Butylbenzene**	auto	47,526	123*,139*				
					399(a)				
3.5-54	98-06-6	***tert*-Butylbenzene**	auto	810,972	123*,139*				
			diesel	138					
			solvent	134					
3.5-55	1075-38-3	**1-*tert*-Butyl-3-methylbenzene**	biomass comb.	597	597				
3.5-56S	824-90-8	**1-Butenylbenzene**			399(a),458(a)				
3.5-57	538-68-1	**Pentylbenzene**	building mtls.	856	601,674				
			rendering	837,879	399(a)				
3.5-58	1077-16-3	**Hexylbenzene**	rendering	837,879	674				
			tobacco smoke	421	399(a),458(a)				
3.5-59	2189-60-8	**Octylbenzene**	tobacco smoke	421					

Table 3.5. *(Continued)*

Species Number	Registry Number	Name	Emission Source	Emission Ref.	Detection Ref.	Code	Bioassay −MA	Bioassay +MA	Bioassay Ref.
3.5-60	1081-77-2	**Nonylbenzene**	refuse comb.	878	399(a)				
3.5-61	104-72-3	**Decylbenzene**	refuse comb.	878	399(a),458(a)				
3.5-62S	92-52-4	**Biphenyl**	asphalt paving	671*(a)	86,553*	SCE	+		B6
			auto	1036	362(a),363(a)	STI00	NEG	NEG	B38
		biomass comb. 597,891(a)t; coal comb. 828t,1159(a); diesel 309(a),685(a);			754*(r)	STI35	NEG	NEG	B38
		phthal. anhyd. mfr. 64*; plastics comb. 181; refuse comb. 788(a),790*(a);			820(I)	STI37	NEG	NEG	B38
		tobacco smoke 339(a), 634(a); wood comb. 814(a)				STI98	NEG	NEG	B38
3.5-63	644-08-6	**4-Methylbiphenyl**	refuse comb.	878	86				
			tobacco smoke	339(a)	361(a),362(a)				
3.5-64	several	**Dimethylbiphenyl**			882(a)				
3.5-65	40529-66-6	**Ethylbiphenyl**			399(a),458(a)				
3.5-66	71277-83-3	**Ethylmethylbiphenyl**	tobacco smoke	130*(a)					
3.5-67S	84-15-1	*o*-**Terphenyl**	refuse comb.	878(a)					
3.5-68	92-06-8	*m*-**Terphenyl**	refuse comb.	878(a)					
3.5-69	92-94-4	*p*-**Terphenyl**	coal comb.	828(a)t					
			refuse comb.	878(a)					
3.5-70S	641-96-3	*o*-**Quaterphenyl**	refuse comb.	878(a)	408(a),832(a)				
3.5-71	1166-18-3	*m*-**Quaterphenyl**	refuse comb.	790(a)t,878(a)					

Species Number	Registry Number	Name	Emission Source	Emission Ref.	Detection Ref.	Code	Bioassay −MA	Bioassay +MA	Bioassay Ref.
3.5-72S	101-81-5	**Diphenylmethane**			364(a) 820(I)				
3.5-73	38888-98-1	**Diphenylethane**	refuse comb.	878(a)	399(a)				
3.5-74	501-65-5	**Diphenylethyne**	coal comb.	828(a)					
3.5-75	1081-75-0	**1,3-Diphenylpropane**	refuse comb.	878(a)					

Me
Me
Me
CH=CH2
CMe:CH2
C:CH

3.5-1 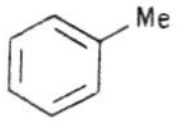 3.5-2 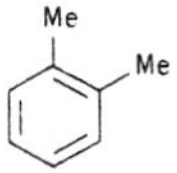 3.5-3 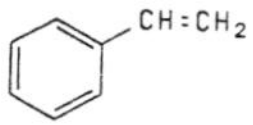 3.5-27 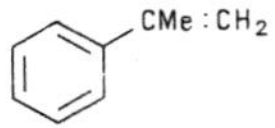 3.5-28 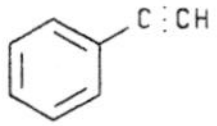 3.5-37

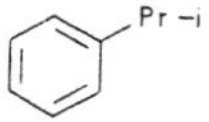

Pr-i
Pr-i
Me
CH2CH:CH2
CH:CHEt

3.5-42 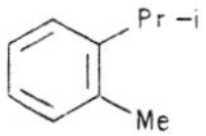 3.5-43 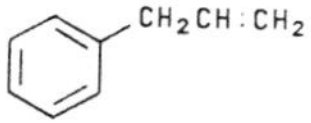 3.5-48 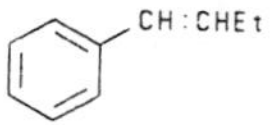 3.5-56 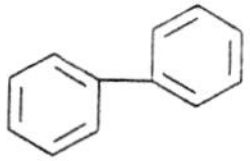 3.5-62

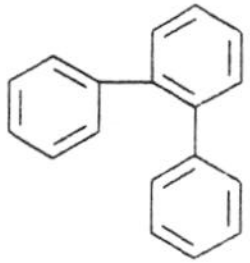

CH2

3.5-67 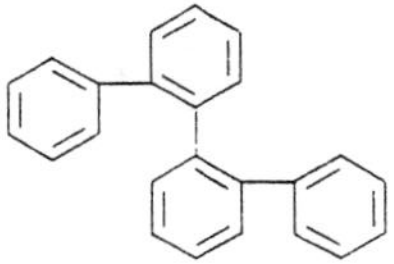 3.5-70 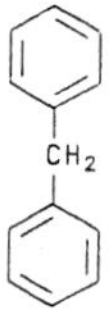 3.5-72

Table 3.6. Bicyclic Arenes

Species Number	Registry Number	Name	Emission Source	Emission Ref.	Detection Ref.	Code	Bioassay −MA	Bioassay +MA	Bioassay Ref.
3.6-1S	496-11-7	**Indan**	auto	1036	200,601 364(a),458(a)				
3.6-2	767-58-8	**1-Methylindan**	auto diesel tobacco smoke	682t 138t,176t 421t	86t,525t 399(a)				
3.6-3	824-63-5	**2-Methylindan**	auto	682t	682				
3.6-4	824-22-6	**4-Methylindan**	auto	682t	682				
3.6-5	874-35-1	**5-Methylindan**	auto	682t	682				
3.6-6	4912-92-9	**Dimethylindan**	biomass comb. diesel	597 138,176	582,597 399(a)				
3.6-7	2613-76-5	**Trimethylindan**	biomass comb. diesel	597 138					
3.6-8	60584-82-9	**Propylindan**			399(a)				
3.6-9	3910-35-8	**1,1,3-Trimethyl-3-phenylindan**	coal comb.	828t	1086(a)				

Species Number	Registry Number	Name	Emission Source	Emission Ref.	Detection Ref.	Bioassay Code	−MA	+MA	Ref.
3.6-10S	95-13-6	**Indene**	auto	1036	200,582				
			biomass comb.	597	364(a),458(a)				
		polymer comb. 853; polymer mfr. 251; tobacco smoke 634,339(a)							
3.6-11	767-59-9	**1-Methylindene**	tobacco smoke	339(a)	364(a),458(a)				
3.6-12	767-60-2	**3-Methylindene**	biomass comb.	597					
3.6-13	29348-63-8	**Dimethylindene**			399(a)				
3.6-14	58924-35-9	**Ethylindene**	tobacco smoke	339(a)	399(a),458(a)				
3.6-15S	91-17-8	**Decalin**	auto	682t	200,682				
			diesel	309(a)	1108(I)				
3.6-16	28258-89-1	**Methyldecalin**	auto	682t	682				
					1108(I)				
3.6-17S	119-64-2	**Tetralin**	auto	682t,880	597,682				
			diesel	138	364(a)				
			river water odor	202					
3.6-18	1559-81-5	**Methyltetralin**	diesel	138,176					
			tobacco smoke	421					
3.6-19	51855-29-9	**Dimethyltetralin**	auto	682t	682				
			diesel	138,309(a)					
3.6-20	40463-15-8	**Trimethyltetralin**	diesel	138					

Table 3.6. *(Continued)*

Species Number	Registry Number	Name	Emission Source	Emission Ref.	Detection Ref.	Bioassay Code	−MA	+MA	Bioassay Ref.
3.6-21	several	**1,2-Dihydromethyl-naphthalene**			628				
3.6-22	several	**1,2-Dihydrotrimethyl-naphthalene**	diesel	309(a)					
3.6-23S	91-20-3	**Naphthalene**	asphalt paving	671*(a)	86,200	STP00	NEG	NEG	B36
			auto	537,880	209(a),553*(a)	STP35	NEG	NEG	B36
		auto 311(a),858*(a); biomass comb. 597,821; brewing 144; coal comb. 828t,828(a)t; diesel 176; diesel 309(a),685(a); phthal anhyd. mfr. 64*; polymer comb. 304,332; refuse comb. 26(a),790(a); river water odor 202; tobacco smoke 406,421,79(a),339(a); volcano 1007; wood comb. 814(a); wood preserv. 1048*			755*(r),876*(r)	STP37	NEG	NEG	B36
					820(I),1108(I)	STP98	NEG	NEG	B36
3.6-24	90-12-0	**1-Methylnaphthalene**	auto	537,1036,858*(a)	86,597				
			brewing	144	755*(r),876*(r)				
		coal comb. 828t,828(a)t; coal tar mfr. 338(a); diesel 176,309(a),685(a);landfill 365; polymer comb. 853; refuse comb. 878,790(a); tobacco smoke 406,421,339(a),634(a); vulcanization 308; wood comb. 814(a)			820(I),1108(I)				
3.6-25	91-57-6	**2-Methylnaphthalene**	auto	1036	86,597				
			coal comb.	828t	399(a),458(a)				
		landfill 365; refuse comb. 878; tobacco smoke 406,421; tobacco smoke 339(a),396(a); wood comb. 814(a)			755*(r),876*(r)				
					820(I),1108(I)				
3.6-26	573-98-8	**1,2-Dimethylnaphthalene**	tobacco smoke	339(a)					
3.6-27	571-58-4	**1,4-Dimethylnaphthalene**			200				

Species Number	Registry Number	Name	Emission Source	Emission Ref.	Detection Ref.	Bioassay Code	−MA	+MA	Ref.
3.6-28	575-43-9	**1,6-Dimethylnaphthalene**	auto	1036	86,200				
			coal comb.	828t,828(a)t	364(a),458(a)				
			tobacco smoke	634(a)					
3.6-29	569-41-5	**1,8-Dimethylnaphthalene**	tobacco smoke	339(a),396(a)	86				
3.6-30	581-40-8	**2,3-Dimethylnaphthalene**	coal tar mfr.	338(a)	200,589(a)				
3.6-31	581-42-0	**2,6-Dimethylnaphthalene**	tobacco smoke	634(a)	86				
			turbine	786*(a)	214*(a)				
			vulcanization	308	876*(r)				
3.6-32	582-16-1	**2,7-Dimethylnaphthalene**	tobacco smoke	634(a)					
3.6-33	28652-77-9	**Trimethylnaphthalene**	diesel	309(a)	214*(a),589(a)				
			refuse comb.	878(a)					
			tobacco smoke	339(a),634(a)					
3.6-34	28652-74-6	**Tetramethylnaphthalene**	tobacco smoke	339(a)					
3.6-35	56908-81-7	**Pentamethylnaphthalene**	tobacco smoke	339(a)					
3.6-36	1127-76-0	**1-Ethylnaphthalene**	diesel	309(a)	200,628				
			refuse comb.	878	399(a)				
		tobacco smoke 339(a); turbine 786*(a)							
3.6-37	939-27-5	**2-Ethylnaphthalene**	vulcanization	308					
3.6-38	2027-17-0	**2-Isopropylnaphthalene**	diesel	309(a)					

Table 3.6. *(Continued)*

Species Number	Registry Number	Name	Emission Source	Emission Ref.	Detection Ref.	Bioassay Code	Bioassay −MA	Bioassay +MA	Bioassay Ref.
3.6-39	483-78-3	**1-Isopropyl-4,7-dimethyl-naphthalene**	biomass comb.	1089(a)					
3.6-40		**Undecylnaphthalene**	auto	311(a)					
3.6-41	605-02-7	**1-Phenylnaphthalene**	biomass comb. coal comb.	1089(a) 828(a)t,1065(a)					
3.6-42	612-94-2	**2-Phenylnaphthalene**	auto biomass comb.	858*(a) 891(a)t	832(a),944(a)				
		diesel 858*(a),903(a); refuse comb. 878(a)							
3.6-43	several	**Methylphenylnaphthalene**	auto diesel	858*(a) 858*(a),903(a)	832(a)				
3.6-44	970-06-9	**1,7-Diphenylnaphthalene**	refuse comb.	878(a)					
3.6-45S	612-78-2	**β,β-Binaphthyl**	coal comb. diesel	828(a)t 309(a)	362(a),363(a) 570(r)				
3.6-46	several	**Methyl-β,β-binaphthyl**			362(a),944(a)				
3.6-47S	7059-70-3	**Trinaphthene benzene**	auto	311(a)					
3.6-48	268-40-6	**Cyclopenta [1,2-*b*] naphthalene**	coal comb. tobacco smoke	361(a) 406	361(a),362(a)				

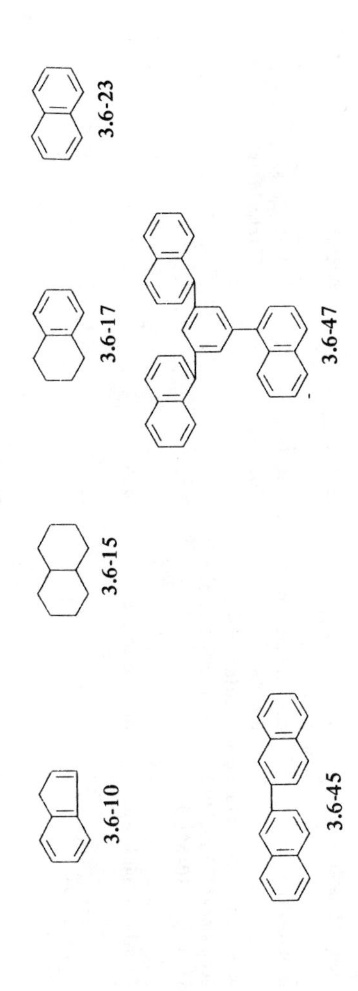
3.6-1
3.6-10
3.6-15
3.6-17
3.6-23
3.6-45
3.6-47

Table 3.7. Tricyclic Arenes

Species Number	Registry Number	Name	Emission Source	Emission Ref.	Detection Ref.	Bioassay Code	−MA	+MA	Ref.
3.7-1S	259-79-0	**Biphenylene**			628				
3.7-2S	83-32-9	**Acenaphthene**	diesel	138,309(a),685(a)	86,200				
			polymer comb.	853	209(a),214(a)				
		refuse comb. 788*(a),840(a); tobacco smoke 396(a),634(a); wood comb. 814(a)			754*(r),755*(r)				
3.7-3	36541-21-6	**Methylacenaphthene**	tobacco smoke	130*(a)					
3.7-4	several	**Dimethylcyclopent-acenaphthene**	auto	311(a)					
3.7-5	several	**Diphenylacenaphthene**	auto	311(a)	408(a)				
3.7-6	208-96-8	**Acenaphthylene**	carbon black mfr.	1048(a)	209(a),528(a)				
			diesel	309(a)	755*(r),876*(r)				
		polymer comb. 853; refuse comb. 26(a),790(a); tobacco smoke 79(a),339(a); wood comb. 814(a); wood preserv. 1048*(a)							
3.7-7	19345-99-4	**1-Methylacenaphthylene**	tobacco smoke	406					

Species Number	Registry Number	Name	Emission Source	Emission Ref.	Detection Ref.	Bioassay Code	Bioassay −MA	Bioassay +MA	Bioassay Ref.
3.7-8S	86-73-7	**Fluorene**	asphalt paving	671*(a)	86,200	STP00	NEG	NEG	B36
			auto	311(a),822(a)	209(a),362(a)	STP35	NEG	NEG	B36
		biomass comb. 891(a)t; coal comb. 361(a),825(a);			754*(r),755*(r)	STP37	NEG	NEG	B36
		diesel 685(a),858*(a); polymer comb. 650*(a),853; refuse comb.				STP98	NEG	NEG	B36
		788*(a); tobacco smoke 79(a),796(a); turbine 786*(a);							
		wood comb. 814(a); wood preserv. 1048*(a)							
3.7-9	41593-21-9	**Dihydrofluorene**	coal comb.	361(a)	362(a),826*(a)				
3.7-10	1730-37-6	**1-Methylfluorene**	auto	858*(a)	362(a)				
			coal comb.	361(a),1065(a)	820(I)t				
		diesel 858*(a),903*(a)t; tobacco smoke 130*(a),396(a); wood comb. 814(a)							
3.7-11	1430-97-3	**2-Methylfluorene**	coal comb.	361(a),1065(a)	362(a),826*(a)				
			refuse comb.	26(a)					
		tobacco smoke 130*(a); wood comb. 814(a)							
3.7-12	2523-37-7	**9-Methylfluorene**	coal comb.	361(a)	362(a)				
			diesel	804(a)					
		tobacco smoke 396(a),634(a); turbine 786*(a)t							
3.7-13	30582-01-5	**Dimethylfluorene**	biomass comb.	1089(a)					
			diesel	309(a)					
3.7-14	30582-02-6	**Trimethylfluorene**	diesel	309(a)					
3.7-15	65319-49-5	**Ethylfluorene**	diesel	309(a)					
3.7-16		**Isopentylfluorene**	auto	311(a)					

Table 3.7. *(Continued)*

Species Number	Registry Number	Name	Emission Source	Emission Ref.	Detection Ref.	Bioassay Code	−MA	+MA	Bioassay Ref.
3.7-17	several	**Phenylfluorene**			832(a)				
3.7-18	238-84-6	**Benzo[*a*]fluorene**	aluminum mfr.	1058*(a)	343(a),676*(a)				
			coal comb.	361(a),982*(a)	570(r)				
			polymer comb.	650*(a)					
		tobacco smoke 130*(a),396(a); wood comb. 814(a),824(a)							
3.7-19	238-79-9	**5*H*-Benzo[*a*]fluorene**	coal comb.	361(a)	918*(a)t				
			tobacco smoke	634(a)					
3.7-20	54811-53-9	**9-Methylbenzo[*a*]fluorene**	tobacco smoke	396(a)					
3.7-21	71265-25-3	**11-Methylbenzo[*a*]fluorene**	tobacco smoke	634(a)					
3.7-22	243-17-4	**Benzo[*b*]fluorene**	aluminum mfr.	1058*(a)	408(a),676*(a)				
			asphalt paving	671*(a)					
		auto 98(a),284(a); coal comb. 361(a),982*(a); coke mfr. 1072(a)t; polymer comb. 650*(a),853; tobacco smoke 130*(a),396(a); turbine 786*(a)t,849(a); wood comb. 814(a),824(a)							
3.7-23	205-12-9	**Benzo[*c*]fluorene**	coal comb.	361(a)	209(a),363(a)				
			tobacco smoke	130*(a),634(a)					
3.7-24	41593-27-5	**Dihydrobenzo[*c*]fluorene**	tobacco smoke	634(a)	362(a),363(a)				
3.7-25	239-60-1	**Dibenzo [*a*,*i*] fluorene**	tobacco smoke	396(a),634(a)					

Species Number	Registry Number	Name	Emission Source	Emission Ref.	Detection Ref.	Code	Bioassay −MA	Bioassay +MA	Bioassay Ref.
3.7-26	220-97-3	**Naphtho [2,1-*a*]fluorrene**	tobacco smoke	396(a),634(a)					
3.7-27S	194-26-3	**Cyclopenta[*cd*]phenalene**	tobacco smoke	130*(a)					
3.7-28	5743-97-5	**Perhydrophenanthrene**	diesel	309(a)					
3.7-29		**4,5-Dimethylperhydrophenanthrene**	diesel	309(a)					
3.7-30	29966-04-9	**Octahydrophenanthrene**	auto coal comb.	311(a) 361(a)					
3.7-31S	57706-44-2	**13-Methylpodocarpa-8,11,13-triene**			955*(a)				
3.7-32S	19407-17-1	**Dehydroabietin**			955*(a)				
3.7-33S	19407-18-2	**19-Norabieta-8,11,13-triene**			955*(a)				
3.7-34	23963-77-1	**19-Norabieta-4,8,11,13-tetraene**			955*(a)				
3.7-35S	19407-28-4	**Dehydroabietane**			955*(a)				

Table 3.7. *(Continued)*

Species Number	Registry Number	Name	Emission Source	Emission Ref.	Detection Ref.	Bioassay Code	Bioassay −MA	Bioassay +MA	Bioassay Ref.
3.7-36	1013-08-7	**Tetrahydrophenanthrene**	auto	311(a)					
3.7-37S	84744-09-2	**Norsimonellite**			955*(a)				
3.7-38S	27530-79-6	**Simonellite**			955*(a)				
3.7-39	26856-35-9	**Dihydrophenanthrene**	coal comb.	361(a)	362(a),625(a)				
3.7-40S	85-01-8	**Phenanthrene**	aluminum mfr.	551*(a),1058*(a)	553*	HMAST		NEG	B13
			asphalt paving	671*(a)	2(a),553*(a)	STP00	NEG	NEG	B36
		auto 284(a),285(a); coal comb. 56(a),1159(a),828t;			754*(r),755*(r)	STP35	NEG	NEG	B36
		coke mfr. 824(a),1072(a); diesel 685(a),804(a);				STP37	NEG	NEG	B36
		polymer comb. 650*(a),853; petroleum mfr. 285(a),338(a);				STP98	NEG	NEG	B36
		refuse cmb. 26(a),661*(a); tobacco smoke 79(a),218(a);				SCE	?		B6
		turbine 786*(a),813(a); volcano 1007;				CCC	I		B37
		wood comb. 814(a),824(a); wood preserv. 1048*(a)				CYC	NEG		B8
						REC		NEG	B9
						ALC	+		B19
						MNT	?/−		B29
						CTV	?		B24
						CTL	NEG		B24
						CTP	NEG		B24

Species Number	Registry Number	Name	Emission Source	Emission Ref.	Detection Ref.	Code	Bioassay −MA	Bioassay +MA	Bioassay Ref.
3.7-41	832-69-9	**1-Methylphenanthrene**	auto	858*(a)	633				
			coal comb.	361(a),825(a)t	362(a),676*(a)				
			coal tar mfr.	338(a)					
		coke mfr. 1072(a)t; diesel 804(a),829(a); polymer comb. 650*(a); tobacco smoke 130*(a),634(a); turbine 786*(a)t; wood comb. 814(a),982*(a)							
3.7-42	2531-84-2	**2-Methylphenanthrene**	coal comb.	1145(a)	408(a),1016(a)				
			diesel	804(a),829(a)					
		tobacco smoke 130*(a),1016(a); wood comb. 814(a),982*(a)							
3.7-43	832-71-3	**3-Methylphenanthrene**	coal comb.	1065(a)	408(a),1016(a)				
			diesel	804(a),829(a)					
		tobacco smoke 130*(a),1016(a); wood comb. 814(a),982*(a)							
3.7-44	832-64-4	**4-Methylphenanthrene**	diesel	804(a),823(a)					
			tobacco smoke	1016(a)					
3.7-45	883-20-5	**9-Methylphenanthrene**	coal comb.	1159(a)					
			coal tar mfr.	338(a)	408(a),1016(a)				
			diesel	804(a),829(a)					
		refuse comb. 878(a); tobacco smoke 130*(a),396(a); wood comb. 814(a)t							
3.7-46	483-87-4	**1,7-Dimethylphenanthrene**	coal comb.	1065(a)					

Table 3.7. *(Continued)*

Species Number	Registry Number	Name	Emission Source	Emission Ref.	Detection Ref.	Code	Bioassay −MA	Bioassay +MA	Bioassay Ref.
3.7-47	3674-66-6	**2,5-Dimethylphenanthrene**	biomass comb.	1089(a)t					
			diesel	309(a)					
			tobacco smoke	396(a),634(a)					
3.7-48	1576-67-6	**3,6-Dimethylphenanthrene**	auto	822(a)					
			coal comb.	1065(a),1145(a)					
3.7-49	30232-26-9	**Trimethylphenanthrene**	biomass comb.	1089(a)					
3.7-50	71607-70-0	**Tetramethylphenanthrene**	biomass comb.	891(a),1089(a)					
3.7-51	30997-38-7	**Ethylphenanthrene**	biomass comb.	891(a)	363(a)				
			coal comb.	361(a)					
			refuse comb.	878					
2.7-53	19353-76-5	**7-Ethyl-1-methylphenanthrene**			955*(a)				
3.7-53	483-65-8	**7-Isopropyl-1-methylphenanthrene**	wood comb.	814(a)	955*(a)				
3.7-54S	203-64-5	**4*H*-Cyclopenta[*def*]phenanthrene**	coal comb.	1065(a)	832(a),1058(a)				
			tobacco smoke	130*(a)					
			wood comb.	814(a),982*(a)					
3.7-55	58548-39-3	**Methyl-4*H*-cyclopenta[*def*] phenathrene**	tobacco smoke	130*(a)	408(a)				

Species Number	Registry Number	Name	Emission Source	Emission Ref.	Detection Ref.	Code	Bioassay −MA	Bioassay +MA	Bioassay Ref.
3.7-56	several	**Dimethyl-4*H*-cyclopenta[*def*]phenanthrene**			944(a)t				
3.7-57	several	**Trimethyl-4*H*-cyclopenta[*def*]phenanthrene**			944(a)t				
3.7-58	65319-51-9	**Ethyl-4*H*-cyclopenta[*def*] phenanthrene**	coal comb. tobacco smoke wood comb.	982*(a) 130*(a) 814(a)t,982*(a)	408(a)				
3.7-59	several	**Ethylmethyl-4*H*-cyclopenta [*def*]phenanthrene**			408(a)				
3.7-60S	82683-68-9	**Ethyldihydromethylene-phenanthrene**	coal comb. wood comb.	982*(a) 982*(a)	408(a),944(a)				
3.7-61S	195-19-7	**Benzo[*c*] phenanthrene**	asphalt paving carbon black mfr.	75(a),338(a) 1148(a)	213(a),676*(a)				
		coal comb. 361(a),825(a); oil comb. 825(a); polymer comb. 825(a); tobacco smoke 396(a),634(a); wood comb. 814(a),830*(a)							
3.7-62	several	**Dihydrobenzo[*c*] phenanthrene**			362(a),363(a)				
3.7-63	73560-82-4	**Methylbenzo[*c*] phenanthrene**			408(a)				

Table 3.7. *(Continued)*

Species Number	Registry Number	Name	Emission Source	Emission Ref.	Detection Ref.	Bioassay Code	Bioassay −MA	Bioassay +MA	Bioassay Ref.
3.7-64	several	**Dimethylbenzo[*c*] phenanthrene**	auto	311(a)					
3.7-65	1079-71-6	**Octahydroanthracene**	coal comb.	361(a)					
3.7-66	613-31-0	**9,10-Dihydroanthracene**	coal comb.	361(a)	826*(a)				
			tobacco smoke	634(a)					
3.7-67S	120-12-7	**Anthracene**	aluminum mfr.	551*(a)	41(a),209(a)	HMAV7		NEG	B13
			asphalt paving	75*(a),285(a)	754*(r),755*(r)	HMAST		+/−	B13
			auto	284(a),285(a)		STP00	NEG	NEG	B36
		coal comb. 56(a),285(a); coke mfr. 824(a),1072(a);				STP35	NEG	NEG	B36
		diesel 685(a),824(a); petroleum mfr. 285(a);				STP37	NEG	NEG	B36
		polymer comb. 650*(a),853; refuse comb. 285(a),661*(a);				STP98	NEG	NEG	B36
		tobacco smoke 79(a),218(a); turbine 786*(a),813(a);				SCE	?		B6
		volcano 1007; wood comb. 814(a),824(a);				CCC	I,N		B37
		wood preserv. 1048*(a)				CYC	NEG		B8
						MNT	NEG		B29
						CTV	?		B24
						CTL	NEG		B24
						CTP	NEG		B24
						SPI	NEG		B25
3.7-68	610-48-0	**1-Methylanthracene**	aluminum mfr.	551*(a)	633				
			auto	822(a)	362(a),363(a)				
			coal comb.	361(a),825(a)t					
		coke mfr. 1072(a)t; oil comb. 825(a)t; polymer comb. 853;							
		refuse comb. 878; tobacco smoke 130*(a),634(a); turbine 786*(a)t							

Species Number	Registry Number	Name	Emission Source	Emission Ref.	Detection Ref.	Code	Bioassay −MA	Bioassay +MA	Bioassay Ref.
3.7-69	613-12-7	**2-Methylanthracene**	auto	822(a)	408(a),676*(a)				
			coal comb.	827(a),1065(a)					
			coal tar mfr. 338*(a); polymer comb. 825(a)t; refuse comb. 878; tobacco smoke 130*(a),396(a); wood comb. 814(a),982*(a)						
3.7-70	779-02-2	**9-Methylanthracene**	auto	822(a)					
			coal comb.	827(a)					
			polymer comb. 650*(a); tobacco smoke 634(a),1016(a)						
3.7-71	781-43-1	**9,10-Dimethylanthracene**	polymer comb.	650*(a)		MNT	?/−		B29
						SPI	NEG		B25
3.7-72	41637-86-9	**Ethylanthracene**	coal comb.	361(a)	363(a)				
3.7-73S	56-55-3	**Benz[*a*]anthracene**	aluminum mfr.	551*(a)	188*(a),676*(a)	HMAST		+	B13
			asphalt paving	671*(a)	570(r)	V790		+	B7
			auto 87(a),312(a); coal comb. 56(a),36(a); coke mfr. 824(a),1072(a); diesel 433*(a),804(a); refuse comb. 661*(a),1093(a); tobacco smoke 79(a),130*(a); turbine 786*(a),813(a),; wood comb. 814(a),824(a); wood preserv. 1048*(a)			V790		NEG	B7
						V79T		+	B7
						V79A	NEG		B7
						STP00		+	B36
						SCE	?		B6
						CCC	SP		B37
						REC		NEG	B9
						CTV	+		B24
						CTL	+/−		B24
						CTP	+		B24
						SRL	?		B30
3.7-74	16434-59-6	**Dihydrobenz[*a*]anthracene**			363(a)				

Table 3.7. *(Continued)*

Species Number	Registry Number	Name	Emission Source	Emission Ref.	Detection Ref.	Bioassay Code	Bioassay −MA	Bioassay +MA	Bioassay Ref.
3.7-75	2498-76-2	**2-Methylbenz[*a*]anthracene**	coal comb.	361(a)	362(a),363(a)	STP00		+(G)	B36
			tobacco smoke	130*(a),1016(a)	570(r)				
			wood comb.	830*(a)t	570(s)				
3.7-76	2498-75-1	**3-Methylbenz[*a*]anthracene**	auto	311(a)	1016(a)	STP00		+(G)	B36
			tobacco smoke	130*(a),396(a)					
3.7-77	316-49-4	**4-Methylbenz[*a*]anthracene**	tobacco smoke	130*(a),1016(a)	1016(a)	STP00		+(G)	B36
3.7-78	2319-96-2	**5-Methylbenz[*a*]anthracene**	tobacco smoke	130*(a),634(a)		STP00		+(G)	B36
3.7-79	316-14-3	**6-Methylbenz[*a*]anthracene**	tobacco smoke	130*(a)		STP00		+(G)	B36
3.7-80	2381-31-9	**8-Methylbenz[*a*]anthracene**	tobacco smoke	130*(a)		STP00		+(G)	B36
						CCC	LP		B37
3.7-81	2381-16-0	**9-Methylbenz[*a*]anthracene**	tobacco smoke	130*(a),1016(a)	1016(a)	STP00		+(G)	B36
3.7-82	2381-15-9	**10-Methylbenz[*a*]anthracene**	tobacco smoke	130*(a),1016(a)	1016(a)	STP00		+(G)	B36
3.7-83	57-97-6	**7,12-Dimethylbenz[*a*] anthracene**	asphalt paving	75(a)		HMAV7		+	B13
			carbon black mfr.	1148(a)		HMAST		NEG	B13
			polymer comb.	650*(a)		V79T	NEG	+	B7
						V79A	NEG	+/−	B7
						STP35		−(C)	B36
						STP38		+	B36
						STP98		+	B36
						STP00		+	B36

Species Number	Registry Number	Name	Emission Source	Emission Ref.	Detection Ref.	Code	Bioassay −MA	Bioassay +MA	Bioassay Ref.
						SCE	+		B6
						CCC	SP		B37
						CYB	+		B8
						SLGM	?		B4
						CYC	+		B8
						CYI	+		B8
						CTV	+		B24
						CTL	+,+,+		B24
						CTP	+		B24
						SRL	?		B30
						NEU	+		B32
						SPI	+		B25
3.7-84	58429-99-5	**9,10-Dimethylbenz[*a*] anthracene**	asphalt paving	671*(a)					
			coal tar mfr.	338*(a)					
			tobacco smoke	396(a),634(a)					
3.7-85		**Benz[*a*]cyclopent[*c*] anthracene**	tobacco smoke	396(a)					
3.7-86S	53-70-3	**Dibenz[*a*,*h*]anthracene**	auto	312*(a),822(a)	213(a),676*(a)	V790		+	B7
			carbon black mfr.	1148(a)t		V79T		+	B7
			coal comb.	361(a),827(a)		V79A	NEG	NEG	B7
		oil comb. 825(a); tobacco smoke 396(a),634(a);				STP00		+	B36
		wood preserv. 1048*(a)				SCE	?		B6
						CCC	SP		B37
						CTV	+		B24
						CTL	+,+,+		B24
						CTP	+,+		B24

Table 3.7. *(Continued)*

Species Number	Registry Number	Name	Emission Source	Emission Ref.	Detection Ref.	Bioassay Code	Bioassay −MA	Bioassay +MA	Bioassay Ref.
						SRL	+		B30
						NEU	+		B32
3.7-87	15595-02-5	**Methyldibenz[*a*,*h*]anthracene**	coal comb.	361(a)	362(a),408(a)				
3.7-88S	224-41-9	**Dibenz[*a*,*j*]anthracene**	coal comb.	1065(a)		STP00		+	B36
3.7-89	215-26-9	**Tribenz[*a*,*c*,*h*]anthracene**	tobacco smoke	634(a)					
3.7-90S	52428-35-0	**7-Diphenylmethylene-1,3,5-cycloheptatriene**	refuse comb.	790(a)					

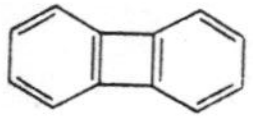

3.7-1

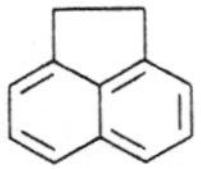

3.7-2

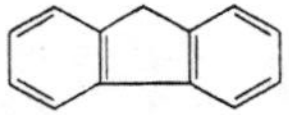

3.7-8

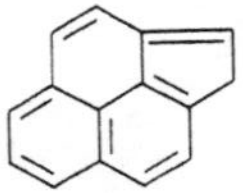

3.7-27

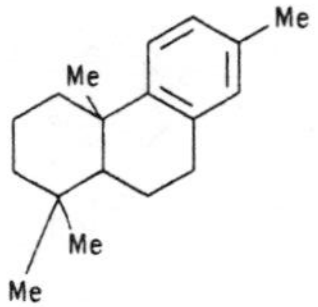

Me Me Me Me
3.7-31

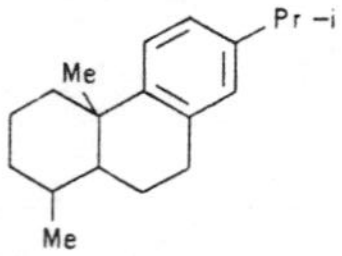

Pr -i Me Me
3.7-32

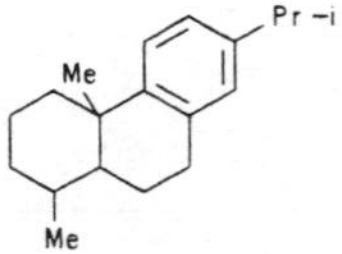

Pr -i Me Me
3.7-33

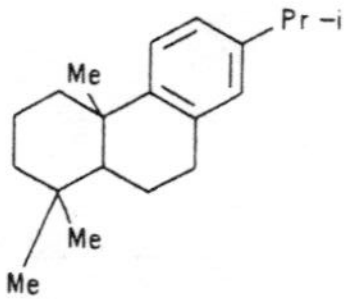

Pr -i Me Me Me
3.7-35

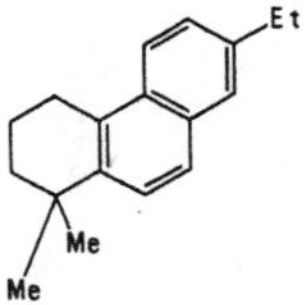

Et Me Me
3.7-37

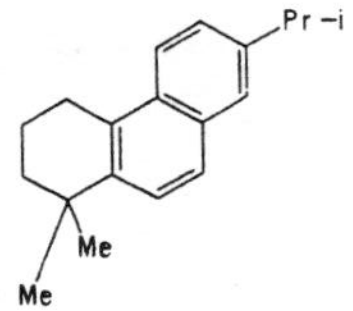

Pr-i
Me
Me

3.7-38

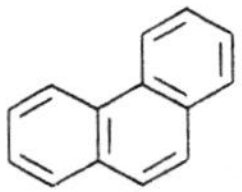

3.7-40

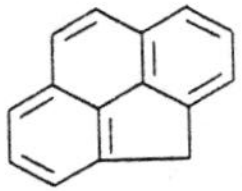

3.7-54

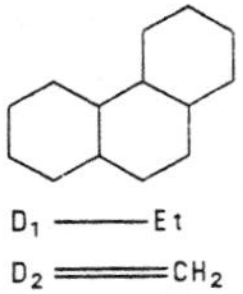

D_1 —— Et
D_2 ══ CH_2

3.7-60

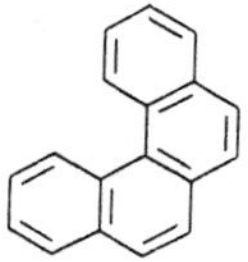

3.7-61

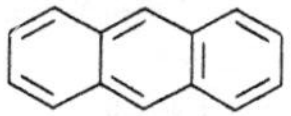

3.7-67

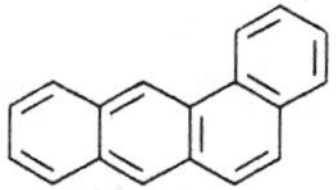

3.7-73

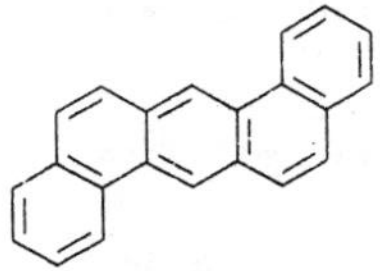

3.7-86

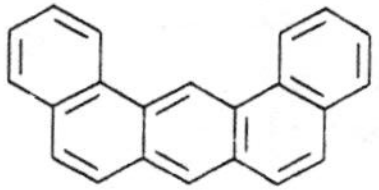

3.7-88

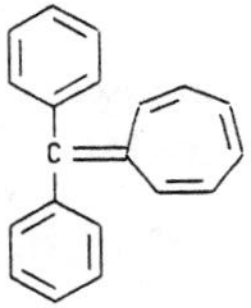

C

3.7-90

Table 3.8. Higher Polycyclic Arenes

Species Number	Registry Number	Name	Emission Source	Emission Ref.	Detection Ref.	Bioassay Code	Bioassay −MA	Bioassay +MA	Bioassay Ref.
3.8-1	41593-22-0	**Octahydrofluoranthene**	coal comb.	361(a)	361(a),362(a)				
3.8-2	41593-24-2	**Dihydrofluoranthene**	coal comb.	361(a)	362(a),826*(a)				
3.8-3S	206-44-0	**Fluoranthene**	aluminum mfr.	551*(a),1058*(a)	553*,633*				
			asphalt paving	285(a),671*(a)	41(a),676*(a)				
		auto 311(a),312*(a); biomass comb. 891(a)t; carbon black mfr. 1148(a);			570(r),754*(r)				
		coal comb. 828t,56(a),285(a); coke mfr. 1072(a); diesel 433*(a),690(a);			570(s),819*(s);				
		oil comb. 825(a); petroleum mfr. 285(a),338*(a);							
		polymer comb. 650*(a),325(a); refuse comb. 341*(a),661*(a);							
		tobacco smoke 79(a); turbine 786*(a),813(a);							
		wood comb. 814(a),824(a); wood preserv. 1048*(a)							
3.8-4	25889-60-5	**1-Methylfluoranthene**	coke mfr.	1072(a)t	408(a),1016(a)				
			tobacco smoke	130*(a),1016(a)					
3.8-5	33543-31-6	**2-Methylfluoranthene**	aluminum mfr.	551*(a)	408(a),1016(a)				
			auto	858*(a)	570(r)				
		coal comb. 1065(a); diesel 690(a),804(a)t; tobacco smoke 130*(a),1016(a)			570(s)				
3.8-6	1706-01-0	**3-Methylfluoranthene**	coal comb.	361(a)	362(a),363(a)				
			tobacco smoke	130*(a),1016(a)					
3.8-7	23339-05-1	**7-Methylflouranthene**	tobacco smoke	130*(a),1016(a)	408(a),1016(a)				

Species Number	Registry Number	Name	Emission Source	Emission Ref.	Detection Ref.	Bioassay Code	Bioassay −MA	Bioassay +MA	Bioassay Ref.
3.8-8	20485-57-8	**8-Methylfluoranthene**	tobacco smoke	130*(a),634(a)	408(a),1016(a)				
3.8-9	60826-74-6	**Dimethylfluoranthene**	tobacco smoke	396(a),634(a)					
3.8-10	203-33-8	**Benzo[*a*]fluoranthene**	coal comb. tobacco smoke	1065(a) 406					
3.8-11	several	**Dihydromethylbenzo[*b*] fluoranthene**			570(r)				
3.8-12	41637-94-9	**Methylbenzo[*b*] fluoranthene**			570(r)				
3.8-13	16135-81-2	**Benzo[*cd*]fluoranthene**	tobacco smoke	396(a)	179*(a)				
3.8-14S	203-12-3	**Benzo[*ghi*]fluoranthene**	auto carbon black mfr.	87(a),312*(a) 1148(a)	209(a),362(a) 570(r)				
		coal comb. 361(a),825(a)t; coke mfr. 824(a); diesel 685(a),824(a); oil comb. 825(a)t; refuse comb. 878(a); tobacco smoke 396(a),634(a); turbine 804(a)t,813(a); wood comb. 814(a),824(a)							
3.8-15	39379-95-8	**Dihydromethylbenzo[*ghi*] fluoranthene**			362(a) 570(r)				
3.8-16	51001-44-6	**Methylbenzo[*ghi*]fluoranthene**	coal comb.	828(a)t	408(a)				
3.8-17	205-82-3	**Benzo [*j*] fluoranthene**	auto coal comb.	87(a),312*(a) 361(a),1065(a)	98(a),362(a)	CCC	LP		B37
		tobacco smoke 130*(a),396(a); turbine 813(a); wood comb. 814(a)							

Table 3.8. *(Continued)*

Species Number	Registry Number	Name	Emission Source	Emission Ref.	Detection Ref.	Bioassay Code	Bioassay −MA	Bioassay +MA	Bioassay Ref.
3.8-18	several	**Methylbenzo[*mno*] fluoranthene**			362(a)				
3.8-19S	193-43-1	**Indeno[1,2,3-*cd*]fluoranthene**	auto	312*(a),824(a)	824(a)				
			coal comb.	1065(a)					
		coke mfr. 824(a); diesel 824(a); tobacco smoke 634(a); wood comb. 824(a)							
3.8-20	2997-45-7	**Dibenzo[*b,e*]fluoranthene**	coal comb.	1065(a)					
3.8-21	several	**Methyldibenzo[*b,k*] fluoranthene**			363(a)				
3.8-22S	201-06-9	**Acephenanthrylene**	biomass comb.	814(a),891(a)t					
			coal comb.	1065(a)					
		tobacco smoke 406(a); turbine 813(a)t,849(a)t							
3.8-23S	205-99-2	**Benz[*e*]acephenathrylene**	aluminum mfr.	551*(a),1058*(a)	98(a),676*(a)	SCE	?		B6
			auto	87(a),312*(a)	570(r)	CCC	SP		B37
		carbon black mfr. 1048(a)t; coal comb. 982*(a),1065(a); coke mfr. 824(a); diesel 209(a),433*(a); refuse comb. 788*(a),790(a); tobacco smoke 130*(a),634(a); turbine 786*(a)t,813(a); wood comb. 814(a),824(a); wood preserv. 1048*(a)t			570(s)				

Species Number	Registry Number	Name	Emission Source	Emission Ref.	Detection Ref.	Bioassay Code	Bioassay −MA	Bioassay +MA	Bioassay Ref.
3.8-24	4766-40-9	**Benz[*j*]acephenanthrylene**	tobacco smoke	396(a)					
3.8-25S	202-03-9	**Aceanthrylene**	turbine	786*(a),813(a)t					
3.8-26S	56-49-5	**3-Methylcholanthrene**	asphalt paving	75(a)	362(a),945*(a)	V79T		+	B7
			carbon black mfr.	1148(a)		V79A	NEG	+/−	B7
			wood comb.	830*(a)		STU38		+(BE)	B36
						STP00		+	B36
						STP35	−ACE	?(E)	B36
						STP37	−(CE)	+	B36
						STP38	−AC	+	B36
						SCE	+		B6
						CCC	SP		B37
						CYC	NEG		B8
						STI00	NEG	?	B38
						STI35	NEG	NEG	B38
						STI37	NEG	NEG	B38
						STI98	?	+	B38
						MNT	?/−		B29
						CTL	+,+,+		B24
						CTP	+,+		B24
						CTV	+		B24
						SRL	?		B30
						NEU	+		B32
						SPI	+		B25
3.8-27S	217-59-4	**Triphenylene**	auto	312*(a),822(a)	832(a)				
			coal comb.	1065(a)					
		refuse comb. 790(a),840(a); turbine 813(a); wood comb. 814(a)							

Table 3.8. *(Continued)*

Species Number	Registry Number	Name	Emission Source	Emission Ref.	Detection Ref.	Bioassay Code	Bioassay −MA	Bioassay +MA	Bioassay Ref.
3.8-28	41637-89-2	**Methyltriphenylene**	coal comb.	361(a)	826*(a)				
3.8-29S	215-58-7	**Benzo[*b*]triphenylene**	asphalt paving	671*(a)	553*	V790		+	B7
			auto	822(a)	362(a),553*(a)	V79T		+	B7
		coal comb. 827(a),1065(a); wood comb. 814(a)t				V79A		+	B7
						STP00		+	B36
						CCC	LP		B37
						CTV	+/−		B24
						CTP	?		B24
3.8-30	55775-16-1	**Octahydropyrene**			361(a),362(a)				
3.8-31	28779-32-0	**Dihydropyrene**	coal comb.	361(a)	362(a),826*(a)				
3.8-32S	129-00-0	**Pyrene**	aluminum mfr.	551*(a),1058*(a)	633	HMAV7		NEG	B13
			asphalt paving	75*(a),671*(a)	209*(a),676*(a)	V790		NEG	B7
		auto 87(a),311(a); biomass comb. 891(a)t;			570(r),754*(r)	V79A		NEG	B7
		carbon black mfr. 1148(a); coal comb. 828t,56(a),285(a);			857*(I)	STP00	NEG	NEG	B36
		coke mfr. 824(a),1072(a); diesel 209(a),433*(a); oil comb. 825(a);				STP35	NEG	NEG	B36
		petroleum mfr. 285(a); polymer comb. 650*(a),825(a); refuse comb.				STP37	NEG	NEG	B36
		26(a),661*(a); tobacco smoke 79(a),218(a); turbine 786*(a),813(a);				STP98	NEG	NEG	B36
		wood comb. 814(a),824(a); wood preserv. 1048*(a)				SCE	?		B6
						CCC	I		B37
						CYC	NEG		B8
						MNT	?/−		B29
						CTV	−,−		B24
						CTP	NEG		B24
						SPI	NEG		B25
						CTL	NEG		B24

Species Number	Registry Number	Name	Emission Source	Emission Ref.	Detection Ref.	Bioassay Code	−MA	+MA	Ref.
3.8-33	2381-21-7	**1-Methylpyrene**	aluminum mfr.	551*(a)	209(a),676*(a)				
			auto	659(a),823(a)	570(r)				
		biomass comb. 891(a)t; coal comb. 361(a),1065(a); coke mfr. 1072(a); diesel 685(a),690(a); polymer comb. 853; tobacco smoke 79(a),130*(a); wood comb. 814(a),982*(a)			570(s)				
3.8-34	3442-78-2	**2-Methylpyrene**	coal comb.	1065(a)	408(a),1016(a)				
			tobacco smoke	130*(a),406(a)					
			wood comb.	814(a),982*(a)					
3.8-35	3353-12-6	**4-Methylpyrene**	auto	659(a),823(a)	408(a),659(a)				
			coal comb.	1065(a)					
		tobacco smoke 130*(a),396(a); wood comb. 814(a),982*(a)							
3.8-36	15679-24-0	**2,7-Dimethylpyrene**			209(a)				
3.8-37S	27208-37-3	**Cyclopenta[*cd*]pyrene**	auto	340(a),556(a)	824(a),1143(a)	STP00		+	B36
			coal comb.	340(a),1065(a)		STP35		?(E)	B36
		diesel 340(a),685(a); turbine 813(a),849(a);				STP37		+	B36
		wood comb. 814(a),824(a)				STP38		+	B36
						STP98		+	B36
						CCC	LP		B37
3.8-38	25732-74-5	**3,4-Dihydrocyclopenta [*cd*]pyrene**	coal comb.	340(a)					
			wood comb.	981*(a)					

Table 3.8. *(Continued)*

Species Number	Registry Number	Name	Emission Source	Emission Ref.	Detection Ref.	Code	Bioassay −MA	Bioassay +MA	Bioassay Ref.
3.8-39S	50-32-8	**Benzo[*a*]pyrene**	aluminum mfr.	551(a),659(a)	553*	TRM	+		B21
			asphalt paving	75(a),671*(a)	113*(a),676*(a)	HMAST		+/−	B13
		auto 87(a),311(a); coal comb. 56(a),285(a);			570(r),755*(r)	HMAV7		+	B13
		coke mfr. 145*(a),824(a); diesel 209(a),433*(a);			570(s),1138*(s)	V790		+	B7
		oil comb. 825(a); petroleum mfr. 285(a);			857*(I),980*(I)	V79T		+	B7
		polymer comb. 825(a),938(a); refuse comb. 26(a),661*(a);				V79A	NEG	+	B7
		rubber abrasion 424(a); steel molds 201(a),277(a);				STU38		+(BE)	B36
		tobacco smoke 4(a),130*(a), turbine 786*(a),813(a);				STI98	−(D)	+(D)	B36
		volcano 1200; wood comb. 814(a),824(a);				STI38	−(E)	+(D)	B36
		wood preserv. 1048*(a)				STP00	−(C)	+	B36
						STP35	−(C)	−(C)	B36
						STP37	−(C)	+	B36
						STP38		+	B36
						STP98	−(C)	+	B36
						SCE	+		B6
						CCC	SP		B37
						SLGM	NEG		B4
						SLPM	?		B4
						MST	+		B5
						MDR	+		B12
						CYC	NEG		B8
						REC		+/?	B9
						STI00	NEG	+	B38
						STI35	NEG	NEG	B38
						STI37	NEG	+	B38
						STI98	NEG	+	B38
						MNT	+		B29
						L5	+?	+	B28
						CTV	+		B24

Species Number	Registry Number	Name	Emission Source	Emission Ref.	Detection Ref.	Code	Bioassay −MA	Bioassay +MA	Ref.
						CTL	+,+,+		B24
						CTP	+,+		B24
						SRL	+		B30
						SPI	+		B25
3.8-40	25167-89-9	**Methylbenzo[*a*]pyrene**	auto	87(a),556(a)	570(r),755*(r)				
			coal comb.	361(a)					
			tobacco smoke	634(a)					
3.8-41	25167-90-2	**Dimethylbenzo[*a*] pyrene**	coal comb.	361(a)					
3.8-42	191-33-3	**Benzo[*cd*]pyrene**	tobacco smoke	396(a)	113*(a),179*(a) 1139*(r) 819*(s)				
3.8-43	86426-53-1	**3,4-Dihydrobenzo [*cd*]pyrene**	tobacco smoke	396(a)					
3.8-44S	192-97-2	**Benzo[*e*]pyrene**	aluminum mfr.	551*(a),1058*(a)	209*(a),676*(a)	V790		NEG	B7
			asphalt paving	75(a),671*(a)	857*(I)	STP00		+	B36
		auto 87(a),312*(a); coal comb. 56(a),285(a); coke mfr. 824(a);				SCE	?		B6
		diesel 433*(a),824(a); oil comb. 825(a); petroleum mfr. 285(a);				CCC	I		B37
		refuse comb. 285(a),661*(a); tobacco smoke 130*(a),634(a);				REC		NEG	B9
		turbine 786*(a),813(a); wood comb. 814(a),824(a)				L5		−?	B28
						CTV	NEG		B24
						CTL	NEG		B24
						CTP	NEG		B24

Table 3.8. *(Continued)*

Species Number	Registry Number	Name	Emission Source	Emission Ref.	Detection Ref.	Code	Bioassay −MA	Bioassay +MA	Bioassay Ref.
3.8-45	41699-04-1	**Methylbenzo[*e*]pyrene**	auto	87(a),556(a)					
3.8-46	29797-12-4	**Dibenzo[*a,b*]pyrene**	auto	284(a)					
3.8-47		**Dibenzo[*a,cd*]pyrene**			375(a),832(a)				
3.8-48	192-65-4	**Dibenzo[*a,e*]pyrene**	auto carbon black mfr. coal comb.	209(a),822(a) 1148(a)t 1065(a)	189(a),375(a)	CCC	SP		B37
3.8-49	several	**Methyldibenzo[*a,e*] pyrene**			363(a)				
3.8-50	191-30-0	**Dibenzo[*a,l*]pyrene**	tobacco smoke	634(a)		CCC	SP		B37
3.8-51S	191-26-4	**Dibenzo[*cd,jk*]pyrene**	auto coal comb.	209(a),312*(a) 56(a),285(a)	189(a),676*(a)				
		coke mfr. 824(a); diesel 209(a),824(a); petroleum mfr. 285(a); refuse comb. 285(a),661*(a); tobacco smoke 79(a),130*(a); wood comb. 830*(a)t,968*(a)							
3.8-52	192-51-8	**Dibenzo[*e,l*]pyrene**	coal comb.	1065(a)	98(a)				

Species Number	Registry Number	Name	Emission Source	Emission Ref.	Detection Ref.	Code	Bioassay −MA	Bioassay +MA	Ref.
3.8-53S	193-39-5	**Indeno[1,2,3-*cd*]pyrene**	aluminum mfr.	1058*(a)	98(a),625(a)	CCC	SP		B37
			auto	312*(a),360(a)	570(r)				
		carbon black mfr. 1148(a); coal comb. 827(a),982*(a);			819*(s)				
		coke mfr. 1072(a); diesel 433*(a); oil comb. 825(a);							
		refuse comb. 341*(a); tobacco smoke 634(a); turbine 813(a);							
		wood comb. 814(a), 981*(a); wood preserv. 1048*(a)							
3.8-54		**Naphtho[1,2,3-*cd*]pyrene**	tobacco smoke	396(a)					
3.8-55	41593-29-7	**Hexahydrochrysene**			363(a)				
					570(r)				
3.8-56	41593-31-1	**Dihydrochrysene**			363(a)				
3.8-57S	218-01-9	**Chrysene**	aluminum mfr.	551*(a)	188*(a),373*(a)	HMAST		NEG	B13
			asphalt paving	671*(a)	754*(r),755*(r)	STP00		+	B36
		auto 87(a),311(a); coal comb. 827(a),1145(a); coke mfr. 1072(a);			857*(I)	SCE	?		B6
		diesel 433*(a),685(a); polymer comb. 853; refuse comb. 26(a),661*(a);				CCC	LP		B37
		tobacco smoke 130*(a),396(a); turbine 786*(a),813(a);				CYS	NEG		B8
		wood comb. 814(a),968*(a); wood preserv. 1048*(a)				REC		NEG	B9
						CTL	NEG		B24
						CTP	+		B24
3.8-58	3351-28-8	**1-Methylchrysene**	coal comb.	361(a),1065(a)	362(a),363(a)	STP00		+(G)	B36
			oil comb.	825(a)t					
			tobacco smoke	130*(a),396(a)					
3.8-59	3351-32-4	**2-Methylchrysene**	coal comb.	1065(a),1145(a)	408(a),1016(a)	STP00		+(G)	B36
			tobacco smoke	130*(a),1016(a)					

Table 3.8. *(Continued)*

Species Number	Registry Number	Name	Emission Source	Emission Ref.	Detection Ref.	Bioassay Code	Bioassay −MA	Bioassay +MA	Bioassay Ref.
3.8-60	3351-31-3	**3-Methylchrysene**	coal comb.	1065(a)	408(a),1016(a)	STP00		+(G)	B36
			tobacco smoke	130*(a),1016(a)					
3.8-61	3351-30-2	**4-Methylchrysene**	coal comb.	1065(a)		STP00		+(G)	B36
3.8-62	3697-24-3	**5-Methylchrysene**	coal comb.	1065(a)		STP00		+(BE)	B36
			tobacco smoke	130*(a)		STP00		+(G)	B36
						CCC	SP		B37
3.8-63	1705-85-7	**6-Methylchrysene**	coal comb.	1065(a)	408(a),1016(a)	STP00		+(G)	B36
			tobacco smoke	130*(a),1016(a)					
3.8-64	41637-92-7	**Dimethylchrysene**	tobacco smoke	396(a),634(a)	362(a)				
					570(r)				
3.8-65	214-17-5	**Benzo[*b*]chrysene**	coal comb.	1065(a)					
3.8-66	216-53-5	**Benzo[*hi*]chrysene**	asphalt paving	671*(a)					
3.8-67	189-64-0	**Dibenzo[*b,def*] chrysene**	asphalt paving	75(a)	98(a),189(a)	CCC	SP		B37
			tobacco smoke	634(a)					
3.8-68S	92-24-0	**Naphthacene**	auto	209(a)					
			turbine 786*(a)						
3.8-69	226-88-0	**Benzo[*a*]naphthacene**	tobacco smoke	396(a),634(a)					
3.8-70	216-00-2	**Dibenzo[*a,c*]naphthacene**	tobacco smoke	396(a),634(a)					

Species Number	Registry Number	Name	Emission Source	Emission Ref.	Detection Ref.	Code	Bioassay −MA	Bioassay +MA	Bioassay Ref.
3.8-71	227-04-3	**Dibenzo[*a,j*]naphthacene**	tobacco smoke	396(a),634(a)					
3.8-72	226-86-8	**Dibenzo[*a,l*]naphthacene**	auto	209(a)	209(a)				
			diesel	209(a)					
3.8-73	193-09-9	**Dibenzo[*de,qr*]naphthacene**	tobacco smoke	396(a)					
3.8-74	196-42-9	**Naphtho[2,1,8-*qra*]naphthacene**	refuse comb.	990*(a)	189(a)				
			wood comb.	990(a)					
3.8-75S	213-46-7	**Picene**	aluminum mfr.	1058*(a)	363(a)				
			auto	822(a)					
			coal comb.	1065(a)					
3.8-76	30283-95-5	**Methylpicene**	coal comb.	1065(a)t					
3.8-77	1242-77-9	**1,2,9-Trimethyl-1,2,3,4-tetrahydropicene**	coal comb.	1065(a)t					
3.8-78	1242-76-8	**2,2,9-Trimethyl-1,2,3,4-tetrahydropicene**	coal comb.	1065(a)t					
3.8-79	189-96-8	**Benzo[*pqr*]picene**	tobacco smoke	634(a)					

Table 3.8. *(Continued)*

Species Number	Registry Number	Name	Emission Source	Emission Ref.	Detection Ref.	Code	Bioassay −MA	Bioassay +MA	Bioassay Ref.
3.8-80S	198-55-0	**Perylene**	aluminum mfr.	1058*(a)	209*(a),676*(a)	HMAV7		NEG	B13
			asphalt paving	671*(a)	570(r)	SCE	NEG		B6
			auto 311(a),312*(a); coal comb. 56(a),285(a); coke mfr. 824(a),1072(a); diesel 804(a),824(a); oil comb. 825(a); petroleum mfr. 285(a); polymer comb. 650*(a); refuse comb. 341(a),661*(a); tobacco smoke 130*(a),396(a); turbine 786*(a),813(a); wood comb. 814(a),824(a)		570(s)				
3.8-81	191-24-2	**Benzo[*ghi*]perylene**	aluminum mfr.	659(a),1058*(a)	188*(a),660*(a)	STP00		+	B36
			asphalt paving	285(a)	570(r),754*(r)				
			auto 87(a),311(a); coal comb. 56(a),285(a); coke mfr. 824(a),1072(a); diesel 209(a),685(a); oil comb. 825(a); petroleum mfr. 285(a); refuse comb. 341(a),661*(a); tobacco smoke 130*(a),396(a); turbine 813(a),849(a); wood comb. 814(a),824(a); wood preserv. 1048*(a)		819*(s)				
3.8-82	41699-09-6	**Methylbenzo[*ghi*]perylene**	auto	556(a)					
			coal comb.	361(a)					
3.8-83S	64503-02-2	**Benzo[*ghi*]cyclopenta [*pqr*]perylene**	auto	824(a)t	824(a)t				
			diesel	824(a)t					
3.8-84	190-95-4	**Dibenzo[*b,pqr*]perylene**	auto	209(a)					
3.8-85S	222-93-5	**Pentaphene**	auto	209(a)					
			diesel	209(a)					
			tobacco smoke	634(a)					

Species Number	Registry Number	Name	Emission Source	Emission Ref.	Detection Ref.	Code	Bioassay −MA	Bioassay +MA	Bioassay Ref.
3.8-86	189-55-9	**Benzo[*rst*]pentaphene**	asphalt paving	75(a)	98(a),189(a)	STP00		+	B36
			coal comb.	1065(a)		CCC	SP		B37
			tobacco smoke	634(a)					
3.8-87S	135-48-8	**Pentacene**			375(a)				
3.8-88S	191-07-1	**Coronene**	aluminum mfr.	659(a)	98(a),188*(a)				
			auto	311(a),312*(a)	754*(r)				
		coal comb. 56(a),285(a); diesel 824(a); oil comb. 825(a)			857*(I)				
		petroleum mfr. 285(a); refuse comb. 341(a),661*(a);							
		tobacco smoke 79(a),396(a);							
		turbine 813(a),849(a); wood comb. 824(a),981*(a)							
3.8-89	190-70-5	**Benzo[*a*]coronene**	auto	822(a)					

3.8-3 3.8-14 3.8-19 3.8-22

3.8-23 3.8-25 3.8-26 3.8-27

Table 3.8. *(Continued)*

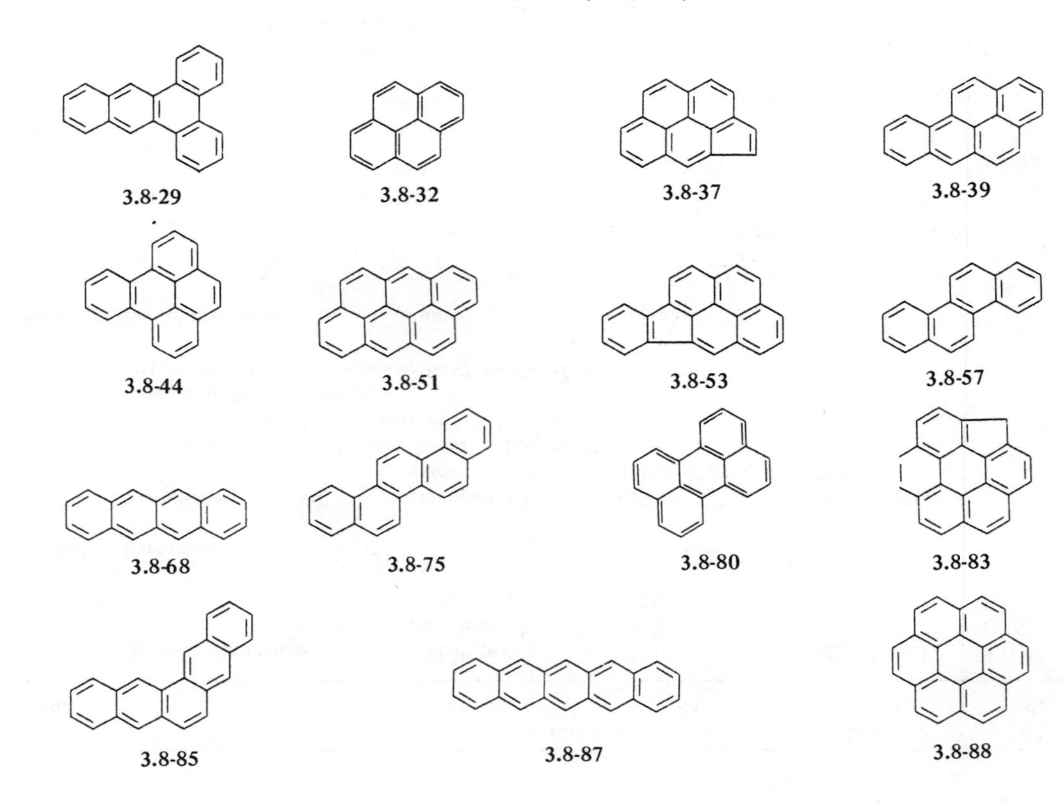

CHAPTER 4

Ethers

4.0 Introduction

The *ethers* are compounds in which two carbon atoms are joined by an oxygen atom, giving them the general formula R-O-R′, where R and R′ represent any organic group (very often one derived from an alkane). The group -O- is called the ether functional group,* the word ether being derived from a Greek word meaning "material filling heavenly space." (It was originally applied to ethyl ether because of that compounds' high volatility.) We divide atmospheric ethers into three groups: (1) alkanic (those derived from alkanes), (2) olefinic (those derived from olefins), and (3) aromatic (those derived from aromatic parent compounds).

* A *functional group* is a group of atoms defining the 'function' or mode of activity of a compound.

4.1 Alkanic Ethers

Eight alkanic ethers have been detected in the atmosphere, all in the gas phase. They are emitted from a variety of sources, none of them strong. The concentrations are low.

The alkanic ethers react with the hydroxyl radical, the process involving a relatively rapid abstraction from a C-H bond,[1221] e.g., for dimethyl ether,

$$CH_3OCH_3 + OH\cdot \rightarrow CH_3OCH_2\cdot + H_2O \qquad (R4.1\text{-}1)$$

with subsequent reactions that mimic those of the alkanes (see Chapter 3).

4.2 Olefinic Ethers

About a dozen unsaturated ethers are known to be emitted into the atmosphere, all in the gas phase. They arise primarily from the incomplete combustion of biomass and petroleum. The main loss reactions are with the OH· radical, and are rapid,[1221] the mechanism involving OH· addition at the olefinic bond. Typical products of these reactions have not been studied, but are expected to be similar to those of the alkenes (see Chapter 3).

4.3 Aromatic Ethers

25 aromatic ethers comprise this group. They originate primarily from incomplete combustion of petroleum fuels and as emittants from vegetation. None of the sources is known to be strong and atmospheric concentrations are low.

The principal loss reaction for the aromatic ethers is reaction with OH·. In the case of anisole, the OH· adds at the ortho position.[1221,1232] A product analysis for anisole or other aromatic ethers in an NO_x-air environment has not been made, but one would anticipate that a variety of benzene derivatives with mixed functional groups would result.

Table 4.1. Alkanic Ethers

Species Number	Registry Number	Name	Emission Source	Emission Ref.	Detection Ref.	Code	Bioassay −MA	Bioassay +MA	Ref.
4.1-1	115-10-6	**Dimethyl ether**	propellant	559	597,628				
4.1-2	540-67-0	**Ethyl methyl ether**			597				
4.1-3	557-17-5	**Methyl propyl ether**	landfill	628	358,597				
4.1-4	1634-04-4	*tert*-**Butyl methyl ether**	gasoline vapor	1088	597				
4.1-5	109-87-5	**Dimethoxymethane**	biomass comb.	597					
4.1-6	60-29-7	**Diethyl ether**	biomass comb.	597	358,591	STD00	NEG	−(D)	B36
			industrial	60		REC	+		B9
		landfill 628; solvent 439				SPI	NEG		B25
4.1-7	462-95-3	**Diethoxyethane**	vegetation	552e					
4.1-8	108-20-3	**Diisopropyl ether**	landfill	628	628				
4.1-9	142-96-1	**Dibutyl ether**			597				

Table 4.2. Olefinic Ethers

Species Number	Registry Number	Name	Emission Source	Emission Ref.	Detection Ref.	Code	Bioassay −MA	Bioassay +MA	Ref.
4.2-1	926-66-9	**Methyl vinyl ether**	landfill	365,628	628				
			turbine	414					
4.2-2	4181-12-8	**Ethyl vinyl ether**	biomass comb.	597	820(I)				
			turbine	414					
4.2-3	1663-35-0	**Ethyl methoxyvinyl ether**			820(I)				
4.2-4	926-65-8	**Isopropyl vinyl ether**	landfill	628	597				
4.2-5	111-34-2	**Butyl vinyl ether**	rendering	1163	358				
			turbine	414					
4.2-6	109-53-5	**Isobutyl vinyl ether**	turbine	414					
4.2-7	5363-64-4	**Hexyl vinyl ether**	turbine	414					
4.2-8	28214-64-4	**Octadecyl vinyl ether**	turbine	414					
4.2-9	109-93-3	**Divinyl ether**	biomass comb.	597					
4.2-10	627-40-7	**Allyl methyl ether**	biomass comb.	597					
4.2-11	557-40-4	**Diallyl ether**	turbine	414					

Table 4.3. Aromatic Ethers

Species Number	Registry Number	Name	Emission Source	Emission Ref.	Detection Ref.	Code	Bioassay −MA	Bioassay +MA	Bioassay Ref.
4.3-1S	100-66-3	**Anisole**	auto	217*	628				
			biomass comb.	597	399(a), 458(a)t				
		diesel 811; vegetation 888							
4.3-2	578-58-5	***o*-Methylanisole**	biomass comb.	597					
4.3-3	104-93-8	***p*-Methylanisole**	diesel	138	358				
			vegetation	507e					
4.3-4	several	**Dimethylanisole**	diesel	138					
4.3-5	several	**Trimethylanisole**	diesel	138					
4.3-6	140-67-0	**4-Allyl-1-methoxy-benzene**	vegetation	531,675					
4.3-7S	104-46-1	**Anethole**	diesel	138					
			vegetation	326, 515e					
4.3-8	91-16-7	**1,2-Dimethoxybenzene**	diesel	138					

Table 4.3. *(Continued)*

Species Number	Registry Number	Name	Emission Source	Emission Ref.	Detection Ref.	Bioassay Code	−MA	+MA	Ref.
4.3-9	93-16-3	**1,2-Dimethoxy-4-(1-propenyl)-benzene**	vegetation	514e,531					
4.3-10	150-78-7	**1,4-Dimethoxybenzene**	vegetation	533		STI00	NEG	NEG	B38
						STI35	NEG	NEG	B38
						STI37	NEG	NEG	B38
						STI98	NEG	NEG	B38
4.3-11	707-07-3	**Methyltrimethoxy-benzene**	biomass comb.	1089(a)					
4.3-12	487-11-6	**5-Allyl-1,2,3-trimethoxy-benzene**	vegetation	531					
4.3-13	487-12-7	**1,2,3-Trimethoxy-5-(1-propenyl) benzene**	vegetation	533					
4.3-14	103-73-1	**Ethoxybenzene**	biomass comb.	891(a)t					
4.3-15	622-85-5	**Propoxybenzene**	refuse comb.	840(a)					
4.3-16S	538-86-3	**Benzyl methyl ether**			628				
4.3-17	539-30-0	**Benzyl ethyl ether**	vegetation	552e					
4.3-18S	3558-60-9	**α-Methoxy-2-phenylethane**	vegetation	531					
4.3-19	4013-37-0	**1,2-Dimethoxy-1-phenylethane**	vegetation	514e					

Species Number	Registry Number	Name	Emission Source	Emission Ref.	Detection Ref.	Bioassay Code	Bioassay −MA	Bioassay +MA	Bioassay Ref.
4.3-20S	101-84-8	**Biphenyl ether**	combustion	431(a)	597	STI00	NEG	NEG	B38
			phthal anhyd. mfr.	6		STI35	NEG	NEG	B38
			turbine	414		STI37	NEG	NEG	B38
						STI98	NEG	NEG	B38
4.3-21	30921-17-6	**Diphenoxybenzene**			597t				
4.3-22	28299-41-4	**Bitolyl ether**			597t				
4.3-23	73493-71-7	**Methoxyfluorene**			832(a)				
4.3-24	61128-87-8	**Methoxyphenanthrene**	diesel	309(a)					
			turbine	786*(a)t					

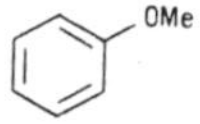

OMe

4.3-1

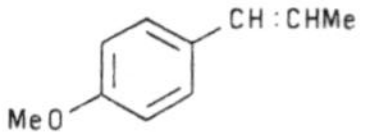

CH : CHMe
MeO

4.3-7

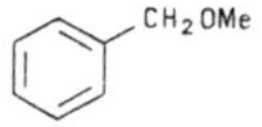

CH_2OMe

4.3-16

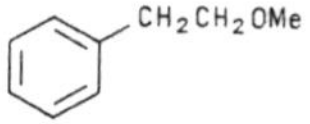

CH_2CH_2OMe

4.3-18

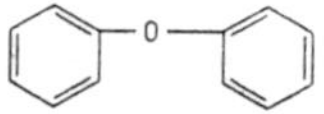

O

4.3-20

CHAPTER 5

Alcohols

5.0 Introduction

Alcohols are organic compounds in which the hydroxy (–OH) functional group is present. The hydroxy group is quite common in atmospheric organic compounds, particularly those from vegetative sources. We divide atmospheric alcohols into four groups: (1) alkanic, (2) olefinic, (3) cyclic, and (4) aromatic.

Some of the alcohols contain molecular groupings which would qualify them for citation as ethers as well as alcohols. For example, biomass combustion produces the compound 1,3-dimethoxy-2-propanol:

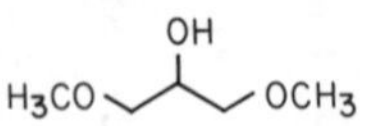

OH

H_3CO OCH_3

The compound is placed in this chapter rather than in Chapter 4 on the basis of its *principal group*, i.e., the functional group with the highest seniority. The seniority in a mixed group compound is specified by CA and IUPAC; for the compounds in this book the order in which the groups have increasing priority for citation as the principal group (and the order in which they are tabulated) is ethers, alcohols, ketones, aldehydes, esters, anhydrides, and acids. The order of seniority is based approximately on reactivity, that is, in a compound with mixed functional groups the functional group most likely to participate in chemical reactions is generally the most senior.

5.1 Alkanic Alcohols

Over seventy alkanic alcohols are present in the atmosphere, some three-quarters of them in the aerosol phase. A number have been detected in indoor air and four of the smaller ones have been found in precipitation. Ionic derivatives of methanol have been tentatively identified in the stratosphere.

The C_1 to C_5 alkanic alcohols have numerous sources. They see wide use in industrial applications and are also emitted from biological processes. It is anticipated that they will be increasingly used as gasoline substitutes or supplements; if so, their atmospheric concentrations will increase. Vegetative sources emit a number of alcohols with carbon numbers as high as eighteen. No direct sources are known for alcohols of higher carbon number, but many are detected; this suggests that they result from the chemical transformation of precursor molecules rather than from direct emission.

The favored gas phase reaction of the alkanic alcohols is with the hydroxyl radical. H atom abstraction occurs at a C-H bond adjacent to the hydroxyl group, followed by reaction with O_2 to abstract H from the now weak O-H bond.[1221] For ethanol, for example,

$$CH_3CH_2OH + OH\cdot \rightarrow CH_3CHOH\cdot + H_2O \qquad \text{(R5.1-1)}$$

$$CH_3CHOH\cdot + O_2 \rightarrow CH_3CHO + HO_2\cdot \qquad \text{(R5.1-2)}$$

5.2 Olefinic Alcohols

More than thirty of these compounds are known to enter the atmosphere in the gas phase, though few have been detected. The smaller compounds are emitted from diesels and turbines as a consequence of incomplete combustion of petroleum fuels. Most of the larger molecules are alkenes with C_6, C_8, C_{10}, and C_{12} backbones that have been emitted by vegetation.

Olefinic alcohols are expected to react in the atmosphere primarily with OH·. A kinetic study of allyl alcohol indicates that the OH· reaction proceeds mainly by addition to the double bond.[1221] Subsequent reactions are expected to be analogous to those of the alkenes (see Chapter 3).

5.3 Cyclic Alcohols

Cyclic alcohols are emitted at relatively high rates from trees and other vegetation. Together with terpenes, they are responsible for the fragrance identified with forests. More than thirty of these compounds may be present in the atmosphere, mostly from vegetation but a few from biomass combustion and other minor sources. About a third of the compounds have been detected in the aerosol phase.

The atmospheric chemistry of the cyclic alcohols is unexplored. It is anticipated that the primary reactions are with OH· and O_3, with the reactions proceeding similarly to those of the monocyclic hydrocarbons and terpenes (see Chapter 3).

5.4 Aromatic Alcohols

There are more than ninety atmospheric aromatic alcohols, a notably large number. The most prolific sources are the combustion of petroleum fuels or biomass, but some of the alcohols are produced by vegetation. Phenol and the cresols have many industrial uses and, as a result, many anthropogenic sources. Not many of the compounds have been detected in ambient air; the majority of those which have been detected are found in the aerosol phase.

The gas phase atmospheric chemistry of phenol and the cresols has been worked out in some detail, despite its complexity. The OH· and NO_3· radicals compete as principal reactants in urban areas. The hydroxyl radical adds to the aromatic ring,[1221,1222] preferably at the site of the existing hydroxyl ligand,[1231] while the nitrate radical abstracts the hydrogen atom from the O-H group.[1234] A tentative reaction sequence is shown in Figure 5.4-1. There are two major routes of reaction. The first begins by OH· addition and favors a sequence ending in ring cleavage to produce small multifunctional organics. Nitrocresol is a minor product. The second route is initiated by NO_3· reaction and proceeds directly to nitrocresol. Similar reaction sequences are likely for all of the aromatic alcohols. A possible indication of the accuracy of these sequences is the detection in the aerosol phase of several of the nitrocresols (see Chapter 11).

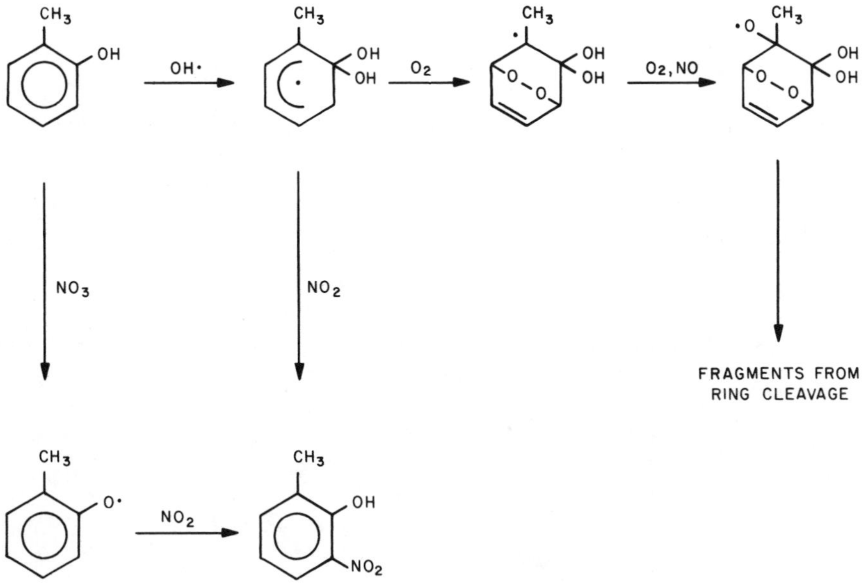

Fig. 5.4-1 A tentative sequence for the gas phase atmospheric reactions of *o*-cresol.

Table 5.1. Alkanic Alcohols

Species Number	Registry Number	Name	Emission Source	Emission Ref.	Detection Ref.	Code	Bioassay −MA	Bioassay +MA	Bioassay Ref.
5.1-1	67-56-1	**Methanol**	acrylonitrile mfr.	626	90*, 376	SCE	NEG		B6
			animal waste	25,80	820(I), 1123*(I)	CTV	NEG		B24
		auto 55,217*; biomass comb. 183,354; charcoal mfr. 58*;				CTP	NEG		B24
		insects 471; microbes 302; paint 695, 1071; petroleum mfr. 58*;				NEU	NEG		B32
		plastics comb. 354; plastics mfr. 695; polymer mfr. 936;							
		printing 543*; refuse comb. 439, 26(a); rendering 52;							
		sewage tmt. 695, 1071; solvent 78, 88; starch mfr. 695,1071;							
		tobacco smoke 396,446; turbine 414; vegetation 552e;							
		volcano 234; wood pulping 19, 267							
5.1-2S	17836-08-7	**Protonated methanol ion**			1167(S)t				
5.1-3S		**Protonated methanol-water ion**			1167(S)t				
5.1-4	64-17-5	**Ethanol**	acrylonitrile mfr.	626	358t,591	TRM	+		B21
			animal waste	25,80	820(I),1123*(I)	CCC	LN		B37
		auto 55,217*; biomass comb. 376,439; building mtls. 856;				STP00	NEG	NEG	B36
		chemical mfr. 1157; fish processing 695,1071; insects 471;				STP35	NEG	NEG	B36
		microbes 302; onion odor 636; paint 695,1071;				STP37	NEG	NEG	B36
		petroleum mfr. 58*; plastics comb. 354; plastics mfr. 695;				STP98	NEG	NEG	B36
		printing 543*; refuse comb. 439,665*, 26(a); rendering 52;				SCE	NEG		B6
		sewage tmt. 695,1071; solvent 88,439; starch mfr. 695,1071;				CYC	NEG		B8
		tobacco smoke 396; turbine 414; vegetation 552e;				VIC	+		B18
		volcano 234; whiskey mfr. 32, 1158; wood pulping 19,267				ARA	NEG		B17
						MNT	?/−		B29

Species Number	Registry Number	Name	Emission Source	Ref.	Detection Ref.	Bioassay Code	−MA	+MA	Ref.
						ASPD	NEG		B10
						CTV	NEG		B24
						CTP	NEG		B24
						NEU	NEG		B32
						SPI	NEG		B25
5.1-5	110-80-5	**2-Ethoxyethanol**	solvent	134	1110*(I)				
5.1-6S	621-63-6	**2-bis(Ethoxy)ethanol**			820(I)				
5.1-7S	111-46-6	**Diethylene glycol**	industrial	60					
			solvent	439					
			tobacco smoke	396					
5.1-8S	112-27-6	**Triethylene glycol**	building mtls.	9*					
			industrial	60					
			tobacco smoke	396					
5.1-9	111-76-2	**2-Butoxyethanol**	industrial	60					
5.1-10		**Dodecoxyethanol**			1086(a)t				
5.1-11	71-23-8	**1-Propanol**	animal waste	25,80	1258*	SCE	NEG		B6
			building mtls.	856					
		diesel 176; microbes 302; onion odor 636; paint 695,1071; plastics comb. 354; printing 543*; rendering 52; sewage tmt. 422*; solvent 78,382; starch mfr. 695,1071; vegetation 552e; volcano 234; whiskey mfr. 32,1158; wood pulping 267,545							

Table 5.1. *(Continued)*

Species Number	Registry Number	Name	Emission Source	Emission Ref.	Detection Ref.	Bioassay Code	Bioassay −MA	Bioassay +MA	Bioassay Ref.
5.1-12	1589-49-7	**3-Methoxy-1-propanol**	polymer comb.	1107(a)					
5.1-13	57-55-6	**1,2-Propanediol**	industrial	60		STI00	NEG	NEG	B38
						STI35	NEG	NEG	B38
						STI37	NEG	NEG	B38
						STI98	NEG	NEG	B38
5.1-14	623-39-2	**3-Methoxy-1,2-propanediol**	biomass comb.	1089(a)					
			polymer comb.	1107(a)					
5.1-15	56-81-5	**1,2,3-Trihydroxypropane**	biomass comb.	597		STI00	?	?	B38
			industrial	60,618		STI35	NEG	NEG	B38
		printing 58*; tobacco smoke 396				STI37	NEG	NEG	B38
						STI98	NEG	NEG	B38
5.1-16	67-63-0	**2-Propanol**	animal waste	25,80	358,591	NEU	NEG		B32
			auto	284	820(I)				
		biomass comb. 597; paint 695,1071; petroleum mfr. 58; plastics comb. 354; printing 543*; sewage tmt. 422; solvent 88,134; starch mfr. 695,1071; vegetation 552e; volcano 234; whiskey mfr. 1158; wood pulping 267,545							
5.1-17	623-69-8	**1,3-Dimethoxy-2-propanol**	biomass comb.	1089(a)					

Species Number	Registry Number	Name	Emission Source	Emission Ref.	Detection Ref.	Code	Bioassay −MA	Bioassay +MA	Ref.
5.1-18	71-36-3	**1-Butanol**	animal waste	25,80	33*,566*	SCE	NEG		B6
			building mtls.	856	820(I),940(I)				
		insects 471; microbes 302; paint 695,1071; rendering 52; sewage tmt. 422*,1071; solvent 88,439; starch mfr. 695,1071; turbine 414; whiskey mfr. 1158; wood pulping 19,267							
5.1-19	78-83-1	**Isobutanol**	animal waste	25,80	597				
			microbes	302	1110*(I)				
		paint 695; petroleum mfr. 58*,221; solvent 88,439; vegetation 552e; whiskey mfr. 32,1158; wood pulping 19,267							
5.1-20	75-65-0	*tert*-**Butyl alcohol**	industrial	60	628,1258*	NEU	NEG		B32
			petroleum mfr.	58*					
			turbine	414					
5.1-21	137-32-6	**2-Methyl-1-butanol**	vegetation	552e					
5.1-22	97-95-0	**2-Ethyl-1-butanol**			820(I)				
5.1-23	78-92-2	**2-Butanol**	auto	284					
			industrial	60					
		paint 695,1071; sewage tmt. 1071; starch mfr. 695,1071; wood pulping 695,1071							
5.1-24	594-60-5	**2,3-Dimethyl-2-butanol**	animal waste	855					
5.1-25	464-07-3	**3,3-Dimethyl-2-butanol**			628				

Table 5.1. *(Continued)*

Species Number	Registry Number	Name	Emission Source	Emission Ref.	Detection Ref.	Bioassay Code	Bioassay −MA	Bioassay +MA	Bioassay Ref.
5.1-26	71-41-0	**1-Pentanol**	animal waste building mtls.	855 856	820(I),1123*(I)				
		paint 695,1071; rendering 52,833; sewage tmt. 1071; solvent 598; turbine 414; vegetation 552e; wood pulping 19							
5.1-27	123-51-3	**Isopentanol**	animal waste industrial	80,855 108	597t				
		microbes 302; turbine 414; vegetation 552e; whiskey mfr. 32							
5.1-28	105-30-6	**2-Methyl-1-pentanol**	turbine	414	820(I)				
5.1-29	589-35-5	**3-Methyl-1-pentanol**	animal waste turbine	855t 414					
5.1-30	106-67-2	**2-Ethyl-4-methyl-1-pentanol**			1113(I)				
5.1-31	19876-64-3	**4-(*p*-Tolyl)-1-pentanol**	wood pulping	267,545					
5.1-32	111-29-5	**1,5-Pentanediol**			532(a)				
5.1-33	6032-29-7	**2-Pentanol**	animal waste vegetation	899 552e	597t				
5.1-34	42027-23-6	**2,3-Pentanediol**	coffee mfr.	598					
5.1-35	584-02-1	**3-Pentanol**	turbine	414					

Species Number	Registry Number	Name	Emission Source	Emission Ref.	Detection Ref.	Code	Bioassay −MA	Bioassay +MA	Ref.
5.1-36	70561-57-8M	**Pentahydroxypentane**			697(a)				
5.1-37	111-27-3	**1-Hexanol**	building mtls.	856	628				
			rendering	52	820(I),1108(I)				
			vegetation	533,552e					
5.1-38	104-76-7	**2-Ethylhexanol**	industrial	60	820(I),1040(I)				
			refuse comb.	840(a)					
		river water odor 202; turbine 414							
5.1-39	40868-73-3	**2-Ethoxyhexanol**	industrial	60					
5.1-40	626-93-7	**2-Hexanol**	vegetation	552e					
5.1-41	629-11-8	**1,6-Hexanediol**			532(a)				
5.1-42	several	**Dimethyl-2,5-hexanediol**	turbine	414					
5.1-43	70561-59-0M	**1,2,3,4,5,6-Hexahydroxy-hexane**			697(a)				
5.1-44	111-70-6	**1-Heptanol**	turbine	414	628				
			vegetation	514e,552e					
5.1-45	10042-59-8	**2-Propyl-1-heptanol**	turbine	414					
5.1-46	543-49-7	**2-Heptanol**	turbine	414					

Table 5.1. *(Continued)*

Species Number	Registry Number	Name	Emission Source	Emission Ref.	Detection Ref.	Bioassay Code	Bioassay −MA	Bioassay +MA	Bioassay Ref.
5.1-47	108-82-7	**2,6-Dimethyl-4-heptanol**	river water odor	202					
5.1-48	629-30-1	**1,7-Heptanediol**			532(a)				
5.1-49	70532-58-0M	**1,2,3,4,5-Pentahydroxy-heptane**			697(a)t				
5.1-50	70532-59-1M	**1,2,3,4,6-Pentahydroxy-heptane**			697(a)t				
5.1-51	70532-60-4M	**1,2,3,5,7-Pentahydroxy-heptane**			697(a)t				
5.1-52	70532-62-6M	**1,2,3,4,5,6-Hexa-hydroxyheptane**			697(a)t				
5.1-53	111-87-5	**1-Octanol**	microbes turbine vegetation	302 414 503e,552e	591,881* 820(I)t,1112(I)t				
5.1-54	20296-29-1	**3-Octanol**	microbes vegetation	302 552e					
5.1-55	143-08-8	**Nonanol**	vegetation	503e,552e					
5.1-56	40589-14-8	**2-Methylnonanol**	industrial	60					

Species Number	Registry Number	Name	Emission Source	Emission Ref.	Detection Ref.	Code	Bioassay −MA	Bioassay +MA	Ref.
5.1-57	36729-58-5	**Decanol**	turbine vegetation	414 503e,552e					
5.1-58	27342-88-7	**Dodecanol**	vegetation	503e	955*(a)				
5.1-59	27196-00-5	**Tetradecanol**	vegetation	514e	955*(a),1073*(a)				
5.1-60	2422-98-2	**5-Pentadecanol**	auto	309(a)					
5.1-61	29354-98-1	**Hexadecanol**			848*(a),1073*(a) 916(I)				
5.1-62	52783-44-5	**Heptadecanol**			848(a),1086(a)				
5.1-63	26762-44-7	**Octadecanol**	vegetation	552e	848*(a),1073*(a) 916(I)				
5.1-64	52783-43-4	**Nonadecanol**			848(a)				
5.1-65	28679-05-2	**Eicosanol**			848*(a),1073*(a)				
5.1-66	15594-90-8	**Heneicosanol**			848(a),1073*(a)				
5.1-67	30303-65-2	**Docosanol**	biomass comb.	1089(a)	848*(a),901*(a)				
5.1-68	3133-01-5	**Tricosanol**			848(a),1073(a)				
5.1-69	52783-45-6	**Tetracosanol**			848*(a),901*(a)				

Table 5.1. *(Continued)*

Species Number	Registry Number	Name	Emission Source	Emission Ref.	Detection Ref.	Code	Bioassay −MA	Bioassay +MA	Bioassay Ref.
5.1-70	26040-98-2	**Pentacosanol**			848(a),1073(a)				
5.1-71	28346-64-7	**Hexacosanol**			848*(a),901*(a)				
5.1-72	70679-23-1	**2-Methylhexacosanol**			901*(a)t				
5.1-73	2004-39-9	**Heptacosanol**			848(a),1073(a)				
5.1-74	68580-63-2	**Octacosanol**			848*(a),901*(a)				
5.1-75	70679-24-2	**2-Methyloctacosanol**			901*(a)t				
5.1-76	25154-56-7	**Nonacosanol**			848(a),1073*(a)				
5.1-77	28351-05-5	**Triacontanol**			848*(a),901*(a)				
5.1-78	26444-39-3	**Hentriacontanol**			1073*(a)				
5.1-79	79554-32-8	**Dotriacontanol**			848*(a),1073*(a)				

$H^+ \cdot CH_3OH$
5.1-2

$H^+ \cdot CH_3OH \cdot (H_2O)_n$
5.1-3

$HOCH_2CH(OEt)_2$
5.1-6

$HOCH_2CH_2OCH_2CH_2OH$
5.1-7

$HOCH_2CH_2OCH_2CH_2OCH_2CH_2OH$
5.1-8

Table 5.2. Olefinic Alcohols

Species Number	Registry Number	Name	Emission Source	Emission Ref.	Detection Ref.	Bioassay Code	Bioassay −MA	Bioassay +MA	Bioassay Ref.
5.2-1S	107-21-1	**Ethylene glycol**	building mtls.	9*		STP00	NEG	NEG	B36
			industrial	60		STP35	NEG	NEG	B36
		solvent 598; tobacco smoke 396				STP37	NEG	NEG	B36
						STP98	NEG	NEG	B36
						NEU	NEG		B32
5.2-2		**2-Ethoxyethen-1-ol**	solvent	88,442					
5.2-3		**2-Butoxyethen-1-ol**	solvent	88					
5.2-4		**Vinyl 1,2-dihydroxyvinyl ether**			628				
5.2-5	107-18-6	**2-Propen-1-ol**	acrylonitrile mfr.	626	1123*(I)				
5.2-6	6117-91-5	**2-Buten-1-ol**	auto	217*					
			turbine	414					
5.2-7	556-82-1	**3-Methyl-2-buten-1-ol**	turbine	414					
5.2-8	115-18-4	**2-Methyl-3-buten-2-ol**	vegetation	552e,888t					
5.2-9	10473-14-0	**3-Methyl-3-buten-2-ol**	turbine	414	628t				
5.2-10	497-06-3	**3-Buten-1,2-diol**	turbine	414					

Table 5.2. *(Continued)*

Species Number	Registry Number	Name	Emission Source	Emission Ref.	Detection Ref.	Code	Bioassay −MA	Bioassay +MA	Bioassay Ref.
5.2-11	821-09-0	**4-Penten-1-ol**	turbine	414					
5.2-12	77-75-8	**3-Methyl-1-pentyn-3-ol**	diesel	1					
5.2-13	70572-99-5M	**3-Hydroxymethyl-4,5,6-trihydroxy-1-hexene**			697(a)t				
5.2-14	2305-21-7	**2-Hexen-1-ol**	turbine vegetation	414 510e,552e					
5.2-15	544-12-7	**3-Hexen-1-ol**	vegetation	533,888					
5.2-16	111-28-4	**2,4-Hexadien-1-ol**	turbine	414					
5.2-17	105-31-7	**1-Hexyn-3-ol**	turbine	414					
5.2-18	22104-78-5	**2-Octen-1-ol**	microbes	302					
5.2-19	106-22-9	**3,7-Dimethyl-6-octen-1-ol**	vegetation	498,512e					
5.2-20	543-39-5	**2-Methyl-6-methylene-7-octen-2-ol**	vegetation	517e					
5.2-21	3391-86-4	**1-Octen-3-ol**	microbes vegetation	302,602 552e					
5.2-22	50306-18-8	*cis*-**1,5-Octadien-3-ol**	microbes	602					

Species Number	Registry Number	Name	Emission Source	Emission Ref.	Detection Ref.	Code	Bioassay −MA	Bioassay +MA	Bioassay Ref.
5.2-23	78-70-6	**3,7-Dimethyl-1,6-octadien-3-ol**	vegetation wood pulping	474,531 267,545	886*				
5.2-24	50306-14-4	*trans* **-1,5-Octadien-3-ol**	microbes	602					
5.2-25	106-25-2	**3,7-Dimethyl-*cis* -2,6-octadien-1-ol**	vegetation	505e,552e					
5.2-26	106-24-1	**3,7-Dimethyl-*trans* -2,6-octadien-1-ol**	vegetation	498,506e					
5.2-27	6994-89-4	*trans* **-2-Methyl-6-methylene-3,7-octadien-2-ol**	vegetation	552e					
5.2-28	70532-57-9M	**1,3,6-Trihydroxy-1-nonene**			697(a)t				
5.2-29	13019-22-2	**9-Decen-1-ol**	vegetation	503e					
5.2-30	4602-84-0	**3,7,11-Trimethyl-2,6,10-dodecatrien-1-ol**	vegetation	514e					
5.2-31	142-50-7	**3,7,11-Trimethyl-1,6,10-dodecatrien-2-ol**	vegetation	514e					
5.2-32S	150-86-7	**Phytol**			848*(a)				

$HOCH_2CH_2OH$

5.2-1

$HOCH_2CH.CMe(CH_2)_3CHMe(CH_2)_3CHMe(CH_2)_3CHMe_2$

5.2-32

Table 5.3. Cyclic Alcohols

Species Number	Registry Number	Name	Emission Source	Emission Ref.	Detection Ref.	Bioassay Code	Bioassay −MA	Bioassay +MA	Bioassay Ref.
5.3-1	96-41-3	**Cyclopentanol**	diesel	1					
5.3-2	27583-37-5	**3-Methyl-1,2-cyclopentanediol**	turbine	414					
5.3-3	108-93-0	**Cyclohexanol**	industrial	60	358	ALC	NEG		B19
			solvent	439		STI00	NEG	NEG	B38
			turbine	414		STI35	NEG	NEG	B38
						STI37	NEG	NEG	B38
						STI98	?	?	B38
5.3-4	25639-42-3	**Methylcyclohexanol**	turbine	414					
5.3-5	70561-58-9M	**1,2,4,5-Tetrahydroxy-cyclohexane**			697(a)				
5.3-6	6917-35-7	**Hexahydroxycyclo-hexane**			697(a)				
5.3-7S	2216-51-5	***l*-Menthol**	vegetation	514e,516e					
5.3-8S	1632-73-1	**Fenchyl alcohol**	tap water odor	555					
			vegetation	531					
			wood pulping	267,545					

Species Number	Registry Number	Name	Emission Source	Emission Ref.	Detection Ref.	Code	Bioassay −MA	Bioassay +MA	Bioassay Ref.
5.3-9S	498-81-7	***p*-Menthan-8-ol**	vegetation	508e					
5.3-10S	586-27-6	***l*-Carvenol**	vegetation	506e					
5.3-11S	89-79-2	**Isopulegol**	vegetation	514e					
5.3-12S	10482-56-1	**α-Terpineol**	vegetation	515e,552e	886*				
			wood pulping	267,545	820(I)t				
5.3-13S	138-87-4	**β-Terpineol**	vegetation	515e,517e					
5.3-14S	586-81-2	**γ-Terpineol**	vegetation	515e,517e					
5.3-15S	562-74-3	**Terpinen-4-ol**	vegetation	552e					
			wood pulping	545					
5.3-16S	115-71-9	**α-Santalol**	vegetation	510e,531					
5.3-17S	18647-78-4	**Civetol**	animal waste	510e					
5.3-18S	489-86-1	**Guaiol**	vegetation	510e,531					
5.3-19S	19700-21-1	**Geosmin**	microbes	664					
			tap water odor	555					
			vegetation	555					
5.3-20S	11070-72-7	**Cadinol**	vegetation	514e					
5.3-21S	77-53-2	**Cedrol**	vegetation	510e					

Table 5.3. *(Continued)*

Species Number	Registry Number	Name	Emission Source	Emission Ref.	Detection Ref.	Code	Bioassay −MA	Bioassay +MA	Bioassay Ref.
5.3-22S	28231-03-0	**Cedrenol**	vegetation	514e					
5.3-23S	472-97-9	**Caryophyllene alcohol**	vegetation	510e					
5.3-24S	57-88-5	**Cholesterol**	tobacco smoke	396(a)	848*(a),851*(a)				
5.3-25S	83-48-7	**Stigmasterol**	tobacco smoke	396(a),542(a)	848*(a),905(a)				
5.3-26S	474-62-4	**Campesterol**			848*(a),905(a)				
5.3-27	474-67-9	**Brassicasterol**			848*(a),851(a)t				
5.3-28S	17605-67-3	**Fucosterol**			851(a)				
5.3-29	18472-36-1	**Avenasterol**			851(a)				
5.3-30S	83-46-5	**β-Sitosterol**	tobacco smoke veneer drying	396(a),542(a) 68(a)	848*(a),851*(a)				
5.3-31S	83-47-6	**γ-Sitosterol**	tobacco smoke	396(a)					

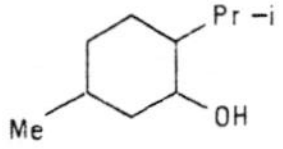

Pr –i
Me
OH

5.3-7

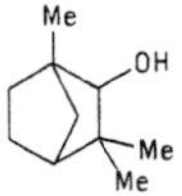

Me
OH
Me
Me

5.3-8

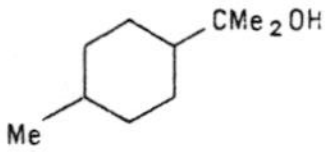

CMe $_2$ OH
Me

5.3-9

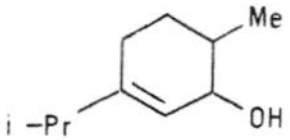

Me
i –Pr
OH

5.3-10

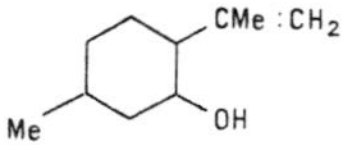

CMe : CH_2
Me
OH

5.3-11

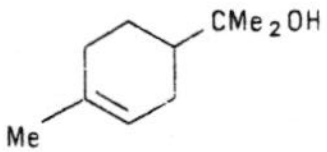

CMe $_2$ OH
Me

5.3-12

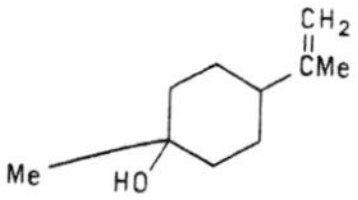

CH_2
||
CMe
Me
HO

5.3-13

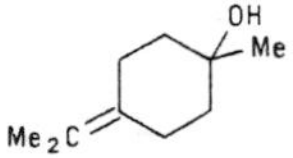

OH
Me
Me_2C

5.3-14

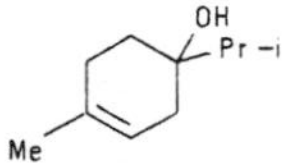

OH
Pr –i
Me

5.3-15

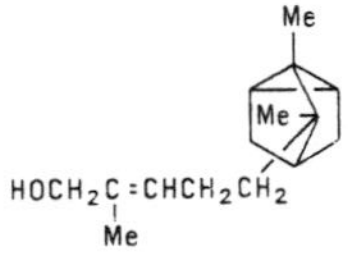

Me
Me
$HOCH_2C{=}CHCH_2CH_2$
Me

5.3-16

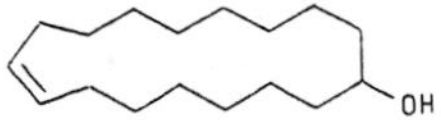

OH

5.3-17

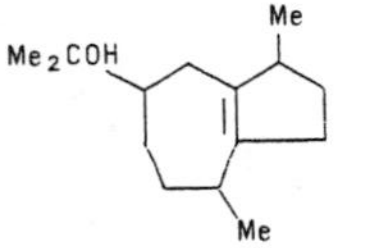

Me
Me_2COH
Me

5.3-18

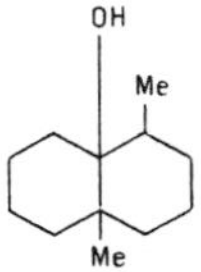

OH
Me
Me

5.3-19

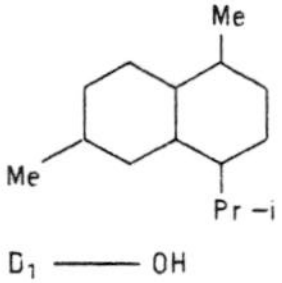

Me
Me
Pr –i
D_1 —— OH

5.3-20

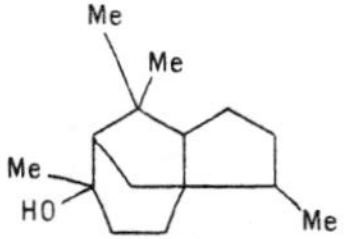

Me
Me
Me
HO
Me

5.3-21

Table 5.3. (*Continued*)

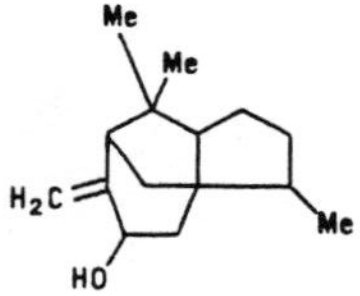

Me Me H_2C HO Me

5.3-22

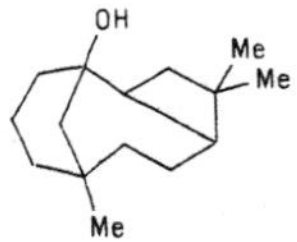

OH Me Me Me

5.3-23

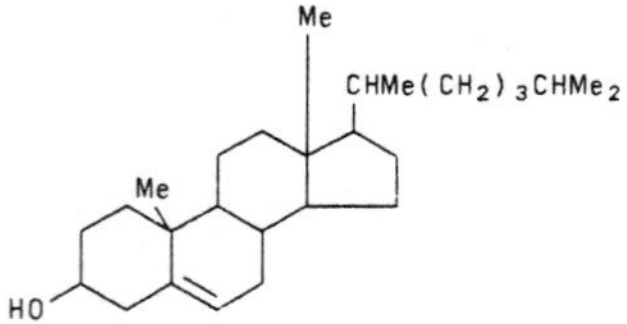

Me $CHMe(CH_2)_3CHMe_2$ Me HO

5.3-24

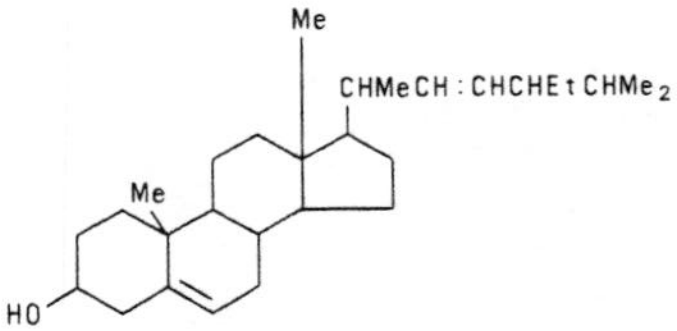

Me $CHMeCH:CHCHEtCHMe_2$ Me HO

5.3-25

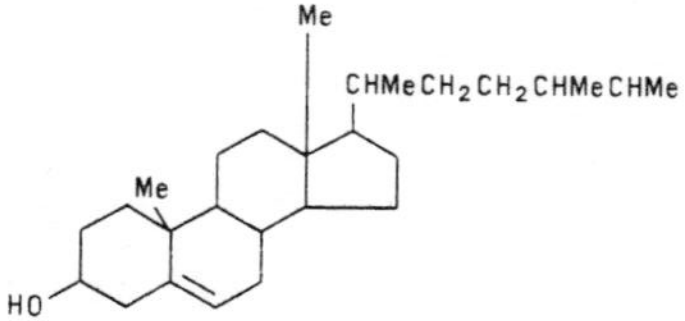

Me $CHMeCH_2CH_2CHMeCHMe$ Me HO

5.3-26

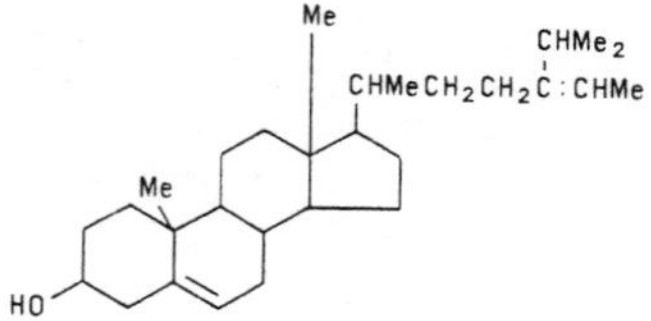

Me $CHMe_2$ $CHMeCH_2CH_2C:CHMe$ Me HO

5.3-28

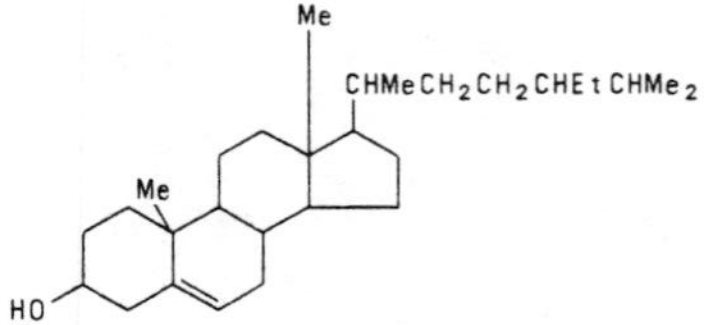

Me $CHMeCH_2CH_2CHEtCHMe_2$ Me HO

5.3-30

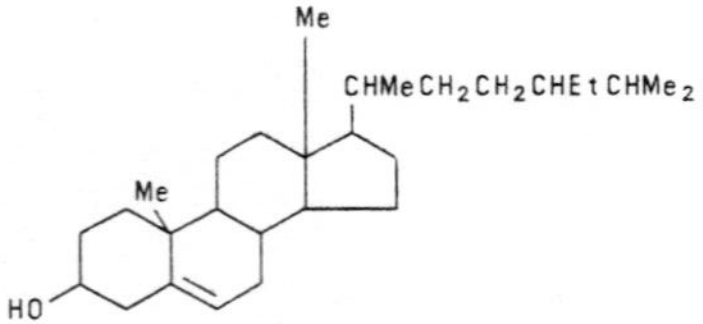

Me $CHMeCH_2CH_2CHEtCHMe_2$ Me HO

5.3-31

Table 5.4. Aromatic Alcohols

Species Number	Registry Number	Name	Emission Source	Emission Ref.	Detection Ref.	Bioassay Code	−MA	+MA	Ref.
5.4-1S	108-95-2	**Phenol**	animal waste	695,855	413*,652*	STI00	NEG	NEG	B38
			auto	95,537,87(a)	214(a),458(a)	STI35	NEG	NEG	B38
		biomass comb. 597,891(a); brewing 144; chemical mfr. 1164;			755*(r)	STI37	NEG	NEG	B38
		coal comb. 1159; diesel 1,137; fish processing 695,1071;			820(I),1123*(I)	STI98	NEG	NEG	B38
		foundry 97*,279*; glass fiber mfr. 122*,1076; lacquer mfr. 427;				NEU	NEG		B32
		oil comb. 624; paint 695,1071; plastics comb. 46;							
		plastics mfr. 695,1076; refuse comb. 26(a),840(a); sewage tmt. 695,1071;							
		solvent 121,439; starch mfr. 695,1071; tap water odor 555;							
		tobacco smoke 4,396; turbine 786(a); wood preserv. 1048*(a);							
		wood pulping 267,545							
5.4-2S	95-48-7	***o*-Cresol**	auto	537,789*,87(a)	633,652*	ALC	+/−		B19
			biomass comb.	597	458(a),973(a)t	STI00	NEG	NEG	B38
		coal comb. 1159; diesel 138,808,809*(a); paint 695,1071;			755*(r)	STI35	NEG	NEG	B38
		starch mfr. 695,1071; tobacco smoke 396,789*; wood pulping 267,545				STI37	NEG	NEG	B38
						STI98	NEG	NEG	B38
5.4-3	108-39-4	***m*-Cresol**	auto	789*,971,87(a)		ALC	NEG		B19
			biomass comb.	891(a)		STI00	NEG	NEG	B38
		coal comb. 1159; diesel 138; glass fiber mfr. 122*;				STI35	NEG	NEG	B38
		tap water odor 555; tobacco smoke 396,789*; wood pulping 267,545				STI37	NEG	NEG	B38
						STI98	NEG	NEG	B38

Table 5.4. *(Continued)*

Species Number	Registry Number	Name	Emission Source	Emission Ref.	Detection Ref.	Bioassay Code	−MA	+MA	Ref.
5.4-4	106-44-5	***p*-Cresol**	animal waste	855,892(a)	1005(a)t	ALC	+/−		B19
			auto	789*,999,87(a)	755*(r)	STI00	NEG	NEG	B38
		brewing 144; coal comb. 1159; diesel 137,808*,809*(a);			820(I)t	STI35	NEG	NEG	B38
		glass fiber mfr. 122*; solvent 598; tobacco smoke 396,789*;				STI37	NEG	NEG	B38
		turbine 786*(a); vegetation 507e; wood pulping 267,545				STI98	NEG	NEG	B38
5.4-5S	526-75-0	**2,3-Xylenol**	auto	537,880t,87(a)	973t				
			brewing	144					
		diesel 137,808,809(a); solvent 598							
5.4-6	105-67-9	**2,4-Xylenol**	auto	999,87(a)					
			diesel	808,809,809(a)					
			tobacco smoke	396(a)					
5.4-7	95-87-4	**2,5-Xylenol**	auto	999,87(a)					
			diesel	808,809,809(a)					
5.4-8	576-26-1	**2,6-Xylenol**	auto	971,999,87(a)					
			paint	695,1071					
5.4-9	95-65-8	**3,4-Xylenol**	auto	971,999,87(a)					
			diesel	808*					
5.4-10	108-68-9	**3,5-Xylenol**	auto	999,87(a)		ALC	+/−		B19
			brewing	144					
		diesel 808,809,809(a); tobacco smoke 396(a)							

Species Number	Registry Number	Name	Emission Source	Emission Ref.	Detection Ref.	Code	Bioassay −MA	Bioassay +MA	Bioassay Ref.
5.4-11	527-60-6	**2,4,6-Trimethylphenol**	auto	537,87(a)	627(a)				
			diesel	138					
			tobacco smoke	396(a)					
5.4-12	66586-93-4	**Tetramethylphenol**	diesel	138					
5.4-13	90-00-6	***o*-Ethylphenol**	animal waste	695,1071,892(a)	628				
			auto	537,971,87(a)					
5.4-14	620-17-7	***m*-Ethylphenol**	animal waste	892(a)					
			auto	971,87(a)					
5.4-15	123-07-9	***p*-Ethylphenol**	animal waste	855					
			auto	971,87(a)					
		biomass comb. 891(a)t; coal comb. 1159; paint 695,1071							
5.4-16	30230-52-5	**Ethylmethylphenol**			628				
					891(a)t				
5.4-17	31019-46-2	**Propylphenol**	auto	537					
5.4-18	99-89-8	**4-Isopropylphenol**			628				
5.4-19	89-83-8	**2-Isopropyl-5-methylphenol**	vegetation	510e,531		ALC	NEG		B19
						SRL	?		B30
5.4-20	499-75-2	**3-Isopropyl-5-methylphenol**	vegetation	510e,531					

Table 5.4. *(Continued)*

Species Number	Registry Number	Name	Emission Source	Emission Ref.	Detection Ref.	Bioassay Code	Bioassay −MA	Bioassay +MA	Bioassay Ref.
5.4-21	501-92-8	**4-Allylphenol**	diesel	138					
			vegetation	505e,531					
5.4-22	several	**Methyl-4-allylphenol**	vegetation	1170					
5.4-23	31195-95-6	**Isobutylphenol**	refuse comb.	840(a)					
5.4-24	88-18-6	**2-*tert*-Butylphenol**	coal comb.	825(a)					
5.4-25	121-00-6	**2-*tert*-Butyl-4-methoxyphenol**	coal comb.	828(a)t					
5.4-26	26746-38-3	**Di-*tert*-butylphenol**			1086(a)				
5.4-27	128-37-0	**2,6-Di-*tert*-butyl-4-methylphenol**	auto	311(a)	358	STP00	−(C)	−(C)	B36
			coal comb.	828t,828(a)t		STP35	−(C)	−(C)	B36
			diesel	309(a)		STP37	−(C)	−(C)	B36
						STP38	−(C)	−(C)	B36
						STP98	−(C)	−(C)	B36
						ALC	+		B19
						MNT	?/−		B29
						SPI	?		B25
5.4-28	4130-42-1	**2,6-Di-*tert*-butyl-4-ethylphenol**	coal comb.	828(a)t					
5.4-29	1322-06-1	**Pentylphenol**	refuse comb.	840(a)					

Species Number	Registry Number	Name	Emission Source	Emission Ref.	Detection Ref.	Bioassay Code	Bioassay −MA	Bioassay +MA	Bioassay Ref.
5.4-30	90-43-7	*o*-**Phenylphenol**	diesel	309(a)		STI00	NEG	NEG	B38
			turbine	786*(a)t		STI35	+	NEG	B38
						STI37	NEG	NEG	B38
			vegetation	552e		STI98	NEG	NEG	B38
5.4-31	580-51-8	*m*-**Phenylphenol**	diesel	309(a)					
5.4-32	92-69-3	*p*-**Phenylphenol**	diesel	309(a)					
5.4-33	90-05-1	*o*-**Methoxyphenol**	animal waste	855		STP38	−ABCE	−ABCE	B36
			biomass comb.	439,597		ALC	+		B19
		brewing 144; diesel 138,227; tobacco smoke 396(a);				STI00	NEG	NEG	B38
		vegetation 552e; wood pulping 267,545(a)				STI35	NEG	NEG	B38
						STI37	NEG	NEG	B38
						STI98	NEG	NEG	B38
5.4-34	150-76-5	*p*-**Methoxyphenol**	animal waste	855		ALC	+		B19
			vegetation	510e		STI00	NEG	NEG	B38
						STI35	NEG	NEG	B38
						STI37	NEG	NEG	B38
						STI98	NEG	NEG	B38
5.4-35	32391-38-1	**Methoxymethylphenol**	diesel	138					
5.4-36S	97-53-0	**Eugenol**	vegetation	326,500		HMAST		+/−	B13
						STI00	NEG	NEG	B38
						STI35	NEG	NEG	B38
						STI37	NEG	NEG	B38
						STI98	NEG	NEG	B38

Table 5.4. *(Continued)*

Species Number	Registry Number	Name	Emission Source	Emission Ref.	Detection Ref.	Code	Bioassay −MA	Bioassay +MA	Ref.
5.4-37S	97-54-1	**Isoeugenol**	vegetation	514e					
5.4-38S	458-35-5	**Coniferyl alcohol**	vegetation	531					
5.4-39S	495-60-3	**Zingiberene**	vegetation	510e					
5.4-40	25154-52-3	**(8-Ethoxynonyl)phenol**	industrial	60					
5.4-41	25155-26-4	**Dimethoxyphenol**	biomass comb.	891(a)t					
5.4-42	29445-64-5	**3-Methoxyeugenol**	biomass comb.	891(a)t					
5.4-43S	122-48-5	**Zingerone**	vegetation	531					
5.4-44S	120-80-9	**Pyrocatechol**	foundry	598	213*(a),891(a)t	ALC	+		B19
			tobacco smoke	396,405		STI00	NEG	NEG	B38
			vegetation	552e		STI35	NEG	NEG	B38
						STI37	NEG	NEG	B38
						STI98	NEG	NEG	B38
5.4-45S	488-17-5	**3-Methylcatechol**	biomass comb.	891(a)t					
			tobacco smoke	405					
5.4-46	452-86-8	**4-Methylcatechol**	tobacco smoke	405					
5.4-47	1124-39-6	**4-Ethylcatechol**	biomass comb.	891(a)t		ALC	NEG		B19
			tobacco smoke	405					

Species Number	Registry Number	Name	Emission Source	Emission Ref.	Detection Ref.	Code	Bioassay −MA	Bioassay +MA	Bioassay Ref.
5.4-48	2525-02-2	**4-Propylcatechol**	tobacco smoke	405					
5.4-49	1126-61-0	**Allylpyrocatechol**	vegetation	531					
5.4-50	92-05-7	**Phenylpyrocatechol**	auto	311(a)					
5.4-51S	108-46-3	**Resorcinol**	tobacco smoke	396(a)		STP00	NEG	NEG	B36
						STP35	NEG	NEG	B36
						STP37	NEG	NEG	B36
						STP98	NEG	NEG	B36
						STI00	NEG	NEG	B38
						STI35	NEG	NEG	B38
						STI37	NEG	NEG	B38
						STI98	NEG	NEG	B38
						ALC	+		B19
						MNT	?/−		B29
						NEU	NEG		B32
5.4-52	608-25-3	**2-Methylresorcinol**	biomass comb.	891(a)t					
5.4-53	504-15-4	**5-Methylresorcinol**	biomass comb.	891(a)t		ALC	NEG		B19
5.4-54S	123-31-9	**Hydroquinone**	diesel	138	213*(a)	ALC	+		B19
			tobacco smoke	396(a)		STI00	NEG	NEG	B38
						STI35	NEG	NEG	B38
						STI37	NEG	NEG	B38
						STI98	NEG	NEG	B38
5.4-55	95-71-6	**2-Methylhydroquinone**	biomass comb.	891(a)t					

Table 5.4. *(Continued)*

Species Number	Registry Number	Name	Emission Source	Emission Ref.	Detection Ref.	Code	Bioassay −MA	Bioassay +MA	Bioassay Ref.
5.4-56		**4,6-Dimethoxypyro-gallol**	biomass comb.	439					
5.4-57S	100-51-6	**Benzyl alcohol**	auto diesel	217*,537 138	532(a)				
		tobacco smoke 396; vegetation 533,552e							
5.4-58	27043-34-1	**Methylbenzyl alcohol**	auto	537					
5.4-59	29718-36-3	**Dimethylbenzyl alcohol**	auto	537					
5.4-60	30584-69-1	**Vinylbenzyl alcohol**	auto	880t					
5.4-61		**Dihydrocuminyl alcohol**	vegetation	531					
5.4-62S	536-60-7	**Cuminyl alcohol**	vegetation	510e					
5.4-63		**2,4,6-Phenylbenzyl alcohol**			944(a)t				
5.4-64	90-01-7	***o*-Hydroxybenzyl alcohol**			973*(a)				
5.4-65	623-05-2	***p*-Hydroxybenzyl alcohol**	vegetation	552e	973*(a)				
5.4-66	60-12-8	**2-Phenylethanol**	animal waste tobacco smoke vegetation	855 396(a) 533,552e					
5.4-67S	98-85-1	**4-Isopropylbenzyl alcohol**	river water odor	202					

Species Number	Registry Number	Name	Emission Source	Emission Ref.	Detection Ref.	Bioassay Code	Bioassay −MA	Bioassay +MA	Bioassay Ref.
5.4-68S	104-54-1	**Cinnamic alcohol**	vegetation	510e,533					
5.4-69S	several	**Methoxy-2,2′-biphenyldiol**	auto	311(a)					
5.4-70S	2467-02-9	**2,2′-Methylenediphenol**			973(a)				
5.4-71	2467-03-0	**2,4′-Methylenediphenol**			973*(a)t				
5.4-72	620-92-8	**4,4′-Methylenediphenol**			973*(a)				
5.4-73		**(3-Hydroxybenzyl)methylenediphenol**			973t				
5.4-74	7559-72-0	**Bis(2-*o*-hydroxyphenyl) propane**	industrial	60					
5.4-75	36643-74-0	**Indanol**	diesel	138					
5.4-76	37480-21-0	**1,2-Dihydronaphthalenol**	biomass comb.	891(a)t					
5.4-77	90-15-3	**1-Naphthalenol**	biomass comb.	891(a)t		STP00	NEG	NEG	B36
			diesel	811		STP35	NEG	NEG	B36
			tobacco smoke	396(a)		STP37	NEG	NEG	B36
						STP98	NEG	NEG	B36
						ALC	+		B19
						MNT	?/−		B29
						NEU	NEG		B32
5.4-78	59534-35-9	**Methyl-1-naphthalenol**	biomass comb.	891(a)t					

Table 5.4. *(Continued)*

Species Number	Registry Number	Name	Emission Source	Emission Ref.	Detection Ref.	Code	Bioassay −MA	Bioassay +MA	Bioassay Ref.
5.4-79	135-19-3	**2-Napthalenol**	tobacco smoke	396(a)	182(a)	ALC	NEG		B19
5.4-80	1689-64-1	**Fluorenol**	diesel	690(a)t					
5.4-81	several	**Methylfluorenol**	diesel	690(a)t					
5.4-82	several	**Dimethylfluorenol**	diesel	690(a)t					
5.4-83	88898-07-1	**Fluorenediol**	diesel	690(a)t					
5.4-84	several	**Methylfluorenediol**	diesel	690(a)t					
5.4-85	several	**Fluorenetriol**	diesel	690(a)t					
5.4-86	30774-95-9	**Phenanthrenol**	biomass comb.	891(a)t					
5.4-87	several	**Dimethylphenanthrenol**	biomass comb.	891(a)t					
5.4-88	several	**Trimethylphenanthrenol**	biomass comb.	891(a)t					
5.4-89	3772-55-2	**1-Phenanthrenemethanol, 1,2,3,4,4a,9,10,10a-octahydro-1,4a-dimethyl-7-(1-methylethyl)-**	biomass comb.	1089(a)					
5.4-90	several	**Dimethylanthracenediol**	diesel	690(a)t					
5.4-91	5315-79-7	**Pyrenol**			897(a)				

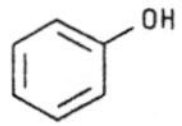

OH
5.4-1

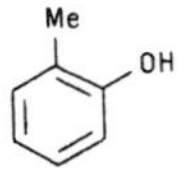

Me OH
5.4-2

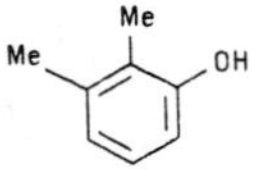

Me Me OH
5.4-5

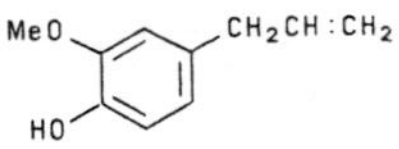

MeO $CH_2CH:CH_2$ HO
5.4-36

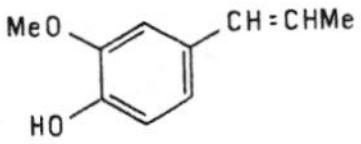

MeO CH=CHMe HO
5.4-37

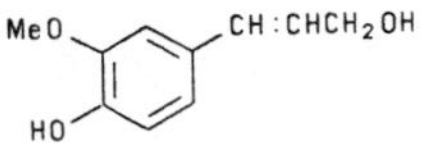

MeO $CH:CHCH_2OH$ HO
5.4-38

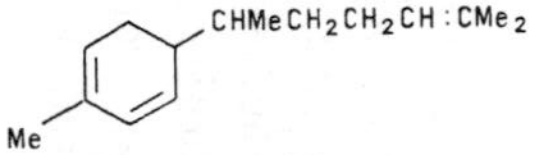

$CHMeCH_2CH_2CH:CMe_2$ Me
5.4-39

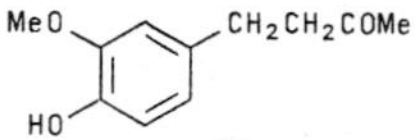

MeO CH_2CH_2COMe HO
5.4-43

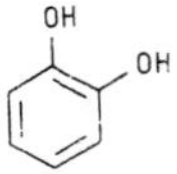

OH OH
5.4-44

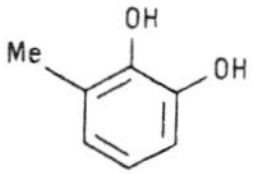

OH Me OH
5.4-45

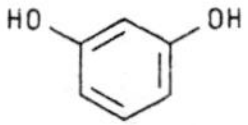

HO OH
5.4-51

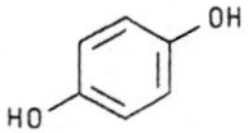

OH HO
5.4-54

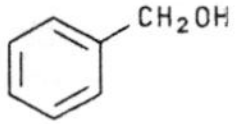

CH_2OH
5.4-57

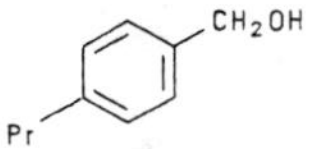

CH_2OH Pr
5.4-62

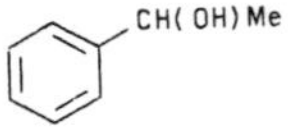

CH(OH)Me
5.4-67

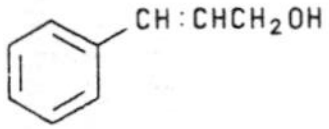

$CH:CHCH_2OH$
5.4-68

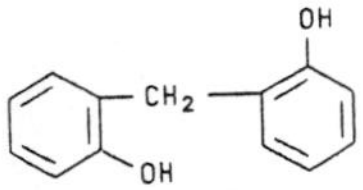

OH CH_2 OH
5.4-70

CHAPTER 6

Ketones

6.0 Introduction

Ketones are organic compounds which contain a carbon atom that is bonded to two other carbon atoms as well as being doubly bonded to an oxygen atom. The simplest example is acetone, $H_3C{-}C({=}O){-}CH_3$, and the ketonic functional group is the doubly-bonded oxygen atom. Combustion processes inject abundant quantities and types of ketones into the atmosphere. We divide ketones into six groups: (1) alkanic ketones, (2) olefinic ketones, (3) cyclic ketones, and (4-6) aromatic ketones of increasing cyclic multiplicity.

6.1 Alkanic Ketones

The simple ketones are common atmospheric constituents. Nearly forty are known to be emitted, and about half of them have been identified in ambient air, mostly in the gas phase. Many of them arise as well from biological processes (sewage treatment, etc.), but the smaller ones also see wide industrial use and are products of fossil fuel combustion. The atmospheric concentrations of most of these compounds are not high, although acetone is often abundant enough (one part per billion or so) to be readily measured.

An important facet of the gas phase atmospheric chemistry of alkanic ketones is their ability to absorb solar photons and dissociate into reactive fragment radicals.[1213] For acetone, the sequence is

$$CH_3C(O)CH_3 \xrightarrow[\lambda<350\ nm]{h\nu} CH_3C(O)\cdot + CH_3\cdot \qquad (R6.1\text{-}1)$$

followed by each product's oxidation:

$$CH_3C(O)\cdot + O_2 \rightarrow CH_3C(O)O_2\cdot \qquad (R6.1\text{-}2)$$

$$CH_3\cdot + O_2 \rightarrow CH_3O_2\cdot \qquad (R6.1\text{-}3)$$

Alternatively, alkanic ketones undergo H abstraction by OH·:[1235]

$$CH_3C(O)CH_3 + OH\cdot \rightarrow CH_3C(O)CH_2\cdot + H_2O \qquad (R6.1\text{-}4)$$

$$CH_3C(O)CH_2\cdot + O_2 \rightarrow CH_3C(O)CH_2O_2\cdot \qquad (R6.1\text{-}5)$$

The peroxy radicals will follow reaction sequences similar to those presented in Chapter 3.

Butanone is the only one of these compounds which has been detected in the aqueous phase in the atmosphere. Its aqueous phase atmospheric chemistry and that of the other atmospheric ketones remain unexplored.

6.2 Olefinic Ketones

A dozen or so olefinic ketones can be present in the atmosphere, mostly as a result of biomass combustion. Few have been detected in ambient air and their concentrations are low. The atmospheric chemistry of these compounds has received little study, but they are known to react rather rapidly with both O_3 and OH·, the reaction in both cases proceeding by addition to the carbon-carbon double bond of the ketone.[1236]

6.3 Cyclic Ketones

The cyclic ketones, more than twenty of them, are produced primarily by vegetation. They have been little studied. It is known that ozone will add at a point of ring unsaturation, though the rate is not rapid.[1236] Reaction with OH· is also anticipated, with reaction sequences similar to those of the cyclic hydrocarbons.

6.4-6.6 Aromatic Ketones

The combustion of petroleum, coal, and biomass results in the emission of large numbers of aromatic ketones to the atmosphere, nearly 150 having been identified. A number of them are found in ambient air, mostly in the aerosol phase. The concentrations are low. Their atmospheric chemistry is unstudied, but should mimic that of the analogous hydrocarbons (see Chapter 3).

Table 6.1. Alkanic Ketones

Species Number	Registry Number	Name	Emission Source	Emission Ref.	Detection Ref.	Bioassay Code	Bioassay −MA	Bioassay +MA	Bioassay Ref.
6.1-1S	67-64-1	**Acetone**	animal waste	436*,695	33*,306*	STP00	NEG	NEG	B36
			auto	15*,217*	820(I),1108(I)	STP35	NEG	NEG	B36
		biomass comb. 354,376; building mtls. 856; chemical mfr. 566*;				STP37	NEG	NEG	B36
		diesel 176,808*; fish processing 110; insects 471;				STP98	NEG	NEG	B36
		microbes 302; onion odor 636; paint 695,1071;				SCE	NEG		B6
		petroleum mfr. 58*; phthalic acid mfr. 535*; plastics comb. 354;				CYC	NEG		B8
		polymer comb. 853; printing 535*,543*; refuse comb. 439,665*,26(a);				ARA	NEG		B17
		solvent 88,134; starch mfr. 695,1071; tobacco smoke 446,453,396(a);				CTV	−,−		B24
		turbine 414; vegetation 552e; volcano 234; whiskey mfr. 1158;				CTP	NEG		B24
		wood comb. 875*; wood pulping 19,267							
6.1-2	43022-03-3	**Protonated acetone ion**			1160(S)t				
6.1-3		**Protonated acetone-water ion**			1160(S)t				
6.1-4	116-09-6	**Hydroxypropanone**	diesel	1					
6.1-5	78-93-3	**Butanone**	animal waste	695,931,892(a)	358,469				
			auto	15*,217*	681*(a),967*(a)				
		biomass comb. 354,535*; building mtls. 856; industrial 133;			748*(c)				
		insects 471; onion odor 636; printing 535*,543*;			748*(i)				
		solvent 88,134; starch mfr. 695,1071; tobacco smoke 396,446;			820(I),1124(I)				
		turbine 414; volcano 234; whiskey mfr. 1158; wood pulping 19,267							

Table 6.1. *(Continued)*

Species Number	Registry Number	Name	Emission Source	Emission Ref.	Detection Ref.	Bioassay Code	Bioassay −MA	Bioassay +MA	Bioassay Ref.
6.1-6	563-80-4	**Methylbutanone**	auto	284	597				
			biomass comb.	597	820(I)				
		building mtls. 856; tobacco smoke 634; wood pulping 19,267							
6.1-7	75-97-8	**Dimethylbutanone**	auto	284					
			solvent	134					
6.1-8	513-86-0	**3-Hydroxybutanone**	animal waste	25,438					
			microbes	302					
6.1-9S	431-03-8	**Biacetyl**	animal waste	25	352,358				
			biomass comb.	439,597					
		rendering 1163t; tobacco smoke 396,634; vegetation 552e							
6.1-10	107-87-9	**2-Pentanone**	animal waste	892(a)	582,597				
			auto	271					
		biomass comb. 354,597; building mtls. 856; polymer comb. 853; solvent 134,598; tobacco smoke 634; turbine 414; wood pulping 19							
6.1-11	565-61-7	**3-Methyl-2-pentanone**	tobacco smoke	634					
6.1-12	108-10-1	**4-Methyl-2-pentanone**	auto	284	358,597				
			building mtls.	856					
		chemical mfr. 1046; landfill 365,628; printing 620; solvent 315,442; tobacco smoke 634; turbine 414; wood pulping 267,545							

Species Number	Registry Number	Name	Emission Source	Emission Ref.	Detection Ref.	Code	Bioassay −MA	Bioassay +MA	Bioassay Ref.
6.1-13	590-50-1	**4,4-Dimethyl-2-pentanone**	turbine	414					
6.1-14	4161-60-8	**4-Hydroxy-2-pentanone**	diesel	138					
6.1-15	123-42-2	**4-Hydroxy-4-methyl-2-pentanone**	biomass comb. solvent	1089(a) 88,134					
6.1-16	600-14-6	**2,3-Pentanedione**	biomass comb. tobacco smoke	354 396,634	1086(a)t				
6.1-17	96-22-0	**3-Pentanone**	animal waste auto	997,1075 284	566*,597 1123*(I)				
		chemical mfr. 566*; solvent 598; tobacco smoke 396,634; wood pulping 267,545							
6.1-18	565-69-5	**2-Methyl-3-pentanone**	biomass comb. chemical mfr.	597 566*					
		tobacco smoke 634; turbine 414							
6.1-19	565-80-0	**2,4-Dimethyl-3-pentanone**	tobacco smoke	634					
6.1-20	591-78-6	**2-Hexanone**	solvent tobacco smoke	134 634	597,628				
6.1-21	589-38-8	**3-Hexanone**	tobacco smoke turbine	634 414	628				

Table 6.1. *(Continued)*

Species Number	Registry Number	Name	Emission Source	Emission Ref.	Detection Ref.	Code	Bioassay −MA	Bioassay +MA	Bioassay Ref.
6.1-22	20633-03-8	**2,2,5-Trimethylhexane-3,4-dione**	turbine	414					
6.1-23	110-43-0	**2-Heptanone**	animal waste	855	628,1086				
			fish processing	946	882(a)				
		vegetation 514e,552e; wood pulping 545							
6.1-24	70713-26-7	**Methyl-2-heptanone**	vegetation	497	628t				
6.1-25	106-35-4	**3-Heptanone**	diesel	414	566*				
6.1-26	123-19-3	**4-Heptanone**	diesel	138					
			solvent	134					
			tobacco smoke	396,634					
6.1-27	108-83-8	**2,6-Dimethyl-4-heptanone**	solvent	134					
6.1-28	111-13-7	**2-Octanone**	animal waste	892(a)	597				
			vegetation	514e,531	882(a)				
6.1-29	106-68-3	**3-Octanone**	animal waste	855t					
			microbes	302					
			vegetation	509e,552e					
6.1-30	821-55-6	**2-Nonanone**	sewage tmt.	687	882(a)				
			vegetation	552e					
6.1-31	925-78-0	**3-Nonanone**			358				

Species Number	Registry Number	Name	Emission		Detection	Bioassay			
			Source	Ref.	Ref.	Code	−MA	+MA	Ref.
6.1-32	693-54-9	**2-Decanone**	rendering	837,879	882(a)				
6.1-33	112-12-9	**2-Undecanone**	sewage tmt. vegetation	687 512e,531	882(a)				
6.1-34	2345-28-0	**2-Pentadecanone**	animal waste	855t					
6.1-35	502-69-2	**6,10, 14-Trimethylpenta-decan-2-one**			882(a),955(a)				
6.1-36	18787-63-8	**2-Hexadecanone**	animal waste tobacco smoke	855t 396					
6.1-37	2922-51-2	**2-Heptadecanone**	biomass comb.	1089(a)					
6.1-38	24724-84-3	**14,16-Hentriacontane-dione**			901*(a)				

MeCOMe
6.1-1

MeCOCOMe
6.1-9

Table 6.2. Olefinic Ketones

Species Number	Registry Number	Name	Emission Source	Emission Ref.	Detection Ref.	Code	Bioassay −MA	Bioassay +MA	Bioassay Ref.
6.2-1S	463-51-4	**Ketene**			820(I)	TRC	+		B20
6.2-2	78-94-4	**3-Buten-2-one**	auto biomass comb.	271,284 597	358t,597				
		tobacco smoke 634; turbine 414; whiskey mfr. 1158							
6.2-3	814-78-8	**3-Methyl-3-buten-2-one**	auto biomass comb. tobacco smoke	271 597 634					
6.2-4	36854-53-2	**2-Methyl-1-oxo-1-butene**			820(I)				
6.2-5	625-33-2	**3-Penten-2-one**	biomass comb.	597					
6.2-6	141-79-7	**4-Methyl-3-penten-2-one**	auto	271	1184(a)t				
6.2-7	13891-87-7	**4-Penten-2-one**	biomass comb. tobacco smoke	597 634	1113(I)				
6.2-8	1629-58-9	**1-Penten-3-one**	tobacco smoke	634					
6.2-9	763-93-9	**3-Hexen-2-one**	refuse comb.	960*(a)					

Species Number	Registry Number	Name	Emission Source	Emission Ref.	Detection Ref.	Code	Bioassay −MA	Bioassay +MA	Ref.
6.2-10	109-49-9	**5-Hexen-2-one**	turbine	414					
6.2-11	110-93-0	**6-Methyl-5-hepten-2-one**	vegetation	514e					
6.2-12	4312-99-6	**1-Octen-3-one**	animal waste	892(a)					
6.2-13	65213-22-8	**1,5-Octadien-3-one**	microbes	602					
6.2-14	2516-30-5	**6,9-Pentadecadien-2-one**	vegetation	552e					

$H_2C:C:O$

6.2-1

Table 6.3. Cyclic Ketones

Species Number	Registry Number	Name	Emission Source	Emission Ref.	Detection Ref.	Code	Bioassay −MA	Bioassay +MA	Bioassay Ref.
6.3-1	765-43-5	**Cyclopropyl methyl ketone**	biomass comb.	597t	628t				
6.3-2	120-92-3	**Cyclopentanone**	tobacco smoke	634					
			turbine	414					
6.3-3	3008-40-0	**1,2-Cyclopentanedione**	vegetation	552e					
6.3-4	108-94-1	**Cyclohexanone**	building mtls.	9*	597	STI00	NEG	NEG	B38
			chemical mfr.	1046	1110*(I)	STI35	NEG	NEG	B38
			solvent	88,134		STI37	NEG	NEG	B38
						STI98	NEG	NEG	B38
6.3-5	2320-30-1	**1,3-Dimethylcyclo-hexan-5-one**	diesel	138					
6.3-6	2408-37-9	**1,3,3-Trimethylcyclo-hexan-2-one**	vegetation	515e					
6.3-7	930-68-7	**Cyclohexen-3-one**	animal waste	598					
6.3-8S	89-80-5	**Menthone**	vegetation	516e,531					
6.3-9S	491-07-6	**Isomenthone**	vegetation	515e					
6.3-10S	1195-79-5	**Fenchone**	wood pulping	267,545					

Species Number	Registry Number	Name	Emission Source	Emission Ref.	Detection Ref.	Code	Bioassay −MA	Bioassay +MA	Bioassay Ref.
6.3-11S	464-49-3	**Camphor**	chemical mfr. vegetation	1164 497e,663	882(a) 1110*(I)				
6.3-12S	546-80-5	**Thujone**	vegetation	510e,514e					
6.3-13S	89-82-7	**Pulegone**	vegetation	510e,531					
6.3-14S	78-59-1	**Isophorone**	solvent	88,134					
6.3-15S	89-81-6	**Piperitone**	vegetation	531					
6.3-16S	2244-16-8	**Carvone**	vegetation	506e,531					
6.3-17	several	**Methylcyclohex-anedione**	diesel	138					
6.3-18	719-22-2	**2,6-Di-*tert*-butyl-2,5-cyclohexadien-1,4-dione**	coal comb.	828(a)t					
6.3-19		**2,5-Bis(1,1-Dimethylpropyl)-2,5-cyclohexadien-1,4-dione**	coal comb.	828(a)t					
6.3-20S	24190-29-2	**α-Ionone**	vegetation	440,531					
6.3-21S	79-77-6	**β-Ionone**	vegetation	440,531					
6.3-22S	79-69-6	**α-Irone**	vegetation	440,513e					

Table 6.3. *(Continued)*

Species Number	Registry Number	Name	Emission Source	Emission Ref.	Detection Ref.	Code	Bioassay −MA	Bioassay +MA	Bioassay Ref.
6.3-23	10363-27-6	**2-Methylcyclo-octanone**	turbine	414					
6.3-24	542-46-1	**Cycloheptadec-9-en-1-one**	animal waste	510e					
6.3-25	70532-56-8M	**2-Ethoxy-5-(1-hydroxy-ethyl)cyclopentanone**			697(a)t				

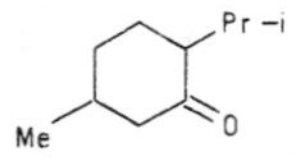

Pr -i
Me
O
6.3-8

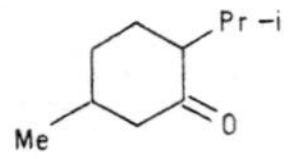

Pr -i
Me
O
6.3-9

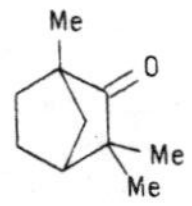

Me
O
Me
Me
6.3-10

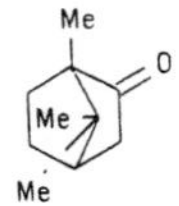

Me
O
Me
Me
6.3-11

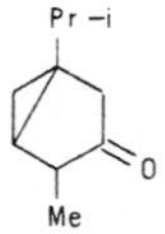

Pr -i
O
Me
6.3-12

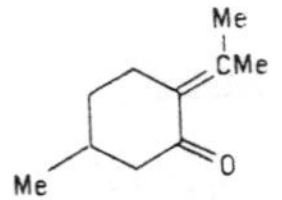

Me
CMe
Me
O
6.3-13

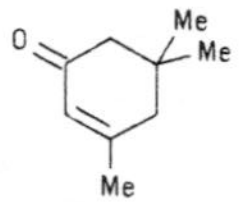

O
Me
Me
Me
6.3-14

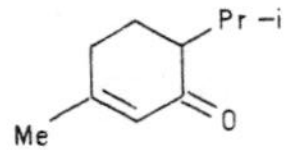

Pr -i
Me
O
6.3-15

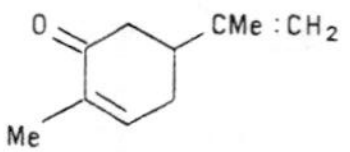

O
CMe : CH₂
Me
6.3-16

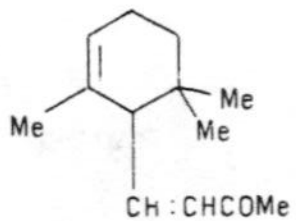

Me
Me
Me
CH : CHCOMe
6.3-20

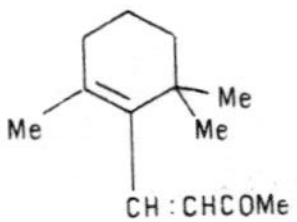

Me
Me
Me
CH : CHCOMe
6.3-21

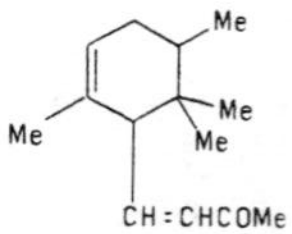

Me
Me
Me
Me
CH : CHCOMe
6.3-22

Table 6.4. Monocyclic Aromatic Ketones

Species Number	Registry Number	Name	Emission Source	Emission Ref.	Detection Ref.	Code	Bioassay −MA	Bioassay +MA	Ref.
6.4-1S	106-51-4	**1,4-Benzoquinone**	tobacco smoke	579(a)	213*(a)	ALC SRL	+ ?		B19 B30
6.4-2	553-97-9	**Methyl-1,4-benzoquinone**	tobacco smoke	579(a)					
6.4-3	526-86-3	**2,3-Dimethyl-1,4-benzoquinone**	tobacco smoke	579(a)					
6.4-4	137-18-8	**2,5-Dimethyl-1,4-benzoquinone**	tobacco smoke	579(a)					
6.4-5	935-92-2	**Trimethyl-1,4-benzoquinone**	tobacco smoke	579(a)					
6.4-6	627-17-3	**Tetramethyl-1,4-benzoquinone**	tobacco smoke	579(a)					
6.4-7	4754-26-1	**Ethyl-1,4-benzoquinone**	biomass comb.	891(a)t					
6.4-8	2460-77-7	**Di-*tert*-butyl-1,4-benzoquinone**	auto biomass comb.	880 1089(a)					

Table 6.4. *(Continued)*

Species Number	Registry Number	Name	Emission Source	Emission Ref.	Detection Ref.	Code	Bioassay −MA	Bioassay +MA	Bioassay Ref.
6.4-9S	98-86-2	**Acetophenone**	auto biomass comb.	217*,880 597	86,202 633(a)	REC	NEG		B9
		coal comb. 1159; diesel 138; river water odor vegetation 503,552e		202;	820(I),1113(I)				
6.4-10	577-16-2	**2-Methylacetophenone**	biomass comb.	597	896				
6.4-11	122-00-9	**4-Methylacetophenone**	vegetation	510e					
6.4-12	89-74-7	**Dimethylacetophenone**	diesel	138					
6.4-13	937-30-4	**4-Ethylacetophenone**	coal comb.	825(a)					
6.4-14	100-06-1	**4-Methoxyacetophenone**	biomass comb. vegetation	1089(a) 510e					
6.4-15	1131-62-0	**3,4-Dimethoxyaceto-phenone**	biomass comb.	1089(a)	628				
6.4-16	121-71-1	**3-Hydroxyacetophenone**	tobacco smoke	396(a)					
6.4-17	99-93-4	**4-Hydroxyacetophenone**	biomass comb. diesel tobacco smoke	891(a)t 138 396(a)					
6.4-18S	498-02-2	**Acetovanillone**	wood pulping	267,545					
6.4-19	23133-83-7	**3-Hydroxy-2,4-dimethoxy-acetophenone**	biomass comb.	891(a)t					

Species Number	Registry Number	Name	Emission Source	Emission Ref.	Detection Ref.	Code	Bioassay −MA	Bioassay +MA	Bioassay Ref.
6.4-20	89-84-9	**Dihydroxyacetophenone**	biomass comb. diesel	891(a)t 138					
6.4-21	93-55-0	**Ethyl phenyl ketone**	biomass comb.	1089(a)					
6.4-22		**Ethyl 2,4-dimethoxyphenyl ketone**	biomass comb.	1089(a)					
6.4-23	768-03-6	**Phenyl vinyl ketone**	diesel	138					
6.4-24	103-79-7	**3-Phenyl-2-propanone**	refuse comb.	878(a)					
6.4-25S	2503-46-0	**1-(4-Hydroxy-3-methoxy-phenyl)-2-propanone**	biomass comb.	891(a)t					
6.4-26	57765-50-1	**(2,4,6-Trihydroxy-3-methylphenyl) propyl ketone**	biomass comb.	891(a)t					
6.4-27	2550-26-7	**4-Phenyl-2-butanone**			358				
6.4-28	63449-79-6	**4-(*p*-Methoxyphenyl)-2-butanone**	vegetation	510e					
6.4-29	several	**Methylphenylhexanone**	coal comb.	828t					
6.4-30S	119-61-9	**Benzophenone**	coal comb. refuse comb.	828(a)t 878(a)	591	REC	NEG		B9

Table 6.4. *(Continued)*

Species Number	Registry Number	Name	Emission Source	Emission Ref.	Detection Ref.	Code	Bioassay −MA	Bioassay +MA	Bioassay Ref.
6.4-31	several	**Hydroxybenzophenone**	turbine	786*(a)t					
6.4-32S	6962-60-3	**2-Tolyl-α-methoxybenz-aldehyde**	biomass comb.	891(a)t					
6.4-33S	119-53-9	**Benzoin**	vegetation	531		CCC	SN		B37
6.4-34S	134-81-6	**Benzil**	coal comb. refuse comb.	825(a) 790(a)					
6.4-35	94-41-7	**1,3-Diphenyl-2-propen-1-one**	refuse comb.	878(a)					

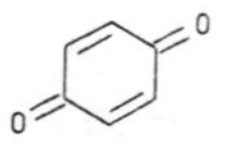

O O

6.4-1

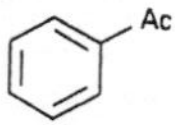

Ac

6.4-9

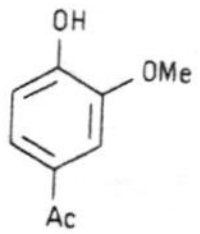

OH OMe Ac

6.4-18

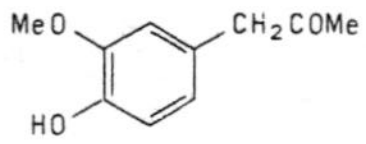

MeO CH$_2$COMe HO

6.4-25

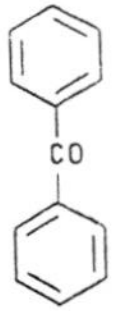

CO

6.4-30

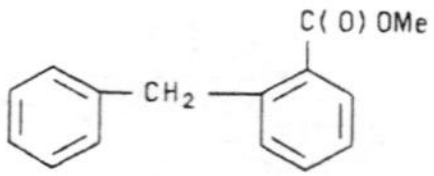

CH$_2$ C(O)OMe

6.4-32

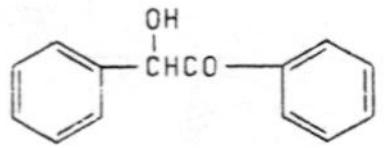

OH CHCO

6.4-33

COCO

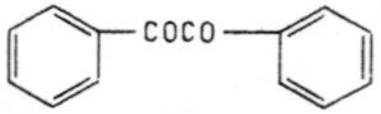

6.4-34

Table 6.5. Bi - and Tri-Cyclic Aromatic Ketones

Species Number	Registry Number	Name	Emission Source	Emission Ref.	Detection Ref.	Bioassay Code	Bioassay −MA	Bioassay +MA	Bioassay Ref.
6.5-1S	83-33-0	**1-Indanone**	biomass comb. diesel tobacco smoke	1089(a) 138,811,685(a) 634(a)	364(a)				
6.5-2	several	**Dimethyl-1-indanone**	diesel	138					
6.5-3	34322-84-0	**3,4,7-Trimethyl-1-indanone**	biomass comb.	1089(a)					
6.5-4		**3,3-Dimethyl-5-*tert*-butylindanone**			364(a)				
6.5-5	several	**Methoxyindanone**	diesel	227					
6.5-6	several	**Hydroxyindanone**	diesel	138,227					
6.5-7	several	**Hydroxymethylindanone**	diesel	138					
6.5-8	several	**Hydroxytrimethyl-indanone**	diesel	138					
6.5-9	several	**Hydroxytetramethyl-indanone**	diesel	138					
6.5-10	615-13-4	**2-Indanone**	diesel	138					

Table 6.5. *(Continued)*

Species Number	Registry Number	Name	Emission Source	Emission Ref.	Detection Ref.	Bioassay Code	Bioassay −MA	Bioassay +MA	Bioassay Ref.
6.5-11	480-90-0	**Indenone**	diesel	138,811					
6.5-12	2887-89-0	**Dimethylindenone**	diesel	138					
6.5-13	several	**Pentamethylindenone**	diesel	138					
6.5-14	several	**Hydroxyindenone**	diesel	138					
6.5-15	several	**Hydroxymethylindenone**	diesel	138					
6.5-16	several	**Hydroxydimethyl-indenone**	diesel	138					
6.5-17S	529-34-0	**1-Tetralone**	diesel	138					
6.5-18	several	**Methyltetralone**	diesel	138					
6.5-19	several	**Methoxytetralone**	diesel	138					
6.5-20S	1333-52-4	**Acetonaphthone**	auto diesel	311(a) 138					
6.5-21S	58-27-5	**2-Methyl-1,4-naphthoquinone**	diesel tobacco smoke	138 579(a)					
6.5-22	20490-42-0	**2,3,6-Trimethyl-1,4-naphthoquinone**	tobacco smoke	579(a),634(a)					

Species Number	Registry Number	Name	Emission Source	Emission Ref.	Detection Ref.	Bioassay Code	Bioassay −MA	Bioassay +MA	Bioassay Ref.
6.5-23	several	**Hydroxy-1,4-naphthoquinone**	coal comb.	828t					
6.5-24	613-20-7	**2,6-Naphthoquinone**	diesel	138	213(a),528(a)				
			phthal anhyd. mfr.	64*					
6.5-25S	486-25-9	**9*H*-Fluoren-9-one**	auto	858*(a)	466(a),749(a)				
			biomass comb.	838(a)	755*(r),876*(r)				
		coal comb. 828t,749(a); diesel 685(a),858*(a); refuse comb. 26(a),749(a); tobacco smoke 634(a); turbine 786*(a); wood comb. 749(a),838(a)							
6.5-26	79147-47-0	**Methylfluorenone**	auto	858*(a)	799(a)t				
			biomass comb.	838(a)					
		diesel 685(a),693(a); wood comb. 838(a)							
6.5-27	several	**Dimethylfluorenone**	diesel	690(a),801(a)t					
6.5-28	several	**Trimethylfluorenone**	diesel	690(a)t,801(a)t					
6.5-29	several	**Tetramethylfluorenone**	diesel	801(a)t					
6.5-30	several	**Hydroxyfluorenone**	diesel	690(a)t					
6.5-31	42523-54-6	**1,4-Fluorenequinone**	diesel	685(a),690(a)					
6.5-32	several	**Methylfluorenequinone**	diesel	690(a)					
6.5-33	several	**Dimethylfluorenequinone**	diesel	690(a)					

Table 6.5. *(Continued)*

Species Number	Registry Number	Name	Emission Source	Emission Ref.	Detection Ref.	Bioassay Code	Bioassay −MA	Bioassay +MA	Bioassay Ref.
6.5-34	479-79-8	**11*H*-Benzo[*a*]fluoren-11-one**	coal comb.	749(a)	749(a),803(a)				
			diesel	693(a),749(a)					
			refuse comb.	749(a)					
			wood comb.	749(a)					
6.5-35	3074-03-1	**11*H*-Benzo[*b*]fluoren-11-one**	coal comb.	749(a)	749(a),803(a)				
			diesel	693(a),749(a)					
			refuse comb.	749(a)					
			wood comb.	749(a)					
6.5-36	6051-98-5	**7*H*-Benzo[*c*]fluoren-7-one**	coal comb.	749(a)	749(a),1142(a)t				
			diesel	749(a)					
			refuse comb.	749(a)					
			wood comb.	749(a)					
6.5-37	63041-47-4	**13*H*-Dibenzo[*a*,*g*]fluoren-13-one**	coal comb.	749(a)t	799(a)t				
			diesel	749(a)t					
			refuse comb.	749(a)t					
			wood comb.	749(a)t,838(a)t					
6.5-38	4599-94-4	**13*H*-Dibenzo[*a*,*h*]fluoren-13-one**	coal comb.	749(a)t	1142(a)t				
			diesel	749(a)t					
			refuse comb.	749(a)t					
			wood comb.	749(a)t					
6.5-39	86854-01-5	**13*H*-Dibenzo[*a*,*i*]fluoren-13-one**	coal comb.	749(a)t					
			diesel	749(a)t					
			refuse comb.	749(a)t					
			wood comb.	749(a)t					

Species Number	Registry Number	Name	Emission Source	Emission Ref.	Detection Ref.	Bioassay Code	Bioassay −MA	Bioassay +MA	Bioassay Ref.
6.5-40	86854-02-6	**7*H*-Dibenzo[*b*,*g*] fluoren-7-one**	coal comb.	749(a)t					
			diesel	749(a)t					
			refuse comb.	749(a)t					
			wood comb.	749(a)t					
6.5-41	53223-75-9	**12*H*-Dibenzo[*b*,*h*] fluoren-12-one**	coal comb.	749(a)t					
			diesel	749(a)t					
			refuse comb.	749(a)t					
			wood comb.	749(a)t					
6.5-42	86853-97-6	**7*H*-Dibenzo[*c*,*g*] fluoren-7-one**	coal comb.	749(a)t					
			diesel	749(a)t					
			refuse comb.	749(a)t					
			wood comb.	749(a)t					
6.5-43	86854-26-4	**7*H*-Naphtho[1,2-*a*]benzo [*h*]fluoren-7-one**	coal comb.	749(a)t					
			diesel	749(a)t					
			refuse comb.	749(a)t					
			wood comb.	749(a)t					
6.5-44	518-85-4	**1*H*-Phenalen-1-one, 2,3-dihydro**			188(a)				
6.5-45S	548-39-0	**1*H*-Phenalen-1-one**	diesel	749(a)	98(a),363(a)				
6.5-46S	84-11-7	**9,10-Phenanthrenequinone**	biomass comb.	891(a)t		ALC	NEG		B19
			diesel	685(a),690(a)					

Table 6.5. *(Continued)*

Species Number	Registry Number	Name	Emission Source	Emission Ref.	Detection Ref.	Bioassay Code	Bioassay −MA	Bioassay +MA	Bioassay Ref.
6.5-47	several	**Methyl-9,10-phenan-threnequinone**	diesel	690(a)					
6.5-48	5737-13-3	**4*H*-Cyclopenta[*def*] phenanthren-4-one**	biomass comb.	838(a)	749(a),799(a)				
			coal comb.	749(a)					
			diesel	685(a),749(a)					
		refuse comb. 749(a),878(a); wood comb. 749(a),838(a)							
6.5-49	several	**Methyl-4*H*-cyclopenta[*def*] phenanthren-4-one**	diesel	684(a)t	1142(a)				
6.5-50	28609-66-7	**8*H*-Dibenzo[*b*,*mn*] phenanthren-8-one**	coal comb.	749(a)t					
			diesel	749(a)t					
			refuse comb.	749(a)t					
			wood comb.	749(a)t					
6.5-51	62716-20-5	**9*H*-Dibenzo[*c*,*mn*] phenanthren-9-one**	coal comb.	749(a)t					
			diesel	749(a)t					
			refuse comb.	749(a)t					
			wood comb.	749(a)t					
6.5-52	86853-95-4	**7*H*-Indeno[2,1-*a*] phenanthren-7-one**	coal comb.	749(a)t					
			diesel	749(a)t					
			refuse comb.	749(a)t					
			wood comb.	749(a)t					

Species Number	Registry Number	Name	Emission Source	Emission Ref.	Detection Ref.	Bioassay Code	Bioassay −MA	Bioassay +MA	Bioassay Ref.
6.5-53	4599-92-2	**11*H*-Indeno[2,1-*a*] phenanthren-11-one**	coal comb.	749(a)t					
			diesel	749(a)t					
			refuse comb.	749(a)t					
			wood comb.	749(a)t					
6.5-54	86854-00-4	**12*H*-Indeno[1,2-*b*] phenanthren-12-one**	coal comb.	749(a)t					
			diesel	749(a)t					
			refuse comb.	749(a)t					
			wood comb.	749(a)t					
6.5-55	86853-98-7	**8*H*-Indeno[2,1-*b*] phenanthren-8-one**	coal comb.	749(a)t					
			diesel	749(a)t					
			refuse comb.	749(a)t					
			wood comb.	749(a)t					
6.5-56	86853-93-2	**9*H*-Indeno[2,1-*c*] phenanthren-9-one**	coal comb.	749(a)t					
			diesel	749(a)t					
			refuse comb.	749(a)t					
			wood comb.	749(a)t					
6.5-57	86854-03-7	**13*H*-Indeno[1,2-*c*] phenanthren-13-one**	coal comb.	749(a)t					
			diesel	749(a)t					
			refuse comb.	749(a)t					
			wood comb.	749(a)t					
6.5-58	86853-96-5	**13*H*-Indeno[1,2-*l*] phenanthren-13-one**	coal comb.	749(a)t					
			diesel	749(a)t					
			refuse comb.	749(a)t					
			wood comb.	749(a)t					

Table 6.5. *(Continued)*

Species Number	Registry Number	Name	Emission Source	Emission Ref.	Detection Ref.	Bioassay Code	−MA	+MA	Ref.
6.5-59S	90-44-8	**9-Anthrone**	diesel	690(a)	1001*(a)				
6.5-60	79075-29-9	**Methyl-9-anthrone**	diesel	690(a)					
6.5-61	84-65-1	**9,10-Anthraquinone**	auto	311(a)	633*				
			biomass comb.	838(a)	190*(a),633*(a)				
		diesel 309(a),684(a); refuse comb. 26(a),790(a); turbine 786*(a); wood comb. 838*(a)			570(r),755*(r) 570(s)				
6.5-62	84-54-8	**2-Methyl-9,10-anthraquinone**	diesel	804(a),829(a)	799(a),803(a)t				
			tobacco smoke	579(a)					
6.5-63	6531-35-7	**2,3-Dimethyl-9,10-anthraquinone**	tobacco smoke	579(a)					
6.5-64S	11067-62-2	**Benz[*a*]anthrone**	auto	311(a)	179(a),188*(a)				
			biomass comb.	838(a)					
		diesel 684(a); wood comb. 838(a)							
6.5-65	69761-08-6	**Benz[*a*]anthracene quinone**	biomass comb.	838(a)	799(a),803(a)				
			diesel	693(a)					
			wood comb.	838(a)					
6.5-66	86853-88-5	**1*H*-Benz[*de*]anthracen-1-one**	coal comb.	749(a)					
			diesel	749(a)					
			refuse comb.	749(a)					
			wood comb.	749(a)					

Species Number	Registry Number	Name	Emission Source	Emission Ref.	Detection Ref.	Code	Bioassay −MA	Bioassay +MA	Bioassay Ref.
6.5-67	80252-14-8	***6H*-Benz[*de*]anthracen-6-one**			1142(a)t				
6.5-68	82-05-3	***7H*-Benz[*de*]anthracen-7-one**	auto	311(a)	98(a),188(a)				
			coal comb.	749(a)					
			diesel	693(a),749(a)					
		refuse comb. 749(a); wood comb. 749(a)							
6.5-69	86854-06-0	***9H*-Dibenz[*a*,*de*]anthracen-9-one**	coal comb.	749(a)t					
			diesel	749(a)t					
			refuse comb.	749(a)t					
			wood comb.	749(a)t					
6.5-70	86854-05-9	***9H*-Dibenz[*b*,*de*]anthracen-9-one**	coal comb.	749(a)t					
			diesel	749(a)t					
			refuse comb.	749(a)t					
			wood comb.	749(a)t					
6.5-71	60848-01-3	***7H*-Dibenz[*de*,*h*]anthracen-7-one**	coal comb.	749(a)t					
			diesel	749(a)t					
			refuse comb.	749(a)t					
			wood comb.	749(a)t					
6.5-72	5623-32-5	**Benzo[*fg*]naphthacen-7-one**	coal comb.	749(a)t	1142(a)t				
			diesel	749(a)t					
			refuse comb.	749(a)t					
			wood comb.	749(a)t					

Table 6.5. *(Continued)*

Species Number	Registry Number	Name	Emission Source	Emission Ref.	Detection Ref.	Code	Bioassay −MA	Bioassay +MA	Bioassay Ref.
6.5-73	80440-44-4	**7*H*-Dibenz[*de*,*j*] anthracen-7-one**	coal comb.	749(a)t					
			diesel	749(a)t					
			refuse comb.	749(a)t					
			wood comb.	749(a)t					
6.5-74	86853-94-3	**8*H*-Inden[2,1-*a*] anthracen-8-one**	coal comb.	749(a)t					
			diesel	749(a)t					
			refuse comb.	749(a)t					
			wood comb.	749(a)t					
6.5-75	86854-04-8	**13*H*-Inden[2,1-*a*] anthracen-13-one**	coal comb.	749(a)t					
			diesel	749(a)t					
			refuse comb.	749(a)t					
			wood comb.	749(a)t					
6.5-76	86853-99-8	**13*H*-Inden[1,2-*b*] anthracen-13-one**	coal comb.	749(a)t					
			diesel	749(a)t					
			refuse comb.	749(a)t					
			wood comb.	749(a)t					

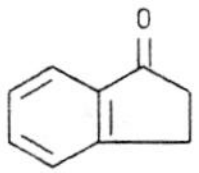

O

6.5-1

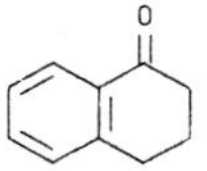

O

6.5-17

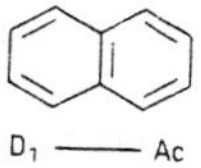

D_1 —— Ac

6.5-20

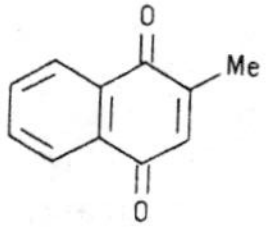

O Me O

6.5-21

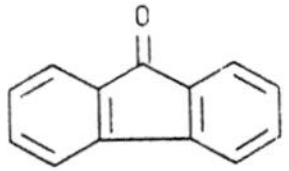

O

6.5-25

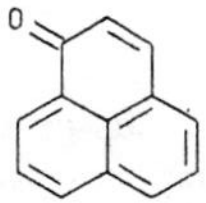

O

6.5-45

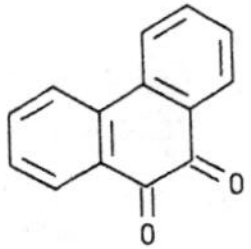

O O

6.5-46

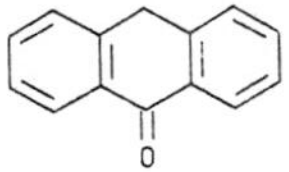

O

6.5-59

O

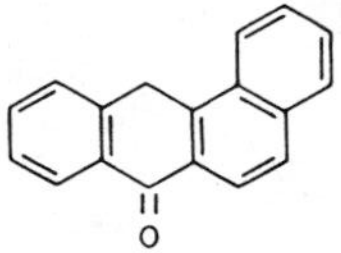

6.5-64

Table 6.6. Higher Polycyclic Aromatic Ketones

Species Number	Registry Number	Name	Emission Source	Emission Ref.	Detection Ref.	Code	Bioassay −MA	Bioassay +MA	Bioassay Ref.
6.6-1S	39407-42-6	**Fluoranthenequinone**	diesel	690(a)					
6.6-2	86853-89-6	**11*H*-Benz[*bc*]ace-anthrylen-11-one**	coal comb.	749(a)t	749(a)t, 1142(a)t				
			diesel	749(a)t					
			refuse comb.	749(a)t					
			wood comb.	749(a)t					
6.6-3	86854-17-3	**13*H*-Dibenz[*bc, j*]ace-anthrylen-13-one**	coal comb.	749(a)t					
			diesel	749(a)t					
			refuse comb.	749(a)t					
			wood comb.	749(a)t					
6.6-4	86854-18-4	**13*H*-Dibenz[*bc, l*]ace-anthrylen-13-one**	coal comb.	749(a)t					
			diesel	749(a)t					
			refuse comb.	749(a)t					
			wood comb.	749(a)t					
6.6-5	86853-90-9	**4*H*-Cyclopenta[*def*]triphenylen-4-one**	coal comb.	749(a)t	749(a)t				
			diesel	749(a)t					
			refuse comb.	749(a)t					
			wood comb.	749(a)t					

Species Number	Registry Number	Name	Emission Source	Emission Ref.	Detection Ref.	Code	Bioassay −MA	Bioassay +MA	Bioassay Ref.
6.6-6	86854-12-8	**13*H*-Benzo[*b*]cyclopenta[*def*] triphenylen-13-one**	coal comb.	749(a)t					
			diesel	749(a)t					
			refuse comb.	749(a)t					
			wood comb.	749(a)t					
6.6-7	86854-13-9	**4*H*-Benzo[*m*]cyclopenta[*cde*] triphenylen-4-one**	coal comb.	749(a)t					
			diesel	749(a)t					
			refuse comb.	749(a)t					
			wood comb.	749(a)t					
6.6-8	86854-14-0	**4*H*-Benzo[*l*]cyclopenta[*cde*] triphenylen-4-one**	coal comb.	749(a)t					
			diesel	749(a)t					
			refuse comb.	749(a)t					
			wood comb.	749(a)t					
6.6-9	86854-25-3	**13*H*-Dibenzo[*b, jkl*]cyclopenta [*def*] triphenylen-13-one**	coal comb.	749(a)t					
			diesel	749(a)t					
			refuse comb.	749(a)t					
			wood comb.	749(a)t					
6.6-10S	79147-51-6	**Pyrenone**	diesel	690(a)					
6.6-11	39461-53-5	**Pyrene quinone**	diesel	690(a)					
6.6-12	57652-57-0	**3*H*-Benzo[*a*]pyren-2-one**	diesel	685(a)					

Table 6.6. *(Continued)*

Species Number	Registry Number	Name	Emission Source	Emission Ref.	Detection Ref.	Bioassay Code	Bioassay −MA	Bioassay +MA	Bioassay Ref.
6.6-13	3067-13-8	**Benzo[*a*]pyrene-1,6-quinone**			190(a)	V79A	NEG		B7
						CCC	I		B37
						STP00	−(B)		B36
						STP38	−(B)		B36
						STP98	−(B)		B36
6.6-14	3067-14-9	**Benzo[*a*]pyrene-3,6-quinone**			190(a)	V79A	+		B7
						CCC	I		B37
						STP00	−(BC)		B36
						STP38	−(BC)		B36
						STP98	−(BC)		B36
6.6-15	3067-12-7	**Benzo[*a*]pyrene-6,12-quinone**			190*(a), 209(a)	V79A	NEG		B7
						CCC	I		B37
						STP00	−(B)		B36
						STP38	−(B)		B36
						STP98	−(B)		B36
6.6-16	4558-16-1	**5*H*-Benzo[*cd*]pyren-5-one**			1142(a)t				
6.6-17	3074-00-8	**6*H*-Benzo[*cd*]pyren-6-one**	biomass comb.	838(a)	749(a), 799(a)				
			coal comb.	749(a)					
			diesel	693(a), 749(a)					
		refuse comb. 749(a), 878(a); wood comb. 749(a), 838(a)							
6.6-18	86854-22-0	**6*H*-Dibenzo[*a,fg*]pyren-6-one**	coal comb.	749(a)t					
			diesel	690(a)t, 749(a)t					
			refuse comb.	749(a)t					
			wood comb.	749(a)t					

Species Number	Registry Number	Name	Emission		Detection		Bioassay		
			Source	Ref.	Ref.	Code	−MA	+MA	Ref.
6.6-19	86854-24-2	**6*H*-Dibenzo[*b, fg*] pyren-6-one**	coal comb.	749(a)t					
			diesel	749(a)t					
			refuse comb.	749(a)t					
			wood comb.	749(a)t					
6.6-20	86854-23-1	**7*H*-Dibenzo[*cd, l*] pyren-7-one**	coal comb.	749(a)t					
			diesel	749(a)t					
			refuse comb.	749(a)t					
			wood comb.	749(a)t					
6.6-21	641-13-4	**Dibenzo[*cd, jk*]pyrene-6,12-quinone**			209(a), 363(a)				
6.6-22	86854-20-8	**7*H*-Indeno[1,2-*a*] pyren-7-one**	coal comb.	749(a)t					
			diesel	749(a)t					
			refuse comb.	749(a)t					
			wood comb.	749(a)t					
6.6-23	7267-90-5	**11*H*-Indeno[2,1-*a*] pyren-11-one**	coal comb.	749(a)t					
			diesel	749(a)t					
			refuse comb.	749(a)t					
			wood comb.	749(a)t					
6.6-24	86854-21-9	**9*H*-Indeno[1,2-*e*] pyren-9-one**	coal comb.	749(a)t					
			diesel	749(a)t					
			refuse comb.	749(a)t					
			wood comb.	749(a)t					

Table 6.6. *(Continued)*

Species Number	Registry Number	Name	Emission Source	Emission Ref.	Detection Ref.	Bioassay Code	Bioassay −MA	Bioassay +MA	Bioassay Ref.
6.6-25	86862-68-2	**11*H*-Indeno[2,1,7-*cde*] pyren-11-one**	coal comb.	749(a)t					
			diesel	749(a)t					
			refuse comb.	749(a)t					
			wood comb.	749(a)t					
6.6-26S	86853-91-0	**4*H*-Cyclopenta[*def*] chrysen-4-one**	coal comb.	749(a)t	749(a)t				
			diesel	749(a)t					
			refuse comb.	749(a)t					
			wood comb.	749(a)t					
6.6-27	86854-08-2	**5*H*-Benzo[*b*]cyclopenta[*def*] chrysen-5-one**	coal comb.	749(a)t					
			diesel	749(a)t					
			refuse comb.	749(a)t					
			wood comb.	749(a)t					
6.6-28	86854-09-3	**4*H*-Benzo[*b*]cyclopenta[*mno*] chrysen-4-one**	coal comb.	749(a)t					
			diesel	749(a)t					
			refuse comb.	749(a)t					
			wood comb.	749(a)t					
6.6-29	86854-10-6	**4*H*-Benzo[*c*]cyclopenta [*mno*]chrysen-4-one**	coal comb.	749(a)t					
			diesel	749(a)t					
			refuse comb.	749(a)t					
			wood comb.	749(a)t					
6.6-30	86854-11-7	**8*H*-Benzo[*g*]cyclopenta [*mno*]chrysen-8-one**	coal comb.	749(a)t					
			diesel	749(a)t					
			refuse comb.	749(a)t					
			wood comb.	749(a)t					

Species Number	Registry Number	Name	Emission Source	Emission Ref.	Detection Ref.	Code	Bioassay −MA	Bioassay +MA	Bioassay Ref.
6.6-31	86853-92-1	**4*H*-Benzo[*def*]cyclopenta [*mno*]chrysen-4-one**	coal comb.	749(a)t	1142(a)t				
			diesel	749(a)t					
			refuse comb.	749(a)t					
			wood comb.	749(a)t					
6.6-32	128-66-5	**Dibenzo[*b,def*]chrysene-7,14-quinone**			190(a)				
6.6-33S	1090-13-7	**Naphthacene-5,12-dione**			803(a), 1142(a)				
6.6-34	86854-19-5	**13*H*-Indeno[2,1,7-*qra*] naphthacen-13-one**	coal comb.	749(a)t					
			diesel	749(a)t					
			refuse comb.	749(a)t					
			wood comb.	749(a)t					
6.6-35S	83484-79-1	**6*H*-Cyclopenta[*ghi*] picen-6-one**	biomass comb.	838(a)	799(a)				
			coal comb.	749(a)t					
			diesel	749(a)t					
		refuse comb. 749(a)t; wood comb. 749(a)t, 838(a)							
6.6-36	86854-16-2	**13*H*-Cyclopenta[*pqr*] picen-13-one**	coal comb.	749(a)t					
			diesel	749(a)t					
			refuse comb.	749(a)t					
			wood comb.	749(a)t					

Table 6.6. *(Continued)*

Species Number	Registry Number	Name	Emission Source	Emission Ref.	Detection Ref.	Code	Bioassay −MA	Bioassay +MA	Bioassay Ref.
6.6-37S	83622-91-7	**11*H*-Cyclopenta[*ghi*] perylen-11-one**	biomass comb.	838(a)	799(a)				
			coal comb.	749(a)t					
			diesel	749(a)t					
		refuse comb. 749(a)t; wood comb. 749(a)t, 838(a)							
6.6-38	86854-07-1	**11*H*-Benzo[*ghi*]cyclopenta [*pqr*]perylen-11-one**	biomass comb.	838(a)	799(a)				
			coal comb.	749(a)					
			diesel	749(a)					
		refuse comb. 749(a); wood comb. 749(a), 838(a)							
6.6-39S	86854-15-1	**13*H*-Cyclopenta[*rst*] pentaphen-13-one**	coal comb.	749(a)t					
			diesel	749(a)t					
			refuse comb.	749(a)t					
			wood comb.	749(a)t					

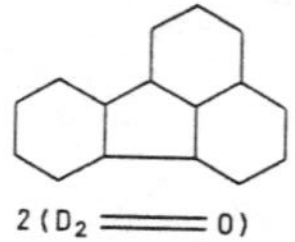

2(D_2 = O)

6.6-1

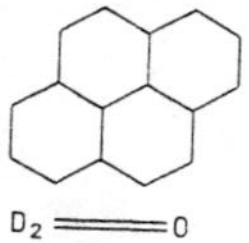

D_2 = O

6.6-10

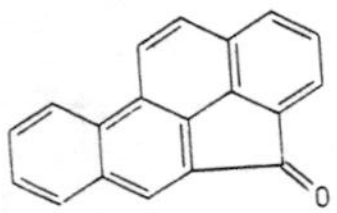

6.6-26

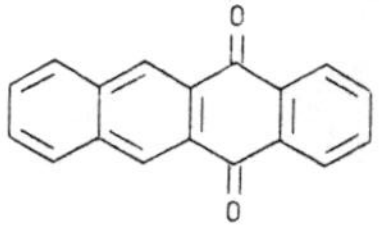

6.6-33

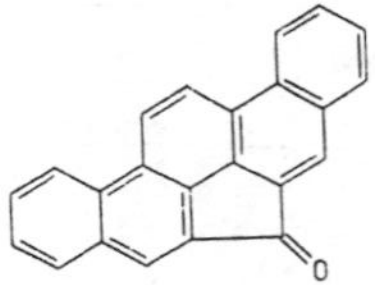

6.6-35

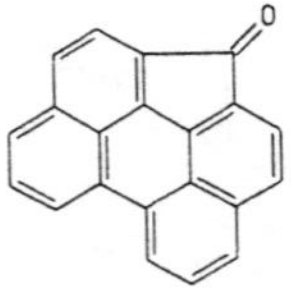

6.6-37

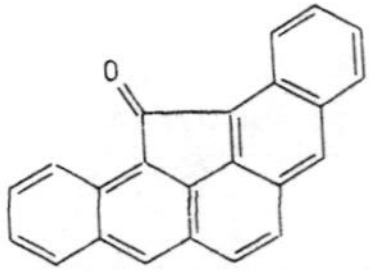

6.6-39

CHAPTER 7

Aldehydes

7.0 Introduction

Aldehydes are organic compounds that contain a carbon atom doubly bonded to an oxygen atom and singly bonded to a hydrogen atom and to another carbon atom. Acetaldehyde, for example, has the structure

$$H_3C-C\begin{matrix}\diagup H\\ \diagdown\!\!\diagdown O\end{matrix}$$

often written in condensed form as CH_3CHO. An aldehyde is often the first stable product of atmospheric oxidation sequences; numerous direct sources exist as well. We divide atmospheric aldehydes into three groups: (1) alkanic, (2) olefinic, and (3) cyclic and aromatic.

7.1 Alkanic Aldehydes

After the hydrocarbons, the alkanic aldehydes are the most abundant and ubiquitous of the organic gases of the atmosphere. Nearly forty appear in Table 7.1. They are commonly detected in both gas and aerosol phases, in precipitation, and indoors.

The C_1-C_5 aldehydes have many sources. The are widely used in industry and are products of the incomplete combustion of petroleum fuels and biomass. As has been seen in previous chapters, the smaller alkanic aldehydes are products of the atmospheric chemistry of hydrocarbons, ethers, alcohols, and other organic compounds. Vegetation and other natural sources emit many of these compounds, both small and large.

The most abundant of the alkanic aldehydes is formaldehyde. It has been measured sufficiently often that a concentration plot can be constructed. As seen in Figure 7.1-1, its urban concentrations sometimes exceed 100 parts per billion. In remote areas, concentrations near one part per billion are typical. This is presumably related to HCHO formation from methane reactions.

Concentrations of several parts per billion HCHO are typical of forests and grassland, where the HCHO is produced at least in part by the atmospheric chemistry of vegetative hydrocarbons. In more remote areas of the troposphere and in the stratosphere much lower HCHO concentrations occur. In these regions, the methane oxidation chain (see Chapter 3) is the only significant source of formaldehyde.

The gas phase alkanic aldehydes are among the few atmospheric compounds that are capable of dissociating upon the absorption of tropospheric solar radiation, and the photolysis of aldehydes represents an important source of free radicals for the initiation of atmospheric chemical cycles. Formaldehyde has two cleavage paths: one to stable species and one to free radicals:[1213]

$$HCHO \xrightarrow[\lambda<360\text{ nm}]{h\nu} H_2 + CO \qquad \text{(R7.1-1)}$$

$$HCHO \xrightarrow[\lambda<335\text{ nm}]{h\nu} H\cdot + CHO\cdot \qquad \text{(R7.1-2)}$$

The larger aldehydes cleave at the bond adjacent to the formyl group:

$$RCHO \xrightarrow{h\nu} R\cdot + CHO\cdot \qquad \text{(R7.1-3)}$$

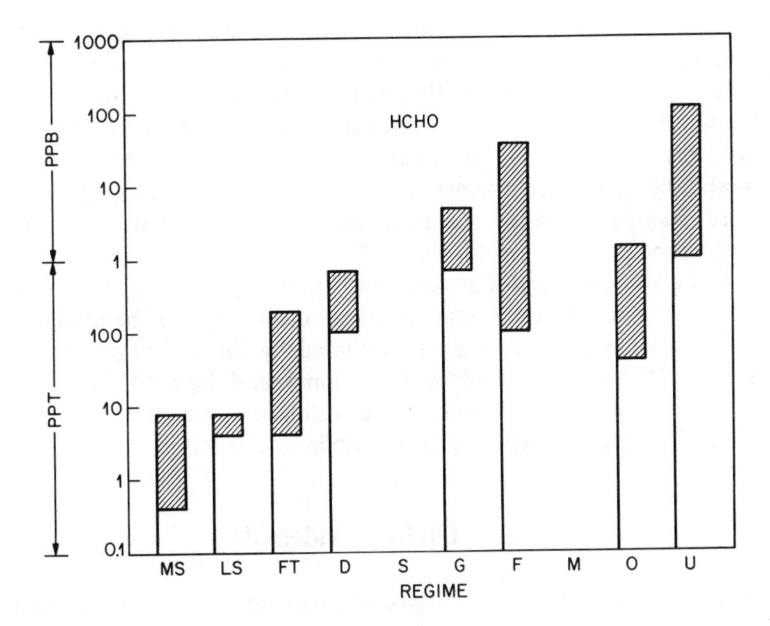

Fig. 7.1-1 Approximate concentration ranges of HCHO is different atmospheric regimes. The regime code is given in the caption of Fig. 2.1-1

Peroxyl radicals are then produced by

$$H\cdot + O_2 \xrightarrow{M} HO_2\cdot \quad (R7.1\text{-}4)$$

$$R\cdot + O_2 \xrightarrow{M} RO_2\cdot \quad (R7.1\text{-}5)$$

$$CHO\cdot + O_2 \rightarrow CO + HO_2\cdot \quad (R7.1\text{-}6)$$

An alternative gas phase sequence for the aldehydes is reaction with OH·. This occurs by H abstraction from the formyl group,[1237] at least for the smaller aldehydes:

$$CH_3CHO + OH\cdot \rightarrow CH_3C(O)\cdot + H_2O \quad (R7.1\text{-}7)$$

followed by

$$CH_3C(O)\cdot + O_2 \rightarrow CH_3C(O)O_2\cdot \quad (R7.1\text{-}8)$$

Kinetic studies show that the rates of OH· attack on glyoxal (OHCCHO) and methylglyoxal ($CH_3C(O)CHO$) are similar to those of formaldehyde and acetaldehyde, indicating that the additional -C(O)- group has little influence on reactivity.[1238] The peroxyl radicals of (R7.1-5) and (R7.1-8) follow the sequences of the alkanes (see Chapter 3). As a result, the atmospheric gas phase chemistry of all the alkanic aldehydes progresses toward fragmentation of the molecule and the ultimate production of carbon monoxide.

Small alkanic aldehydes are common in atmospheric droplets. It is clear that the products of their aqueous phase chemistry will be organic acids. These oxidation processes may be initiated by the hydroxyl radical[1328] or may be catalyzed by dissolved transition metal ions.[1326] The details of these processes, which remain to be specified, probably play important roles in the aqueous chemistry of atmospheric droplets.

7.2 Olefinic Aldehydes

Nearly twenty unsaturated aliphatic aldehydes occur in the atmosphere. The sources are varied and include biomass and petroleum combustion, biological decay processes, and vegetation. The smaller of these compounds have often been detected in urban areas in the gas phase.

The olefinic aldehydes react readily with OH·. This occurs largely by addition to the double bond,[1237] but H abstraction from the formyl group is significant at least for methacrolein.[1239] The result in either case will be the formation of small multifunctional organic molecules. Unsaturated dicarbonyls (such as species 7.2-6) may be subject to photolysis, though the mechanisms and products are not yet well established.[1329] The compounds react with ozone as well,[1236] but not rapidly enough for the process to be important in ambient air.

7.3 Cyclic and Aromatic Aldehydes

Fifty atmospheric compounds possess formyl groups attached to cyclic or aromatic systems. Most of these compounds are found in motor vehicle exhaust; a few are emitted from vegetation. Benzaldehyde, the simplest, has many industrial and combustion sources and is a common constitutent of urban air.

also common in urban atmospheres:

$$C_6H_5CHO \xrightarrow{OH\cdot} C_6H_5C(O)\cdot \xrightarrow{O_2} C_6H_5C(O)O_2\cdot \xrightarrow{NO_2} C_6H_5C(O)O_2NO_2 \quad \text{(R7.3-9)}$$

For more complex cyclic or aromatic aldehydes, OH· addition to a ring will be increasingly favored and a multiplicity of products are expected in the atmosphere.

Table 7.1. Alkanic Aldehydes

Species Number	Registry Number	Name	Emission Source	Emission Ref.	Detection Ref.	Bioassay Code	Bioassay −MA	Bioassay +MA	Bioassay Ref.
7.1-1	50-00-0	**Formaldehyde**	animal waste	25,80	23*,89	CCC	SP		B37
			auto	15*,217*	748*(c),767*(c)	MDR	+		B12
		biomass comb. 354; building mtls. 9*,270*; casting resin 97*;			746*(f),748*(f)	REC	+		B9
		charcoal mfr. 58*; chemical mfr. 1164; coffee mfr. 133;			748*(i)	STI00	?	+	B38
		diesel 1,330; H CHO mfr. 535*; lacquer mfr. 427,428;			748*(r),761*(r)	STI35	NEG	NEG	B38
		microbes 302; paint 133*,643*; petroleum mfr. 221*,390;			632(I),980*(I)	STI37	NEG	NEG	B38
		phthalic acid mfr. 535*; plastics comb. 354;				STI98	?	?	B38
		plastics mfr. 643*; polymer comb. 853;				ASPD	+		B10
		printing 133*,535*; refuse comb. 571,661*; tobacco smoke 4,554*;				SRL	+		B30
		turbine 594,836*; vegetation 552e; volcano 1007;				NEU	+		B32
		wood comb. 875*,890							
7.1-2S	17030-74-9	**Formyl ion**			1167(S)t				
7.1-3	75-07-0	**Acetaldehyde**	acrylonitrile mfr.	626	171*,550*	SCE	+		B6
			animal waste	18,25	399(a),681*(a)	REC	+		B9
		auto 15*,217*; biomass comb. 183,354; casting resin 97*;			748*(c)				
		charcoal mfr. 58*; chemical mfr. 1147*,1164; coffee mfr. 133*,598;			748*(f)				
		diesel 1,176; fish processing 110,255*; H CHO mfr. 535*;			748*(i)				
		insects 471; landfill 628; microbes 302; onion odor 636;			748*(r)				
		paint 643*,695; petroleum mfr. 111,221*; phthalic acid mfr. 535*;			632(I),1123*(I)				
		plastics comb. 354; plastics mfr. 643*,695; polymer comb. 853;							
		printing 133*,535*; refuse comb. 439,665*; rendering 992*;							
		sewage tmt. 422*,695; skunk odor 1162; starch mfr. 695,1071;							
		tobacco smoke 446,448,396(a),554(a); turbine 273,414;							

Species Number	Registry Number	Name	Emission Source	Emission Ref.	Detection Ref.	Code	Bioassay −MA	Bioassay +MA	Bioassay Ref.
		vegetation 552e; volcano 234,1007; whiskey mfr. 1158; wood comb. 875*,890; wood pulping 695,1071							
7.1-4	4746-86-5	**Hydroxyethanal**	turbine	414					
7.1-5	123-38-6	**Propanal**	acrylonitrile mfr.	626	358,380				
			animal waste	25,80	681*(a),967*(a)				
		auto 15*,217*; biomass comb. 376,535*; coffee mfr. 133; diesel 176,381; fish processing 254*,946; insects 471; microbes 302; onion odor 636; paint 133,643*; petroleum mfr. 221*,390; printing 133,535*; rendering 102,833; sewage tmt. 422*; skunk odor 1162; tobacco smoke 613,643*,396(a); turbine 273,414; vegetation 552e; volcano 234; whiskey mfr. 1158; wood comb. 875*			748*(r)				
7.1-6	123-72-8	**Butanal**	animal waste	18,695	165,582				
			auto	132,643*	681*(a),967*(a)				
		biomass comb. 376; coffee mfr. 133; diesel 176,330; fish processing 254*; lacquer mfr. 427; microbes 302; onion odor 636; paint 695,1071; petroleum mfr. 390; printing 133; rendering 833,837; tobacco smoke 643*,396(a); turbine 414; vegetation 552e; whiskey mfr. 1158; wood comb. 875*			748*(c) 748*(f)				
7.1-7	78-84-2	**Isobutanal**	animal waste	25,80	162,582				
			auto	55,132					
		biomass comb. 376; diesel 686,808*; fish processing 254*; industrial 133; insects 471; microbes 302; rendering 833; tobacco smoke 634,396(a); turbine 414; vegetation 552e							

Table 7.1. *(Continued)*

Species Number	Registry Number	Name	Emission Source	Emission Ref.	Detection Ref.	Code	Bioassay −MA	Bioassay +MA	Bioassay Ref.
7.1-8	96-17-3	**2-Methylbutanal**	microbes	302					
			rendering	833					
		tobacco smoke 634; whiskey mfr. 1158							
7.1-9	97-96-1	**2-Ethylbutanal**	auto	519					
7.1-10	110-62-3	**Pentanal**	animal waste	25,80	886				
			auto	132,519	748*(c)				
		building mtls. 856; diesel 176,381; microbes 302;			748*(f)				
		paint 695,1071; plastics mfr. 643*; rendering 833,837;			748*(i)				
		tobacco smoke 634; turbine 414; vegetation 552e; wood comb. 875*			1110*(I)				
7.1-11	590-86-3	**Isopentanal**	animal waste	695,1071,892(a)					
			auto	67*,643*					
		biomass comb. 597; diesel 686; microbes 302; plastics mfr. 643*; rendering 833,837; starch mfr. 695,1071; tobacco smoke 634; turbine 414; vegetation 552e							
7.1-12	630-19-3	**Neopentanal**	auto	284,313					
			turbine	414					
7.1-13	123-15-9	**2-Methylpentanal**	tobacco smoke	634					
7.1-14	4221-03-8	**5-Hydroxypentanal**			214*(a)				

Species Number	Registry Number	Name	Emission Source	Emission Ref.	Detection Ref.	Code	Bioassay −MA	Bioassay +MA	Bioassay Ref.
7.1-15	111-30-8	**1,5-Pentanedial**			214*(a)	STI00	?	?	B38
						STI35	NEG	NEG	B38
						STI37	NEG	NEG	B38
						STI98	?	?	B38
7.1-16	66-25-1	**Hexanal**	animal waste	997,1075,892(a)	886				
			auto	807*,881	748*(f)				
		building mtls. 856; diesel 176,330; microbes 302;			748*(r)				
		rendering 833,837; tobacco smoke 634,643*; turbine 273,414;			1110*(I),1124(I)				
		vegetation 503e,552e; wood comb. 875*							
7.1-17	925-54-2	**2-Methylhexanal**	animal waste	892(a)					
			biomass comb.	597					
7.1-18	55320-57-5	**3,3-Dimethylhexanal**			1113(I)				
7.1-19	several	**Ethylhexanal**			820(I)				
7.1-20	34067-76-0	**6-Hydroxyhexanal**			214*(a)				
7.1-21S	26566-61-0	**Galactose**	tobacco smoke	542(a)					
7.1-22	1072-21-5	**1,6-Hexanedial**			214*(a)				
7.1-23	111-71-7	**Heptanal**	animal waste	25,80	886				
			diesel	176,381	1110*(I),1112(I)				
		microbes 302; rendering 833,837; turbine 414;							
		vegetation 514e,552e							

Table 7.1. *(Continued)*

Species Number	Registry Number	Name	Emission Source	Emission Ref.	Detection Ref.	Bioassay Code	−MA	+MA	Ref.
7.1-24	22054-14-3	**7-Hydroxyheptanal**			214*(a)				
7.1-25	53185-69-6	**1,7-Heptanedial**			214*(a)				
7.1-26	4359-57-3	**Octanal**	animal waste	25,80	86,886				
			diesel	176,381	1110*(I),1112(I)				
		microbes 302; rendering 833,837; vegetation 503e,552e							
7.1-27	124-19-6	**Nonanal**	biomass comb.	821	886,894				
			microbes	302	882(a)				
		rendering 833,837; vegetation 503e,552e			820(I),1108(I)				
7.1-28	112-31-2	**Decanal**	animal waste	25,80	886,894				
			auto	67*	882(a)				
		rendering 833,837; vegetation 504e,552e			820(I),1108(I)				
7.1-29	112-44-7	**Undecanal**	vegetation	504e,552e	886,894				
7.1-30	112-54-9	**Dodecanal**	vegetation	504e,531	886,894				
7.1-31	10486-19-8	**Tridecanal**	vegetation	504e	894				
7.1-32	124-25-4	**Tetradecanal**	vegetation	504e	894				
7.1-33	2765-11-9	**Pentadecanal**			894				
7.1-34	629-80-1	**Hexadecanal**			894				
7.1-35	629-90-3	**Heptadecanal**			894				

Species Number	Registry Number	Name	Emission Source	Emission Ref.	Detection Ref.	Code	Bioassay −MA	Bioassay +MA	Bioassay Ref.
7.1-36	26627-85-0	**Hexacosanal**			901*(a)				
7.1-37	22725-64-0	**Octacosanal**			901*(a)				
7.1-38	22725-63-9	**Triacontanal**			901*(a)				
7.1-39	78-98-8	**2-Oxopropanal**	tobacco smoke	396					

$^{+}HC{=}O$

7.1-2

$HOCH_2CH(OH)CH(OH)CH(OH)CH(OH)CHO$

7.1-21

Table 7.2. Olefinic Aldehydes

Species Number	Registry Number	Name	Emission Source	Emission Ref.	Detection Ref.	Bioassay Code	−MA	+MA	Ref.
7.2-1S	107-02-8	**Acrolein**	animal waste	695,1071	5*,411	STI00	NEG	?	B38
			auto	15*,243*	632(I)	STI35	NEG	+	B38
		biomass comb. 354,376; chemical mfr. 1164; coffee mfr. 133*,598; diesel 314*,432*; fish processing 110,241*; lacquer mfr. 8,427; meat cooking 598; paint 133*,1071; plastics comb. 46; polymer comb. 612; printing 58*,133*; refuse comb. 665; starch mfr. 695,1071; syn. rubber mfr. 58*; tobacco smoke 448,643*,396(a),554(a); turbine 273,414; vegetation 635; whiskey mfr. 1158; wood comb. 875*				STI37	NEG	NEG	B38
						STI98	NEG	NEG	B38
7.2-2S	78-85-3	**Methacrolein**	auto	67*,519	597				
			biomass comb.	597					
		tobacco smoke 634; volcano 234							
7.2-3S	4170-30-3	**Crotonaldehyde**	auto	15*,217*	869	STI00	+	+	B38
			diesel	1,314*	820(I)	STI35	NEG	NEG	B38
		polymer comb. 853; tobacco smoke 643*,396(a),554(a); turbine 414; volcano 234; wood comb. 875*				STI37	NEG	NEG	B38
						STI98	NEG	NEG	B38
7.2-4	497-03-0	**2-Methylcrotonaldehyde**	auto	217*					
7.2-5	13153-14-5	**2,3-Dimethylcrotonaldehyde**	solvent	134					
7.2-6		**2-Methyl-4-pentene-1,5-dial**			532(a)				

Species Number	Registry Number	Name	Emission Source	Emission Ref.	Detection Ref.	Code	Bioassay −MA	Bioassay +MA	Bioassay Ref.
7.2-7	1070-66-2	**2-Methylenehexanal**	vegetation	512e					
7.2-8	6728-26-3	*trans*-**2-Hexenal**	microbes vegetation	302 552e					
7.2-9	142-83-6	**2,4-Hexadienal**	biomass comb.	597					
7.2-10	5095-43-2	**6-Hydroxy-4-methyl-2,4-hexadienal**			532(a)				
7.2-11	2463-63-0	**2-Heptenal**	animal waste	892(a)					
7.2-12	5910-85-0	**2,4-Heptadienal**	animal waste	892(a)					
7.2-13	106-23-0	**3,7-Dimethyl-6-octenal**	vegetation	498,506e					
7.2-14	several	**4-Hydroxy-3,7-dimethyl-6-octenal**	vegetation	498					
7.2-15	38743-20-3	**Octadienal**	rendering	879t					
7.2-16	5392-40-5	**3,7-Dimethyl-2,6-octadienal**	vegetation	498,505e					
7.2-17	6750-03-4	**2,4-Nonadienal**	animal waste	892(a)					
7.2-18	26370-28-5	**2,6-Nonadienal**	rendering vegetation	1163t 531					

Table 7.2. *(Continued)*

Species Number	Registry Number	Name	Emission Source	Ref.	Detection Ref.	Code	Bioassay −MA	+MA	Ref.
7.2-19	2363-88-4	**2,4-Decadienal**	animal waste	892(a)					

$OCHCH:CH_2$
7.2-1

$OCHCMe:CH_2$
7.2-2

$MeCH:CHCHO$
7.2-3

Table 7.3. Cyclic and Aromatic Aldehydes

Species Number	Registry Number	Name	Emission Source	Emission Ref.	Detection Ref.	Bioassay Code	Bioassay −MA	Bioassay +MA	Bioassay Ref.
7.3-1S	100-52-7	**Benzaldehyde**	animal waste	855,892(a)	358,681*	STI00	NEG	NEG	B38
			auto	67*,217*	633(a)	STI35	NEG	NEG	B38
		biomass comb. 597,1018*; building mtls. 856; coal comb. 828t,1159; diesel 1,176; plastics mfr. 643*; refuse comb. 26(a),840(a); rendering 837; tobacco smoke 634,643*,396(a); turbine 273,414; wood comb. 875*			748*(c)	STI37	NEG	NEG	B38
					748*(i)	STI98	NEG	NEG	B38
					570(r),748*(r)				
					570(s)				
					820(I),1110*(I)				
7.3-2	529-20-4	**2-Methylbenzaldehyde**	auto	48,217*	86,525				
			diesel	137					
7.3-3	620-23-5	**3-Methylbenzaldehyde**	auto	537,643*	86,525				
7.3-4	104-87-0	**4-Methylbenzaldehyde**	auto	537,643*	358,455*				
			diesel	686					
		turbine 414; wood comb. 875*							
7.3-5	28351-09-9	**Dimethylbenzaldehyde**	auto	537	86,896				
			diesel	138					
7.3-6S	116-26-7	**Safranal**	vegetation	531					
7.3-7	53951-50-1	**Ethylbenzaldehyde**	auto	217*,1154	896				
			diesel	176					

Table 7.3. *(Continued)*

Species Number	Registry Number	Name	Emission Source	Emission Ref.	Detection Ref.	Bioassay Code	−MA	+MA	Ref.
7.3-8	several	**Ethyltetramethyl-benzaldehyde**	diesel	138					
7.3-9	43145-54-6	**Vinylbenzaldehyde**	coal comb.	828(a)t					
7.3-10	28785-06-0	**Propylbenzaldehyde**	diesel	138					
7.3-11	62708-42-3	**Allylbenzaldehyde**			880t				
7.3-12	several	**Dihydrocuminaldehyde**	vegetation	531					
7.3-13	122-03-2	**Cuminaldehyde**	vegetation	507e,531					
7.3-14	several	**Butylbenzaldehyde**	diesel	138					
7.3-15	123-11-5	**4-Methoxybenzaldehyde**	biomass comb.	1089(a)					
7.3-16S	50984-52-6	**Anisaldehyde**	diesel microbes vegetation	138 302 510e,533					
7.3-17S	120-14-9	**Veratraldehyde**	biomass comb. vegetation	1089(a) 514e					
7.3-18S	90-02-8	**Salicylaldehyde**	auto diesel	217* 808,809,809(a)	973*(a)				
7.3-19	698-27-1	**2-Hydroxy-4-methyl-benzaldehyde**			973t				

Species Number	Registry Number	Name	Emission Source	Emission Ref.	Detection Ref.	Bioassay Code	−MA	+MA	Ref.
7.3-20	123-08-0	**4-Hydroxybenzaldehyde**	vegetation	552e	973*(a)				
7.3-21S	121-33-5	**Vanillin**	biomass comb. diesel	1089(a) 138					
		vegetation 501,531; wood comb. 344*(a); wood pulping 219,267							
7.3-22S	134-96-3	**Syringicaldehyde**	wood comb.	344(a)					
7.3-23	several	**Allylhydroxybenz-aldehyde**	biomass comb. diesel	1089(a) 138					
7.3-24	81051-20-9	**2,3-Dihydroxy-6-methylbenz-aldehyde**	biomass comb.	891(a)t					
7.3-25S	643-79-8	**Phthalaldehyde**	diesel	138					
7.3-26S	623-27-8	**Terephthalaldehyde**	auto	311(a)	880t				
7.3-27	122-78-1	**Phenylacetaldehyde**			358 820(I)				
7.3-28S	104-55-2	**Cinnamaldehyde**	vegetation	510e					
7.3-29	1504-74-1	***o*-Methoxycinnam-aldehyde**	vegetation	514e					
7.3-30S	104-53-0	**β-Phenylpropanal**	vegetation	510e					

Table 7.3. *(Continued)*

Species Number	Registry Number	Name	Emission Source	Emission Ref.	Detection Ref.	Bioassay Code	Bioassay −MA	Bioassay +MA	Bioassay Ref.
7.3-31S	458-36-6	**Coniferlaldehyde**	vegetation wood comb.	531 344(a)					
7.3-32S	4206-58-0	**Sinapylaldehyde**	wood comb.	344(a)					
7.3-33	36884-28-3	**Phenylpentanal**	diesel	962t					
7.3-34	3218-36-8	**Biphenylcarbaldehyde**	auto diesel	858*(a) 684(a)t,804(a)					
7.3-35	several	**Methylbiphenylcarbaldehyde**	auto diesel	858*(a) 802(a),804(a)					
7.3-36	66-77-3	**1-Naphthalenecarbaldehyde**	auto biomass comb.	858*(a) 838(a)					
		diesel 138,811,684(a),690(a); refuse comb. 878(a); turbine 786*(a); wood comb. 838(a)							
7.3-37	66-99-9	**2-Naphthalenecarbaldehyde**	diesel	690(a),804(a)					
7.3-38	5084-46-8	**6-Methyl-2-naphthalene-carbaldehyde**	auto diesel wood comb.	858*(a)t 138,684(a),858*(a)t 838(a)					
7.3-39	several	**Dimethylnaphthalene-carbaldehyde**	diesel	690(a)					

Species Number	Registry Number	Name	Emission Source	Emission Ref.	Detection Ref.	Bioassay Code	Bioassay −MA	Bioassay +MA	Bioassay Ref.
7.3-40	several	**Trimethylnaphthalene-carbaldehyde**	diesel	690(a)					
7.3-41	several	**Methoxynaphthalene-carbaldehyde**	diesel	138					
7.3-42	several	**Naphthalenedicarbaldehyde**	diesel	690(a)					
7.3-43	several	**Acenaphthenecarbaldehyde**	biomass comb.	838(a)					
7.3-44	20615-64-9	**Fluorenecarbaldehyde**	diesel	690(a)					
7.3-45S	13601-88-2	**1-Phenanthrenecarbaldehyde, 1,2,3,4,4a,9,10,10a-octahydro-1,4a-dimethyl-7-(1-methylethyl)-**	biomass comb.	1089(a)					
7.3-46	26842-00-2	**2-Phenanthrenecarbaldehyde**	auto diesel	858*(a)t 802(a),804(a)					
7.3-47	several	**Methylphenanthrene-carbaldehyde**	auto diesel	858*(a)t 802(a),804(a)					
7.3-48	642-31-9	**9-Anthracenecarbaldehyde**	biomass comb. diesel wood comb.	838(a) 684(a) 838(a)	466(a)				
7.3-49	7072-00-6	**Methylanthracene-carbaldehyde**	diesel	684(a)t					

Table 7.3. *(Continued)*

Species Number	Registry Number	Name	Emission Source	Emission Ref.	Detection Ref.	Bioassay Code	Bioassay −MA	Bioassay +MA	Bioassay Ref.
7.3-50	68967-09-9	**Pyrenecarbaldehyde**	wood comb.	838(a)					

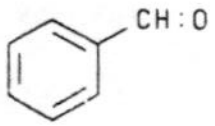

7.3-1 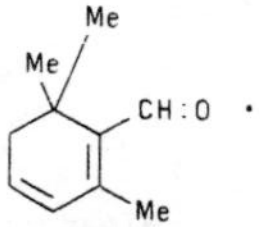 7.3-6 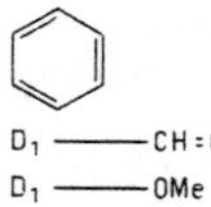 7.3-16 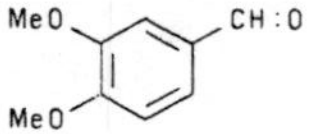 7.3-17

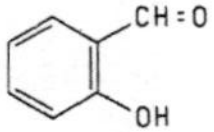

7.3-18 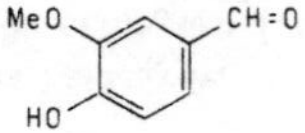 7.3-21 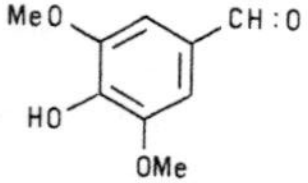 7.3-22 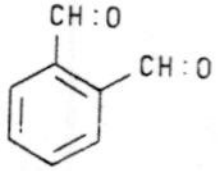 7.3-25

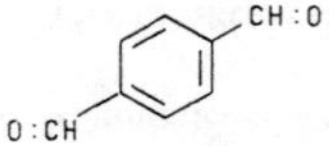

7.3-26 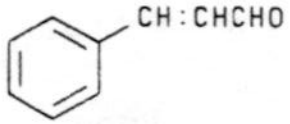 7.3-28 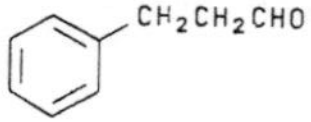 7.3-30

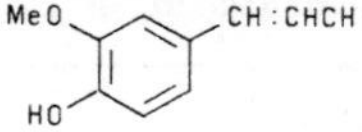

7.3-31 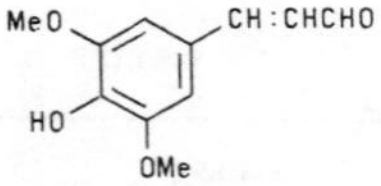 7.3-32 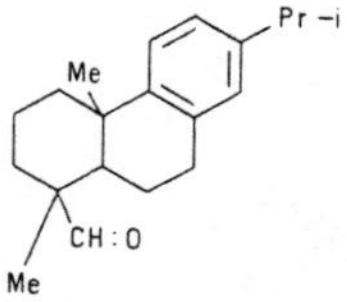 7.3-45

CHAPTER 8

Carboxylic Acid Derivatives

8.0 Introduction

Carboxylic acids are organic compounds which transfer protons to bases more readily than does water and which include the *carboxyl* group -C(O)OH. A compound which can be thought of as representing the combination (with loss of a water molecule) of two carboxylic acid molecules is termed an *anhydride*. For example, acetic acid has the structure

$$H_3C{-}C{\overset{\displaystyle O}{\underset{\displaystyle OH}{<}}}$$

and acetic anhydride has the structure

$$H_3C{-}C(O){-}O{-}C(O){-}CH_3$$

Anhydrides are relatively unstable in the atmosphere and thus are rarely found there. Much more common are the *esters*, in which the hydrogen atom of the carboxyl group is replaced by an organic group, as in methyl acetate:

$$H_3C-C(=O)-OCH_3$$

Among the related compounds important for atmospheric chemistry are the peroxides, which contain either the group -OOH or -OOR (as in methylhydroperoxide, CH_3OOH).

We divide the carboxylic acid derivatives into five groups: (1) anhydrides, (2) alkanic esters, (3) olefinic esters, (4) cyclic and aromatic esters, and (5) phthalates and peroxides. The phthalates are diesters of benzenedicarboxylic acid, and are very common industrial chemicals.

8.1 Anhydrides

Eleven anhydrides are listed in Table 8.1. Three are emitted during their widespread industrial use, while a few others are products of diesel combustion. The larger anhydrides have been identified by several research groups as minor constituents of aerosol particles, so their ocassional presence there seems well established.

Anhydrides are expected to have short lifetimes in the atmosphere, but their chemistry in such environments has not been studied.

8.2-8.4 Esters

Nearly 200 esters are known to be emitted into or to have been identified in the atmosphere. The majority are found as constituents of aerosol particles, but many of the smaller compounds occur in the gas phase. Most of the atmospheric esters have no known source except vegetation; many of the enticing smells of nature involve esters as odor components. The total fluxes from vegetation may be substantial, but have not been assessed. Some of the simpler esters are emitted from animal waste and the acetates are emitted from a variety of industrial processes; the fluxes in each case appear to be small.

Several esters have been identified in raindrops and indoors. No concentrations have been determined. For the gas phase species, however, it is possible to make a rough concentration estimate since the simple esters are customarily detected by humans at an olfactory threshold of 1-100 ppb.[1286,1287] Thus, ester concentrations of a few ppb are likely near vegetation, since the aromas of trees, flowers, and other plants are often readily detected.

The rate constants for the gas phase reactions of several simple esters with the hydroxyl radical are known from laboratory studies. The esters react a bit more rapidly than do the corresponding alkanes, indicating that the C-H bonds are somewhat weaker.[588] The preferred site for attack is the alkoxyl end of the molecule rather than the acyl end, with hydrogen abstraction being followed in the atmosphere by peroxyl radical formation. For ethyl formate, for example,

$$OHC{-}O{-}CH_2CH_3 + OH\cdot \rightarrow OHC{-}O{-}\dot{C}HCH_3 + H_2O \quad (R8.2\text{-}1)$$

$$OHC{-}O{-}\dot{C}HCH_3 + O_2 \xrightarrow{M} OHC{-}O{-}C(\dot{O}_2)HCH_3 \quad (R8.2\text{-}2)$$

The probability of 1:5 intramolecular H transfer in this system is large.[439]

$$OHC{-}O{-}C(\dot{O}_2)HCH_3 \rightarrow O\dot{C}{-}O{-}C(O_2H)HCH_3 \quad (R8.2\text{-}3)$$

The transformed molecule will then undergo β scission:

$$O\dot{C}{-}O{-}C(O_2H)HCH_3 \rightarrow CO_2 + CH_3CHO + OH\cdot \quad (R8.2\text{-}4)$$

Ester chemistry in the atmosphere can thus be expected to proceed along lines similar to that of the alkanes, with the ultimate products being small oxygen-containing organic and inorganic molecules.

8.5 Phthalates and Peroxides

Phthalates see extensive use as polymer plasticizers and in other industrial applications. To a lesser degree, they are also products of combustion. They have moderate vapor pressures and long lifetimes. The result is that they are present in small quantities throughout the troposphere. They are readily detected in the gas phase, as aerosol particle constituents, in rain and snow, and indoors. There are no important atmospheric degradation processes for the phthalates.

A second group of compounds in Table 8.5 is the peroxides. Those known to be emitted into the atmosphere are generated by polymer combustion or by vegetation. Simple peroxides are thought to be formed in the gas phase in the atmosphere by disproportionation reactions such as

$$CH_3O_2\cdot + HO_2\cdot \rightarrow CH_3OOH + O_2 \qquad \text{(R8.5-1)}$$

The O-O bond in peroxides is susceptible to cleavage by solar photons, and the lifetime of peroxides in the atmosphere is thus short.

Table 8.1. Anhydrides

Species Number	Registry Number	Name	Emission Source	Emission Ref.	Detection Ref.	Bioassay Code	Bioassay −MA	Bioassay +MA	Bioassay Ref.
8.1-1S	108-24-7	**Acetic anhydride**	chemical mfr.	1122*					
			industrial	60					
8.1-2S	108-31-6	**Maleic anhydride**	maleic anhyd. mfr.	351		STI00	NEG	NEG	B38
			phthal anhyd. mfr.	58*,64*		STI35	NEG	NEG	B38
						STI37	NEG	NEG	B38
						STI98	NEG	NEG	B38
8.1-3S	85-44-9	**Phthalic anhydride**	building mtls.	9*	799(a),1086(a)				
			lacquer mfr.	428					
			phthal anhyd. mfr.	58*,64*					
8.1-4	30140-42-2	**Methylphthalic anhydride**			799(a)				
8.1-5	several	**Dimethylphthalic anhydride**			799(a)				
8.1-6	several	**Trimethylphthalic anhydride**			799(a)				
8.1-7S	81-84-5	**Naphthalene-1,8-dicarboxylic acid anhydride**	diesel	690(a)	799(a),803(a)				
8.1-8	several	**Methylnaphthalene dicarboxylic acid anhydride**	diesel	690(a)	799(a)t				

Table 8.1. *(Continued)*

Species Number	Registry Number	Name	Emission Source	Emission Ref.	Detection Ref.	Bioassay Code	Bioassay −MA	Bioassay +MA	Bioassay Ref.
8.1-9	several	**Dimethylnaphthalene dicarboxylic acid anhydride**	diesel	690(a)					
8.1-10	several	**Hydroxynaphthalene dicarboxylic acid anhydride**	diesel	690(a)					
8.1-11S	76895-43-7	**Pyrene-3,4-dicarboxylic acid anhydride**			689(a),799(a)t				

O(Ac) 2

8.1-1

8.1-2

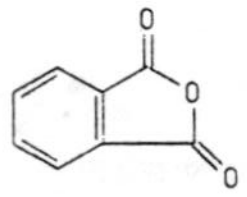

8.1-3

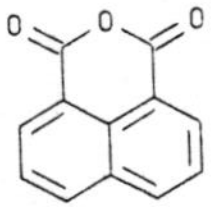

8.1-7

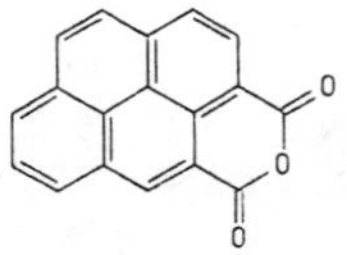

8.1-11

Table 8.2. Alkanic Esters

Species Number	Registry Number	Name	Emission Source	Emission Ref.	Detection Ref.	Bioassay Code	Bioassay −MA	Bioassay +MA	Bioassay Ref.
8.2-1S	107-31-3	**Methyl formate**	animal waste	166,1075					
			auto	217*					
		biomass comb. 354; solvent 598; tobacco smoke		613;					
		turbine 414; vegetation 552e							
8.2-2	109-94-4	**Ethyl formate**	animal waste	166					
			tobacco smoke	634					
8.2-3	592-84-7	**Butyl formate**	solvent	598					
8.2-4	762-75-4	*tert*-**Butyl formate**	building mtls.	856					
8.2-5S	105-85-1	**Citronellyl formate**	vegetation	506e					
8.2-6S	2142-94-1	**Neryl formate**	vegetation	510e,517e					
8.2-7S	115-99-1	**Linalyl formate**	vegetation	514e					
8.2-8S	104-57-4	**Benzyl formate**	vegetation	552e					
8.2-9	104-62-1	**Phenylethyl formate**	vegetation	552e					

Table 8.2. *(Continued)*

Species Number	Registry Number	Name	Emission Source	Emission Ref.	Detection Ref.	Code	Bioassay −MA	Bioassay +MA	Bioassay Ref.
8.2-10S	79-20-9	**Methyl acetate**	animal waste	166,436*	591				
			biomass comb.	354	1123*(I)				
		lacquer mfr. 428; petroleum mfr. 103,111; solvent 58*,439; tobacco smoke 613,396(a)							
8.2-11	141-78-6	**Ethyl acetate**	adhesives	11	358,591				
			animal waste	166,436*	820(I),1108(I)				
		biomass comb. 969*; building mtls. 856; chemical mfr. 1160; landfill 628; microbes 302; petroleum mfr. 103,221; solvent 598; tobacco smoke 396(a); vegetation 509e,552e; whiskey mfr. 32							
8.2-12	109-60-4	**Propyl acetate**	animal waste	327,1075	358t,597				
			solvent	598	820(I)				
			vegetation	552e					
8.2-13	591-87-7	**Allyl acetate**			628t				
8.2-14	108-21-4	**Isopropyl acetate**	animal waste	166,1075	597,628				
			biomass comb.	597t	973(a)				
		solvent 58*,88; vegetation 552e							
8.2-15	123-86-4	**Butyl acetate**	animal waste	327,1075	591,601				
			building mtls.	856					
		chemical mfr. 462; polymer mfr. 103; solvent 78,598; vegetation 552e							

Species Number	Registry Number	Name	Emission Source	Emission Ref.	Detection Ref.	Code	Bioassay −MA	Bioassay +MA	Bioassay Ref.
8.2-16	110-19-0	**Isobutyl acetate**	animal waste	166,1075	597				
			microbes	302					
			solvent	88,134					
8.2-17	105-46-4	*sec*-**Butyl acetate**	solvent	588					
8.2-18	540-88-5	*tert*-**Butyl acetate**	building mtls.	856					
8.2-19	628-63-7	**Pentyl acetate**	vegetation	552e	358				
8.2-20	123-92-2	**Isopentyl acetate**	vegetation	513e,552e	591				
			whiskey mfr.	32					
8.2-21	142-92-7	**Hexyl acetate**	vegetation	510e,888					
8.2-22S	103-09-3	**1-Ethylhexyl acetate**	vegetation	552e					
8.2-23	112-06-1	**Heptyl acetate**	turbine	414					
8.2-24	112-14-1	**Octyl acetate**	turbine	414					
			vegetation	510e					
8.2-25	143-13-5	**Nonyl acetate**	vegetation	552e					
8.2-26	112-17-4	**Decyl acetate**	vegetation	514e,552e					
8.2-27	1731-81-3	**Undecyl acetate**	vegetation	552e					
8.2-28		**Propyltridecyl acetate**			597t				

Table 8.2. *(Continued)*

Species Number	Registry Number	Name	Emission Source	Emission Ref.	Detection Ref.	Bioassay Code	Bioassay −MA	Bioassay +MA	Bioassay Ref.
8.2-29	629-70-9	**Hexadecyl acetate**	vegetation	552e					
8.2-30	822-23-1	**Octadecyl acetate**	vegetation	552e					
8.2-31	108-05-4	**Vinyl acetate**	biomass comb.	354	383,597	CCC	I		B37
			industrial	60		STP00	NEG	NEG	B36
						STP35	NEG	NEG	B36
						STP37	NEG	NEG	B36
						STP98	NEG	NEG	B36
8.2-32	2497-18-9	*trans*-**2-Hexenyl acetate**	vegetation	552e					
8.2-33	1708-82-3	**3-Hexenyl acetate**	vegetation	512e					
8.2-34	2380-48-5	**3-Octenyl acetate**	vegetation	552e					
8.2-35	150-84-5	**Citronellyl acetate**	vegetation	506e					
8.2-36S	105-87-3	**Geranyl acetate**	vegetation	531					
8.2-37	141-12-8	**Neryl acetate**	vegetation	510e,531					
8.2-38	115-95-7	**Linalyl acetate**	vegetation	516e,531					
8.2-39S	30232-11-2	**Methylcyclohexyl acetate**	acetyl. mfr.	128					
8.2-40S	2623-23-6	*l*-**Menthyl acetate**	vegetation	516e					

Species Number	Registry Number	Name	Emission Source	Emission Ref.	Detection Ref.	Code	Bioassay −MA	Bioassay +MA	Ref.
8.2-41S	76-49-3	**Bornyl acetate**	vegetation	506e,1066					
8.2-42S	80-26-2	**α-Terpinyl acetate**	vegetation	510e					
8.2-43S	10198-23-9	**β-Terpinyl acetate**	vegetation	510e					
8.2-44S	10235-63-9	**γ-Terpinyl acetate**	vegetation	510e					
8.2-45	122-79-2	**Phenyl acetate**	solvent	134	880t				
8.2-46	140-39-6	**4-Methylphenyl acetate**	vegetation	507e					
8.2-47	140-11-4	**Benzyl acetate**	tobacco smoke vegetation	634 500,501	820(I)	CCC	SP		B37
8.2-48	93-92-5	**α-Methylbenzyl acetate**	vegetation	517e,552e					
8.2-49	3056-59-5	**2-Phenylethyl acetate**	vegetation	501,510e					
8.2-50	4606-15-9	**Phenylpropyl acetate**	vegetation	510e					
8.2-51S	103-54-8	**Cinnamyl acetate**	vegetation	506e					
8.2-52	110-49-6	**2-Methoxyethyl acetate**			597				
8.2-53	111-15-9	**2-Ethoxyethyl acetate**	building mtls. solvent	856 88,442	1110*(I)				
8.2-54	93-28-7	**4-Allyl-2-methoxyphenyl acetate**	vegetation	510e					

Table 8.2. *(Continued)*

Species Number	Registry Number	Name	Emission Source	Emission Ref.	Detection Ref.	Code	Bioassay −MA	Bioassay +MA	Bioassay Ref.
8.2-55	68750-23-2	**5-Pentanol acetate**	vegetation	552e					
8.2-56	592-20-1	**2-Oxopropyl acetate**	vegetation	552e					
8.2-57S	108-59-8	**Dimethyl malonate**	vegetation	552e	1086(a)				
8.2-58	554-12-1	**Methyl propionate**	biomass comb.	597t					
8.2-59	105-37-3	**Ethyl propionate**	tobacco smoke vegetation	396(a) 552e					
8.2-60	637-78-5	**Isopropyl propionate**	animal waste	166,1075					
8.2-61	105-68-0	**Isopentyl propionate**	vegetation	514e					
8.2-62	2408-20-0	**Allyl propionate**	turbine	414					
8.2-63	105-91-9	**Neryl propionate**	vegetation	510e					
8.2-64	144-39-8	**Linalyl propionate**	vegetation	514e					
8.2-65	80-27-3	**α-Terpinyl propionate**	vegetation	510e					
8.2-66		**β-Terpinyl propionate**	vegetation	510e					
8.2-67		**γ-Terpinyl propionate**	vegetation	510e					
8.2-68S	106-65-0	**Dimethyl succinate**			1086(a)				

Species Number	Registry Number	Name	Emission Source	Emission Ref.	Detection Ref.	Code	Bioassay −MA	Bioassay +MA	Bioassay Ref.
8.2-69	1604-11-1	**Dimethyl methylsuccinate**			1086(a)				
8.2-70	623-42-7	**Methyl butyrate**	vegetation	552e					
8.2-71	105-54-4	**Ethyl butyrate**	tobacco smoke vegetation	396(a) 509e,552e					
8.2-72	109-21-7	**Butyl butyrate**	vegetation	552e					
8.2-73	2639-63-6	**Hexyl butyrate**	vegetation	552e					
8.2-74	62549-24-0	**3-Octenyl butyrate**	vegetation	552e					
8.2-75	141-16-2	**Citronellyl butyrate**	vegetation	506e					
8.2-76	999-40-6	**Neryl butyrate**	vegetation	510e					
8.2-77	29674-47-3	**Methyl β-hydroxybutyrate**	vegetation	552e					
8.2-78	97-62-1	**Ethyl isobutyrate**	vegetation	552e					
8.2-79	97-87-0	**Butyl isobutyrate**	vegetation	514e					
8.2-80	109-15-9	**Octyl isobutyrate**	vegetation	514e					
8.2-81	2345-26-8	**Geranyl isobutyrate**	vegetation	512e					
8.2-82	78-35-3	**Linalyl isobutyrate**	vegetation	514e					

Table 8.2. *(Continued)*

Species Number	Registry Number	Name	Emission Source	Emission Ref.	Detection Ref.	Code	Bioassay −MA	Bioassay +MA	Bioassay Ref.
8.2-83	2445-78-5	**2-Methylbutyl 2-methylbutyrate**	vegetation	552e					
8.2-84S	1119-40-0	**Dimethyl glutarate**			1086(a)				
8.2-85	14035-94-0	**Dimethyl 2-methylglutarate**			1086(a)				
8.2-86	539-82-2	**Ethyl pentanoate**	vegetation	552e					
8.2-87	10588-10-0	**Isobutyl pentanoate**	vegetation	552e					
8.2-88	108-64-5	**Ethyl isopentanoate**	tobacco smoke	396(a)					
8.2-89S	7779-73-9	**Bornyl isopentanoate**	vegetation	510e					
8.2-90	39255-32-8	**Ethyl 2-methylpentanoate**	tobacco smoke	396(a)					
8.2-91	624-45-3	**Methyl 4-oxopentanoate**			1086(a)				
8.2-92S	77-94-1	**Tributylcitrate**			570(r)				
8.2-93	123-66-0	**Ethyl hexanoate**	tobacco smoke vegetation	396(a) 552e					
8.2-94	2177-82-4	**Methyl 4-methylhexanoate**	vegetation	552e					
8.2-95	2177-83-5	**Methyl 5-methylhexanoate**	vegetation	552e					
8.2-96	several	**Methyl β-hydroxyhexanoate**	vegetation	552e					

Species Number	Registry Number	Name	Emission Source	Emission Ref.	Detection Ref.	Code	Bioassay −MA	Bioassay +MA	Bioassay Ref.
8.2-97	several	**Ethyl β-hydroxyhexanoate**	vegetation	552e					
8.2-98	106-73-0	**Methyl heptanoate**	vegetation	552e					
8.2-99	106-30-9	**Ethyl heptanoate**	vegetation	514e					
8.2-100S	123-79-5	**Dioctyl adipate**	PVC mfr.	521*	758(r)				
8.2-101	103-23-1	**bis(2-Ethylhexyl)adipate**			916(I)				
8.2-102	849-99-0	**Dicyclohexyl adipate**	refuse comb.	790(a)					
8.2-103S	1732-10-1	**Dimethyl azelate**			1086(a)				
8.2-104	103-24-2	**bis(2-Ethylhexyl)azelate**			1109*(I)				
8.2-105	110-42-9	**Methyl decanoate**			614(a)				
8.2-106		**Docosyl decanoate**			955(a)				
8.2-107	91283-94-2	**Tetracosyl decanoate**			955(a)				
8.2-108		**Hexacosyl decanoate**			955(a)				
8.2-109		**Octacosyl decanoate**			955(a)				
8.2-110		**Triacontyl decanoate**			955(a)				

Table 8.2. *(Continued)*

Species Number	Registry Number	Name	Emission Source	Emission Ref.	Detection Ref.	Code	Bioassay −MA	Bioassay +MA	Bioassay Ref.
8.2-111	1731-86-8	**Methyl undecanoate**			1103(a)				
8.2-112	111-82-0	**Methyl dodecanoate**			614(a),1086(a)				
8.2-113	106-33-2	**Ethyl dodecanoate**	vegetation	514e					
8.2-114	36617-18-2	**Eicosyl dodecanoate**			955(a)				
8.2-115	42231-82-3	**Docosyl dodecanoate**			955(a)				
8.2-116	68231-26-5	**Tetracosyl dodecanoate**			955(a)				
8.2-117		**Hexacosyl dodecanoate**			955(a)				
8.2-118	84744-07-0	**Octacosyl dodecanoate**			955(a)				
8.2-119	124-10-7	**Methyl tetradecanoate**	auto	682t	682 614(a),888(a) 916(I)				
8.2-120	7132-64-1	**Methyl pentadecanoate**			614(a),1086(a)				
8.2-121	5129-60-2	**Methyl 14-methylpentadecanoate**			888(a)				
8.2-122	112-39-0	**Methyl hexadecanoate**			614(a),1084(a) 916(I)				
8.2-123	628-97-7	**Ethyl hexadecanoate**			570(a)				

Species Number	Registry Number	Name	Emission Source	Emission Ref.	Detection Ref.	Code	Bioassay −MA	Bioassay +MA	Bioassay Ref.
8.2-124	43211-62-7	**Acetyl hexadecanoate**	vegetation	552e					
8.2-125	6929-04-0	**Methyl 15-methylhexadecanoate**			888(a)				
8.2-126	1731-92-6	**Methyl heptadecanoate**			614(a)				
8.2-127	5129-61-3	**Methyl 16-methylheptadecanoate**			888(a)				
8.2-128	112-61-8	**Methyl octadecanoate**			614(a),1084(a) 916(I)				
8.2-129	111-61-5	**Ethyl octadecanoate**			570(a)				
8.2-130	55124-97-5	**Methyl 17-methyloctadecanoate**			888(a)				
8.2-131	18281-04-4	**Ethyl nonadecanoate**			570(a)				
8.2-132	1120-28-1	**Methyl eicosanoate**			614(a)				
8.2-133	18281-05-5	**Ethyl eicosanoate**			570(a)				
8.2-134	28898-67-1	**Ethyl heneicosanoate**			570(a)				
8.2-135	929-77-1	**Methyl docosanoate**			614(a)				
8.2-136	2442-49-1	**Methyl tetracosanoate**			614(a)				
8.2-137	5802-82-4	**Methyl hexacosanoate**			614(a)				

Table 8.2. *(Continued)*

$MeOCH:O$
8.2-1

$O:CHOCH_2CH_2CHMeCH_2CH_2CH:CMe_2$
8.2-5

$O:CHOCH_2CH:CMeCH_2CH_2CH:CMe_2$
8.2-6

$Me_2C:CHCH_2CH_2CMe(CH:CH_2)OCH:O$
8.2-7

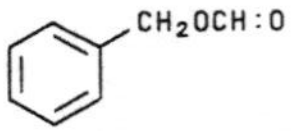

$C_6H_5CH_2OCH:O$
8.2-8

$MeOAc$
8.2-10

$AcOCH_2CHEt(CH_2)_3Me$
8.2-22

$MeC(Me)=CHCH_2CH_2C(Me)=CHCH_2OAc$
8.2-36

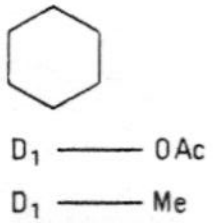

D_1 —— OAc
D_1 —— Me
8.2-39

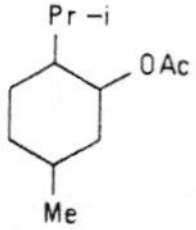

Pr-i, OAc, Me
8.2-40

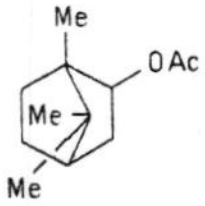

Me, OAc, Me, Me
8.2-41

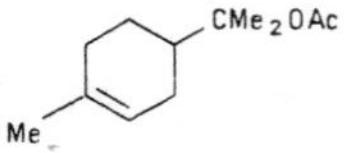

CMe_2OAc, Me
8.2-42

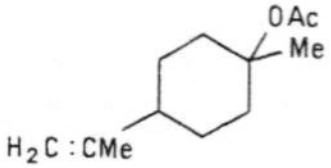

OAc, Me, $H_2C:CMe$
8.2-43

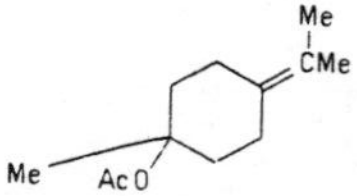

Me, CMe, Me, AcO
8.2-44

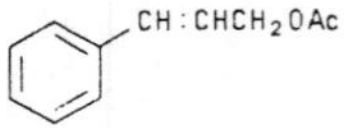

$CH:CHCH_2OAc$
8.2-51

$MeOC(O)CH_2C(O)OMe$
8.2-57

$MeOC(O)CH_2CH_2C(O)OMe$
8.2-68

$MeOC(O)(CH_2)_3C(O)OMe$
8.2-84

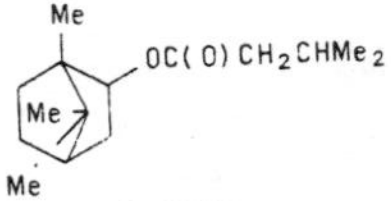

Me, $OC(O)CH_2CHMe_2$, Me, Me
8.2-89

$Me(CH_2)_3OC(O)CH_2C(OH)(C(O)O(CH_2)_3Me)CH_2C(O)O(CH_2)_3Me$
8.2-92

$Me(CH_2)_7OC(O)(CH_2)_4C(O)O(CH_2)_7Me$
8.2-100

$MeOC(O)(CH_2)_7C(O)OMe$
8.2-103

Table 8.3. Olefinic Esters

Species Number	Registry Number	Name	Emission Source	Emission Ref.	Detection Ref.	Code	Bioassay −MA	Bioassay +MA	Ref.
8.3-1S	96-33-3	**Methyl acrylate**			597				
8.3-2	140-88-5	**Ethyl acrylate**	vegetation	514e	597,628	STI00	NEG	?	B38
						STI35	NEG	NEG	B38
						STI37	NEG	NEG	B38
						STI98	NEG	NEG	B38
8.3-3	141-32-2	**Butyl acrylate**			597				
8.3-4S	80-62-6	**Methyl methacrylate**	industrial	60	597,628				
			polymer mfr.	447*, 936	399(a)				
8.3-5	97-88-1	**Butyl methacrylate**			820(I)				
8.3-6	142-09-6	**Hexyl methacrylate**			628				
8.3-7S	623-43-8	**Methyl crotonate**	vegetation	552e					
8.3-8	623-70-1	**Ethyl crotonate**	vegetation	552e					
8.3-9	18060-77-0	**Isopropyl crotonate**	vegetation	552e					
8.3-10	589-66-2	**Isobutyl crotonate**	vegetation	552e					
8.3-11S	623-91-6	**Diethyl fumarate**			1086(a)				

Table 8.3. *(Continued)*

Species Number	Registry Number	Name	Emission Source	Emission Ref.	Detection Ref.	Code	Bioassay −MA	Bioassay +MA	Bioassay Ref.
8.3-12S	7785-64-0	**Butyl angelate**	vegetation	510e					
8.3-13	10482-55-0	**Isopentyl angelate**	vegetation	510e					
8.3-14		**Neryl angelate**	vegetation	510e					
8.3-15S	7493-71-2	**Allyl tiglate**			399(a)				
8.3-16S	7568-58-3	**Tributyl aconitate**			570(r) 570(s)				
8.3-17	1552-67-6	**Ethyl 2-hexenoate**	vegetation	552e					
8.3-18	1732-00-9	**Methyl-*cis*-4-octenoate**	vegetation	552e					
8.3-19	31422-28-3	**Methyl pentadecenoate**			614(a)				
8.3-20	2462-82-2	**Methyl 9-octadecenoate**			614(a)				
8.3-21	28555-06-8	**Ethyl octadecenoate**			570(a)				

$H_2C{:}CHC(O)OMe$
8.3-1

$H_2C{:}CMeC(O)OMe$
8.3-4

$MeCH{:}CHC(O)OMe$
8.3-7

$EtOC(O)CH{:}CHC(O)OEt$
8.3-11

$Me(CH_2)_3OC(O)CMe{:}CHMe$
8.3-12

$H_2C{:}CHCH_2OC(O)CMe{:}CHMe$
8.3-15

$Me(CH_2)_3OC(O)CH_2C(C(O)O(CH_2)_3Me){:}CHC(O)O(CH_2)_3Me$
8.3-16

Table 8.4. Cyclic and Aromatic Esters

Species Number	Registry Number	Name	Emission Source	Emission Ref.	Detection Ref.	Code	Bioassay −MA	Bioassay +MA	Bioassay Ref.
8.4-1S	54598-10-6	**Methyl thujate**	vegetation	531					
8.4-2S	30968-45-7	**Methyl dihydroabietate**	coal comb.	825(a)					
8.4-3S	1334-76-5	**Methyl furoate**	microbes	302					
8.4-4S	93-58-3	**Methyl benzoate**	auto	880	597t				
			microbes	302	1103(a)				
		solvent 134; vegetation 499, 552e							
8.4-5	93-89-0	**Ethyl benzoate**	vegetation	552e					
8.4-6	2049-96-9	**Pentyl benzoate**	vegetation	504e					
8.4-7	94-48-4	**Neryl benzoate**	vegetation	510e					
8.4-8	126-64-7	**Linalyl benzoate**	vegetation	516e					
8.4-9	93-99-2	**Phenyl benzoate**	refuse comb.	878(a)					
8.4-10	120-51-4	**Benzyl benzoate**	vegetation	506e, 510e					
8.4-11	94-47-3	**Phenylethyl benzoate**	vegetation	514e					
8.4-12S	4159-29-9	**Coniferyl benzoate**	vegetation	531					

Table 8.4. *(Continued)*

Species Number	Registry Number	Name	Emission Source	Emission Ref.	Detection Ref.	Code	Bioassay −MA	Bioassay +MA	Bioassay Ref.
8.4-13	several	**Methyl methylbenzoate**			597				
8.4-14	121-98-2	**Methyl *p*-methoxybenzoate**	vegetation	516e					
8.4-15S	119-36-8	**Methyl salicylate**	refuse comb. vegetation	26(a) 501,518					
8.4-16	118-58-1	**Benzyl salicylate**	vegetation	506e					
8.4-17	99-76-3	**Methyl *p*-hydroxybenzoate**	biomass comb.	891(a)t					
8.4-18	1866-39-3	**Methyl cinnamate**	vegetation	500,531					
8.4-19	103-36-6	**Ethyl cinnamate**	vegetation	509e					
8.4-20	103-41-3	**Benzyl cinnamate**	vegetation	506e, 510e					
8.4-21	122-68-9	**Phenylpropyl cinnamate**	vegetation	510e					
8.4-22	122-69-0	**Cinnamyl cinnamate**	vegetation	514e					
8.4-23	several	**Ethyl naphthoate**	auto	311(a)					

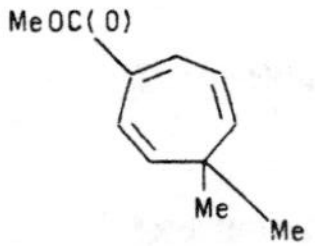

MeOC(O)
Me
Me
8.4-1

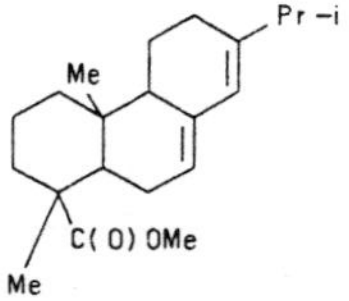

Me
Pr-i
C(O)OMe
Me
8.4-2

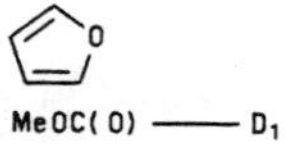

O
MeOC(O) —— D_1
8.4-3

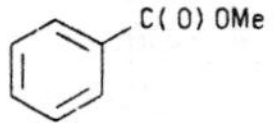

C(O)OMe
8.4-4

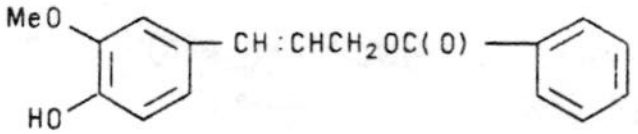

MeO
HO
CH:CHCH_2OC(O) ——
8.4-12

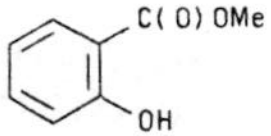

C(O)OMe
OH
8.4-15

Table 8.5. Phthalates and Peroxides

Species Number	Registry Number	Name	Emission Source	Emission Ref.	Detection Ref.	Bioassay Code	Bioassay −MA	Bioassay +MA	Bioassay Ref.
8.5-1S	131-11-3	**Dimethylphthalate**			1084(a),1086(a)t 570(r) 570(s)				
8.5-2	34006-76-3	**Butylmethylphthalate**			570(s)				
8.5-3	84-66-2	**Diethylphthalate**	auto coal comb.	311(a) 825(a)	633 363(a),443(a)				
		polymer mfr. 84; refuse comb. 790(a),878(a); tobacco smoke 421			570(s)				
8.5-4	7299-93-6	**Butylethylphthalate**			570(s)				
8.5-5	84-74-2	**Dibutylphthalate**	auto coal comb.	311(a) 825(a)	633*,1027* 98(a),638*(a)	SCE	?		B6
		diesel 693(a); PVC mfr. 521*; refuse comb. 26(a),790(a)t; water treatment 248(a)			570(r),1028*(r) 570(s) 916(I),1109*(I)				
8.5-6	84-69-5	**Diisobutylphthalate**	auto coal comb.	311(a) 825(a)	633* 363(a),633*(a)				
8.5-7	4489-61-6	**Di-*sec*-butylphthalate**			633* 363(a)				
8.5-8	30448-43-2	**Di-*tert*-butylphthalate**			1103(a)				

Species Number	Registry Number	Name	Emission Source	Emission Ref.	Detection Ref.	Bioassay Code	Bioassay −MA	Bioassay +MA	Bioassay Ref.
8.5-9	84-75-3	**Dihexylphthalate**			916(I)t				
8.5-10	117-81-7	**Di (2-ethylhexyl) phthalate**	industrial	60	1027*,1028*	SCE	?		B6
			refuse comb.	960*(a)	363(a),443(a)	CCC	SP		B37
			water treatment	248(a)	1028*(r)				
					916(I),1109*(I)				
8.5-11	3648-21-3	**Diheptylphthalate**			1109*(I)				
8.5-12	117-84-0	**Dioctylphthalate**	auto	311(a)	633*				
			PVC mfr.	92,521*	633*(a),638*(a)				
			refuse comb.	790(a)t,840(a)	570(r),758(r)				
					570(s)				
8.5-13	27554-26-3	**Diisooctylphthalate**	coal comb.	825(a)	888(a)				
			refuse comb.	878(a)					
8.5-14	119-07-3	**Decyloctylphthalate**	industrial	60					
8.5-15	84-76-4	**Dinonylphthalate**			916(I),1109*(I)				
8.5-16	84-77-5	**Didecylphthalate**			916(I),1109*(I)				
8.5-17	26761-40-0	**Diisodecylphthalate**	industrial	60					
			PVC mfr.	521*					
8.5-18	84-61-7	**Dicyclohexylphthalate**	coal comb.	825(a)					

Table 8.5. *(Continued)*

Species Number	Registry Number	Name	Emission Source	Emission Ref.	Detection Ref.	Bioassay Code	Bioassay −MA	Bioassay +MA	Bioassay Ref.
8.5-19S	85-68-7	**Benzylbutylphthalate**	auto refuse comb.	311(a) 790(a),840(a)	633 363(a),443(a) 1109*(I)				
8.5-20	2528-16-7	**Dibenzylphthalate**	PVC mfr.	521*					
8.5-21S	120-61-6	**Dimethylterephthalate**	industrial tobacco smoke	60 421	614(a),1086(a)				
8.5-22	several	**Ethylmethylterephthalate**			614(a)				
8.5-23S	85-70-1	**Butyl-4 (2-hydroxyethoxycar-bonyl) butylphthalate**	water treatment	248(a)					
8.5-24S	3229-98-9	**Dodecanyl peroxide**	polymer mfr.	407*					
8.5-25S	1561-49-5	**Dicyclohexyl peroxide carbonate**	polymer mfr.	407*					
8.5-26S	94-36-0	**Benzoyl peroxide**	polymer mfr.	407*					
8.5-27		**Evernicaldehyde**	vegetation	531					

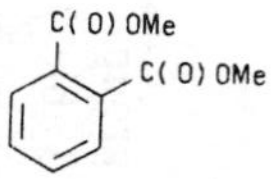

C(O)OMe
C(O)OMe

8.5-1

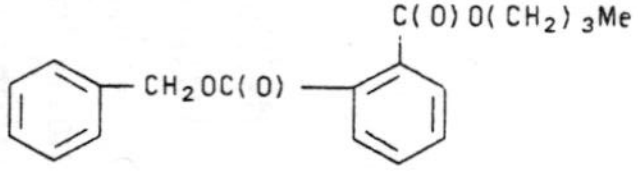

$CH_2OC(O)$ — C(O)O$(CH_2)_3$Me

8.5-19

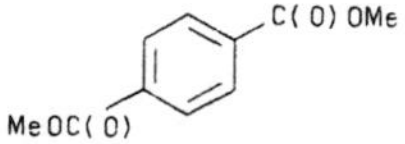

MeOC(O) — C(O)OMe

8.5-21

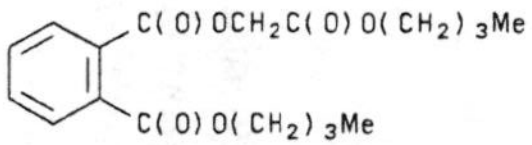

C(O)OCH_2C(O)O$(CH_2)_3$Me
C(O)O$(CH_2)_3$Me

8.5-23

HOO$(CH_2)_{11}$Me

8.5-24

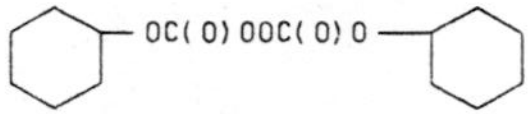

OC(O)OOC(O)O

8.5-25

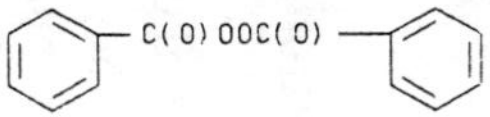

C(O)OOC(O)

8.5-26

CHAPTER 9

Carboxylic Acids

9.0 Introduction

As discussed in 8.0, carboxylic acids are those organic compounds that contain the -C(O)OH group. While they appear to be relatively uncommon in the atmosphere in the gas phase, it is becoming widely recognized that condensed phase atmospheric systems (aerosol particles, raindrops, etc.) contain a variety of carboxylic acids, often at reasonably high concentrations. We divide the atmospheric carboxylic acids into four groups: (1) alkanic, (2) olefinic, (3) cyclic, and (4) aromatic.

9.1 Alkanic Acids

Alkanic acids are one of the largest chemically distinct groups of atmospheric compounds, more than a hundred appearing in Table 9.1. The majority are found in aerosol particles, but a number occur in both gas and condensed phases.

Sources of atmospheric alkanic acids are numerous. Many of the sources involve biological decay, although a variety of industrial processes and biomass combustion also deserve mention. Even in combination, these sources are not particularly prolific, however, and it seems likely that chemical production from less highly oxidized precursors is responsible for the presence of a large fraction of the atmospheric alkanic acids. This surmise is consistent with the identification of the acids in juxtaposition with other lipids* of similar size and structure in aerosols.

In addition to their occurrence in gaseous and particulate phases, alkanic acids are very commonly identified in rain, snow, and indoors.

The gas phase chemistry of alkanic acids in the atmosphere has received little study, in part because their aqueous solubility renders them good candidates for prompt deposition to surfaces. In the aqueous phase, reaction with OH· and NO_3· can occur, with the result being the formation of smaller oxygenated fragment species.[1326]

9.2 Olefinic Acids

Less than twenty compounds in this group have been identified, all as aerosol particle constituents. The known sources are diverse and seem unlikely to explain the occurrence of the compounds; we thus infer that the olefinic acids arise principally as products of chemical reactions taking place in the condensed phase in the atmosphere. Subsequent atmospheric chemical processes involving the olefinic acids have received no study.

9.3 Cyclic Acids

This group of ten compounds arises entirely from emissions from trees, the combustion of trees, or the use of trees in various aspects of the forest products industry. They are large molecules which occur only in the condensed phase. The first three compounds are derivatives of pinene, and speculative chemical routes for their production (in the gas phase) have been presented,[1259] the implication being that their production is followed by deposition onto particles.

None of these compounds is abundant in the atmosphere and they seem unlikely to play important roles in atmospheric chemistry.

* *Lipid* is an inclusive term for fats and fat-derived materials.

9.4 Aromatic Acids

Nearly forty aromatic acids comprise Table 9.4. They are largely present in the condensed phase, being common in aerosol particles and in precipitation. A few sources are known, biomass combustion being one of the most often encountered. With the exception of benzoic acid (which receives fairly extensive industrial use), the sources are quite insufficient to explain the frequent appearance of the aromatic acids in aerosol particles. This suggests that these acids as well as the non-aromatic acids are products of atmospheric chemistry, a theory supported by the ubiquitous presence of reasonable acid precursors. For example, note that phenanthrene has many sources and is almost universally found in aerosol particles. Phenanthrenecarboxylic acid has no known sources, but has been identified by at least two studies of aerosol particle composition. As with simpler organic structures, the evidence for organic acids as products of the atmospheric chemistry of related precursors seems quite conclusive.

Table 9.1. Alkanic Acids

Species Number	Registry Number	Name	Emission Source	Emission Ref.	Detection Ref.	Code	Bioassay −MA	Bioassay +MA	Bioassay Ref.
9.1-1	64-18-6	**Formic acid**	biomass comb.	354	90*, 342*				
			diesel	863	1120*(f)				
		lacquer mfr. 8; plastics comb. 46,354; refuse comb. 26(a); tobacco smoke 396(a); vegetation 552e			761*(r), 764(r)				
					874(s)				
9.1-2	71-47-6	**Formate ion**			1085(r)				
9.1-3	64-19-7	**Acetic acid**	animal waste	25,436*	842*,994*	STP00	NEG	NEG	B36
			auto body comb.	58*	1120*(f)	STP35	NEG	NEG	B36
		biomass comb. 183,354; chemical mfr. 1160; diesel 1,176; explosives mfr. 1092; fish processing 695,946; food decay 1074; lacquer mfr. 8,427; plastics comb. 46,354; refuse comb. 249,661*,26(a); rendering 833; sewage tmt. 422*,695*; starch mfr. 149,695; tobacco smoke 396(a); vegetation 552e; wood pulping 695,1071			761*(r),764(r)	STP37	NEG	NEG	B36
					874(s)	STP98	NEG	NEG	B36
					1110*(I),1123*(I)				
9.1-4	338-70-5	**Oxalate ion**			747(a)				
					770(c)				
					747(r), 770(r)				
9.1-5S	79-14-1	**Glycolic acid**			761*(r), 764(r)				
9.1-6	666-14-8	**Glycolate ion**			1085(r)				
9.1-7S	298-12-4	**Glyoxylic acid**	tobacco smoke	634, 396(a)	1198(r)				

Table 9.1. *(Continued)*

Species Number	Registry Number	Name	Emission Source	Emission Ref.	Detection Ref.	Bioassay Code	−MA	+MA	Ref.
9.1-8S	144-62-7	**Oxalic acid**	rendering	833	882(a)1198(a)	STI00	NEG	NEG	B38
			tobacco smoke	396(a)	761*(r),1198(r)	STI35	NEG	NEG	B38
						STI37	NEG	NEG	B38
						STI98	NEG	NEG	B38
9.1-9	79-09-4	**Propionic acid**	animal waste	25, 436*	843*				
			diesel	176	1198(a)				
		fish processing 946; food decay 1074; lacquer mfr. 427; paint 695, 1071; plastics comb. 354; starch mfr. 149, 695; tobacco smoke 396(a); vegetation 552e; whiskey mfr. 1158; wood pulping 695, 1071			1120*(f) 761*(r),764(r) 874(s)				
9.1-10	72-03-7	**Propionate ion**			1085(r)				
9.1-11S	598-82-3	**Lactic acid**	tobacco smoke	396(a)	761*(r),1198(r)				
9.1-12	113-21-3	**Lactate ion**			1085(r)				
9.1-13S	127-17-3	**Pyruvic acid**	tobacco smoke	396(a)	1199 1198(r)				
9.1-14S	80-72-8	**Reductic acid**	tobacco smoke	634, 396(a)					
9.1-15S	141-82-2	**Malonic acid**	tobacco smoke	396(a)	480*(a), 882(a) 1198(r)t				
9.1-16	1636-27-7	**2,2-Dipropylmalonic acid**			882(a)				

Species Number	Registry Number	Name	Emission Source	Emission Ref.	Detection Ref.	Bioassay Code	Bioassay −MA	Bioassay +MA	Bioassay Ref.
9.1-17S	80-69-3	**Tartronic acid**			1198(a)				
					1198(r)				
9.1-18	107-92-6	**Butyric acid**	animal waste	25, 436*	843*				
			chemical mfr.	1146*, 1147*	1120*(f)				
		diesel 176; fish processing 695, 1071; food decay 1074;			761*(r), 764(r)				
		food processing 62; industrial 156; microbes 540, 664;			1123*(I)				
		paint 695, 1071; rendering 833, 1077; sewage tmt. 695, 1071;							
		starch mfr. 149, 695; sweat glands 610; tobacco smoke 396(a);							
		vegetation 552e							
9.1-19	14287-61-7	**2,3-Dimethylbutyric acid**	sweat glands	598					
			vegetation	552e					
9.1-20S	110-15-6	**Succinic acid**	biomass comb.	1089(a)	480*(a), 697(a)				
			tobacco smoke	396(a)	761*(r), 764(r)				
9.1-21S	6915-15-7	**Malic acid**	tobacco smoke	396(a)	1198(a)t				
					761*(r),1198(r)t				
9.1-22	79-31-2	**Isobutyric acid**	animal waste	25, 695	843*				
			fish processing	695, 1071	1120*(f)				
		food decay 1074; microbes 302; rendering 833;			1120*(r)				
		starch mfr. 695, 1071; tobacco smoke 396(a)							
9.1-23		**Isobutanedioic acid**			480*(a)				

Table 9.1. *(Continued)*

Species Number	Registry Number	Name	Emission Source	Emission Ref.	Detection Ref.	Code	Bioassay −MA	Bioassay +MA	Bioassay Ref.
9.1-24	109-52-4	**Valeric acid**	animal waste	25, 695	843*				
			chemical mfr.	1147*, 1164	1120*(f)				
		diesel 176; food decay 1074; food processing 62; microbes 540; paint 695, 1071; starch mfr. 695, 1071; tobacco smoke 396(a); wood pulping 695, 1071			761*(r), 1120*(r)				
					874(s)				
9.1-25	97-61-0	**2-Methylvaleric acid**	rendering	833					
			tobacco smoke	396(a)					
9.1-26	646-07-1	**4-Methylvaleric acid**	animal waste	855					
9.1-27	13392-69-3	**5-Hydroxyvaleric acid**			214*(a), 532(a)				
9.1-28S	123-76-2	**Levulinic acid**	biomass comb.	1089(a)					
			tobacco smoke	396(a)					
9.1-29	5746-02-1	**5-Formylbutyric acid**			214*(a), 532(a)				
9.1-30S	110-94-1	**Glutaric acid**	rendering	833	213*(a), 654*(a)				
			tobacco smoke	396(a)					
9.1-31	617-62-9	**2-Methylpentanedioic acid**			480*(a)				
9.1-32	626-51-7	**3-Methylpentanedioic acid**			480*(a)				
9.1-33	17179-91-8	**2,3-Dimethylpentanedioic acid**			480*(a)				
9.1-34	328-50-7	**2-Oxoglutaric acid**	tobacco smoke	396(a)					

Species Number	Registry Number	Name	Emission Source	Emission Ref.	Detection Ref.	Code	Bioassay −MA	Bioassay +MA	Bioassay Ref.
9.1-35	320-77-4	**1-Hydroxy-1,2,3-propanetricarboxylic acid**			761*(r), 764(r)				
9.1-36S	77-92-9	**Citric acid**			761*(r)				
9.1-37	126-44-3	**Citrate ion**			1085(r)				
9.1-38	503-74-2	**Isovaleric acid**	animal waste	25, 695	843*				
			chemical mfr.	1146*	761*(r)				
		food decay 1074; microbes 302; rendering 833; starch mfr. 695, 1071; vegetation 552e							
9.1-39S	498-21-5	**Isopentanedioic acid**			480*(a), 882(a)				
9.1-40	142-62-1	**Hexanoic acid**	animal waste	855, 892(a)	1120*(f)				
			diesel	176	758(r), 761*(r)				
		food decay 1074; food processing 62; rendering 833; tobacco smoke 396(a); vegetation 552e							
9.1-41	4536-23-6	**2-Methylhexanoic acid**	vegetation	552e					
9.1-42	1191-25-9	**6-Hydroxyhexanoic acid**			214*(a), 532(a)				
9.1-43	928-81-4	**5-Formylvaleric acid**			214*(a), 532(a)				
9.1-44	124-04-9	**Hexanedioic acid**	biomass comb.	1089(a)	214*(a), 532(a)	WPU	NEG		B1
			industrial	60	874*(r)				
			tobacco smoke	396(a)	874(s)				

Table 9.1. *(Continued)*

Species Number	Registry Number	Name	Emission Source	Emission Ref.	Detection Ref.	Code	Bioassay −MA	Bioassay +MA	Bioassay Ref.
9.1-45	18667-07-7	**Hexanedioate ion**			1085(r)				
9.1-46	626-70-0	**2-Methylhexanedioic acid**			480*(a)				
9.1-47	3058-01-3	**3-Methylhexanedioic acid**			480*(a)				
9.1-48	18294-85-4	**2-Hydroxy-1,6-hexanedioic acid**			697(a)				
9.1-49	111-14-8	**Heptanoic acid**	tobacco smoke vegetation	396(a) 552e	633				
9.1-50	1188-02-9	**2-Methylheptanoic acid**	vegetation	552e					
9.1-51	3710-42-7	**7-Hydroxyheptanoic acid**			214*(a), 532(a)				
9.1-52	35923-65-0	**6-Formylhexanoic acid**			214*(a), 532(a)				
9.1-53	111-16-0	**Heptanedioic acid**	animal waste	892(a)	214*(a), 532(a)				
9.1-54	124-07-2	**Octanoic acid** tobacco smoke 396(a); vegetation 552e	animal waste biomass comb.	892*(a) 991*(a), 1089(a)	633 991*(a), 1023(a) 758(r), 761*(r) 570(s)				
9.1-55	505-48-6	**Octanedioic acid**	biomass comb.	1089(a)	480*(a) 570(r) 570(s)				
9.1-56	34284-35-0	**3-Methyloctanedioic acid**			882(a)				

Species Number	Registry Number	Name	Emission Source	Emission Ref.	Detection Ref.	Bioassay Code	−MA	+MA	Ref.
9.1-57	112-05-0	**Nonanoic acid**	animal waste	892(a)	633				
			biomass comb.	991*(a), 1089(a)	528(a)				
		tobacco smoke 396(a); vegetation 552e			570(r), 758(r)				
					570(s)				
9.1-58	24323-21-5	**2-Methylnonanoic acid**	vegetation	552e					
9.1-59	123-99-9	**Nonanedioic acid**	biomass comb.	1089(a)	480*(a), 697(a)				
					570(r), 874*(r)				
					570(s), 874(s)				
9.1-60	334-48-5	**Decanoic acid**	animal waste	892(a)	633				
			biomass comb.	991*(a)	528(a)				
			vegetation	552e	758(r), 761*(r)				
					570(s)				
9.1-61	1679-53-4	**10-Hydroxydecanoic acid**			955(a)				
9.2-62	111-20-6	**Decanedioic acid**			480*(a)				
					570(r)				
9.1-63	112-37-8	**Undecanoic acid**	animal waste	892(a)	633				
			biomass comb.	991*(a), 1089(a)	528(a)				
			vegetation	552e	570(r)				
					570(s)				
9.1-64	24323-25-9	**2-Methylundecanoic acid**	vegetation	552e					

Table 9.1. *(Continued)*

Species Number	Registry Number	Name	Emission Source	Emission Ref.	Detection Ref.	Bioassay Code	−MA	+MA	Ref.
9.1-65	1852-04-6	**Undecanedioic acid**			1082(a)				
					570(r)				
9.1-66	143-07-7	**Dodecanoic acid**	animal waste	892(a)	633				
			biomass comb.	991*(a), 1089(a)	420*(a), 633*(a)				
		tobacco smoke 396(a), 542(a); vegetation 552e			570(r), 758(r)				
					570(s)				
					916(I), 1109*(I)				
9.1-67	several	**Methyldodecanoic acid**			570(r)				
9.1-68	505-95-3	**12-Hydroxydodecanoic acid**			955(a)				
9.1-69	693-23-2	**Dodecanedioic acid**			955*(a), 1082(a)				
9.1-70	505-52-2	**Tridecanoic acid**	animal waste	892(a)	633				
			tobacco smoke	542(a)	420*(a), 528(a)				
		vegetation 552e; wood comb. 991*(a), 1089(a)			570(r), 762*(r)				
					570(s), 762*(s)				
					1109*(I)				
9.1-71	71987-34-3	**Methyltridecanoic acid**			570(r)				
9.1-72	544-63-8	**Tetradecanoic acid**	animal waste	892(a)	633*				
			tobacco smoke	542(a)	420*(a), 633*(a)				
		vegetation 531, 552e; wood comb. 991*(a), 1089(a)			570(r), 762*(r)				
					570(s), 762*(s)				
					1109*(I)				

Species Number	Registry Number	Name	Emission Source	Emission Ref.	Detection Ref.	Bioassay Code	−MA	+MA	Ref.
9.1-73	several	**Methyltetradecanoic acid**			570(r) 570(s)				
9.1-74	17278-74-9	**14-Hydroxytetradecanoic acid**			955(a)				
9.1-75	821-38-5	**Tetradecanedioic acid**			955*(a)				
9.1-76	1002-84-2	**Pentadecanoic acid**	tobacco smoke wood comb.	542(a) 991*(a), 1089(a)	633* 420*(a), 633*(a) 570(r), 762*(r) 570(s), 762*(s) 1109*(I)				
9.1-77	72000-71-6	**Methylpentadecanoic acid**			570(r)				
9.1-78	57-10-3	**Hexadecanoic acid**	refuse comb. tobacco smoke vegetation 531, 552e; veneer drying 68, 791*(a); wood comb. 991*(a), 1089(a)	26(a) 396(a), 542(a)	633* 420*(a), 633*(a) 570(r), 762*(r) 570(s), 762*(s) 916(I), 1109*(I)				
9.1-79	28801-93-6	**Methylhexadecanoic acid**			570(r) 570(s)				
9.1-80	506-13-8	**16-Hydroxyhexadecanoic acid**			955(a)				
9.1-81	505-54-4	**Hexadecanedioic acid**			955*(a)				

Table 9.1. *(Continued)*

Species Number	Registry Number	Name	Emission Source	Emission Ref.	Detection Ref.	Bioassay Code	−MA	+MA	Ref.
9.1-82	506-12-7	**Heptadecanoic acid**	tobacco smoke	542(a)	633*				
			wood comb.	991*(a), 1089(a)	420*(a), 633*(a)				
					570(r), 762*(r)				
					570(s), 762*(s)				
					1109*(I)				
9.1-83	several	**Methylheptadecanoic acid**			570(r)				
9.1-84	57-11-4	**Octadecanoic acid**	refuse comb.	26(a), 192(a)	633*				
			tobacco smoke	396(a), 542(a)	420*(a), 633*(a)				
		vegetation 552e; veneer drying 791*(a); wood comb. 991*(a), 1089(a)			570(r), 762*(r)				
					570(s), 762*(s)				
					1109*(I)				
9.1-85	7217-83-6	**2-Methyloctadecanoic acid**	veneer drying	791*(a)	570(r)				
9.1-86	3155-42-8	**18-Hydroxyoctadecanoic acid**			955(a)				
9.1-87	871-70-5	**Octadecanedioic acid**			955*(a)				
9.1-88	646-30-0	**Nonadecanoic acid**	tobacco smoke	542(a)	633*				
			wood comb.	991*(a), 1089(a)	528(a), 633*(a)				
					570(r), 762*(r)				
					570(s)				
					1109*(I)				

Species Number	Registry Number	Name	Emission Source	Emission Ref.	Detection Ref.	Bioassay Code	−MA	+MA	Ref.
9.1-89	506-30-9	**Eicosanoic acid**	tobacco smoke veneer drying wood comb.	396(a), 542(a) 791*(a) 991*(a), 1089(a)	633* 269(a), 633*(a) 570(r), 762*(r) 570(s), 762*(s) 1109*(I)				
9.1-90	62643-46-3	**20-Hydroxyeicosanoic acid**			955(a)				
9.1-91	2424-92-2	**Eicosanedioic acid**			955*(a)				
9.1-92	2363-71-5	**Heneicosanoic acid**	tobacco smoke wood comb.	542(a) 991*(a), 1089(a)	633* 363(a), 633*(a) 570(r), 762*(r) 570(s), 762*(s) 1109*(I)				
9.1-93	112-85-6	**Docosanoic acid**	tobacco smoke wood comb.	542(a) 991*(a), 1089(a)	98(a),633*(a) 570(r), 762*(r) 570(s), 762*(s) 1109*(I)				
9.1-94	506-45-6	**22-Hydroxydocosanoic acid**			955(a)				
9.1-95	505-56-6	**Docosanedioic acid**			955*(a)				
9.1-96	2433-96-7	**Tricosanoic acid**	tobacco smoke wood comb.	542(a) 991*(a), 1089(a)	633*(a), 991*(a) 570(r), 762*(r) 570(s), 762*(s) 1109*(I)				

Table 9.1. *(Continued)*

Species Number	Registry Number	Name	Emission Source	Emission Ref.	Detection Ref.	Bioassay Code	−MA	+MA	Ref.
9.1-97	557-59-5	**Tetracosanoic acid**	tobacco smoke wood comb.	542(a) 991*(a), 1089(a)	443(a), 633*(a) 570(r), 762*(r) 570(s), 762*(s) 1109*(I)				
9.1-98	75912-18-4	**24-Hydroxytetracosanoic acid**			955(a)				
9.1-99	2450-31-9	**Tetracosanedioic acid**			955*(a)				
9.1-100	506-38-7	**Pentacosanoic acid**	tobacco smoke wood comb.	542(a) 991*(a)	443(a), 633*(a) 570(r), 762*(r) 1109*(I)				
9.1-101	506-46-7	**Hexacosanoic acid**	tobacco smoke wood comb.	396(a), 542(a) 991*(a), 1089(a)	443(a), 951*(a) 570(r), 762*(r) 570(s), 762*(s) 1109*(I)				
9.1-102	7138-40-1	**Heptacosanoic acid**	tobacco smoke wood comb.	542(a) 991*(a)	887(a), 951*(a) 570(r)				
9.1-103	506-48-9	**Octacosanoic acid**	tobacco smoke wood comb.	542(a) 991*(a), 1089(a)	887(a), 991*(a) 570(r), 762*(r) 762*(s)				
9.1-104	4250-38-8	**Nonacosanoic acid**	tobacco smoke wood comb.	542(a) 991*(a)	887(a), 1023(a) 570(r)				

Species Number	Registry Number	Name	Emission Source	Emission Ref.	Detection Ref.	Bioassay Code	−MA	+MA	Ref.
9.1-105	506-50-3	**Triacontanoic acid**	tobacco smoke wood comb.	542(a) 991*(a)	991*(a), 1023(a) 570(r)				
9.1-106	38232-01-8	**Hentriacontanoic acid**	tobacco smoke wood comb.	542(a) 991*(a)	1023(a), 1073(a)				
9.1-107	3625-52-3	**Dotriacontanoic acid**	tobacco smoke wood comb.	542(a) 991*(a)	1023(a), 1073(a)				
9.1-108	38232-03-0	**Tritriacontanoic acid**	wood comb.	991*(a)	1023(a)				
9.1-109	38232-04-1	**Tetratriacontanoic acid**	wood comb.	991*(a)	1023(a)				

$HOCH_2CO_2H$
9.1-5

HO_2CCHO
9.1-7

HO_2CCO_2H
9.1-8

$MeCH(OH)CO_2H$
9.1-11

HO_2CCOMe
9.1-13

$HOCH_2COCO_2H$
9.1-14

$CH_2(CO_2H)_2$
9.1-15

HO_2C_2CHOH
9.1-17

$HO_2CCH_2CH_2CO_2H$
9.1-20

$HO_2CCH_2CH(OH)CO_2H$
9.1-21

$HO_2CCH_2CH_2COMe$
9.1-28

$HO_2C(CH_2)_3CO_2H$
9.1-30

$HO_2CCH_2C(OH)(CO_2H)CH_2CO_2H$
9.1-36

$HO_2CCH_2CHMeCO_2H$
9.1-39

Table 9.2. Olefinic Acids

Species Number	Registry Number	Name	Emission Source	Emission Ref.	Detection Ref.	Bioassay Code	Bioassay −MA	Bioassay +MA	Bioassay Ref.
9.2-1S	79-10-7	**Acrylic acid**	industrial	60					
9.2-2S	3724-65-0	**Crotonic acid**	diesel lacquer mfr.	1 428					
9.2-3	13201-46-2	**2-Methyl-2-butenoic acid**	animal waste	855t					
9.2-4		**2-Formyl-4-hydroxy-2-butenoic acid**			532(a)				
9.2-5	6915-18-0	**Butenedioic acid**			761*(r)				
9.2-6		**2-Formyl-5-hydroxy-2-pentenoic acid**			532(a)				
9.2-7	1289-45-8	**Dodecenoic acid**			955*(a)				
9.2-8	28555-21-7	**Tridecenoic acid**			420*(a)				
9.2-9	26444-03-1	**Tetradecenoic acid**			955*(a)				
9.2-10	21444-04-2	**Pentadecenoic acid**			570(a)				

Species Number	Registry Number	Name	Emission Source	Emission Ref.	Detection Ref.	Bioassay Code	Bioassay −MA	Bioassay +MA	Bioassay Ref.
9.2-11	2091-29-4	**9-Hexadecenoic acid**	veneer drying	791*(a)	420*(a),955*(a) 762*(r) 762*(s)				
9.2-12	26265-99-6	**Heptadecenoic acid**			570(a)				
9.2-13	4712-34-9	**6-Octadecenoic acid**			882(a)				
9.2-14	26764-26-1	**9-Octadecenoic acid**	biomass comb. tobacco smoke veneer drying	1089(a) 396(a),451(a) 68(a),791*(a)	420*(a),633*(a) 762*(r) 762*(s) 1109*(I)				
9.2-15	2197-37-7	**Octadec-9,12-dienoic acid**	tobacco smoke	396(a),451(a)	98(a),1001*(a)				
9.2-16	1955-33-5	**Octadec-9,12,15-trienoic acid**	tobacco smoke	396(a),451(a)	98(a),1001*(a)				
9.2-17	26764-41-0	**Eicosenoic acid**			420*(a),955*(a)				
9.2-18	25378-26-1	**Docosenoic acid**			955*(a)				

$HO_2CCH:CH_2$
9.2-1

$MeCH:CHCO_2H$
9.2-2

Table 9.3. Cyclic Acids

Species Number	Registry Number	Name	Emission		Detection	Bioassay			
			Source	Ref.	Ref.	Code	−MA	+MA	Ref.
9.3-1S	473-72-3	**Pinonic acid**			269(a),532(a)				
9.3-2S	473-68-7	**Nor-pinonic acid**			532(a),627*(a)t				
9.3-3S	473-73-4	**Pinic acid**			532(a),627*(a)t				
9.3-4S	1231-35-2	**Elliotinoic acid**	veneer drying	791*(a)					
9.3-5S	514-10-3	**Abietic acid**	biomass comb.	1089(a)	1109*(I)				
			refuse comb.	192(a)					
			veneer drying	68(a),197(a)					
9.3-6S	19407-37-5	**Dihydroabietic acid**	biomass comb.	1089(a)	955*(a)				
			veneer drying	68(a)					
9.3-7S	1945-53-5	**Palustric acid**	veneer drying	791*(a)					
9.3-8S	127-27-5	**α-Pimaric acid**	veneer drying	68(a)					
9.3-9S	79-54-9	**β-Pimaric acid**	veneer drying	68(a),791*(a)					
9.3-10S	471-74-9	**Sandaracopimaric acid**	veneer drying	68(a)					

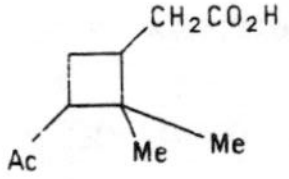

CH2CO2H
Ac Me Me

9.3-1

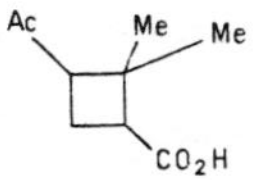

Ac Me Me
CO2H

9.3-2

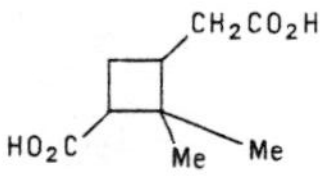

CH2CO2H
HO2C Me Me

9.3-3

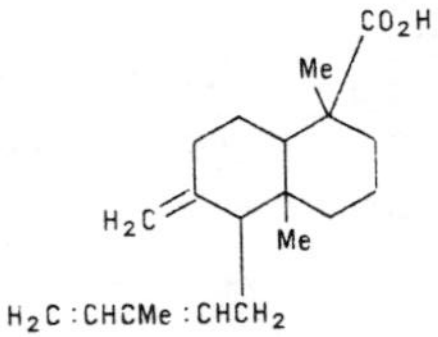

CO2H
Me
H2C
Me
H2C:CHCMe:CHCH2

9.3-4

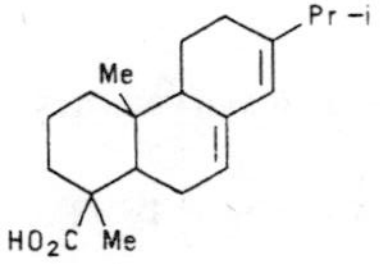

Pr-i
Me
HO2C Me

9.3-5

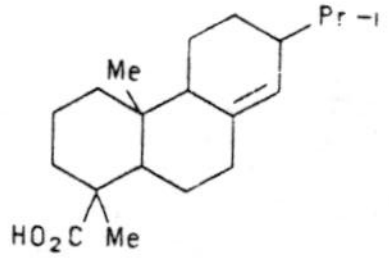

Pr-i
Me
HO2C Me

9.3-6

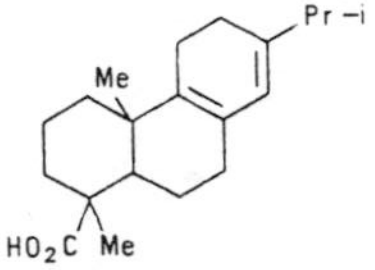

Pr-i
Me
HO2C Me

9.3-7

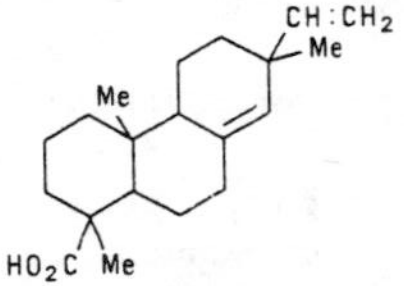

CH:CH2
Me
Me
HO2C Me

9.3-8

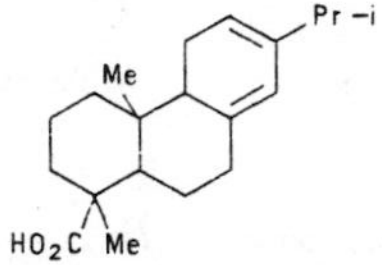

Pr-i
Me
HO2C Me

9.3-9

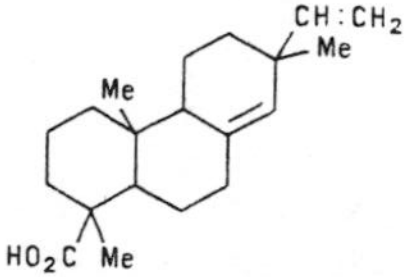

CH:CH2
Me
Me
HO2C Me

9.3-10

Table 9.4. Aromatic Acids

Species Number	Registry Number	Name	Emission Source	Emission Ref.	Detection Ref.	Bioassay Code	−MA	+MA	Ref.
9.4-1S	65-85-0	**Benzoic acid**	animal waste	855,892(a)	307,633	STP00	NEG	NEG	B36
			auto	269(a)	213*(a),532(a)	STP35	NEG	NEG	B36
		biomass comb. 1089(a); coal comb. 825(a); diesel 137;			570(r),761*(r)	STP37	NEG	NEG	B36
		phthal anhyd. mfr. 64*; refuse comb. 26(a),878(a); tobacco smoke 396(a);			570(s)	STP38	NEG	NEG	B36
		vegetation 531,552e				STP98	NEG	NEG	B36
9.4-2	99-04-7	**3-Methylbenzoic acid**	biomass comb.	1089(a)	633				
9.4-3	25567-10-6	**4-Methylbenzoic acid**	biomass comb.	1089(a)	633				
					363(a),627(a)t				
9.4-4	27458-15-7	**Trimethylbenzoic acid**			627(a)				
9.4-5	28134-31-8	**Ethylbenzoic acid**			627(a)				
					570(r)				
9.4-6	1335-08-6	**Methoxybenzoic acid**	biomass comb.	1089(a)					
9.4-7	93-07-2	**3,4-Dimethoxybenzoic acid**	biomass comb.	1089(a)					
9.4-8	69-72-7	**2-Hydroxybenzoic acid**	biomass comb.	1089(a)	633	STP00	NEG	NEG	B36
			refuse comb.	26(a)	363(a),1023(a)	STP35	NEG	NEG	B36
					761*(r)	STP37	NEG	NEG	B36
					570(s)	STP98	NEG	NEG	B36

Species Number	Registry Number	Name	Emission Source	Emission Ref.	Detection Ref.	Code	Bioassay −MA	Bioassay +MA	Bioassay Ref.
9.4-9	99-06-9	**3-Hydroxybenzoic acid**			633 363(a),1023(a)				
9.4-10	99-96-7	**4-Hydroxybenzoic acid**			633 897(a),1023(a)				
9.4-11	99-50-3	**3,4-Dihydroxybenzoic acid**			897(a),1023(a)				
9.4-12S	121-34-6	**Vanillic acid**			1023(a)				
9.4-13S	530-57-4	**Syringic acid**			1023(a)				
9.4-14	619-66-9	**4-Formylbenzoic acid**	auto	309(a)					
9.4-15S	88-99-3	**Phthalic acid**	tobacco smoke vinyl chloride mfr.	396(a) 70	363(a),897(a) 570(r),761*(r) 570(s)				
9.4-16	several	**Methylphthalic acid**			363(a),882(a)				
9.4-17	several	**Butylphthalic acid**			570(r) 570(s)				
9.4-18	601-97-8	**3-Hydroxyphthalic acid**			696(a)				
9.4-19S	121-91-5	**Isophthalic acid**			633 363(a),897(a)				

Table 9.4. *(Continued)*

Species Number	Registry Number	Name	Emission Source	Emission Ref.	Detection Ref.	Code	Bioassay −MA	Bioassay +MA	Bioassay Ref.
9.4-20S	100-21-0	**Terephthalic acid**	industrial	60	60,633 363(a),897(a)				
9.4-21	569-51-7	**1,2,3-Benzenetricarboxylic acid**			697(a)				
9.4-22S	103-82-2	**Phenylacetic acid**	animal waste auto biomass comb. 1089(a); vegetation	855,892(a) 269(a) 514e,531	214*(a)t,532(a)				
9.4-23	637-27-4	**Phenylpropionic acid**	animal waste	855	214*(a)t,532(a)				
9.4-24S	501-52-0	**Hydrocinnamic acid**	animal waste	892(a)					
9.4-25S	621-82-9	**Cinnamic acid**	vegetation	531					
9.4-26	2316-26-9	**3,4-Dimethoxycinnamic acid**	biomass comb.	1089(a)					
9.4-27	7400-08-0	**4-Hydroxycinnamic acid**			1023(a)				
9.4-28	1135-24-6	**4-Hydroxy-3-methoxycinnamic acid**			1023(a)				
9.4-29	90-27-7	**Phenylbutyric acid**			214*(a)t,532(a)				
9.4-30	7226-83-7	**2-Hydroxy-4-phenylbutyric acid**			697(a)				
9.4-31S	86-55-5	**2-Naphathalenecarboxylic acid**	biomass comb.	1089(a)	633* 363(a)				

Species Number	Registry Number	Name	Emission Source	Emission Ref.	Detection Ref.	Code	Bioassay −MA	Bioassay +MA	Bioassay Ref.
9.4-32S		**16,17-Bisnordehydroabietic acid**			955*(a)				
9.4-33S	1740-19-8	**Dehydroabietic acid**	veneer drying	68(a),791*(a)	955*(a) 570(r) 570(s)				
9.4-34S	6040-04-6	**13-Isopropyl-5α-podocarpa-6,8,11,13-tetraen-15-oic acid**	veneer drying	791*(a)	955*(a)				
9.4-35	27875-89-4	**Phenanthrenecarboxylic acid**			363(a),897(a)				
9.4-36	607-42-1	**Anthracenecarboxylic acid**			363(a)				
9.4-37	19694-02-1	**Pyrenecarboxylic acid**			363(a)				

Table 9.4. *(Continued)*

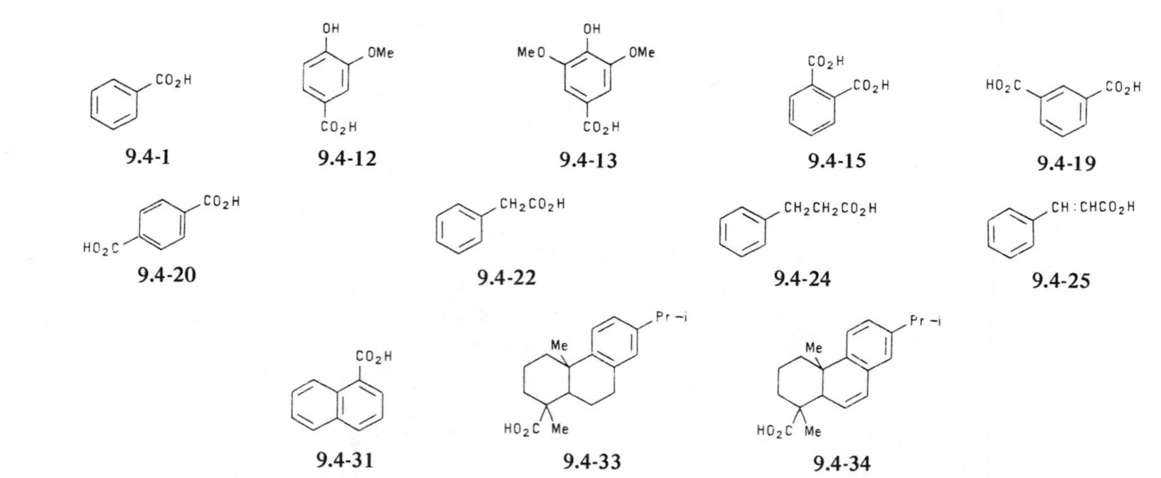

CHAPTER 10

Heterocyclic Oxygen Compounds

10.0 Introduction

Organic compounds that contain rings made up of more than one kind of atom are termed *heterocyclic*. This chapter is comprised of heterocyclic compounds in which one, or at most two, oxygen atoms take the place of carbon atoms in a ring. Thus one has, for example, the analogous compounds

cyclohexane 1,4-dioxane

An associated group of compounds are the *lactones*, intramolecular cyclic esters in which the hydroxyl group and the carboxyl group of a hydroxy acid are considered to have interacted with loss of water. Thus one has, for example, the related compounds

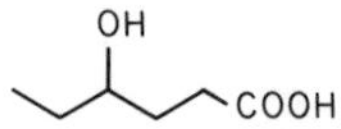

OH COOH

4-hydroxyhexanoic acid

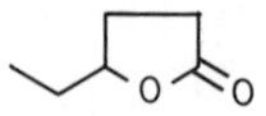

γ-hexalactone

The locant γ indicates that ring closure occurs with the carbon atom third down the chain from the carboxylic carbon atom.

The compounds in this chapter are divided into three groups: (1) lactones, (2) nonaromatic heterocyclic oxygen compounds, and (3) aromatic heterocyclic oxygen compounds.

10.1 Lactones

Lactones are uncommon in the atmosphere. They are emitted to some degree from vegetation, forming a portion of the aroma of certain plants. A few have been identified in combustion emissions from diesels and turbines. Their atmospheric chemistry has not been studied and it seems unlikely that they play any significant role in atmospheric processes.

10.2-3 Heterocyclic Oxygen Compounds

There are more than eighty of these compounds in the atmosphere. Their concentrations are low, but their presence is frequent. The nonaromatic compounds occur almost entirely as gases and are emitted from vegetation and a variety of incomplete combustion processes: biomass, auto, diesel, etc. The aromatic compounds tend to be present in condensed form rather than as gases. Vegetation is the principal source of the smaller aromatics; the larger ones arise from combustion of biomass, fossil fuels, and refuse. The chemistry of two gas phase heterocyclic oxygen compounds has been studied. OH· appears to be the principal reactant with each. For tetrahydrofuran, a saturated compound, the reaction proceeds by H abstraction:[1240]

$$\text{C}_4\text{H}_8\text{O} + \text{OH}\cdot \longrightarrow \text{C}_4\text{H}_7\text{O}\cdot + H_2O \qquad \text{(R10.2-1)}$$

For furan, an unsaturated compound, addition to the ring is favored:[1240]

$$C_4H_4O + OH\cdot \longrightarrow C_4H_4O(OH)\cdot \qquad \text{(R10.2-2)}$$

In each case, subsequent reactions proceed as do those of analogous arenes. The arenes are also expected to be reasonable models for the chemistry of other heterocyclic oxygen compounds.

Table 10.1. Lactones

Species Number	Registry Number	Name	Emission Source	Emission Ref.	Detection Ref.	Code	Bioassay −MA	Bioassay +MA	Bioassay Ref.
10.1-1S	1955-45-9	**β,β-Dimethylpropiolactone**	turbine	414					
10.1-2S	96-48-0	**γ-Butyrolactone**	vegetation	552e	628t	STI00	NEG	NEG	B38
						STI35	NEG	NEG	B38
						STI37	NEG	NEG	B38
						STI98	NEG	NEG	B38
						MNT	?/−		B29
						SPI	NEG		B25
10.1-3S	108-29-2	**γ-Valerolactone**	diesel	1					
10.1-4	695-06-7	**γ-Hexalactone**	turbine	414					
10.1-5	104-50-7	**γ-Octalactone**	vegetation	552e					
10.1-6	104-61-0	**γ-Nonalactone**	vegetation	514e					
10.1-7	104-67-6	**γ-Undecalactone**	vegetation	514e					
10.1-8S	542-28-9	**δ-Valerolactone**	diesel	138					
10.1-9S	698-76-0	**δ-Octalactone**	vegetation	552e					
10.1-10	106-02-5	**Cyclopentadecalactone**	vegetation	514e					

Species Number	Registry Number	Name	Emission Source	Emission Ref.	Detection Ref.	Code	Bioassay −MA	Bioassay +MA	Bioassay Ref.
10.1-11	109-29-5	**ω-Hexadecenlactone**	vegetation	514e					
10.1-12	469-61-4	**Cedrene**	vegetation	510e					
10.1-13	several	**Methylphthalide**			358				
10.1-14	6066-49-5	**Butylphthalide**	vegetation	531					
10.1-15S	551-08-6	***n*-Butylidenephthalide**	vegetation	531					

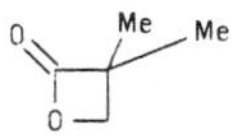

O Me Me O

10.1-1

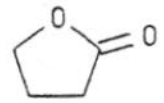

O O

10.1-2

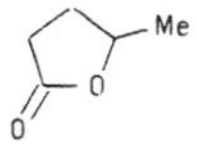

Me O O

10.1-3

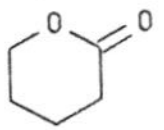

O O

10.1-8

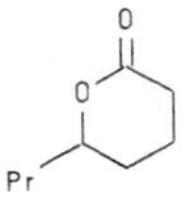

O O Pr

10.1-9

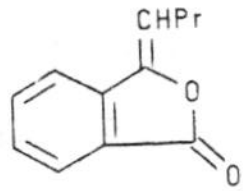

CHPr O O

10.1-15

Table 10.2. Heterocyclic Oxygen Compounds (Nonaromatic)

Species Number	Registry Number	Name	Emission Source	Emission Ref.	Detection Ref.	Code	Bioassay −MA	Bioassay +MA	Ref.
10.2-1S	75-21-8	**Epoxyethane**	auto	217*	525t,597t	CCC	I,P		B37
			chemical mfr.	1157	399(a)	STU35	+(B)		B36
		diesel 261; industrial 60				TRC	+		B20
						SRL	+		B30
						NEU	+		B32
10.2-2	75-56-9	**1,2-Epoxypropane**	auto	284	597t	CCC	LP		B37
			diesel	261		YEA	+		B27
		petroleum mfr. 111; turbine 414				SRL	+		B30
						NEU	+		B32
10.2-3	10317-17-6	**1,3-Epoxy-2-isopropylpropane**	turbine	414					
10.2-4	1758-33-4	*cis*-**2,3-Epoxybutane**	turbine	414					
10.2-5	1758-32-3	*trans*-**2,3-Epoxybutane**	turbine	414					
10.2-6	1192-31-0	**2,3-Epoxy-4-methylpentane**	coal comb.	828t					
10.2-7S	109-99-9	**Tetrahydrofuran**	solvent	134,598	833				
					820(I)				
10.2-8	96-47-9	**2-Methyltetrahydrofuran**	turbine	414					
10.2-9	1003-38-9	**2,5-Dimethyltetrahydrofuran**	biomass comb.	597					

Species Number	Registry Number	Name	Emission Source	Emission Ref.	Detection Ref.	Code	Bioassay −MA	Bioassay +MA	Bioassay Ref.
10.2-10	3358-28-9	**2,2,4,4-Tetramethyltetra-hydrofuran**	auto	217*					
			turbine	414					
10.2-11S	33081-34-4	**Lilac alcohol-a**	vegetation	533					
10.2-12S	33081-35-5	**Lilac alcohol-b**	vegetation	533,552e					
10.2-13	1191-99-7	**2,3-Dihydrofuran**	biomass comb.	597					
10.2-14	1487-15-6	**2-Methyl-4,5-dihydrofuran**	biomass comb.	597					
10.2-15S	110-00-9	**Furan**	acrylonitrile mfr.	626	358,597				
			auto	284	820(I)				
		biomass comb. 354,821; coffee mfr. 598; diesel 1,811; fish processing 110; tobacco smoke 396							
10.2-16	534-22-5	**2-Methylfuran**	biomass comb.	354,597	358,582				
			diesel	1	820(I)t				
		tobacco smoke 396; wood pulping 19,267							
10.2-17	930-27-8	**3-Methylfuran**			597				
					208(r)				
10.2-18	3710-43-8	**2,4-Dimethylfuran**	biomass comb.	597	597,1086				

Table 10.2. *(Continued)*

Species Number	Registry Number	Name	Emission Source	Emission Ref.	Detection Ref.	Code	Bioassay −MA	Bioassay +MA	Bioassay Ref.
10.2-19	625-86-5	**2,5-Dimethylfuran**	auto	284	358t,597				
			biomass comb.	354,597					
			diesel	138					
10.2-20	74430-19-6	**Trimethylfuran**	biomass comb.	597					
10.2-21	27252-25-1	**Ethylfuran**	biomass comb.	597t	591				
10.2-22	several	**Diethylfuran**	biomass comb.	597	358t				
10.2-23	10599-59-4	**2-Isopropylfuran**	biomass comb.	597					
10.2-24	10504-05-9	**2-Isopropyl-5-methylfuran**	biomass comb.	597					
10.2-25	3777-69-3	**2-Pentylfuran**	rendering	837	833				
10.2-26	1487-18-9	**2-Vinylfuran**	biomass comb.	597					
10.2-27	10504-13-9	**5-Methyl-2-vinylfuran**	biomass comb.	597					
10.2-28S	10599-55-0	**2-Propenylfuran**	biomass comb.	597					
10.2-29	10599-66-3	**2-Isopropenyl-5-methylfuran**	biomass comb.	597					
10.2-30	several	**Methoxyfuran**	refuse comb.	26(a)					
10.2-31	several	**Hydroxyfuran**	coffee mfr.	598					

Species Number	Registry Number	Name	Emission Source	Emission Ref.	Detection Ref.	Code	Bioassay −MA	Bioassay +MA	Bioassay Ref.
10.2-32S	98-01-1	**Furfural**	biomass comb.	354,597	597	STP00	+(B)	+(B)	B36
			brewing	144					
		coffee mfr. 598; diesel 137,176; microbes 302; tobacco smoke 396,554,579(a); vegetation 552e; wood comb. 875*							
10.2-33	620-02-0	**5-Methylfurfural**	biomass comb.	354,597					
			brewing	144					
		diesel 137; tobacco smoke 634							
10.2-34	59173-59-0	**Hydroxyfurfural**	diesel	137					
10.2-35	67-47-0	**5-Hydroxymethylfurfural**	tobacco smoke	554(a),634(a)					
10.2-36S	4437-22-3	**Difurfuryl ether**	rendering	1163					
10.2-37	49612-49-9	**2-(3-Formylpropyl) furan**	biomass comb.	597	628				
10.2-38	26447-28-0	**Furancarboxylic acid**	biomass comb.	821	1113(I)				
			tobacco smoke	396(a)					
10.2-39	27987-03-7	**Methylfurancarboxylic acid**	diesel	137					
10.2-40	50-99-7	**Glucose**	tobacco smoke	542(a)		MNT	?/−		B29
						SPI	NEG		B25
10.2-41S	110-87-2	**Dihydropyran**	biomass comb.	597t	628				
10.2-42S	118-71-8	**Maltol**	animal waste	855					
			vegetation	514e					

Table 10.2. *(Continued)*

Species Number	Registry Number	Name	Emission Source	Emission Ref.	Detection Ref.	Code	Bioassay −MA	Bioassay +MA	Bioassay Ref.
10.2-43S	470-82-6	**1,8-Cineol**	vegetation	500,501		STI00	NEG	NEG	B38
						STI35	NEG	NEG	B38
						STI37	NEG	NEG	B38
						STI98	NEG	NEG	B38
10.2-44S	470-67-7	**2,6-Cineol**	vegetation	198					
10.2-45S	123-91-1	**1,4-Dioxane**	solvent	175,598	601,797*	CCC	SP		B37
					820(I),1108(I)	STI00	NEG	NEG	B38
						STI35	NEG	NEG	B38
						STI37	NEG	NEG	B38
						STI98	NEG	NEG	B38

10.2-1  10.2-7 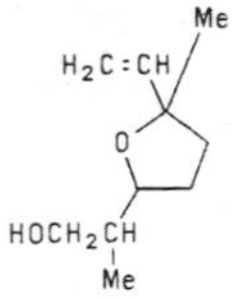 10.2-11 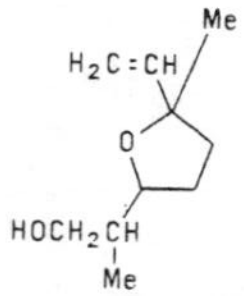 10.2-12 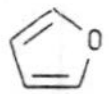 10.2-15

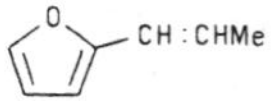

10.2-28 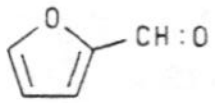 10.2-32 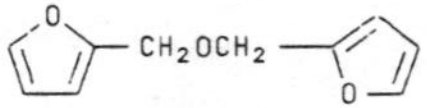 10.2-36  10.2-41

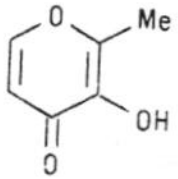

10.2-42 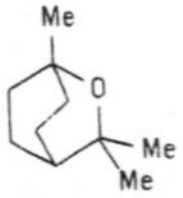 10.2-43  10.2-44 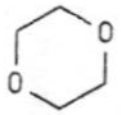 10.2-45

Table 10.3. Heterocyclic Oxygen Compounds (Aromatic)

Species Number	Registry Number	Name	Emission Source	Emission Ref.	Detection Ref.	Bioassay Code	Bioassay −MA	Bioassay +MA	Bioassay Ref.
10.3-1S	96-09-3	**Phenylepoxyethane**			358t	HMASC		+	B13
						V79A	+		B7
						WPU	+		B1
						CCC	LP		B37
						STP98	−(B)	−(B)	B36
						STS00	+		B36
						STU00	+		B36
						STP00	+	+	B36
						STP35	+	+	B36
						STP37	−(B)	−(B)	B36
						STP38	−(B)	+(B)	B36
						YEA	+	NEG	B27
						MNT	?/−		B29
						SRL	+		B30
10.3-2S	94-59-7	**Safrole**	vegetation	326,497e		TRM	+		B21
						HMAST		+	B13
						CCC	SP		B37
						STP37	NEG	NEG	B36
						STP98	NEG	NEG	B36
						STP00	NEG	NEG	B36
						STP35	NEG	NEG	B36
						REC	+		B9
						MNT	?/−		B29
						SRL	NEG		B30

Table 10.3. *(Continued)*

Species Number	Registry Number	Name	Emission Source	Emission Ref.	Detection Ref.	Bioassay Code	Bioassay −MA	Bioassay +MA	Bioassay Ref.
						SPI	?		B25
10.3-3S	120-58-1	**3,4-Methylenedioxy-1-propenylbenzene**	vegetation	515e		CCC	SP		B37
						STP00	−(CG)	−(CG)	B36
						STP35	−(CG)	−(CG)	B36
10.3-4S	607-91-0	**Myristicin**	vegetation	517e,531					
10.3-5S	523-80-8	**Apiole**	vegetation	514e					
10.3-6		**Mericyl alcohol**	vegetation	531					
10.3-7S	120-57-0	**Piperonal**	diesel	138		STI00	NEG	NEGF	B38
			vegetation	510e		STI35	NEG	NEG	B38
						STI37	NEG	NEG	B38
						STI98	NEG	NEG	B38
10.3-8S	91-64-5	**2-Coumarin**	diesel	138	528(a),1001*(a)	ALC	+		B19
			refuse comb.	26(a)		STI00	NEG	+	B38
						STI35	NEG	NEG	B38
						STI37	NEG	NEG	B38
						STI98	NEG	NEG	B38
10.3-9	1333-47-7	**Methylcoumarin**	refuse comb.	26(a)					
			vegetation	517e					
10.3-10	55599-95-6	**Maraniol**	vegetation	531					
10.3-11	64927-40-8	**Dihydroxycoumarin**	refuse comb.	26(a)					

Species Number	Registry Number	Name	Emission Source	Emission Ref.	Detection Ref.	Bioassay Code	Bioassay −MA	Bioassay +MA	Bioassay Ref.
10.3-12S	491-31-6	**3-Coumarin**	diesel	138					
			vegetation	507e					
10.3-13S	271-89-6	**Benzofuran**	auto	217*	86	STI00	NEG	NEG	B38
			biomass comb.	597,891(a)	1086(a)t	STI35	NEG	NEG	B38
		coal comb. 828t; diesel 811; polymer mfr. 251;				STI37	NEG	NEG	B38
		tobacco smoke 339(a)				STI98	NEG	NEG	B38
10.3-14	4265-25-2	**2-Methylbenzofuran**	biomass comb.	597	358				
			tobacco smoke	339(a)					
10.3-15	17059-52-8	**7-Methylbenzofuran**	biomass comb.	597					
			diesel	685(a)					
10.3-16	24410-50-2	**3,6-Dimethylbenzofuran**	vegetation	531					
10.3-17	3782-00-1	**4,7-Dimethylbenzofuran**	biomass comb.	597					
			diesel	138					
			refuse comb.	878(a)					
10.3-18	3131-63-3	**Ethylbenzofuran**	tobacco smoke	339(a)					
10.3-19	several	**Hydroxybenzofuran**	diesel	138					
10.3-20S	210-79-7	**Benzo[*b*]furano[5,6-*b*]furan**	biomass comb.	891(a)t					
10.3-21	5656-82-6	**Naphtho[2,3-*b*]furan-4,9-dione**	diesel	690(a)					

Table 10.3. *(Continued)*

Species Number	Registry Number	Name	Emission Source	Emission Ref.	Detection Ref.	Bioassay Code	Bioassay −MA	Bioassay +MA	Bioassay Ref.
10.3-22S	132-64-9	**Dibenzofuran**	biomass comb.	891(a)	86				
			coal comb.	828,828(a)t,1065(a)t	209(a),466(a)				
		diesel 685(a); refuse comb. 790(a),878(a); tobacco smoke 130*(a),406(a)							
10.3-23	60826-62-2	**Methyldibenzofuran**	coal comb.	828t					
10.3-24S	239-30-5	**Benzo[*b*]naphtho[2,1-*d*]furan**	coal comb.	1065(a)					
10.3-25	243-42-5	**Benzo[*b*]naphtho[2,3-*d*]furan**	coal comb.	1065(a)					
10.3-26S	92-61-5	**Scopoletin**	tobacco smoke	396(a)		ALC	+		B19
10.3-27S	525-82-6	**2-Phenylchromone**			1001*(a)				
10.3-28S	90-46-0	**Hydroxyxanthene**	diesel	690(a)t					
10.3-29S	90-47-1	**Xanthen-9-one**	refuse comb.	790(a),878(a)	98(a),213(a)				
			turbine	786*(a)t					
10.3-30	several	**Hydroxyxanthen-9-one**	diesel	690(a)					
10.3-31S	191-37-7	**1,8,9-Perinaphthoxanthene**	tobacco smoke	396(a)					
10.3-32S	1162-65-8	**Aflatoxin B_1**			871*(a)	HMAST		+	B13
						V79T		+	B7
						V790		+	B7
						CCC	SP		B37
						STP35		−(C)	B36
						STP38		+	B36

Species Number	Registry Number	Name	Emission Source	Emission Ref.	Detection Ref.	Bioassay Code	Bioassay −MA	Bioassay +MA	Bioassay Ref.
						STS00		+	B36
						STS98		+	B36
						STU00	−BCEG		B36
						STU98	−BCEG		B36
						STP98		+(GH)	B36
						STP00	+(B)	+	B36
						STP98	+(B)	+(B)	B36
						SCE	+		B6
						MDR	+		B12
						CYG	NEG		B8
						CYC	+		B8
						REC	+		B9
						MNT	+		B29
						CTV	+		B24
						CTP	+,+		B24
						SRL	+		B30
						NEU		+	B32
						SPI	NEG		B25
10.3-33S	7220-81-7	**Aflatoxin B_2**			871*(a)	HMAST		NEG	B13
						V79T		+	B7
						CCC	LP		B37
						STP98		−ACEH	B36
						STU00	NE,BC		B36
						STU98	NE,BC		B36
						REC		NEG	B9

Table 10.3. *(Continued)*

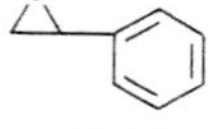

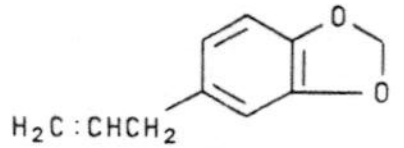

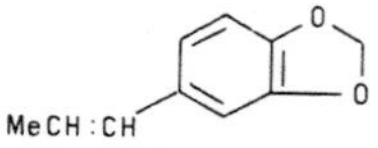

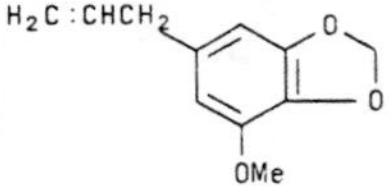

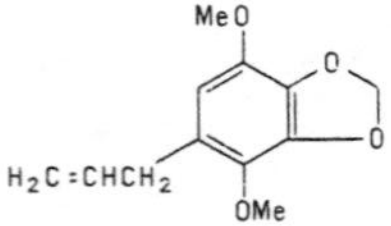

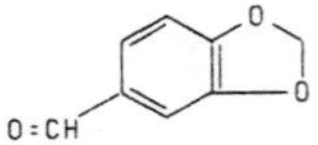

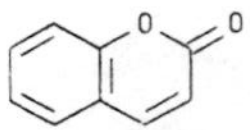

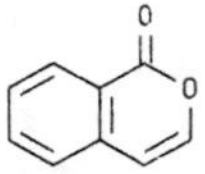

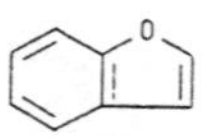

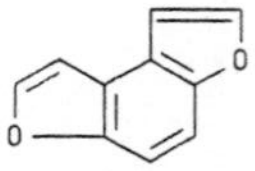

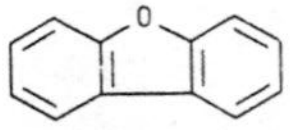

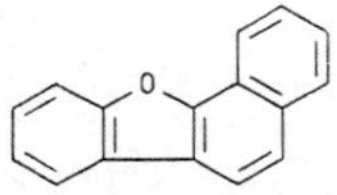

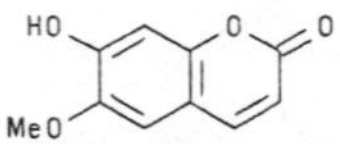

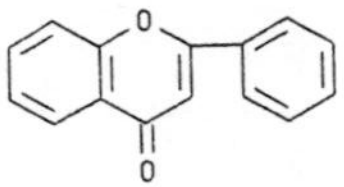

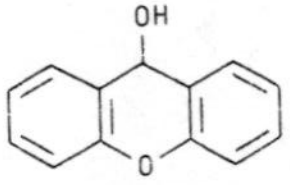

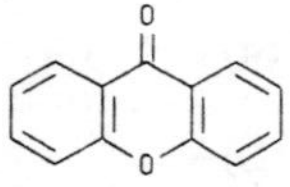

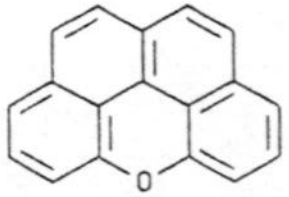

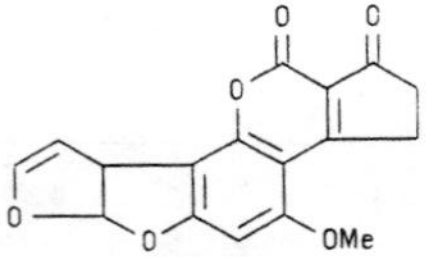

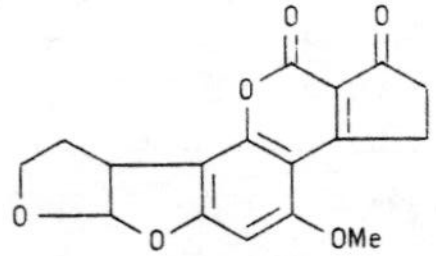

CHAPTER 11

Nitrogen – Containing Organic Compounds

11.0 Introduction

Organic acyclic oxygen compounds can be considered as derivatives of the water molecule in which one or more of the hydrogen atoms are replaced by other groups. Similarly, many organic acyclic nitrogen compounds may be thought of as derivatives of the ammonia molecule in which one or more hydrogen atoms have been replaced by other groups. This leads to

NH_3	ammonia
RNH_2	an *amine*
$RC(O)NH_2$	an *amide*

The nitrogen atom can also occur triply bonded to a carbon atom. These compounds are termed *nitriles* (or *cyanides*), and occasionally occur in the atmosphere. A group of much broader atmospheric interest is the *nitro compounds* of general formula RNO_2, with the nitro (NO_2) group being linked directly to a carbon atom. The active atmospheric chemistry of NO_2 has resulted in a surprisingly large group of atmospheric nitro compounds.

Heterocyclic compounds in which nitrogen atoms replace carbon atoms within a ring structure occur frequently in the atmosphere. Their nomenclature involves a degree of complexity similar to that of the arenes.

The nitrogen-containing organic compounds are divided here into five groups: (1) nitriles, (2) amines and amides, (3) aliphatic nitro compounds, (4) aromatic nitro compounds, and (5) heterocyclic nitrogen compounds.

11.1 Nitriles

Some thirty different nitriles are known to be atmospheric constituents. They occur about equally in gas and condensed phases and are frequently present in indoor air. Nitriles are produced as byproducts of incomplete combustion, especially when the combustion involves polymeric materials containing nitrogen. Biomass combustion also results in the emission of a number of nitriles.

HCN loss in the atmosphere occurs as a result of reaction with OH·. The initial step is the formation of an HO-HCN adduct, followed by collisional stabilization and subsequent reaction.[1325] The atmospheric chemical reactions of the higher nitriles generally involve the hydrocarbon moieties of the molecules rather than the -CN groups. For saturated systems, this consists of OH· abstraction of a hydrogen atom.[1241] For propanenitrile, for example,

$$CH_3CH_2CN + OH\cdot \rightarrow CH_3\dot{C}H\,CN + H_2O \qquad (R11.1\text{-}1)$$

The nitrile-OH· reactions are not very rapid, and some hydrogen cyanide and ethanenitrile, at least, are able to live long enough to be transported to the stratosphere.

If the molecule is unsaturated, addition of OH· to a $-C=C-$ bond will occur, as for acrylonitrile:[1241]

$$CH_2{=}CHCN + OH\cdot \rightarrow \cdot CH_2CH(OH)CN\,. \qquad (R11.1\text{-}2)$$

In each case, the nitrile-containing product will react further by the mechanisms outlined in Chapter 3 for hydrocarbon compounds.

11.2 Amines, Amides

More than seventy amines and amides have been identified by atmospheric chemists. In addition, reduced nitrogen is abundant in aerosol particles and unidentified amines and amides are major fractions of the reduced nitrogen.[1243] The most common of these compounds in the gas phase are the lower alkanic amines, which are produced naturally by a wide variety of biological decay processes involving microorganisms. Aromatic members of this class, on the other hand, are almost entirely anthropogenic, arising from chemical manufacturing or combustion of polymeric materials. The atmospheric concentrations of the amines have not been determined, but based on their olfactory thresholds and the fact that they are commonly noticed in the vicinity of such sources as sewage treatment facilities and decaying fish, the concentrations are probably a part per billion or so near sources. Amides are rare in the atmosphere; a few pesticides formulated by substitutions to an amine substrate are more common.

The atmospheric chemistry of the amines is complicated by the possibility of OH· abstraction from either a C-H or an N-H bond.[1221] The branching ratio for reactions in these channels varies with the compound involved. In Figure 11.2-1, we show a probable reaction sequence for dimethylamine, a compound in which the two channels are competitive. In channel A, abstraction takes place at the N-H bond, the product being dimethylnitramine (I) in the presence of 300 ppb NO_x. Channel B, the channel involving abstraction at a C-H bond, leads to a radical that may decay to produce a amide (Channel C, product II) or an aldehyde and a smaller radical (Channel D, product III). In any case, hydroxyl attack on amines is relatively rapid and atmospheric lifetimes of a few hours are expected.[1244] Mention should also be made of other compounds in Table 11.2 which are present in the atmosphere in low concentrations and whose chemistry has been studied. The hydrazines (compounds 11.2-1 and 11.2-2) are components of rocket fuel. They react rapidly with OH·, the mechanism being H abstraction from the weak N-H bonds.[1246] A second group is the hydroxyamines (compounds 11.2-22, 23, 24). In these compounds, abstraction by OH· from an α-carbon is rapid and the products are hydroxy derivative analogues of the unsubstituted amines.[1242]

Finally, we consider the carcinogenic nitrosamines (compounds 11.2-43, 44, 63). These are formed in the gas phase in the atmosphere by the reaction of amines with nitrous acid:[1245]

$$R_2NH + HNO_2 \rightarrow R_2NNO + H_2O \qquad \text{(R11.2-1)}$$

The principle fate is photodissociation to yield nitric oxide:[1330]

$$R_2NNO \xrightarrow{h\nu} R_2N\cdot + NO \qquad \text{(R11.2-2)}$$

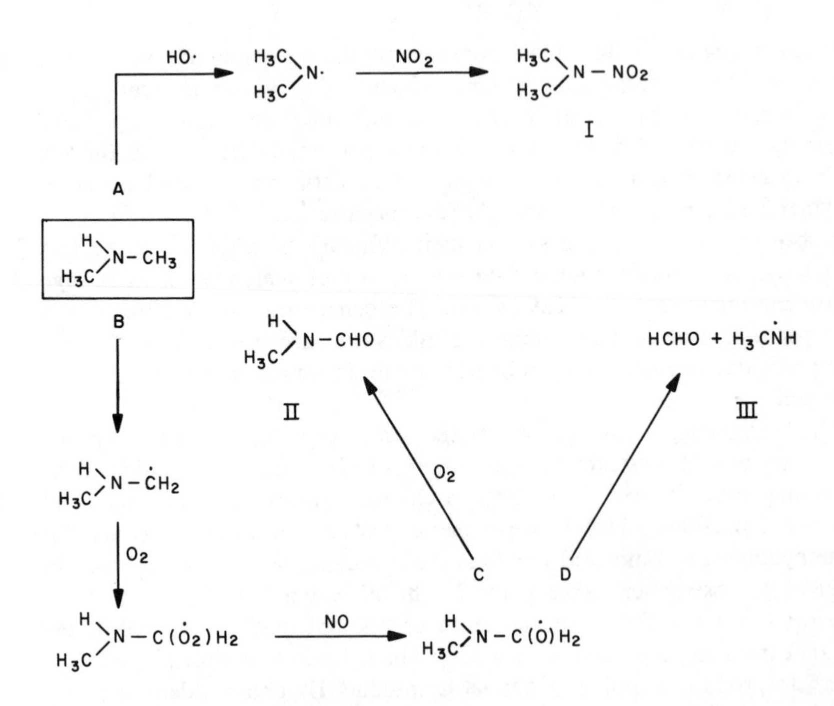

Fig. 11.2-1 Gas phase atmospheric reaction sequences for dimethylamine. See text for discussion. This figure was drawn from information contained in Atkinson et al.[1244] and Pitts et al.[1245]

Since HNO_2 is photolyzed in daylight and thus unavailable to form the nitrosamines, nitrosamine concentrations are quite low except near their sources and at reduced light levels.

11.3 Aliphatic Nitro Compounds

More than thirty compounds appear in Table 11.3. They are largely emitted in the gas phase and largely detected on aerosol particles in urban regions. A few of these compounds are known to come from autos and diesels, but most must be present as a result of atmospheric chemical reactions. One probable sequence involves the alkanes (see Chapter 3), the final nitrate production reactions being

$$RO_2\cdot + NO \xrightarrow{M} RONO_2 \qquad (R11.3\text{-}1)$$

$$RO\cdot + NO_2 \xrightarrow{M} RONO_2 \qquad (R11.3\text{-}2)$$

Direct addition of NO_2 to hydrocarbon radicals will form not nitrates, but nitro compounds.

The gas phase concentrations of the nitro compounds, particularly the peroxynitro compounds, have received extensive study, as the compounds may be a major reservoir for reactive atmospheric nitrogen.[1281] Background oceanic concentrations are of order 20-40 ppt and rural concentrations are perhaps ten times higher. The major component is peroxyacetylnitrate (PAN); peroxypropionylnitrate (PPN) concentrations are a few tenths of the PAN concentrations.

Molecules containing oxygen-nitrogen bonds are rather long-lived in the atmosphere. Three possible chemical fates are known. The first is abstraction of H from a C-H bond by OH·. The second is photolysis at an N-O bond. The third is unimolecular dissociation. In the case of PAN and PPN, photolysis occurs less rapidly than dissociation, which plays the major role.[1247] Photolysis is also slow for the alkyl nitrates,[1250] whose loss is controlled by relatively inefficient reactions with OH·.[1249] Gas phase nitro compounds may thus be largely lost as a result of deposition onto particle surfaces or to the ground.

11.4 Aromatic Nitro Compounds

Aromatic nitro compounds are surprisingly numerous, more than a hundred being present in the atmosphere at detectable concentrations. Some of these are present in the gas phase, but most are found instead as constituents of aerosol particles. Virtually all of the compounds are found in diesel exhaust and many as a byproduct of aluminum manufacture. It is likely that other processes involving high temperature combustion generate them as well. In addition, some are known to be products of atmospheric chemistry, as exemplified by the discussion of cresol chemistry in Chapter 5.

Virtually nothing is known of the reactions of the aromatic nitro compounds, particularly in the condensed phase. Their ubiquity suggests that the lifetimes are relatively long, however, an important fact in view of the carcinogenic potential of many of them.

11.5 Heterocyclic Nitrogen Compounds

Nearly a hundred and fifty of these compounds occur in the atmosphere, mostly in aerosol particles. Some are found indoors. They are products of incomplete combustion of fossil fuels and biomass of various types. Coal combustion and tobacco smoke are especially prolific sources and polymer combustion also produces a number of nitrogen heterocycles.

The reactions of these compounds are thought to be generally similar to those of the analogous arenes, though they have received little study. A common process is OH· addition at a point of ring unsaturation, as seen in the gas phase for pyrrole[1251] and in the liquid phase for pyridone.[1252] Both reactions are facile. The presence of heterocyclic nitrogen compounds in the aerosol phase, their relatively high concentrations there,[1243] and the low OH· concentrations expected in aerosols suggest long atmospheric lifetimes for these compounds.

Table 11.1. Nitriles

Species Number	Registry Number	Name	Emission Source	Emission Ref.	Detection Ref.	Bioassay Code	Bioassay −MA	Bioassay +MA	Bioassay Ref.
11.1-1	74-90-8	**Hydrogen cyanide**	acrylonitrile mfr.	626	242,986*				
			auto	205,415	726*(S)				
		coal comb. 895; diesel 772*,895; electronics mfr. 616; foundry 279*; microbes 302; petroleum mfr. 58*; plastics comb. 46,242; polymer comb. 611,853; refuse comb. 13,416*,26(a); sewage tmt. 1013*; steel mfr. 105*; tobacco smoke 4,396; turbine 594,666*; vegetation 531,583							
11.1-2S	75-05-8	**Ethanenitrile**	acrylonitrile mfr.	626	815*,816*				
			auto	895,974	817(S)				
		industrial 58*,60; polymer comb. 853,1169; syn. rubber mfr. 58*; turbine 414,895			820(I)				
11.1-3	2074-87-5	**Cyanogen**	auto	415,807	389*				
			refuse comb.	13					
			tobacco smoke	396					
11.1-4	107-12-0	**Propanenitrile**	auto	895					
			polymer comb.	853					
			turbine	414,895					
11.1-5	18936-17-9	**2-Methylbutanenitrile**			597				
11.1-6	110-59-8	**Pentanenitrile**	turbine	414,895					

Table 11.1. *(Continued)*

Species Number	Registry Number	Name	Emission Source	Emission Ref.	Detection Ref.	Bioassay Code	Bioassay −MA	Bioassay +MA	Bioassay Ref.
11.1-7	542-54-1	**4-Methylpentanenitrile**			820(I)				
11.1-8S	107-13-1	**Acrylonitrile**	acrylonitrile mfr.	626,1156*	820(I)	WPU	+		B1
			auto	217*,895		WP2	+		B1
		industrial 58*,60; plastics mfr. 104; polymer comb. 853,1169;				CCC	SP		B37
		syn. rubber mfr. 58*				STD38		?(G)	B36
						STI35		?(G)	B36
						STD35		+(G)	B36
						CTP	+		B24
						CTV	+		B24
						SRL	?		B30
11.1-9	126-98-7	**2-Methylacrylonitrile**	industrial	60					
11.1-10	4786-20-3	**2-Butenenitrile**	polymer comb.	853					
11.1-11	17656-09-6	**2-Butenedinitrile**	polymer comb.	853					
11.1-12	109-75-1	**3-Butenenitrile**	polymer comb.	853					
11.1-13	107-16-4	**2,3-Dihydroxypropanenitrile**	turbine	414,895					
11.1-14	5264-33-5	**5-Cyanopentanoic acid**	polymer comb.	853					
11.1-15	19060-13-0	**3-Cyanopropenoic acid**	acrylonitrile mfr.	626					
11.1-16S	624-83-9	**Methylisocyanate**	tobacco smoke	634					
11.1-17S	1531-37-9	**Azodiisobutylnitrile**	polymer mfr.	407*					

Species Number	Registry Number	Name	Emission Source	Emission Ref.	Detection Ref.	Code	Bioassay −MA	Bioassay +MA	Bioassay Ref.
11.1-18S	45043-50-3	**Azo-bis(succinonitrile)**	plastics mfr.	46					
11.1-19S	100-47-0	**Benzonitrile**	auto	895	358,597t				
			polymer comb.	853	399(a),458(a)				
			refuse comb.	13	820(I),1113(I)				
11.1-20	32074-25-2	**Dicyanobenzene**	polymer comb.	853					
11.1-21	25550-22-5	**Toluonitrile**	polymer comb.	1169	458(a)				
11.1-22	140-29-4	**Phenylethanenitrile**	vegetation	531	820(I)				
11.1-23S	86-53-3	**Naphthalene-1-carbonitrile**	biomass comb.	838(a)		ALC	+		B19
			wood comb.	838(a)					
11.1-24	613-46-7	**Naphthalene-2-carbonitrile**	biomass comb.	838(a)		ALC	+		B19
			wood comb.	838(a)					
11.1-25	83536-56-5	**Acenaphthenecarbonitrile**	biomass comb.	838(a)	799(a)				
			wood comb.	838(a)					
11.1-26	2523-48-0	**Fluorenecarbonitrile**	coal comb.	361(a)	362(a)				
11.1-27	3752-42-9	**Anthracenecarbonitrile**	biomass comb.	838(a)	799(a)				
			wood comb.	838(a)					
11.1-28	4107-64-6	**Pyrenecarbonitrile**	biomass comb.	838(a)	799(a)				
			wood comb.	838(a)					

Table 11.1. *(Continued)*

Species Number	Registry Number	Name	Emission Source	Emission Ref.	Detection Ref.	Bioassay Code	Bioassay −MA	Bioassay +MA	Bioassay Ref.
11.1-29S	103-71-9	**Phenylisocyanate**	polymer comb.	1169					
11.1-30	584-84-9	**Toluene-2,4-diisocyanate**	industrial	60					
11.1-31	91-08-7	**Toluene-2,6-diisocyanate**	industrial	60					

NCMe
11.1-2

$NCCH:CH_2$
11.1-8

OCNMe
11.1-16

$-Bu_2NCN$
11.1-17

$NC(CH_2)_3N:N(CH_2)_3CN$
11.1-18

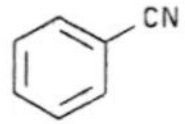

CN
11.1-19

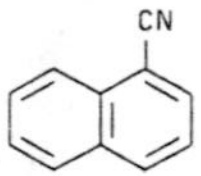

CN
11.1-23

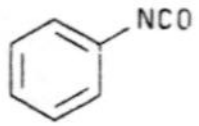

NCO
11.1-29

Table 11.2. Amines, Amides

Species Number	Registry Number	Name	Emission Source	Emission Ref.	Detection Ref.	Code	Bioassay −MA	Bioassay +MA	Ref.
11.2-1S	302-01-2	**Hydrazine**	rocket	536		CCC	SP		B37
						STU35		+	B36
						SRL	?		B30
11.2-2	57-14-7	**Dimethylhydrazine**	rocket	536		CCC	SP		B37
						STP00	−(C)	−(CD)	B36
						STP35	−(C)	−(CD)	B36
						STP37	−(C)	−(CD)	B36
						STP38	−(C)	−(CD)	B36
						STP98	−(C)	+(D)	B36
						MNT	?/−		B29
						L5	NEG	NEG	B28
						SPI	NEG		B25
11.2-3S	74-89-5	**Methylamine**	animal waste	80,160	628t				
			auto	792*,895					
		fish processing 110,255*; tobacco smoke 396,534							
11.2-4	124-40-3	**Dimethylamine**	animal waste	160					
			auto	792*,895					
		chemical mfr. 1046,1164; fish processing 255*; tobacco smoke 396,449							

Table 11.2. *(Continued)*

Species Number	Registry Number	Name	Emission Source	Emission Ref.	Detection Ref.	Code	Bioassay −MA	Bioassay +MA	Bioassay Ref.
11.2-5	75-50-3	**Trimethylamine**	animal waste	157,436*					
			chemical mfr.	1147*,1164					
		fish processing 58*,110; leather mfr. 1151*,1152; microbes 302; rendering 325,384; sewage tmt. 174*; starch mfr. 695,1071; tobacco smoke 396,534							
11.2-6	75-04-7	**Ethylamine**	animal waste	80,160	591				
			auto	792*,807*					
		sewage tmt. 422*; tobacco smoke 396,534							
11.2-7	109-89-7	**Diethylamine**	chemical mfr.	1046					
			fish processing	254*,255*					
		solvent 134; tobacco smoke 534,634							
11.2-8	121-44-8	**Triethylamine**	animal waste	80,900					
			sewage tmt.	422*					
			solvent	134					
11.2-9	107-10-8	**Propylamine**	animal waste	80,160					
			fish processing	254*,255*					
11.2-10	75-31-0	**Isopropylamine**	animal waste	80,160					
11.2-11	108-18-9	**Diisopropylamine**	sewage tmt.	422*					
11.2-12	109-73-9	**Butylamine**	animal waste	160,898	358t				
			fertilizer mfr.	180					
		fish processing 254*; rendering 325,1077; sewage tmt. 422*							

Species Number	Registry Number	Name	Emission Source	Emission Ref.	Detection Ref.	Code	Bioassay −MA	Bioassay +MA	Ref.
11.2-13	111-92-2	**Dibutylamine**	sewage tmt.	422*		SCE	+		B6
11.2-14	110-96-3	**Diisobutylamine**	sewage tmt.	422*					
11.2-15	13952-84-6	*sec*-**Butylamine**	animal waste	160					
11.2-16S	110-60-1	**Putrescine**	food processing	62					
			rendering	325,637					
11.2-17	110-58-7	**1-Pentylamine**	animal waste	80,160					
11.2-18	625-30-9	**2-Pentylamine**			597	SCE	+		B6
11.2-19S	462-94-2	**Cadaverine**	food processing	62					
			rendering	325,637					
11.2-20	111-26-2	**Hexylamine**	sewage tmt.	422*					
11.2-21	124-09-4	**1,6-Hexanediamine**	industrial	60					
11.2-22	141-43-5	**2-Hydroxyethylamine**	industrial	60					
			natural gas proc.	423*					
11.2-23	111-42-2	**bis(2-Hydroxyethyl) amine**	industrial	60		STI00	NEG	NEG	B38
						STI35	NEG	NEG	B38
						STI37	NEG	NEG	B38
						STI98	NEG	NEG	B38
11.2-24	102-71-6	**tris(2-Hydroxyethyl) amine**	industrial	60		ALC	+		B19

Table 11.2. *(Continued)*

Species Number	Registry Number	Name	Emission Source	Emission Ref.	Detection Ref.	Code	Bioassay −MA	Bioassay +MA	Ref.
11.2-25S	56-40-6	**2-Aminoacetic acid**	volcano	1007	757*(r)	STI00	NEG	NEG	B38
						STI35	NEG	NEG	B38
						STI37	NEG	NEG	B38
						STI98	NEG	NEG	B38
11.2-26	56-41-1	**2-Aminopropionic acid**			757*(r)				
11.2-27	56-45-1	**2-Amino-3-hydroxypropionic acid**	volcano	1007					
11.2-28	56-12-2	**4-Aminobutyric acid**	volcano	1007					
11.2-29	72-19-5	**2-Amino-3-hydroxybutyric acid**	volcano	1007					
11.2-30	56-84-8	**2-Amino-1,4-butanedioic acid**			757*(r)	ALC	+		B19
						SPH	+		B26
11.2-31	61-90-5	**2-Amino-4-methylpentanoic acid**			757*(r)				
11.2-32	617-65-2	**2-Amino-1,5-pentanedioic acid**	tobacco smoke	396(a)	757*(r)				
11.2-33	929-17-9	**7-Aminoheptanoic acid**	volcano	1007					
11.2-34S	6899-04-3	**Glutamine**	tobacco smoke	396(a)					
11.2-35S	68-12-2	**Dimethylformamide**	industrial	156	597,628	MNT	?/−		B29
			printing	620		SPI	NEG		B25
			solvent	442					

Species Number	Registry Number	Name	Emission Source	Emission Ref.	Detection Ref.	Bioassay Code	Bioassay −MA	Bioassay +MA	Bioassay Ref.
11.2-36S	60-35-5	**Acetamide**	tobacco smoke	1177(a)		HMAST		NEG	B13
						CCC	SP		B37
						STP00	NEG	NEG	B36
						STP35	NEG	NEG	B36
						STP37	NEG	NEG	B36
						STP98	NEG	NEG	B36
						REC		NEG	B9
						STI00	NEG	NEG	B38
						STI35	NEG	NEG	B38
						STI37	NEG	NEG	B38
						STI98	NEG	NEG	B38
						CTP	+		B24
						SPI	NEG		B25
						CTV	+,+		B24
11.2-37S	127-19-5	**N,N-Dimethylacetamide**	tobacco smoke	1177(a)					
11.2-38	625-50-3	**N-Ethylacetamide**	tobacco smoke	1177(a)					
11.2-39	79-05-0	**Propionamide**	tobacco smoke	1177(a)					
11.2-40	541-35-5	**Butyramide**	tobacco smoke	1177(a)					
11.2-41	563-83-7	**Isobutyramide**	tobacco smoke	1177(a)					
11.2-42	541-46-8	**Isopentanamide**	tobacco smoke	1177(a)					

Table 11.2. *(Continued)*

Species Number	Registry Number	Name	Emission Source	Emission Ref.	Detection Ref.	Code	Bioassay −MA	Bioassay +MA	Ref.
11.2-43S	62-75-9	**N-Nitrosodimethylamine**	amine mfr.	368,377	368,378*	HMAML		+	B13
			auto	807t		HMANC		+	B13
		diesel 872*,921*; tobacco smoke 404,450				HMAV7		+	B13
						HMACH		+	B13
						HMASC		+	B13
						HMASM		+/−	B13
						HMAEC		+	B13
						HMAST		+	B13
						V790		+	B7
						V79T		+	B7
						V79A		+	B7
						WP2	+		B1
						WPU	+		B1
						STI00		+(A)	B36
						STU35		+	B36
						SCE	+		B6
						CCC	SP		B37
						CHOM		+	B3
						MST	NEG		B5
						MDR	+		B12
						CYG	NEG		B8
						REC		+	B9
						STI00	NEG	+	B38
						STI35	NEG	+	B38
						STI37	NEG	NEG	B38
						STI98	NEG	NEG	B38
						ARA	+		B17
						YEA		+	B27
						MNT	+/−		B29

Species Number	Registry Number	Name	Emission Source	Emission Ref.	Detection Ref.	Code	Bioassay −MA	Bioassay +MA	Bioassay Ref.
						L5		+	B28
						CTV	?		B24
						CTL	+		B24
						CTP	?		B24
						SRL	+		B30
						NEU		+	B32
						SPI	NEG		B25
11.2-44	55-18-5	**N-Nitrosodiethylamine**			378,924*	V790		+	B7
						V79A		+	B7
						STP35		+	B36
						STU35		+	B36
						STI98	NEG	+(A)	B36
						STI00		+(A)	B36
						STI35		+	B36
						SCE	+		B6
						CCC	SP		B37
						SLPM	?		B4
						SLGM	NEG		B4
						MST	NEG		B5
						MDR	+		B12
						CYC		+	B8
						REC	NEG	+	B9
						L5		+	B28
						CTV	+		B24
						CTP	+		B24
						SRL	+		B30
						NEU	+(D)		B32
						MNT	?/−		B29

Table 11.2. *(Continued)*

Species Number	Registry Number	Name	Emission Source	Emission Ref.	Detection Ref.	Code	Bioassay −MA	Bioassay +MA	Bioassay Ref.
11.2-45S	123-33-1	**Maleic hydrazide**	chemical mfr.	1046		CCC	LN		B37
						STP00	NEG	NEG	B36
						STP35	NEG	NEG	B36
						STP37	NEG	NEG	B36
						STP98	NEG	NEG	B36
						SCE	NEG		B6
						VIC	+		B18
						ALC	+		B19
						STI00	NEG	NEG	B38
						STI35	NEG	NEG	B38
						STI37	NEG	NEG	B38
						STI98	NEG	NEG	B38
						ARA	NEG		B17
						MNT	?/−		B29
						SRL	+		B30
						NEU	NEG		B32
11.2-46	108-91-8	**Cyclohexylamine**	biomass comb.	597		HMAST		NEG	B13
						HMASM		+/−	B13
						CCC	I,P		B37
						CYB	NEG		B8
						CYS	+		B8
						CYG	NEG		B8
						CYI	NEG		B8
						SRL	NEG		B30

Species Number	Registry Number	Name	Emission Source	Emission Ref.	Detection Ref.	Code	Bioassay −MA	Bioassay +MA	Bioassay Ref.
11.2-47S	62-53-3	**Aniline**	industrial	60,250	597	CCC	I		B37
			plastics comb.	354	528(a)	STP00	NEG	NEG	B36
		polymer comb. 1169,1107(a); tobacco smoke 534(a)				STP35	NEG	NEG	B36
						STP37	NEG	NEG	B36
						STP98	NEG	NEG	B36
						REC	+/−		B9
						ALC	+/−		B19
						STI00	NEG	NEG	B38
						STI35	NEG	NEG	B38
						STI37	NEG	NEG	B38
						STI98	NEG	NEG	B38
						CTL	+		B24
						CTP	NEG		B24
11.2-48S	106-49-0	***p*-Toluidine**	industrial	250					
			plastics comb.	354					
		polymer comb. 1169,1107(a); tobacco smoke 534							
11.2-49	95-64-7	**3,4-Dimethylaniline**	polymer comb.	1107(a)	274,628t				
11.2-50	several	**Tetramethylaniline**	polymer comb.	1107(a)					
11.2-51	100-61-8	**N-Methylaniline**			597				
11.2-52	6068-69-5	**N-*sec*-Butylaniline**	vulcanization	308					
11.2-53	620-84-8	**N-Phenyl-*p*-toluidine**	polymer comb.	1107(a)					
11.2-54	122-39-4	**Diphenylamine**	chemical mfr.	1046*	570(r)				

Table 11.2. *(Continued)*

Species Number	Registry Number	Name	Emission Source	Emission Ref.	Detection Ref.	Code	Bioassay −MA	Bioassay +MA	Bioassay Ref.
11.2-55	25265-76-3	**Benzenediamine**	chemical mfr.	1046					
11.2-56	several	**Biphenylamine**	refuse comb.	878(a)					
11.2-57S	101-77-9	**4,4'-di(Aminophenyl)methane**	polymer comb.	1107(a)					
11.2-58	several	**Methyl-4,4'-di(amino-phenyl)methane**	polymer comb.	1107(a)					
11.2-59	several	**Dimethyl-4,4'-di(amino-phenyl)methane**	polymer comb.	1107(a)					
11.2-60	several	**Trimethyl-4,4'-di(amino-phenyl)methane**	polymer comb.	1107(a)					
11.2-61	134-32-7	**2-Naphthenamine**			597	TRM	+		B21
					528(a)	HMAST		+	B13
						CCC	LN		B37
						STP00		+	B36
						MNT	+/−		B29
						CTV	−,+		B24
						CTP	NEG		B24
						NEU	+/−		B32
						SPI	?		B25
11.2-62	370-14-9	**4-Hydroxy-N-methylaniline**	vegetation	533					
11.2-63S	59-89-2	**N-Nitrosomorpholine**	diesel	872		HMAST		+	B13

Species Number	Registry Number	Name	Emission Source	Emission Ref.	Detection Ref.	Bioassay Code	Bioassay −MA	Bioassay +MA	Bioassay Ref.
						V79A		+	B7
						STP35		+ABH	B36
						CCC	SP		B37
						MDR	+		B12
						CYI	+		B8
						MNT	+/−		B29
						SRL	+		B30
						SPI	NEG		B25
11.2-64S	134-20-3	**Methylanthranilate**	vegetation	531,552e					
11.2-65	several	**Dimethylanthranilate**	vegetation	514e					
11.2-66	87-25-2	**Ethylanthranilate**	vegetation	552e					
11.2-67S	114-26-1	**Propoxur**	pesticide	154*,573*		WP2	NEG		B1
11.2-68S	1563-66-2	**Carbofuran**	pesticide	1118		WPU	NEG		B1
						STP00	NEG	NEG	B36
						STP35	NEG	NEG	B36
						STP37	NEG	NEG	B36
						STP38	NEG	NEG	B36
						STP98	NEG	NEG	B36
						SRL	NEG		B30

Table 11.2. *(Continued)*

Species Number	Registry Number	Name	Emission Source	Emission Ref.	Detection Ref.	Code	Bioassay −MA	Bioassay +MA	Ref.
11.2-69S	63-25-2	**Carbaryl**	pesticide	126,347(a)		V790	+		B7
						WP2	NEG		B1
						STP00	NEG	NEG	B36
						STP38	−(C)	−(C)	B36
						STP98	NEG	NEG	B36
						STU38	−(E)	+(E)	B36
						STP35	NEG	NEG	B36
						STP37	NEG	NEG	B36
						MDR	+		B12
						VIC	+		B18
						ALC	+		B19
						SRL	?		B30
						SPL	+		B25
11.2-70	150-30-1	**2-Amino-3-phenylpropionic acid**	volcano	1007					
11.2-71S	949-87-1	**4-Methylazobenzene**	biomass comb.	597					
11.2-72S	26444-20-2	**bis(Methylphenyl)diazene**	refuse comb.	878(a)					

H_2NNH_2 11.2-1 — $MeNH_2$ 11.2-3 — $H_2N(CH_2)_4NH_2$ 11.2-16 — $H_2N(CH_2)_5NH_2$ 11.2-19 — $HO_2CCH_2NH_2$ 11.2-25

$HO_2CCH(NH_2)CH_2CH_2CONH_2$ 11.2-34 — $Me_2NCH:O$ 11.2-35 — $AcNH_2$ 11.2-36 — Me_2NAc 11.2-37 — Me_2NNO 11.2-43

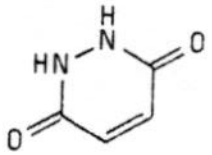

11.2-45

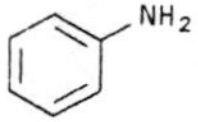

11.2-47

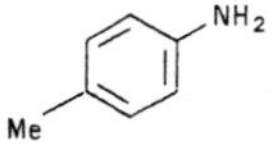

11.2-48

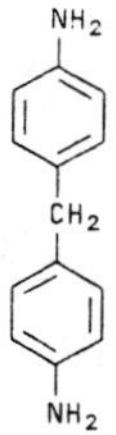

11.2-57

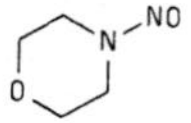

11.2-63

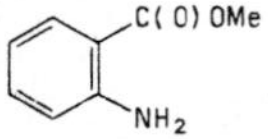

11.2-64

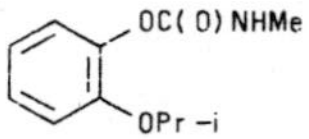

11.2-67

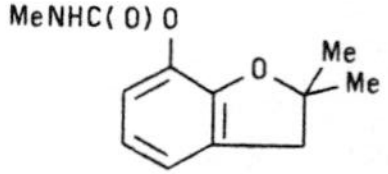

11.2-68

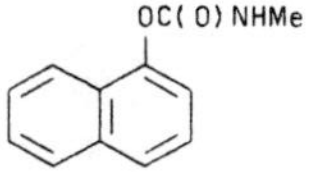

11.2-69

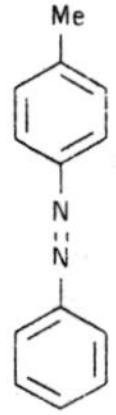

11.2-71

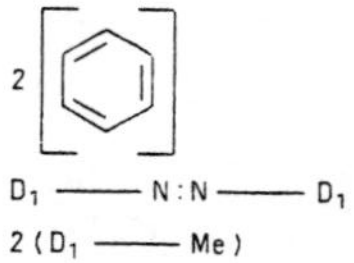

11.2-72

Table 11.3. Aliphatic Nitro Compounds

Species Number	Registry Number	Name	Emission Source	Emission Ref.	Detection Ref.	Code	Bioassay −MA	Bioassay +MA	Bioassay Ref.
11.3-1S	75-52-5	**Nitromethane**	auto	217*,415					
			diesel	692					
		explosives mfr. 1092; tobacco smoke 396; turbine 414,895							
11.3-2	509-14-8	**Tetranitromethane**	explosives mfr.	1092					
11.3-3S	598-58-3	**Methylnitrate**	explosives mfr.	1092	173t				
11.3-4	79-24-3	**Nitroethane**	auto	415,895		MNT	?/−		B29
11.3-5	625-58-1	**Ethylnitrate**			186,263				
11.3-6S	2278-22-0	**Peroxyacetylnitrate**			231*,329 963*(I)				
11.3-7	79-46-9	**2-Nitropropane**	solvent	88,134		CCC	SP		B37
						STI00	+	+	B38
						STI35	+/−	+/−	B38
						STI37	+/−	+/−	B38
						STI98	+	+	B38
						MNT	?/−		B29
11.3-8	594-70-7	**2-Methyl-2-nitropropane**	turbine	414,895					
11.3-9	627-13-4	**Propylnitrate**			307				

Species Number	Registry Number	Name	Emission Source	Emission Ref.	Detection Ref.	Code	Bioassay −MA	Bioassay +MA	Bioassay Ref.
11.3-10	5796-89-4	**Peroxypropionylnitrate**			173,231*				
11.3-11	928-45-0	**Butylnitrate**			307				
11.3-12		**Neopentylnitrate**	turbine	414					
11.3-13	1606-31-1	**2-Methyl-1-nitropropene**	auto	895					
11.3-14	1606-30-0	**2-Methyl-1-nitro-2-propene**	auto	895					
11.3-15	19031-80-2	**2-Methyl-3-nitro-2-butene**	auto	895					
11.3-16	4812-22-0	**3-Nitro-3-hexene**	auto	895					
11.3-17S		**2-Oxooctylnitrate**	turbine	414,895					
11.3-18S		**4-Formylbutylnitrate**			214*(a)t				
11.3-19		**5-Formylpentylnitrate**			214*(a)t				
11.3-20		**6-Formylhexylnitrate**			214*(a)t				
11.3-21S		**4-Formyl-2-oxobutylnitrite**			214*(a),532(a)				
11.3-22		**5-Formyl-2-oxopentylnitrite**			214*(a),532(a)				
11.3-23		**6-Formyl-2-oxopentylnitrite**			532(a)				

Table 11.3. *(Continued)*

Species Number	Registry Number	Name	Emission Source	Emission Ref.	Detection Ref.	Code	Bioassay −MA	Bioassay +MA	Bioassay Ref.
11.3-24S		**4-Formyl-2-oxobutylnitrate**			214*(a)t				
11.3-25		**5-Formyl-2-oxopentylnitrate**			214*(a)t				
11.3-26		**6-Formyl-2-oxohexylnitrate**			214*(a)t				
11.3-27S	74754-56-6	**5-Nitratopentanoic acid**			214*(a),532(a)				
11.3-28		**5-Nitrato-5-oxopentanoic acid**			532(a)				
11.3-29	74754-55-5	**6-Nitratohexanoic acid**			214*(a),532(a)				
11.3-30		**6-Nitrato-6-oxohexanoic acid**			214*(a),532(a)				
11.3-31		**7-Nitratoheptanoic acid**			532(a)				
11.3-32		**7-Nitrato-7-oxoheptanoic acid**			214*(a),532(a)				
11.3-33S	4164-28-7	**Dimethylnitroamine**	tobacco smoke	924					

O_2NMe
11.3-1

$MeONO_2$
11.3-3

O_2NOOAc
11.3-6

$CH_3(CH_2)_5COCH_2ONO_2$
11.3-17

$OHC(CH_2)_4ONO_2$
11.3-18

$OHC(CH_2)_3C(O)ONO$
11.3-21

$OHC(CH_2)_3C(O)ONO_2$
11.3-24

$O_2NO(CH_2)_4CO_2H$
11.3-27

Me_2NNO_2
11.3-33

Table 11.4. Aromatic Nitro Compounds

Species Number	Registry Number	Name	Emission Source	Emission Ref.	Detection Ref.	Bioassay Code	Bioassay −MA	Bioassay +MA	Bioassay Ref.
11.4-1S	98-95-3	**Nitrobenzene**	industrial	60	797*	STI00	NEG	NEG	B38
			solvent	439	925(a)	STI35	NEG	NEG	B38
						STI37	NEG	NEG	B38
						STI98	NEG	NEG	B38
11.4-2S	88-72-2	***o*-Nitrotoluene**	explosives mfr.	1092	925(a)t	STI00	NEG	NEG	B38
						STI35	NEG	NEG	B38
						STI37	NEG	NEG	B38
						STI98	NEG	NEG	B38
11.4-3	99-08-1	***m*-Nitrotoluene**	explosives mfr.	1092		STI00	NEG	NEG	B38
						STI35	NEG	NEG	B38
						STI37	NEG	NEG	B38
						STI98	NEG	NEG	B38
11.4-4	99-99-0	***p*-Nitrotoluene**	explosives mfr.	1092		STI00	NEG	NEG	B38
						STI35	NEG	NEG	B38
						STI37	NEG	NEG	B38
						STI98	NEG	NEG	B38
11.4-5	25168-04-1	**Dimethylnitrobenzene**			925(a)				
11.4-6S	86-00-0	**2-Nitrobiphenyl**	aluminum mfr.	850(a)					
			diesel	800(a)					

Table 11.4. *(Continued)*

Species Number	Registry Number	Name	Emission Source	Emission Ref.	Detection Ref.	Bioassay Code	Bioassay −MA	Bioassay +MA	Bioassay Ref.
11.4-7	2113-58-8	**3-Nitrobiphenyl**	aluminum mfr.	850(a)					
			diesel	688(a),800(a)					
11.4-8	92-93-3	**4-Nitrobiphenyl**	aluminum mfr.	850(a)		HMAST		+/−	B13
			diesel	800(a)		STP38	NE,AB	NE,AB	B36
						STP00	+	+	B36
						CCC	LP		B37
						CTV	+		B24
						CTP	+		B24
						NEU	NEG		B32
11.4-9	80182-39-4	**Methylnitrobiphenyl**	diesel	688(a)					
11.4-10	2436-96-6	**2,2'-Dinitrobiphenyl**	aluminum mfr.	850(a)					
11.4-11	86695-75-2	**Nitroterphenyl**	diesel	800(a)					
11.4-12S	86-57-7	**1-Nitronaphthalene**	diesel	800(a)	845(a)	CCC	SN		B37
11.4-13	881-03-8	**2-Methyl-1-nitronaphthalene**	diesel	800(a)					
11.4-14	581-89-5	**2-Nitronaphthalene**	aluminum mfr.	850(a)	799(a)t,845(a)	HMAST		+	B13
			diesel	800(a)		STP00	+		B36
						REC	+		B9
11.4-15	80182-40-7	**Trimethylnitronaphthalene**	diesel	688(a),800(a)					

Species Number	Registry Number	Name	Emission Source	Emission Ref.	Detection Ref.	Code	Bioassay −MA	Bioassay +MA	Bioassay Ref.
11.4-16	605-71-0	**1,5-Dinitronaphthalene**	aluminum mfr.	850(a)					
			diesel	800(a)					
11.4-17	602-38-0	**1,8-Dinitronaphthalene**	aluminum mfr.	850(a)					
			diesel	800(a)					
11.4-18	several	**Nitroacenaphthene**	diesel	688(a),928(a)					
11.4-19	several	**Methylnitroacenaphthene**	diesel	688(a)					
11.4-20	several	**Nitroacenaphthylene**	diesel	688(a)					
11.4-21S	55345-04-5	**Nitrofluorene**	aluminum mfr.	850(a)		WPU	+		B1
			diesel	688(a),800(a)					
11.4-22	several	**Methylnitrofluorene**	diesel	688(a)					
11.4-23	15110-74-4	**2,5-Dinitrofluorene**	aluminum mfr.	850(a)					
			diesel	800(a)					
11.4-24	5405-53-8	**2,7-Dinitrofluorene**	aluminum mfr.	850(a)					
			diesel	800(a),985*(a)					
11.4-25S	17024-18-9	**2-Nitrophenanthrene**	diesel	688(a),800(a)					
11.4-26	80191-44-2	**Methylnitrophenanthrene**	diesel	688(a)					
11.4-27	several	**Dimethylnitrophenanthrene**	diesel	688(a)					

Table 11.4. *(Continued)*

Species Number	Registry Number	Name	Emission Source	Emission Ref.	Detection Ref.	Code	Bioassay −MA	Bioassay +MA	Bioassay Ref.
11.4-28	several	**4,5-Methylenenitro-phenanthrene**			1180(a)				
11.4-29S	36925-31-2	**1-Nitroanthracene**			799(a)				
11.4-30	3586-69-4	**2-Nitroanthracene**	diesel	684(a),688(a)	845(a)t				
11.4-31	602-60-8	**9-Nitroanthracene**	aluminum mfr. diesel	850(a) 985*(a),1136*(a)	1143(a),1180(a)				
11.4-32	86695-76-3	**1-Methyl-9-nitroanthracene**	diesel	684(a),688(a)	799(a)t,845(a)t				
11.4-33	86689-95-4	**1-Methyl-10-nitroanthracene**	diesel	800(a),1136*(a)					
11.4-34	84457-22-7	**9-Methyl-10-nitroanthracene**	diesel	1136*(a)					
11.4-35	80191-45-3	**Dimethylnitroanthracene**	diesel	684(a),688(a)					
11.4-36	86689-92-1	**Trimethylnitroanthracene**	diesel	684(a)t,800(a)					
11.4-37S	20268-51-3	**7-Nitrobenz[*a*]anthracene**	diesel	800(a)					
11.4-38	13209-09-1	**10-Nitrobenz[*a*]anthracene**			1143(a),1180(a)				
11.4-39	several	**Methylnitrobenz[*a*] anthracene**	diesel	688(a)					
11.4-40S	13177-28-1	**1-Nitrofluoranthene**	diesel	800(a)					

Species Number	Registry Number	Name	Emission Source	Emission Ref.	Detection Ref.	Code	Bioassay −MA	Bioassay +MA	Bioassay Ref.
11.4-41	13177-29-2	**2-Nitrofluoranthene**	diesel	688(a),800(a)	699(a)				
11.4-42	892-21-7	**3-Nitrofluoranthene**	diesel	800(a),928(a)	799(a),845(a)				
11.4-43	13177-31-6	**7-Nitrofluoranthene**	diesel	800(a)					
11.4-44	13177-32-7	**8-Nitrofluoranthene**	diesel	800(a),1144(a)	1143(a)t,1180(a)t				
11.4-45	80182-29-2	**Methylnitrofluoranthene**	diesel	688(a)					
11.4-46S	several	**Methylnitrotriphenylene**	diesel	688(a)					
11.4-47S	5522-43-0	**1-Nitropyrene**	aluminum mfr.	850(a)	799(a),831*(a)				
			auto	831*(a)					
		coal comb. 1136*(a); diesel 800(a),831*(a); wood comb. 831*(a)							
11.4-48	789-07-1	**2-Nitropyrene**	diesel	684(a),688(a)	1180(a)				
11.4-49	86689-96-5	**1-Methyl-3-nitropyrene**	diesel	688(a),800(a)					
11.4-50	57835-92-4	**4-Nitropyrene**	diesel	800(a)					
11.4-51	86689-97-6	**1-Methyl-6-nitropyrene**	diesel	800(a)					
11.4-52	89198-47-0	**1-Methyl-8-nitropyrene**	diesel	800(a)					
11.4-53	75321-20-9	**1,3-Dinitropyrene**	diesel	800(a),985*(a)					
11.4-54	42397-64-8	**1,6-Dinitropyrene**	diesel	800(a),985*(a)					

Table 11.4. *(Continued)*

Species Number	Registry Number	Name	Emission Source	Emission Ref.	Detection Ref.	Bioassay Code	Bioassay −MA	Bioassay +MA	Bioassay Ref.
11.4-55	42397-65-9	**1,8-Dinitropyrene**	diesel	800(a),985*(a)	926*(a)				
11.4-56S	86689-98-7	**6-Nitrocyclopenta[*cd*] pyrene**	diesel	800(a)					
11.4-57S	70021-99-7	**1-Nitrobenzo[*a*]pyrene**	diesel	800(a)					
11.4-58	70021-98-6	**3-Nitrobenzo[*a*]pyrene**	diesel	688(a),800(a)	699(a)				
11.4-59	63041-90-7	**6-Nitrobenzo[*a*]pyrene**	auto	831*(a)	831*(a),870(a)t	STP00	+	+	B36
			diesel	800(a),831*(a)					
			wood comb.	831*(a)					
11.4-60S	70021-42-0	**Nitrobenzo[*e*]pyrene**	diesel	688(a)					
11.4-61S	81316-77-0	**1-Nitrochrysene**	diesel	800(a)					
11.4-62	80182-33-8	**Methylnitrochrysene**	diesel	688(a)					
11.4-63S	20589-63-3	**3-Nitroperylene**	diesel	688(a),800(a)t		STP38	+	+	B36
11.4-64S	88-75-5	***o*-Nitrophenol**			927(a)	STI00	NEG	NEG	B38
						STI35	NEG	NEG	B38
						STI37	NEG	NEG	B38
						STI98	NEG	NEG	B38
						ALC	NEG		B19

Species Number	Registry Number	Name	Emission Source	Emission Ref.	Detection Ref.	Bioassay Code	Bioassay −MA	Bioassay +MA	Bioassay Ref.
11.4-65	100-02-7	***p*-Nitrophenol**			307,411	HMAST		NEG	B13
					755*(r),1006(r)	HMASM		+/−	B13
						STP00	NEG	NEG	B36
						STP35	NEG	NEG	B36
						STP37	NEG	NEG	B36
						STP98	NEG	NEG	B36
						STI00	NEG	NEG	B38
						STI35	NEG	NEG	B38
						STI37	NEG	NEG	B38
						STI98	NEG	NEG	B38
						ALC	NEG		B19
11.4-66	119-33-5	**4-Methyl-2-nitrophenol**	tobacco smoke	447*	411				
11.4-67	32021-53-7	**3,4,5-Trimethyl-2-nitrophenol**	coal comb.	1165(a)					
11.4-68	99-53-6	**2-Methyl-4-nitrophenol**			927(a) 1006(r)				
11.4-69	2581-34-2	**3-Methyl-4-nitrophenol**			927(a)	ALC	+		B19
11.4-70	2423-71-4	**2,6-Dimethyl-4-nitrophenol**	coal comb.	1165(a)					
11.4-71	32021-54-8	**2-Ethyl-6-methyl-4-nitrophenol**	coal comb.	1165(a)					

Table 11.4. *(Continued)*

Species Number	Registry Number	Name	Emission Source	Emission Ref.	Detection Ref.	Bioassay Code	Bioassay −MA	Bioassay +MA	Bioassay Ref.
11.4-72	13073-29-5	**2-Methyl-6-nitrophenol**			927(a) 1006(r)				
11.4-73	51-28-5	**2,4-Dinitrophenol**			927(a)	ALC ARA	+ NEG		B19 B17
11.4-74	573-56-5	**2,6-Dinitrophenol**			927(a)				
11.4-75	several	**Dihydroxynitronaphthalene**	diesel	688(a)					
11.4-76	several	**Hydroxynitrofluorene**	diesel	690(a),800(a)					
11.4-77	86674-49-9	**1-Hydroxy-3-nitropyrene**	diesel	800(a)					
11.4-78	1767-28-8	**1-Hydroxy-6-nitropyrene**	diesel	800(a)					
11.4-79	1732-29-2	**1-Hydroxy-8-nitropyrene**	diesel	800(a)					
11.4-80S	80267-67-0	**Nitronaphthoquinone**	diesel	688(a)					
11.4-81S	3096-52-4	**2-Nitro-9-fluorenone**	diesel	985*(a)					
11.4-82	42135-22-8	**3-Nitro-9-fluorenone**	diesel	688(a),800(a)					
11.4-83	53197-58-3	**2,5-Dinitro-9-fluorenone**	diesel	985*(a)					
11.4-84	31551-45-8	**2,7-Dinitro-9-fluorenone**	diesel	800(a)					
11.4-85	129-79-3	**2,4,7-Trinitro-9-fluorenone**	diesel	985*(a)					

Species Number	Registry Number	Name	Emission Source	Emission Ref.	Detection Ref.	Code	Bioassay −MA	Bioassay +MA	Bioassay Ref.
11.4-86S	80267-69-2	**Nitrophenanthrone**	diesel	688(a)					
11.4-87	80267-42-0	**Nitroanthrone**	diesel	688(a)					
11.4-88S	80267-73-8	**Nitrofluoranthone**	diesel	688(a)					
11.4-89S	80267-77-2	**Nitrofluoranthenequinone**	diesel	688(a)					
11.4-90S	80267-71-6	**Nitropyrone**	diesel	688(a)1100(a)					
11.4-91	86689-94-3	**Nitropyrene-3,6-quinone**	diesel	688(a),800(a)					
11.4-92	several	**Dimethylnitrophenanthrene-carbaldehyde**	diesel	688(a)					
11.4-93	several	**Dimethylnitroanthracene-carbaldehyde**	diesel	688(a)					
11.4-94S	3027-38-1	**3-Nitro-1,8-naphthalic acid anhydride**	diesel diesel	985*(a) 985*(a)					
11.4-95	80191-41-9	**Nitronaphthalenecarboxylic acid**	diesel diesel	688(a) 688(a)					

Table 11.4. *(Continued)*

Species Number	Registry Number	Name	Emission Source	Emission Ref.	Detection Ref.	Bioassay Code	Bioassay −MA	Bioassay +MA	Bioassay Ref.
11.4-96S	100-01-6	***p*-Nitroaniline**	industrial	250		STI00	NEG	NEG	B38
						STI35	NEG	NEG	B38
						STI37	NEG	NEG	B38
						STI98	+	+	B38
						SPI	NEG		B25
11.4-97S	32368-69-7	**Peroxybenzoylnitrate**			549*				
11.4-98S	607-34-1	**5-Nitroquinoline**	diesel	800(a)		STP98	+		B36
						STP00	+		B36
11.4-99	607-35-2	**8-Nitroquinoline**	diesel	800(a)		STI00	−(B)	+(B)	B36
						STI98	−(BC)	+(B)	B36
						STP00	−(C)	+	B36
						STP37	+	+	B36
						STP98	+	+	B36
11.4-100	several	**Nitrobenzoquinoline**	diesel	800(a)					
11.4-101S	76025-15-5	**Nitroacridine**	diesel	800(a)					
11.4-102	50764-83-5	**Dinitroacridine**	diesel	800(a)					
11.4-103	several	**Nitrobenzothiophene**	diesel	800(a)					

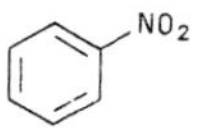

NO_2

11.4-1

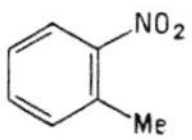

NO_2 Me

11.4-2

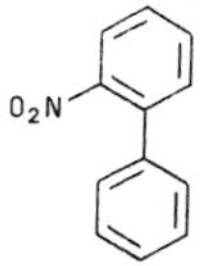

O_2N

11.4-6

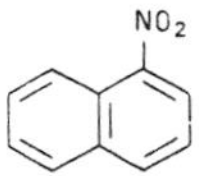

NO_2

11.4-12

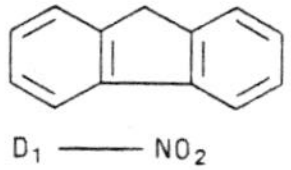

D_1 —— NO_2

11.4-21

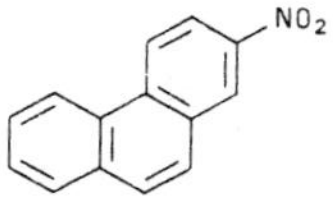

NO_2

11.4-25

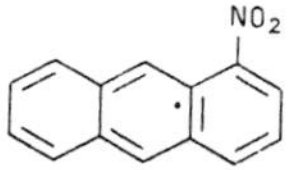

NO_2

11.4-29

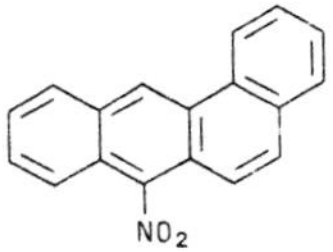

NO_2

11.4-37

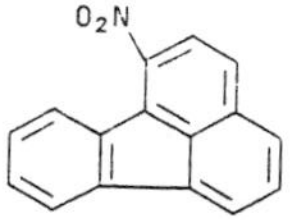

O_2N

11.4-40

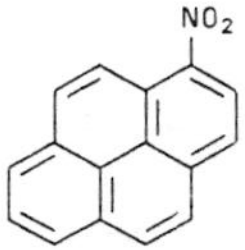

NO_2

11.4-47

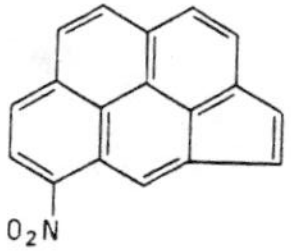

O_2N

11.4-56

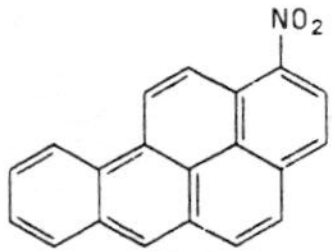

NO_2

11.4-57

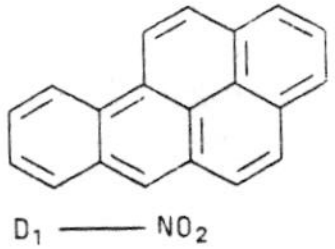

D_1 —— NO_2

11.4-60

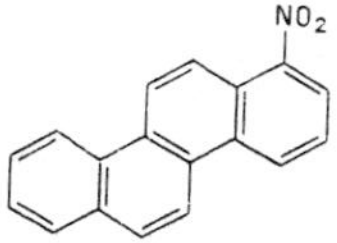

NO_2

11.4-61

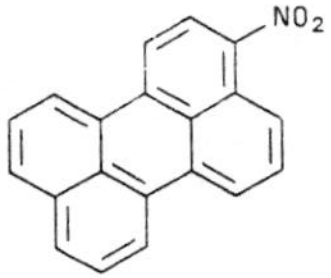

NO_2

11.4-63

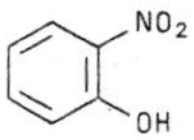

NO_2 OH

11.4-64

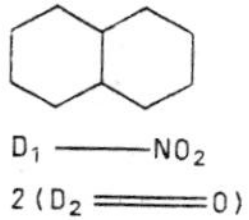

D_1 —— NO_2
2 (D_2 ══ O)

11.4-80

Table 11.4. *(Continued)*

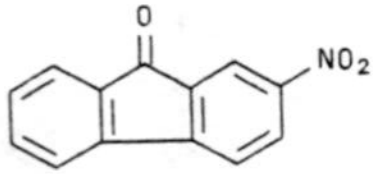

O
NO_2

11.4-81

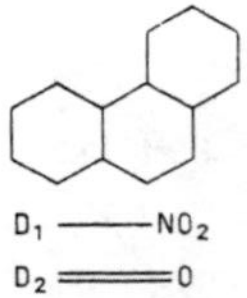

D_1 —— NO_2
D_2 ═══ O

11.4-86

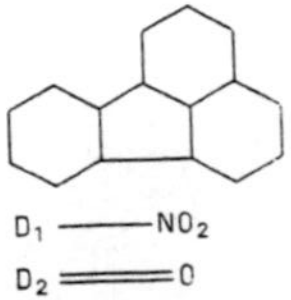

D_1 —— NO_2
D_2 ═══ O

11.4-88

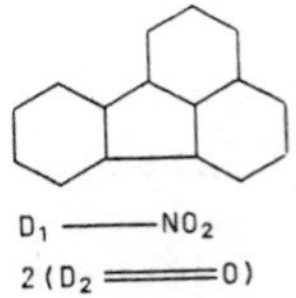

D_1 —— NO_2
2 (D_2 ═══ O)

11.4-89

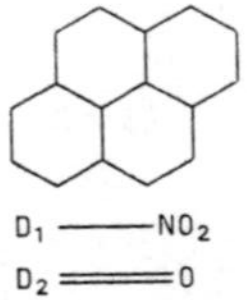

D_1 —— NO_2
D_2 ═══ O

11.4-90

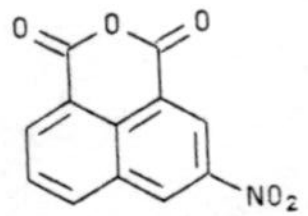

O O O
NO_2

11.4-94

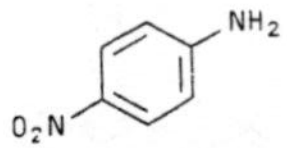

NH_2
O_2N

11.4-96

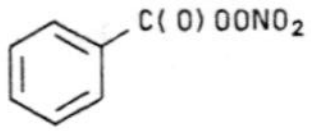

C(O)OONO$_2$

11.4-97

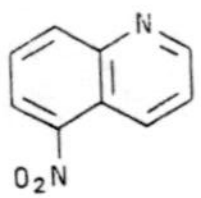

N
O_2N

11.4-98

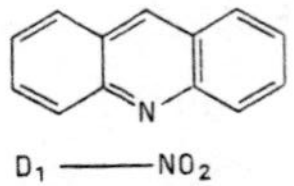

N
D_1 —— NO_2

11.4-101

Table 11.5. Heterocyclic Nitrogen Compounds

Species Number	Registry Number	Name	Emission Source	Emission Ref.	Detection Ref.	Code	Bioassay −MA	Bioassay +MA	Bioassay Ref.
11.5-1S	109-97-7	**Pyrrole**	tobacco smoke	396	399(a),458(a) 820(I)				
11.5-2	27417-39-6	**Methylpyrrole**			458(a)				
11.5-3	3274-56-4	**2,4-Diphenylpyrrole**	biomass comb.	891(a)t					
11.5-4S	288-32-4	**Imidazole**	turbine	414,895					
11.5-5	616-47-7	**1-Methylimidazole**	tobacco smoke	1177(a)					
11.5-6	693-98-1	**2-Methylimidazole**	tobacco smoke	1177(a)					
11.5-7	1739-84-0	**1,2-Dimethylimidazole**	tobacco smoke	1177(a)					
11.5-8	40356-65-8	**2-Ethylimidazole**	tobacco smoke	1177(a)	399(a)				
11.5-9S	930-61-0	**2,4-Dimethylimidazoline**	turbine	414,895					
11.5-10S	288-13-1	**Pyrazole**	turbine	895					
11.5-11	67771-72-6	**Dimethylpyrazole**			399(a)				
11.5-12	several	**Ethylpyrazole**			399(a)				

Table 11.5. *(Continued)*

Species Number	Registry Number	Name	Emission Source	Emission Ref.	Detection Ref.	Bioassay Code	Bioassay −MA	Bioassay +MA	Bioassay Ref.
11.5-13S	110-89-4	**Piperidine**			627(a)	HMAST		NEG	B13
11.5-14	52642-16-7	**Phenylpiperidine**			214*(a)				
11.5-15S	110-86-1	**Pyridine**	acrylonitrile mfr.	626	597t	SCE	+		B6
			chemical mfr.	1164	528(a),627*(a)	STI00	NEG	NEG	B38
		coffee mfr. 598; coke oven 387; polymer comb. 853,1169;			820(I)	STI35	NEG	NEG	B38
		sewage tmt. 1161; tobacco smoke 396,1177(a)				STI37	NEG	NEG	B38
						STI98	NEG	NEG	B38
11.5-16	109-06-8	**2-Methylpyridine**	coal comb.	1166(a)t					
			tobacco smoke	396(a),1177(a)					
11.5-17	108-99-6	**3-Methylpyridine**	tobacco smoke	396(a),1177(a)	214(a)	STI00	NEG	NEG	B38
						STI35	NEG	NEG	B38
						STI37	NEG	NEG	B38
						STI98	NEG	NEG	B38
11.5-18	108-89-4	**4-Methylpyridine**	tobacco smoke	647(a),1177(a)					
11.5-19	583-61-9	**2,3-Dimethylpyridine**	tobacco smoke	647(a),1177(a)					
11.5-20	108-47-4	**2,4-Dimethylpyridine**	tobacco smoke	647(a),1177(a)					
11.5-21	589-93-5	**2,5-Dimethylpyridine**	tobacco smoke	647(a),1177(a)					
11.5-22	108-48-5	**2,6-Dimethylpyridine**	coal comb.	1166(a)t					
			tobacco smoke	647(a),1177(a)					

Species Number	Registry Number	Name	Emission Source	Emission Ref.	Detection Ref.	Code	Bioassay −MA	Bioassay +MA	Bioassay Ref.
11.5-23	583-58-4	**3,4-Dimethylpyridine**	tobacco smoke	647(a),1177(a)					
11.5-24	591-22-0	**3,5-Dimethylpyridine**	tobacco smoke	647(a),1177(a)					
11.5-25	29611-84-5	**Trimethylpyridine**	coal comb.	1166(a)					
			tobacco smoke	647(a),1177(a)					
11.5-26	53123-74-3	**Tetramethylpyridine**	coal comb.	1166(a)t					
11.5-27	3748-83-2	**Pentamethylpyridine**	coal comb.	1166(a)t					
11.5-28	28631-77-8	**2-Ethylpyridine**	tobacco smoke	647(a),1177(a)					
11.5-29	536-78-7	**3-Ethylpyridine**	tobacco smoke	647(a),1177(a)					
11.5-30	536-75-4	**4-Ethylpyridine**	tobacco smoke	647(a),1177(a)	458(a)				
11.5-31	104-90-5	**2-Ethyl-5-methylpyridine**			202	STD00	NEG	NEG	B36
						STD35	NEG	NEG	B36
						STD37	NEG	NEG	B36
						STD38	NEG	NEG	B36
						STD98	NEG	NEG	B36
						STP00	NEG	NEG	B36
						STP35	NEG	NEG	B36
						STP37	NEG	NEG	B36
						STP38	NEG	NEG	B36
						STP98	NEG	NEG	B36
11.5-32	529-21-5	**3-Ethyl-4-methylpyridine**	tobacco smoke	647(a)					

Table 11.5. *(Continued)*

Species Number	Registry Number	Name	Emission Source	Emission Ref.	Detection Ref.	Bioassay Code	Bioassay −MA	Bioassay +MA	Bioassay Ref.
11.5-33	536-88-9	**4-Ethyl-2-methylpyridine**	tobacco smoke	396(a)					
11.5-34	4673-31-8	**3-Propylpyridine**	tobacco smoke	1177(a)					
11.5-35	3978-81-2	**4-*tert*-Butylpyridine**	tobacco smoke	647(a)					
11.5-36	2294-76-0	**2-Pentylpyridine**	rendering	837,879					
11.5-37	7399-50-0	**2-(3-Pentyl) pyridine**	tobacco smoke	647(a)					
11.5-38	1008-88-4	**3-Phenylpyridine**	coal comb.	828(a)t					
			tobacco smoke	1177(a)					
11.5-39	26274-35-1	**2,4-Diphenylpyridine**	biomass comb.	891(a)t					
11.5-40	1121-55-7	**3-Vinylpyridine**	tobacco smoke	647(a),1177(a)					
11.5-41S	54-11-5	**Nicotine**	tobacco smoke	356,421,1177(a)	1109*(I)	STP00	NEG	NEG	B36
						STP35	NEG	NEG	B36
						STP37	NEG	NEG	B36
						STP98	NEG	NEG	B36
						NEU	NEG		B32
11.5-42S	494-97-3	**Nornicotine**	tobacco smoke	396(a),449(a)					
11.5-43S	487-19-4	**Nicotyrine**	tobacco smoke	396(a),534(a)					
11.5-44S	494-98-4	**Nornicotyrine**	tobacco smoke	396(a),534(a)					

Species Number	Registry Number	Name	Emission Source	Emission Ref.	Detection Ref.	Code	Bioassay −MA	Bioassay +MA	Bioassay Ref.
11.5-45S	494-52-0	**Anabasine**	tobacco smoke	396(a),534(a)					
11.5-46S	581-49-7	**Anatabine**	tobacco smoke	396(a),534(a)					
11.5-47S	581-50-0	**2,3′-Bipyridyl**	tobacco smoke	1177(a)					
11.5-48	26844-80-4	**5-Methyl-2,3′-bipyridyl**	tobacco smoke	1177(a)					
11.5-49	553-26-4	**4,4′-Bipyridyl**	tobacco smoke	1177(a)					
11.5-50S	110-85-0	**Piperazine**			214*(a)	STI00	NEG	NEG	B38
						STI35	NEG	NEG	B38
						STI37	NEG	NEG	B38
						STI98	NEG	NEG	B38
11.5-51	106-55-8	**2,5-Dimethylpiperazine**	diesel	1					
11.5-52S	290-37-9	**Pyrazine**	tobacco smoke	1177(a)					
11.5-53	109-08-0	**Methylpyrazine**	rendering	1077					
			sewage tmt.	687					
			tobacco smoke	647(a),1177(a)					
11.5-54	5910-89-4	**2,3-Dimethylpyrazine**	tobacco smoke	647(a),1177(a)					
11.5-55	123-32-0	**2,5-Dimethylpyrazine**	rendering	1078,1163t					
			tobacco smoke	1177(a)					

Table 11.5. *(Continued)*

Species Number	Registry Number	Name	Emission Source	Emission Ref.	Detection Ref.	Code	Bioassay −MA	Bioassay +MA	Bioassay Ref.
11.5-56	108-50-9	**2,6-Dimethylpyrazine**	tobacco smoke	1177(a)					
11.5-57	14667-55-1	**Trimethylpyrazine**	rendering sewage tmt.	1163t 687					
11.5-58	13925-00-3	**Ethylpyrazine**	sewage tmt. tobacco smoke	687 1177(a)					
11.5-59	71607-73-3	**Ethyldimethylpyrazine**	rendering sewage tmt.	1163t 687					
11.5-60	39723-60-9	**Diethylmethylpryazine**	sewage tmt.	687					
11.5-61S	13717-92-5	**3,6-Dipropyl-1,2,4,5-tetrazine**	turbine	414,895					
11.5-62S	100-97-0	**Formin**	explosives mfr.	60		REC	+		B9
11.5-63	several	**Methylindolizine**	coal comb.	828(a)					
11.5-64S	120-72-9	**Indole**	animal waste coal comb.	166,240 1159	1001*(a) 820(I)				
		food processing 62; paint 695; sewage tmt. 687; starch mfr. 1071; tobacco smoke 534(a),1177(a); vegetation 510e; wood pulping 695,1071							
11.5-65	95-20-5	**2-Methylindole**	animal waste	695,1071	1000(a),1001*(a)				

Species Number	Registry Number	Name	Emission Source	Emission Ref.	Detection Ref.	Code	Bioassay −MA	Bioassay +MA	Bioassay Ref.
11.5-66	83-34-1	**3-Methylindole**	animal waste	25,166					
			drug mfr.	193					
		food processing 62; rendering 325,637; sewage tmt. 1161; tobacco smoke 130*(a),1177(a)							
11.5-67	614-96-0	**5-Methylindole**	animal waste	695,1071					
			starch mfr.	695,1071					
11.5-68	933-67-5	**7-Methylindole**			1001*(a)				
11.5-69	875-79-6	**1,2-Dimethylindole**	animal waste	695,1071					
			wood pulping	695,1071					
11.5-70	91-55-4	**2,3-Dimethylindole**	animal waste	695,1071					
11.5-71	1196-79-8	**2,5-Dimethylindole**			1001*(a)				
11.5-72		**1,7-Methylene-2,3-dimethyl-indole**	polymer comb.	1107(a)t					
11.5-73	several	**Ethylindole**	tobacco smoke	130*(a)					
11.5-74	64844-52-6	**Methylphenylindole**	refuse comb.	878(a)					
11.5-75S	271-63-6	**7-Azaindole**	tobacco smoke	1177(a)					
11.5-76S	119-65-3	**Isoquinoline**	tobacco smoke	1177(a)	363(a),488*(a)	STI98	−(B)	−(B)	B36
						STI00	−(B)	−(B)	B36

Table 11.5. *(Continued)*

Species Number	Registry Number	Name	Emission Source	Emission Ref.	Detection Ref.	Code	Bioassay −MA	Bioassay +MA	Bioassay Ref.
11.5-77	1721-93-3	**1-Methylisoquinoline**	tobacco smoke	1177(a)	488(a),884(a)				
11.5-78	several	**Dimethylisoquinoline**			488(a)				
11.5-79	64828-51-9	**Ethylisoquinoline**			488(a)				
11.5-80	229-67-4	**Benzo[*f*]isoquinoline**			488*(a),884(a)				
11.5-81S	91-22-5	**Quinoline**	coal comb.	1166(a)	363(a),488*(a)	STI00	−(BC)	+	B36
			rendering	1078,1163		STI98	−(BC)	+	B36
			tobacco smoke	396(a),634(a)		STP00	−(CE)	+	B36
						STP35	−(CE)	−(CE)	B36
						STP98	−(CE)	−(CE)	B36
						STP00	NEG	+	B36
						STP37	−(C)	+	B36
						SCE	NEG		B6
						STI00	NEG	+	B38
						STI35	NEG	?	B38
						STI37	NEG	?	B38
						STI98	NEG	+	B38
11.5-82	91-63-4	**2-Methylquinoline**	tobacco smoke	1177(a)		STP00		+	B36
11.5-83	491-35-0	**4-Methylquinoline**	coal comb.	828(a)t,1166(a)t	488(a),897(a)	STP00		+	B36
			polymer comb.	1107(a)t		STI00	−(B)	+(B)	B36
		skunk odor 1162t; tobacco smoke 1177(a)				STI98	−(B)	+(B)	B36

Species Number	Registry Number	Name	Emission Source	Emission Ref.	Detection Ref.	Code	Bioassay −MA	Bioassay +MA	Bioassay Ref.
11.5-84	612-60-2	**7-Methylquinoline**	tobacco smoke	1177(a)		STP00		+	B36
						STI00	−(BC)	+(B)	B36
						STI98	−(BC)	+(B)	B36
11.5-85	1198-37-4	**2,4-Dimethylquinoline**			488(a)t,1001*(a)				
11.5-86	877-43-0	**2,6-Dimethylquinoline**	coal comb.	1166(a)t	1001*(a)				
			polymer comb.	1107(a)					
11.5-87	1463-17-8	**2,8-Dimethylquinoline**			1001*(a)				
11.5-88	51366-52-0	**Trimethylquinoline**	polymer comb.	1107(a)					
11.5-89	1613-34-9	**2-Ethylquinoline**			597				
					831(a)t				
11.5-90	19020-26-9	**4-Ethylquinoline**	polymer comb.	1107(a)					
11.5-91	85-02-9	**Benzo[*f*]quinoline**	coal comb.	98(a),361(a)	213(a),488*(a)	STP00		+	B36
11.5-92	230-27-3	**Benzo[*h*]quinoline**	coal comb.	98(a)	213(a),488*(a)				
			polymer comb.	1107(a)					
11.5-93	37062-82-1	**3-Methylbenzo[*h*]quinoline**	polymer comb.	1107(a)					
11.5-94	several	**Dimethylbenzo[*h*]quinoline**	polymer comb.	1107(a)					
11.5-95	several	**Trimethylbenzo[*h*]quinoline**	polymer comb.	1107(a)					

Table 11.5. *(Continued)*

Species Number	Registry Number	Name	Emission Source	Emission Ref.	Detection Ref.	Code	Bioassay −MA	Bioassay +MA	Bioassay Ref.
11.5-96	243-51-6	**11-*H*-Indeno[1,2-*b*]quinoline**	coal comb.	98(a)	98(a),488*(a)				
11.5-97S	230-17-1	**Benzo[*c*]cinnoline**	refuse comb.	878(a)					
11.5-98	several	**Nitrobenzo[*c*]cinnoline**	diesel	800(a)t					
11.5-99S	86-74-8	**9*H*-Carbazole**	aluminum mfr. tobacco smoke	551*,850(a) 130*(a),534(a)	209(a),214*(a)				
11.5-100	27323-29-1	**9-Methylcarbazole**	aluminum mfr. diesel tobacco smoke	850(a) 800(a) 130*(a)					
11.5-101	30642-38-7	**Dimethylcarbazole**	aluminum mfr.	850(a)					
11.5-102	64844-51-5	**Trimethylcarbazole**	aluminum mfr.	850(a)					
11.5-103	239-01-0	**Benzo[*a*]carbazole**			182(a),363(a)				
11.5-104	205-25-4	**Benzo[*c*]carbazole**	aluminum mfr.	850(a)	363(a)				
11.5-105	203-65-6	**Benzo[*def*]carbazole**	aluminum mfr. biomass comb.	850(a) 891(a)t					
11.5-106	several	**Methylbenzo[*def*]carbazole**	aluminum mfr.	850(a)					
11.5-107		**Dibenzo[*b,def*]carbazole**	aluminum mfr.	850(a)					
11.5-108	194-59-2	**Dibenzo[*c,g*]carbazole**	carbon black mfr.	1148(a)		CCC	SP		B37

Species Number	Registry Number	Name	Emission Source	Emission Ref.	Detection Ref.	Code	Bioassay −MA	Bioassay +MA	Ref.
11.5-109S	244-99-5	**4-Azafluorene**			488*(a),884(a)				
11.5-110	27799-79-7	**5,6-Dihydrophenanthridine**	coal comb.	828(a)					
11.5-111S	229-87-8	**Phenanthridine**	coal comb.	98(a)	488*(a),528(a)	STP00		+	B36
11.5-112S	229-70-9	**3,8-Diazaphenanthrene**	polymer comb.	1107(a)					
11.5-113S	260-94-6	**Acridine**	coal comb. refuse comb.	98(a),361(a) 26(a)	362(a),488*(a)				
11.5-114	54116-90-4	**Methylacridine**	refuse comb.	26(a)	488(a)				
11.5-115	225-11-6	**Benz[*a*]acridine**	coal comb. turbine	98(a) 786*(a)t	98(a),343(a)				
11.5-116	225-51-4	**Benz[*c*]acridine**	auto coal comb.	311(a) 98(a)	98(a)				
11.5-117	226-36-8	**Dibenz[*a,h*]acridine**	coal comb.	98(a)	98(a),363(a)	CCC	SP		B37
11.5-118	224-42-0	**Dibenz[*a,j*]acridine**	coal comb.	98(a)	98(a),363(a)	CCC STP00	SP	 +	B37 B36
11.5-119S	206-56-4	**1-Azafluoranthene**	coal comb.	98(a)	488*(a),884(a)				
11.5-120S	313-80-4	**1-Azapyrene**			882(a)				

Table 11.5. *(Continued)*

Species Number	Registry Number	Name	Emission Source	Emission Ref.	Detection Ref.	Code	Bioassay −MA	Bioassay +MA	Bioassay Ref.
11.5-121	194-03-6	**4-Azapyrene**	coal comb.	98(a)	488*(a),884(a)				
11.5-122S	236-02-2	**Phenanthro[9,10-*d*]imidazole**	coal comb.	828(a)					
11.5-123	54751-98-3	**1-Ethyl-6-methyl-3-piperidinol**	biomass comb. biomass comb.	1089(a) 1089(a)					
11.5-124S	13190-97-1	**Solanesol**	tobacco smoke	396(a)					
11.5-125S	4030-18-6	**1-Acetylpyrrolidine**	diesel	962t					
11.5-126	616-45-5	**2-Pyrrolidone**	polymer comb. tobacco smoke	1169 1177(a)					
11.5-127	51013-18-4	**Methylpyrrolidone**	solvent	134					
11.5-128	872-50-4	**N-Methylpyrrolidone**	tobacco smoke	1177(a)					
11.5-129	350-03-8	**3-Acetylpyridine**	tobacco smoke	396(a),1177(a)					
11.5-130	1570-48-5	**3-Propionylpyridine**	tobacco smoke	396(a)					
11.5-131	1701-70-8	**3-Pyridyl propyl ketone**	tobacco smoke	396(a)					
11.5-132S	486-56-6	**Cotinine**	tobacco smoke	1177(a)					
11.5-133	491-26-9	**Nicotine-1′-oxide**	tobacco smoke	1177(a)					
11.5-134S	66393-65-5	**1,3-Dioxoisoindole**	biomass comb.	891(a)t					

Species Number	Registry Number	Name	Emission Source	Emission Ref.	Detection Ref.	Bioassay Code	Bioassay −MA	Bioassay +MA	Bioassay Ref.
11.5-135S	58-08-2	**Caffeine**	refuse comb.	878(a)	546*(a),548(a)	TRM	NEG		B21
						V790	NEG	NEG	B7
						V79A	NEG		B7
						WPU	NEG		B1
						STP00	NEG	NEG	B36
						STP35	NEG	NEG	B36
						STP37	NEG	NEG	B36
						STI98	NEG	NEG	B36
						SCE	+		B6
						CCC	I,N		B37
						SLGM	NEG		B4
						SLFM	?		B4
						MST	?		B5
						CYG	NEG		B8
						CYC	+		B8
						HTM	NEG		B2
						REC	+/?		B9
						VIC	+/−		B18
						ALC	+		B19
						MNT	?/−		B29
						ASPD	NEG		B10
						CTV	+		B24
						CTP	NEG		B24
						SRL	+		B30
						NEU	+		B32
						SPI	NEG		B25
11.5-136	500-22-1	**Pyridine-3-carbaldehyde**	tobacco smoke	396(a)					

Table 11.5. *(Continued)*

Species Number	Registry Number	Name	Emission Source	Emission Ref.	Detection Ref.	Code	Bioassay −MA	Bioassay +MA	Bioassay Ref.
11.5-137S		**5-Oxopyrrolidine-2-carboxylic acid**	volcano volcano	1007t 1007t					
11.5-138	82231-51-4	**3-Methylpyrazole-4-carboxylic acid**	diesel diesel	962t 962t					
11.5-139	59-67-6	**3-Pyridinecarboxylic acid**	tobacco smoke	396(a)					
11.5-140S	94-62-2	**Piperine**	vegetation	531					
11.5-141	100-70-9	**2-Pyridinecarbonitrile**	tobacco smoke	1177(a)	358				
11.5-142	504-29-0	**2-Aminopyridine**	tobacco smoke	1177(a)					
11.5-143	462-08-8	**3-Aminopyridine**	tobacco smoke	1177(a)					
11.5-144S	23255-20-1	**Nicotinamide**	tobacco smoke	396(a),634(a)					
11.5-145	114-33-0	**N-Methylnicotinamide**	tobacco smoke	1177(a)					

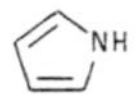

11.5-1 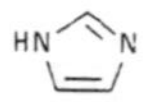 11.5-4 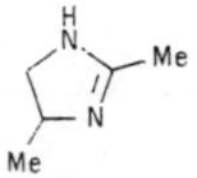 11.5-9 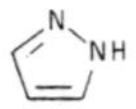 11.5-10 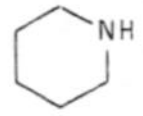 11.5-13  11.5-15

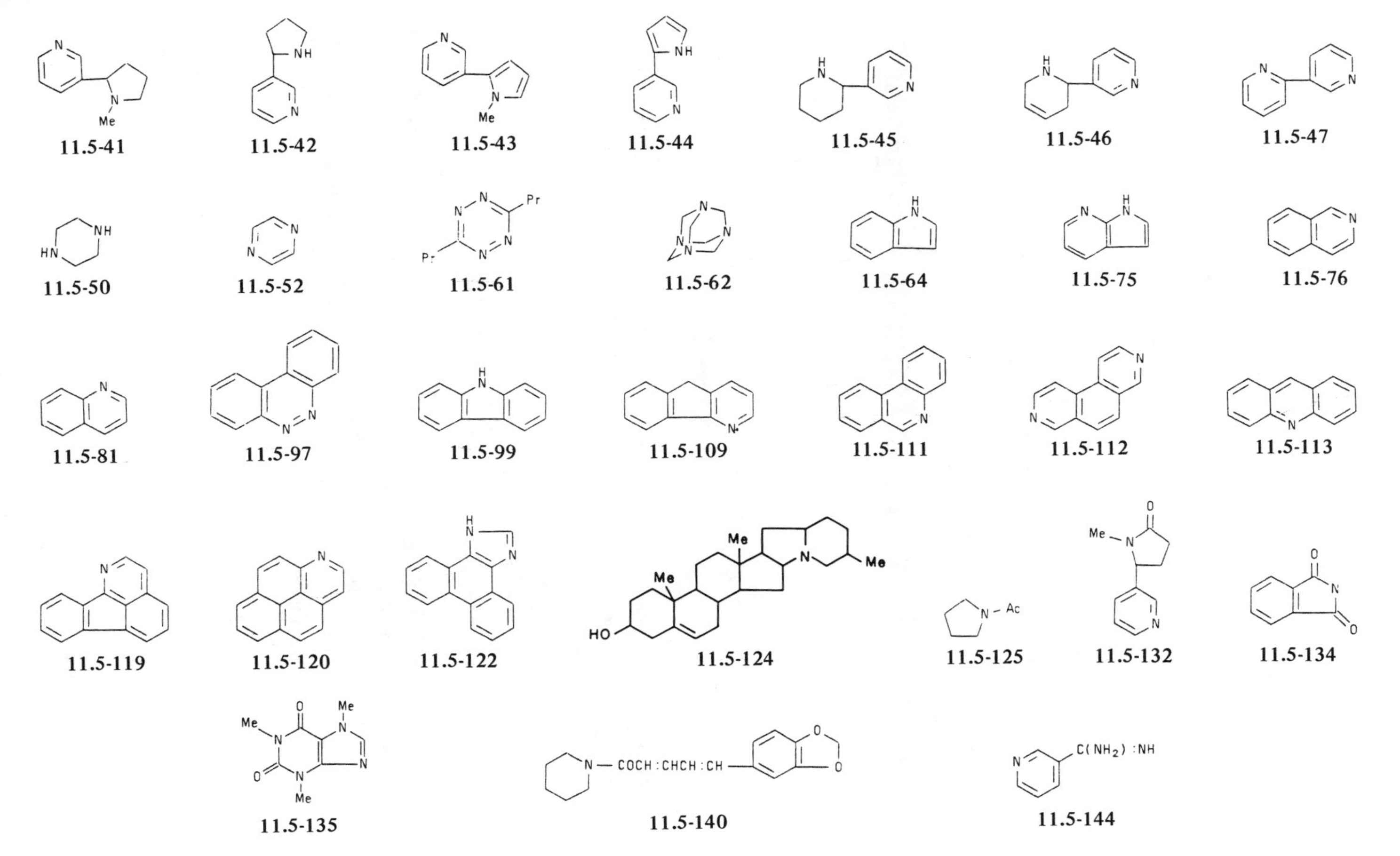
11.5-41
11.5-42
11.5-43
11.5-44
11.5-45
11.5-46
11.5-47
11.5-50
11.5-52
11.5-61
11.5-62
11.5-64
11.5-75
11.5-76
11.5-81
11.5-97
11.5-99
11.5-109
11.5-111
11.5-112
11.5-113
11.5-119
11.5-120
11.5-122
11.5-124
11.5-125
11.5-132
11.5-134
11.5-135
11.5-140
11.5-144

CHAPTER 12

Sulfur – Containing Organic Compounds

12.0 Introduction

Sulfur and oxygen both occur in Group VIA of the periodic table, and some of the common organic sulfur compounds are analogous to organic oxygen compounds:

RSH	a *thiol*[*]	(thio alcohol)
RSR′	a *sulfide*	(thio ether)
RSSR′	a *disulfide*	("persulfide").

A number of heterocyclic compounds in which sulfur atoms replace carbon atoms are also found in the atmosphere, though many fewer than those with nitrogen. Finally, a few compounds in which the extra pair of electrons on the sulfur atom permit the formation of sulfates and other sulfur-containing ligands are of interest.

* Thiols are alternatively termed *mercaptans*

We divide these compounds into four groups: (1) thiols, (2) sulfides and sulfates, (3) heterocyclic sulfur compounds, and (4) miscellaneous sulfur compounds.

12.1 Thiols

More than twenty thiols are emitted into the atmosphere. Methanethiol, the simplest, has been detected under ambient conditions. The thiol sources are mostly natural, involving either biological decay processes initiated by microorganisms or the venting of gases from sulfur-rich petroleum. However, two common thiol sources are anthropogenic. One is the emission of thiols (and other sulfur gases) from wood pulping and paper-making processes. The other is the addition of thiols to natural gas. The presence of the thiols provides early warning of gas leaks, because the olfactory threshold of thiols is very low. In the ambient atmosphere, in contrast to the situation near sources, thiol concentrations are well below 1 ppb and are generally undetectable.

The principal atmospheric fate of the thiols is reaction with OH·. For methanethiol the initial step is addition to the sulfur atom, followed promptly by unimolecular decomposition:[1253]

$$CH_3SH + OH\cdot \rightarrow [CH_3SOH_2\cdot]^* \rightarrow CH_3S\cdot + H_2O\ . \qquad (R12.1\text{-}1)$$

Alternate reaction paths are possible for the $CH_3S\cdot$ radical. One possibility is the addition of an oxygen molecule followed by unimolecular cleavage

$$CH_3S\cdot + O_2 \rightarrow [CH_3SO_2\cdot] \rightarrow CH_3\cdot + SO_2\ ; \qquad (R12.1\text{-}2)$$

another is reaction with OH· to produce methyl sulfonic acid

$$[CH_3SO_2\cdot] + OH\cdot \xrightarrow{M} CH_3SO_3H \qquad (R12.1\text{-}3)$$

In environments with very high concentrations of oxides of nitrogen, it may be possible for the following reaction to compete[1254] with R12.1-2:

$$CH_3S\cdot \rightarrow NO_2 \rightarrow CH_3SNO_2 \qquad (R12.1\text{-}4)$$

Sulfur dioxide and methyl sulfonic acid, the products of R12.1-2 and R12.1-3, have both been identified in the atmosphere, but the product of R12.1-4 has not.

12.2 Sulfides, Sulfates

Nearly forty of these compounds appear in Table 12.2. They are produced rather profusely by microbial processes in the soil, in oceans, and in decaying biomass. Wood pulping and other industrial processes are also sources of sulfides and disulfides. A few of the smaller molecules are detected fairly routinely in the atmosphere, though at concentrations generally less than 1 ppb.

The atmospheric chemistry of a few of the simple compounds in this group has been worked out in some detail. For dimethyl sulfide, the most abundant, the initial step is the addition of OH·:

$$CH_3SCH_3 + OH\cdot \rightarrow [CH_3\dot{S}(OH)CH_3]^{\bullet} \qquad (R12.2\text{-}1)$$

The excited complex can undergo two different unimolecular decomposition processes,[1253,1254] with the first being favored:

$$[CH_3\dot{S}(OH)CH_3]^{\bullet} \rightarrow CH_3SCH_2\cdot + H_2O \qquad (R12.2\text{-}2)$$

$$[CH_3\dot{S}(OH)CH_3]^{\bullet} \rightarrow CH_3\cdot + CH_3SOH \qquad (R12.2\text{-}3)$$

In the first case, the probable chain carrier reactions (at least at high NO_x) are[1254]

$$CH_3SCH_2\cdot + O_2 \rightarrow CH_3SCH_2O_2\cdot \qquad (R12.2\text{-}4)$$

$$CH_3SCH_2O_2\cdot + NO \rightarrow CH_3SCH_2O\cdot + NO_2 \qquad (R12.2\text{-}5)$$

$$CH_3SCH_2O\cdot \rightarrow CH_3S\cdot + HCHO \qquad (R12.2\text{-}6)$$

The $CH_3S\cdot$ product then reacts as in R12.1-2. In the case of the product of R12.2-3, methylsulfonic acid is produced by[1253]

$$CH_3SOH + O_2 \rightarrow CH_3SO_3H \qquad (R12.2\text{-}7)$$

The reaction sequences for the higher sulfides (and di- and tri-sulfides) are expected to be similar.

Organic sulfates have recently been found to be emitted in aerosol particle form by fossil fuel combustion and are present in ambient particles as well. They appear unlikely to participate in any significant chemical processes.

12.3 Heterocyclic Sulfur Compounds

Thirty five heterocyclic sulfur compounds make up Table 12.3. The smaller ones arise from microbial processes, the larger from combustion of sulfur-containing fossil fuels. Although several have been detected in the ambient atmosphere as gases or as aerosol constituents, none is abundant.

Little is known of the atmospheric chemistry of these compounds. Thiophene and tetrahydrothiophene are known to react rapidly with OH·, though whether addition occurs at the sulfur atom or across a $-C=C-$ bond is not certain.[1240] The reactions of structurally similar arenes probably provide good models for the chemistry of sulfur heterocycles.

12.4 Miscellaneous Sulfur Compounds

This table lists five compounds known to be emitted into the atmosphere in small quantities. None has been detected in ambient air and the chemical reactions of the compounds in the atmosphere have not been investigated.

Table 12.1. Thiols

Species Number	Registry Number	Name	Emission Source	Emission Ref.	Detection Ref.	Bioassay Code	Bioassay −MA	Bioassay +MA	Bioassay Ref.
12.1-1	74-93-1	**Methanethiol**	animal waste	25,157	165,566*				
			fish processing 695,1071						
		food decay 1074; leather mfr. 1151*,1152; microbes 210,302; natural gas 262; paint 695,1071; petroleum mfr. 252*,1164; rendering 325,833; sewage tmt. 174*,204; starch mfr. 523*,695; vegetation 584,635; wood pulping 19,58*							
12.1-2	75-08-1	**Ethanethiol**	animal waste	25,695					
			auto	771*					
		microbes 302; natural gas 262,357; petroleum mfr. 252*,1164; sewage tmt. 422*,1161; starch mfr. 695,1071; wood pulping 238							
12.1-3	107-03-9	**Propane-1-thiol**	animal waste	25	364(a)				
			fish processing	385*					
		natural gas 262,357; onion odor 326,636; petroleum mfr. 1164; sewage tmt. 174*,422*; skunk odor 1162; wood pulping 238,623							
12.1-4	513-44-0	**2-Methylpropane-1-thiol**	natural gas	262,357					
			skunk odor	1162					
12.1-5	1679-08-9	**2,2-Dimethylpropane-1-thiol**	sewage tmt.	422*					
12.1-6	75-33-2	**Propane-2-thiol**	natural gas	262,357					
			wood pulping	623					

Species Number	Registry Number	Name	Emission Source	Ref.	Detection Ref.	Code	Bioassay −MA	+MA	Ref.
12.1-7	75-66-1	**2-Methylpropane-2-thiol**	natural gas	262,357					
			sewage tmt.	422*					
12.1-8	870-23-5	**Propene-3-thiol**	petroleum mfr.	1164					
			sewage tmt.	1161					
12.1-9	109-79-5	**Butanethiol**	industrial	618					
			natural gas	262					
			skunk odor	326,1162					
12.1-10	541-31-1	**3-Methylbutane-1-thiol**	skunk odor	1162					
12.1-11	5954-72-3	**2-Butene-1-thiol**	sewage tmt.	1161					
			skunk odor	1162					
12.1-12	110-66-7	**Pentanethiol**	sewage tmt.	174*,422*					
12.1-13	1633-89-2	**2-Methylpentane-1-thiol**	turbine	414					
12.1-14	1633-97-2	**2-Methylpentane-2-thiol**	thiol mfr.	143					
12.1-15	111-31-9	**Hexanethiol**	skunk odor	1162					
12.1-16	111-88-6	**Octanethiol**	turbine	414					
12.1-17	108-98-5	**Benzenethiol**	sewage tmt.	1161					
12.1-18	26445-03-4	**Methylbenzenethiol**	sewage tmt.	1161					

Table 12.1. *(Continued)*

Species Number	Registry Number	Name	Emission Source	Ref.	Detection Ref.	Code	Bioassay −MA	+MA	Ref.
12.1-19S	100-53-8	**Phenylmethanethiol**	petroleum mfr.	1164					
			sewage tmt.	1161					
			skunk odor	1162					
12.1-20	6263-65-6	**1-Phenylethane-1-thiol**	skunk odor	1162					
12.1-21	13129-35-6	**Furanthiol**	coffee mfr.	598					

CH_2SH

12.1-19

Table 12.2. Sulfides, Sulfates

Species Number	Registry Number	Name	Emission Source	Emission Ref.	Detection Ref.	Code	Bioassay −MA	Bioassay +MA	Bioassay Ref.
12.2-1S	75-18-3	**Dimethyl sulfide**	algae	464	140,491*				
			animal waste	25,436*	399(a)				
		fish processing 1071; food decay 1074; leather mfr. 1151*,1152; marshland 1049*,1056*; meat cooking 598; microbes 302,670*; natural gas 262,357; ocean 199; paint 695,1071; petroleum mfr. 252*,1164; rendering 325,833; sewage tmt. 174*,695; SO_2 scrubbing 1104; soil 1049*; starch mfr. 523*,1071; vegetation 140,635; wood pulping 19,111*							
12.2-2	624-89-5	**Ethyl methyl suflide**	wood pulping	238					
12.2-3	352-93-2	**Diethyl sulfide**	animal waste	327,1075	199,597				
			auto	807*					
		natural gas 262; petroleum mfr. 1164; sewage tmt. 1161; wood pulping 238,623							
12.2-4	111-47-7	**Dipropyl sulfide**	natural gas	262					
			wood pulping	623					
12.2-5	592-88-1	**Diallyl sulfide**	industrial	618					
			onion odor	598					
		rendering 1163t; sewage tmt. 422*; vegetation 584							
12.2-6	625-80-9	**Diisopropyl sulfide**	natural gas	262					

Table 12.2. *(Continued)*

Species Number	Registry Number	Name	Emission Source	Emission Ref.	Detection Ref.	Bioassay Code	Bioassay −MA	Bioassay +MA	Bioassay Ref.
12.2-7	35976-82-0	**2-Butenyl ethyl sulfide**	skunk odor	1162					
12.2-8	83688-95-3	**2-Butenyl propyl sulfide**	skunk odor	1162					
12.2-9	5622-73-1	**Di-2-butenyl sulfide**	skunk odor	1162					
12.2-10	83688-96-4	**2-Butenyl isopentyl sulfide**	skunk odor	1162					
12.2-11	2432-38-4	**Isopentyl 1-oxoethyl sulfide**	skunk odor	1162t					
12.2-12	139-66-2	**Diphenyl sulfide**	sewage tmt.	1161					
12.2-13S	624-92-0	**Dimethyl disulfide**	animal waste fish processing	436*,437 695	199,263				
		food decay 1074; microbes 210,302; natural gas 262; oceans 199; onion odor 636; rendering 833,837; sewage tmt. 687,1071; SO_2 scrubbing 1104; soil 1049*; starch mfr. 695; vegetation 584; whiskey mfr. 1158; wood pulping 19,58*							
12.2-14	20333-39-5	**Ethyl methyl disulfide**	natural gas	262					
12.2-15	110-81-6	**Diethyl disulfide**	natural gas	262	399(a)				
12.2-16	2179-60-4	**Methyl propyl disulfide**	onion odor rendering	636 833					
12.2-17	629-19-6	**Dipropyl disulfide**	onion odor	636					

Species Number	Registry Number	Name	Emission Source	Emission Ref.	Detection Ref.	Bioassay Code	Bioassay −MA	Bioassay +MA	Bioassay Ref.
12.2-18S	24645-67-8	**Cystine**	volcano	1007					
12.2-19	2179-58-0	**Allyl methyl disulfide**	vegetation	584					
12.2-20	27817-67-0	**Allyl propyl disulfide**	garlic odor	598					
12.2-21	2179-57-9	**Diallyl disulfide**	petroleum mfr.	1164					
			vegetation	584					
12.2-22	83688-94-2	**2-Butenyl methyl disulfide**	skunk odor	1162					
12.2-23	83688-98-6	**2-Butenyl butyl disulfide**	skunk odor	1162					
12.2-24	36889-28-8	**Di-2-butenyl disulfide**	skunk odor	1162					
12.2-25	75679-09-3	**Butyl isopentyl disulfide**	skunk odor	1162					
12.2-26	83688-97-5	**2-Butenyl isopentyl disulfide**	skunk odor	1162					
12.2-27	2051-04-9	**Diisopentyl disulfide**	skunk odor	1162					
12.2-28	3658-80-8	**Dimethyl trisulfide**	animal waste	1071					
			natural gas	262					
		onion odor 636; rendering 833,837; sewage tmt. 689; vegetation 584; wood pulping 1071							
12.2-29	3600-24-6	**Diethyl trisulfide**	natural gas	262					

Table 12.2. *(Continued)*

Species Number	Registry Number	Name	Emission Source	Emission Ref.	Detection Ref.	Bioassay Code	Bioassay −MA	Bioassay +MA	Bioassay Ref.
12.2-30	17619-36-2	**Methyl propyl trisulfide**	onion odor	636					
			vegetation	584					
12.2-31	6028-61-1	**Dipropyl trisulfide**	onion odor	636					
12.2-32	34135-85-8	**Allyl methyl trisulfide**	vegetation	584					
12.2-33	5756-24-1	**Dimethyl tetrasulfide**	sewage tmt.	687					
12.2-34S	75-75-2	**Methylsulfonic acid**			793*(a),862(a)	SRL	?		B30
					794*(r),873*(r)				
12.2-35S	67-71-0	**Dimethylsulfone**	animal waste	855					
12.2-36	87954-49-2	**bis-(Hydroxymethyl)sulfone**			1083(a)				
12.2-37S	77-78-1	**Dimethylsulfate**	coal comb.	694*(a),1063*(a)	1063*(a)	CCC	SP		B37
			oil comb.	694*(a),1153(a)		STS35	−(C)		B36
						STS37	−(C)		B36
						STS38	−(C)		B36
						SCE	+		B6
						CHOM	+		B3
						MDR	+		B12
						REC	+		B9
						ARA	+		B17
						YEA	+		B27
						ASPH	+		B11
						ASPD	+		B10

Species Number	Registry Number	Name	Emission Source	Emission Ref.	Detection Ref.	Code	Bioassay −MA	Bioassay +MA	Bioassay Ref.
12.2-38S	21228-90-0	**Methylsulfate ion**	coal comb.	694*(a),889(a)	1063*(a)				
			oil comb.	694*(a),1153(a)					

SMe_2 12.2-1

$MeSSMe$ 12.2-13

$HO_2CCH(NH_2)CH_2SSCH_2CH(NH_2)CO_2H$ 12.2-18

HO_3SMe 12.2-34

$MeSO_2Me$ 12.2-35

$MeOSO_2OMe$ 12.2-37

$MeOSO_3$ 12.2-38

Table 12.3. Heterocyclic Sulfur Compounds

Species Number	Registry Number	Name	Emission Source	Emission Ref.	Detection Ref.	Bioassay Code	Bioassay −MA	Bioassay +MA	Bioassay Ref.
12.3-1	110-01-0	**2,3,4,5-Tetrahydrothiophene**	natural gas	262,357					
12.3-2	several	**Methyltetrahydrothiophene**	refuse comb.	840(a)					
12.3-3S	110-02-1	**Thiophene**	industrial	618	833				
			natural gas	454*					
		petroleum mfr. 252*; rendering 992*; wood pulping 19,267							
12.3-4	25154-40-9	**Methylthiophene**	sewage tmt.	687	833				
12.3-5	28632-15-7	**Dimethylthiophene**	sewage tmt.	687	833				
12.3-6	82530-87-8	**Trimethylthiophene**	diesel	176					
12.3-7	52006-63-0	**Ethylthiophene**	auto	682t	682t				
12.3-8	1551-27-5	**2-Propylthiophene**	rendering	837,879					
12.3-9	several	**Methylpropylthiophene**	sewage tmt.	687					
12.3-10	several	**Ethylpropylthiophene**	sewage tmt.	687					
12.3-11	1455-20-5	**2-Butylthiophene**	rendering	837,879					
12.3-12	4861-58-9	**2-Pentylthiophene**	rendering	837,879					

Species Number	Registry Number	Name	Emission Source	Emission Ref.	Detection Ref.	Code	Bioassay −MA	Bioassay +MA	Bioassay Ref.
12.3-13		**2-Ethyl-5-isopentylthiophene**	coal comb.	825(a)					
12.3-14	18794-77-9	**2-Hexylthiophene**	rendering	837,879					
12.3-15	several	**Phenylthiophene**	coal comb.	828(a)t					
12.3-16	42140-26-1	**Diphenylthiophene**			1086(a)				
12.3-17	31468-30-9	**2-Formyl-5-propynylthiophene**	microbes	302					
12.3-18S	88-15-3	**(1-Oxoethyl) thiophene**	sewage tmt.	687					
12.3-19	62656-49-9	**(1-Oxopropyl) thiophene**	sewage tmt.	687					
12.3-20	several	**Dihydrobenzo[*b*]thiophene**	coal comb.	828(a)t					
12.3-21S	95-15-8	**Benzo[*b*]thiophene**	coal comb.	828t,828(a)t	566*				
12.3-22	31393-23-4	**Methylbenzo[*b*]thiophene**	coal comb.	828(a)t	1113(I)				
12.3-23	30027-44-2	**Dimethylbenzo[*b*]thiophene**	diesel	137					
12.3-24	132-65-0	**Dibenzothiophene**	carbon black mfr.	1148(a)	209(a),408(a)				
			coal comb.	828t,828(a)t,1065(a)					
		diesel 804(a)t,829(a); refuse comb. 878(a); turbine 786*(a)							
12.3-25	30995-64-3	**Methyldibenzothiophene**			408(a)				

Table 12.3. *(Continued)*

Species Number	Registry Number	Name	Emission Source	Emission Ref.	Detection Ref.	Bioassay Code	Bioassay −MA	Bioassay +MA	Bioassay Ref.
12.3-26	79313-22-7	**Ethyldibenzothiophene**			408(a)				
12.3-27S	205-43-6	**Benzo[*b*]naphtho[1,2-*d*] thiophene**	coal comb.	1065(a)					
12.3-28	239-35-0	**Benzo[*b*]naphtho[2,1-*d*] thiophene**	coal comb.	828(a)t,1065(a)	408(a),466(a)				
			diesel	804(a),829(a)					
12.3-29	several	**Methylbenzo[*b*]naphtho [2,1-*d*]thiophene**			408(a),944(a)				
12.3-30	243-46-9	**Benzo[*b*]naphtho[2,3-*d*] thiophene**	coal comb.	1065(a)					
12.3-31S	52006-64-1	**Methylthiazole**	chemical mfr.	1046					
12.3-32	95-16-9	**Benzothiazole**	chemical mfr.	1046*	86,597				
			sewage tmt.	687	488(a)				
			vulcanization	308	820(I)				
12.3-33	several	**Methylbenzothiazole**	chemical mfr.	1046					
12.3-34S	492-22-8	**9*H*-Thioxanthen-9-one**	auto	858*(a)					
			diesel	690(a),693(a)					
12.3-35	65587-68-0	**Methyl-9*H*-thioxanthen-9-one**	auto	858*(a)					
			diesel	802(a),858*(a)					

S

12.3-3

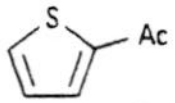

S
Ac

12.3-18

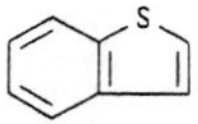

S

12.3-21

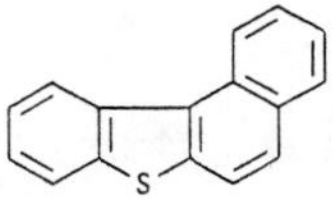

S

12.3-27

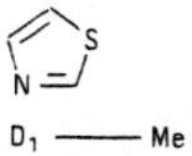

S
N
D_1 —— Me

12.3-31

O
S

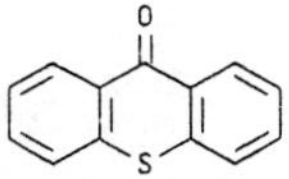

12.3-34

Table 12.4. Miscellaneous Sulfur Compounds

Species Number	Registry Number	Name	Emission Source	Emission Ref.	Detection Ref.	Code	Bioassay −MA	Bioassay +MA	Bioassay Ref.
12.4-1S	68-11-1	**Mercaptoacetic acid**	food processing	62					
12.4-2S	463-56-9	**Hydrogen thiocyanate**	tobacco smoke volcano	396(a) 913t					
12.4-3S	505-14-6	**Thiocyanogen**	tobacco smoke	396(a)					
12.4-4S	57-06-7	**Allylisothiocyanate**	food processing vegetation	62 531		SRL	+		B30
12.4-5S	759-94-4	**EPTC**	herbicide	1034,1118		NEU ZMS	NEG NEG		B32 B22

$HSCH_2CO_2H$
12.4-1

NCSH
12.4-2

NCSSCN
12.4-3

$SCNCH_2CH:CH_2$
12.4-4

EtSC(O)N(Pr)Pr
12.4-5

CHAPTER 13

Halogen – Containing Organic Compounds

13.0 Introduction

Halogenated compounds are common in the atmosphere, both because of their natural presence in marine air, volcanic emissions, and fossil fuels and because of their wide industrial use. Because halogen atoms are monovalent, they substitute for hydrogen in organic compounds. We divide them into four groups: (1) alkanic halogenated compounds, (2) olefinic halogenated compounds, (3) cyclic and aromatic halogenated compounds, and (4) halogen-containing pesticides.

13.1 Alkanic Halogenated Organics

Nearly seventy compounds comprise Table 13.1. A few are naturally produced, mostly by oceanic microbial action, but the vast majority are anthropogenic. The compounds occur almost entirely in the gas phase, both in the troposphere and stratosphere. Many have been detected in indoor air and a few are found in precipitation.

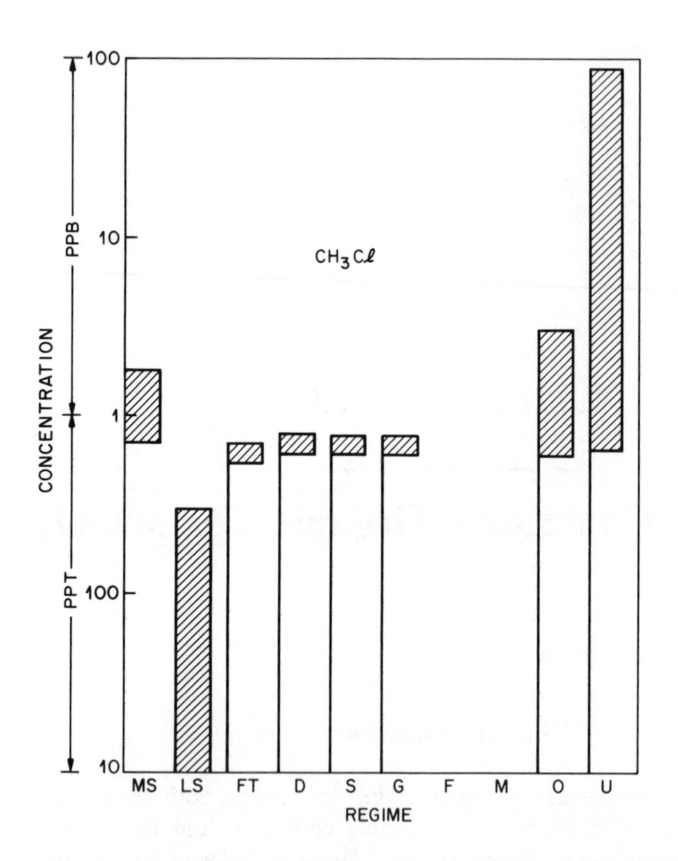

Fig. 13.1-1 Approximate concentration ranges of CH_3Cl in different atmospheric regimes. The regime code is given in the caption of Figure 2.1-1. (Adapted from Graedel.[1257])

The most abundant of the organic halogenates in the troposphere is methyl chloride. Its concentration plot (Figure 13.1-1) indicates that industrial and marine sources are dominant. Away from sources, the concentrations diminish as methyl chloride is consumed by atmospheric reactions.

Extensive concentration data are also available for CF_2Cl_2 and $CFCl_3$, two anthropogenic compounds used extensively as refrigerants and propellants. Their concentration patterns are shown in Figure 13.1-2. The figure shows the highest concentrations in urban areas (where they are emitted), relatively uniform concentrations throughout the rest of the troposphere (where they are well-mixed and not lost in chemical reactions or surface deposition), and lower concentrations in the stratosphere (where they are destroyed by high-energy solar photons).

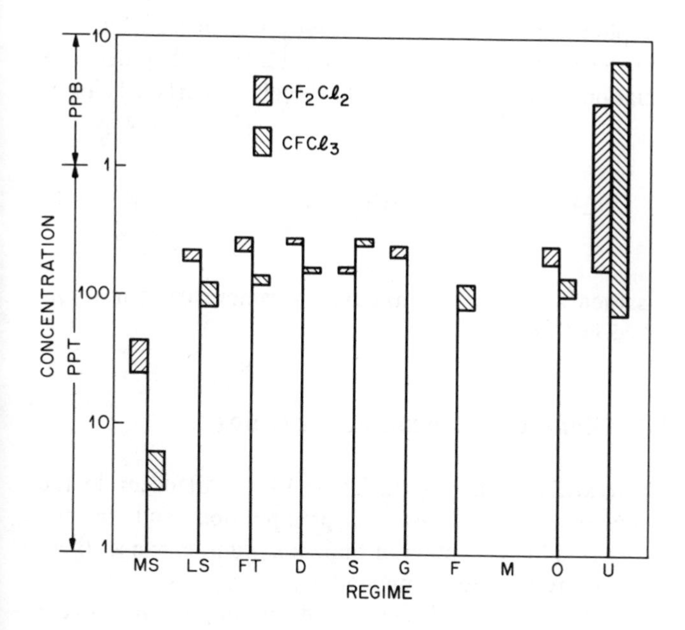

Fig. 13.1-2 Approximate concentration ranges of CF_2Cl_2 and $CFCl_3$ in different atmospheric regimes. The regime code is given in the caption of Figure 2.1-1. The data are from Graedel.[1257]

The atmospheric chemistry of the haloalkanes is determined by the presence or absence of hydrogen atoms in the molecules. If a molecule contains one or more hydrogens, it will react with OH· by H atom abstraction. The reaction rate constants are modest.[1221] The reactions subsequent to OH· radical attack have not been specified, but in the case of the halomethanes it appears likely that the radicals formed will initially react with O_2 to form peroxyl radicals. The ultimate products of the chemical chains are likely to be ClO· or analogous radicals or Cl or analogous atoms. Similar chemical sequences are proposed for chlorinated ethanes, although the greater complexity of the molecules produces a greater diversity of products.[1282]

Fully halogenated molecules do not react with OH·, since abstraction of halogenated atoms is an endothermic* process.[1221] As a result, they are

* An *endothermic* process is one which will occur only if heat is supplied. An *exothermic* process is one in which heat is liberated.

retained in the atmosphere until they are eventually transported into the stratosphere by atmospheric motions. Above 20 km altitude, they begin to encounter solar radiation energetic enough to break a carbon-halogen bond, e.g.,

$$CFCl_3 \xrightarrow[\lambda<230\ \text{nm}]{h\nu} CFCl_2\cdot + Cl\cdot \qquad (R13.1\text{-}1)$$

The halogen atom is then available to catalytically remove oxygen atoms and ozone, as described in Chapter 2.

13.2 Olefinic Halogenated Organics

More than twenty haloalkenes appear in Table 13.2. All occur in the gas phase, many indoors, and a few in precipitation and in the stratosphere. With the possible exception of emission from volcanos, there are no natural sources of any of these compounds.

Haloalkenes react rapidly with OH·, the radical adding across the double bond. It is thought that the subsequent reactions will involve the elimination of halogen atoms from the adducts,[1221] but detailed reaction mechanisms have not yet been determined.

13.3 Halogenated Cyclic and Aromatic Compounds

Nearly a hundred of these compounds are emitted into or detected in the atmosphere, the smaller molecules in the gas phase and the larger ones in aerosol particles. Many are present indoors and a few have been found in precipitation. They are entirely anthropogenic in origin.

These compounds react as do their hydrocarbon analogs. In the case of halogenated aromatics, for example, OH· adds to the aromatic ring.[1283] Subsequent reactions will eventually produce oxygenated ring compounds, perhaps by elimination of the halogen atom present.

13.4 Halogenated Pesticides

More than three dozen compounds are included in this chemical group, which will doubtless become larger as more sensitive and specific detection methods are applied to atmospheric samples. There are no natural sources for these compounds, which have much greater molecular weights (typically >200 amu) than do most other species present in the atmosphere. The structures are similar: they consist of one or more aromatic rings with substituted chlorine atoms and sometimes with added epoxide or ester groups.

Because these compounds are all anthropogenically produced and because of their toxic influence on food chains, their study has become a subspecialty in the atmospheric sciences. Volatilization, both from soil[567] and water,[1305] is a general property of these pesticides, as is their transport on small aerosol particles; the compounds are present in ambient air as far as 2000 km from their point of application.[91] As a result, although the pesticides are generally applied in aerosol form, they have been frequently found in both gas and aerosol phases.

The chemistry of the chlorinated pesticides in the air environment is complex and incompletely understood. They absorb photons at atmospheric wavelengths, and subsequent isomerization[1306] and oxidation[568] processes have been identified. The degree of participation of photochemical products in interactive atmospheric chemistry is less certain. Evidence has been presented for a first-order photochemical decomposition of trifluralin with a 70-sec half life[1307] and for photocleavage of carbon-chlorine bonds in *o*-chlorobiphenyl.[569] In this latter case, hydrogen chloride and phosgene were among the identified products. It thus appears that the atmospheric chemistry of the halogenated pesticides may interact to some degree in the general cycles of atmospheric species.

Table 13.1. Alkanic Halogenated Organics

Species Number	Registry Number	Name	Emission Source	Emission Ref.	Detection Ref.	Bioassay Code	Bioassay −MA	Bioassay +MA	Bioassay Ref.
13.1-1	75-73-0	**Carbon tetrafluoride**	aluminum mfr.	256*,1172	72*,639*				
			electronics mfr.	616	714*(S),978*(S)				
13.1-2	74-87-3	**Methyl chloride**	biomass comb.	495*,655*	85*,321*	STD35	+(B)	+(B)	B36
			chemical mfr.	559,560	713*(S),722*(S)	TRC	+		B20
		ocean 1173*; polymer comb. 211,304; propellant 562; refuse comb. 26(a); solvent 439; tobacco smoke 298,396; turbine 414; vegetation 547; volcano 234,783*; wood comb. 1171			820(I)				
13.1-3S	75-09-2	**Methylene chloride**	biomass comb.	969*	358,392	STD98	+(B)	+(B)	B36
			chemical mfr.	462*,559	820(I),1108(I)	STD00	+(B)	+(B)	B36
		foaming agent 562; landfill 365; solvent 50,60				SRL	NEG		B30
13.1-4S	67-66-3	**Chloroform**	auto	993*	85*,135*	V79A	NEG		B7
			biomass comb.	597	213(a)	CCC	SP		B37
		chemical mfr. 559; landfill 365,628; refuse comb. 26(a); sewage tmt. 204; solvent 60,559; vegetation 547; water treatment 457,487			983*(r)	MNT	?/−		B29
					983*(s)	SPI	?		B25
					713*(S)				
					820(I),952(I)				

Species Number	Registry Number	Name	Emission Source	Emission Ref.	Detection Ref.	Code	Bioassay −MA	Bioassay +MA	Bioassay Ref.
13.1-5	56-23-5	**Carbon tetrachloride**	biomass comb.	597	89*,224*	CCC	SP		B37
			chemical mfr.	559	213(a)	STP00	NEG	NEG	B36
		fumigating agent 560; petroleum mfr. 58*; refuse comb. 26(a);			983*(r)	STP35	NEG	NEG	B36
		solvent 60,439; vegetation 547; volcano 1149			983*(s)	STP37	NEG	NEG	B36
					713*(S),729*(S)	STP98	NEG	NEG	B36
					952(I),1108(I)	SPI	NEG		B25
13.1-6	74-83-9	**Methyl bromide**	fumigating agent	560,595	373,603				
			ocean	1173*	714*(S),1009*(S)				
		turbine 414; volcano 1091							
13.1-7	74-95-3	**Methylene bromide**	landfill	628	597,628				
					1008*(S)				
13.1-8	75-25-2	**Bromoform**			358,462	STI00	+/−	+/−	B38
					755*(r),876*(r)	STI35	NEG	NEG	B38
					1009*(S)	STI37	NEG	NEG	B38
						STI98	NEG	NEG	B38
13.1-9	74-88-4	**Methyl iodide**	ocean	1173*	135*,322	CCC	SP		B37
			volcano	1091,1149	983*(r)	REC	+		B9
					983*(s)	L5	+	+	B28
13.1-10	593-70-4	**Chlorofluoromethane**			358,597				
13.1-11	373-52-4	**Bromofluoromethane**			628				

Table 13.1. *(Continued)*

Species Number	Registry Number	Name	Emission Source	Emission Ref.	Detection Ref.	Code	Bioassay −MA	Bioassay +MA	Bioassay Ref.
13.1-12	75-43-4	**Dichlorofluoromethane**	air cond.	323	358,645				
			plastics comb.	354					
			volcano	234					
13.1-13	75-45-6	**Chlorodifluoromethane**	propellant	559	658	TRM	+		B21
			refrigeration	323,560	714*(S)				
			volcano	234					
13.1-14	75-69-4	**Trichlorofluoromethane**	air cond.	323	85*,135*				
			foaming agent	559,560	983*(r)				
		plastics comb. 354; propellant 559,560; refrigeration 559,560; volcano 234,1149			983*(s)				
					713*(S),718*(S)				
					1108(I)				
13.1-15	75-71-8	**Dichlorodifluoromethane**	air cond.	323	85*,135*	TRM	NEG		B21
			biomass comb.	969*	713*(S),718*(S)				
		foaming agent 559,560; propellant 559,560; refrigeration 559,560							
13.1-16	75-72-9	**Chlorotrifluoromethane**	refrigeration	562	691*				
13.1-17	75-63-8	**Bromotrifluoromethane**	refrigeration	563	691*,1011*				
					714*(S)				
13.1-18	74-97-5	**Bromochloromethane**			1010*				

Species Number	Registry Number	Name	Emission Source	Emission Ref.	Detection Ref.	Code	Bioassay −MA	Bioassay +MA	Bioassay Ref.
13.1-19	75-27-4	**Bromodichloromethane**			952(I)	STD00	+	+	B36
						STD35	+	+	B36
						STP00	NEG	NEG	B36
						STP35	NEG	NEG	B36
						STP37	NEG	NEG	B36
						STP38	NEG	NEG	B36
						STP98	NEG	NEG	B36
13.1-20	124-48-1	**Dibromochloromethane**			628				
13.1-21	594-18-3	**Dibromodichloromethane**			597t				
13.1-22	353-55-9	**Dibromochlorofluoromethane**			628				
13.1-23	353-59-3	**Bromochlorodifluoromethane**			1010*				
13.1-24	75-37-6	**1,1-Difluoroethane**	propellant	562		SRL	?		B30
13.1-25	76-16-4	**Perfluoroethane**	aluminum mfr.	256*,691	691*				
			electronics mfr.	616	714*(S)				
13.1-26	75-00-3	**Ethyl chloride**	biomass comb.	354	213,358				
			chemical mfr.	559					
		plastics comb. 354; refuse comb. 26(a); solvent 60							

Table 13.1. *(Continued)*

Species Number	Registry Number	Name	Emission Source	Emission Ref.	Detection Ref.	Bioassay Code	Bioassay −MA	Bioassay +MA	Bioassay Ref.
13.1-27	107-06-2	**1,2-Dichloroethane**	auto	867*	597,867*	CCC	SP		B37
			landfill	628	952(I)	STP35		+(B)	B36
						STP00	?(F)		B36
						REC	+		B9
						SRL	+		B30
13.1-28	71-55-6	**1,1,1-Trichloroethane**	biomass comb.	597,969*	89,135*	CCC	I		B37
			chemical mfr.	559	213(a),973(a)	STI00	NEG	NEG	B38
		landfill 365,628; solvent 60,369; volcano 1149			983(r)	STI35	NEG	NEG	B38
					983(s)	STI37	NEG	NEG	B38
					713*(S),714*(S)	STI98	NEG	NEG	B38
					820(I),952(I)	MNT	−?		B29
						SPI	NEG		B25
13.1-29	79-00-5	**1,1,2-Trichloroethane**	industrial	462	462,658				
13.1-30	79-34-5	**1,1,2,2-Tetrachloroethane**			797*,919*	CCC	LP		B37
						REC	+		B9
						STI00	NEG	NEG	B38
						STI35	NEG	NEG	B38
						STI37	NEG	NEG	B38
						STI98	NEG	NEG	B38
13.1-31	630-20-6	**1,1,1,2-Tetrachloroethane**			919*	STI00	NEG	NEG	B38
						STI35	NEG	NEG	B38
						STI37	NEG	NEG	B38
						STI98	NEG	NEG	B38

Species Number	Registry Number	Name	Emission Source	Emission Ref.	Detection Ref.	Code	Bioassay −MA	Bioassay +MA	Bioassay Ref.
13.1-32	74-96-4	**Ethyl bromide**	landfill	628	628	STI00	NEG	NEG	B38
						STI35	NEG	NEG	B38
						STI37	NEG	NEG	B38
						STI98	NEG	NEG	B38
						SRL	?		B30
13.1-33	106-93-4	**1,2-Dibromoethane**	auto	867*	867*,919*	TRM	+		B21
						HMASM		NEG	B13
						HMAST		+	B13
						CCC	SP		B37
						STP00	?		B36
						REC	+		B9
						L5	+	+	B28
						TRC	+		B20
						ASPH	+		B11
						SRL	+		B30
						NEU	+		B32
13.1-34	75-68-3	**1-Chloro-1,1-difluoroethane**	propellant	562					
13.1-35	76-15-3	**Chloropentafluoroethane**	propellant	559,562	691*				
			refrigeration	563	714*(S)				
13.1-36	76-14-2	**1,2-Dichlorotetrafluoro-ethane**	foaming agent	559,561	658				
			propellant	559,562	713*(S),714*(S)				
			refrigeration	559,561					

Table 13.1. *(Continued)*

Species Number	Registry Number	Name	Emission Source	Emission Ref.	Detection Ref.	Code	Bioassay −MA	Bioassay +MA	Bioassay Ref.
13.1-37	76-13-1	**1,2,2-Trichloro-1,1,2-trifluoroethane**	foaming agent	559	135*,373*				
			refrigeration	561	713*(S),714*(S)				
			solvent	264,559	820(I)				
		turbine 414; volcano 234							
13.1-38	354-58-5	**1,1,1-Trichloro-2,2,2-trifluoroethane**			628				
13.1-39	28605-74-5	**Tetrachloro-1,2-difluoroethane**	dry cleaning	561					
			solvent	561					
13.1-40	107-04-0	**1-Bromo-2-chloroethane**	landfill	628	628	REC	+		B9
13.1-41	353-61-7	**2-Fluoro-2-methylpropane**			358				
13.1-42	540-54-5	**1-Chloropropane**	refuse comb.	26(a)					
13.1-43	75-29-6	**2-Chloropropane**	biomass comb.	597	597t	STD00	+	+	B36
						STP00	NEG	NEG	B36
						STP35	NEG	NEG	B36
						STP37	NEG	NEG	B36
						STP38	NEG	NEG	B36
						STP98	NEG	NEG	B36
13.1-44	78-99-9	**1,1-Dichloropropane**			597				

Species Number	Registry Number	Name	Emission Source	Emission Ref.	Detection Ref.	Code	Bioassay −MA	Bioassay +MA	Bioassay Ref.
13.1-45	78-87-5	**1,2-Dichloropropane**			920*	STI00	+?	?+	B38
					1095*(I),1110*(I)	STI35	NEG	NEG	B38
						STI37	NEG	NEG	B38
						STI98	NEG	NEG	B38
13.1-46	142-28-9	**1,3-Dichloropropane**			566*	STP00	+	+	B36
13.1-47	598-76-5	**1,1-Dichloro-2-methylpropane**			628t				
13.1-48	96-18-4	**1,2,3-Trichloropropane**			628t	STI00	NEG	+	B38
						STI35	NEG	+	B38
						STI37	NEG	NEG	B38
						STI98	NEG	NEG	B38
13.1-49	598-77-6	**1,1,2-Trichloropropane**	biomass comb.	597t	973(a)t				
13.1-50	54833-05-5	**1,1,3,3-Tetrachloro-2-methylpropane**			628t				
13.1-51	25512-47-4	**Fluorobutane**	turbine	414					
13.1-52	78-86-4	**2-Chlorobutane**	turbine	414t	263				
13.1-53	107-84-6	**1-Chloro-3-methylbutane**	turbine	414					
13.1-54	631-65-2	**2-Chloro-3-methylbutane**	turbine	414					
13.1-55	7581-97-7	**2,3-Dichlorobutane**			628				

Table 13.1. *(Continued)*

Species Number	Registry Number	Name	Emission Source	Emission Ref.	Detection Ref.	Bioassay Code	Bioassay −MA	Bioassay +MA	Bioassay Ref.
13.1-56	544-10-5	**1-Chlorohexane**	turbine	414					
13.1-57	661-11-0	**1-Fluoroheptane**	turbine	414					
13.1-58	1002-69-3	**1-Chlorodecane**			597				
13.1-59	2425-54-9	**1-Chlorotetradecane**	vegetation	888t					
13.1-60S	353-50-4	**Carbonyl fluoride**	electronics mfr. polymer mfr.	616 124					
13.1-61S	75-44-5	**Phosgene**	chemical mfr. plastics comb. refuse comb.	167,177 46,184 31	394*,919*				
13.1-62S	75-87-6	**Chloral**			597,841*				
13.1-63	107-30-2	**Chloromethyl methyl ether**	ion xch. resin mfr.	38		CCC	LP		B37
13.1-64	542-88-1	**bis(Chloromethyl) ether**	ion xch. resin mfr.	38		CCC	SP		B37
13.1-65	111-44-4	**bis(2-Chloroethyl) ether**	river water odor	202					
13.1-66S	542-58-5	**2-Chloroethylacetate**			597				
13.1-67S	17640-08-3	**Pentyl-2,2-dichloropropionate**	turbine	414					
13.1-68S	506-77-4	**Chlorocyanide**	industrial	1150					

CH_2Cl_2
13.1-3

$CHCl_3$
13.1-4

FCOF
13.1-60

ClCOCl
13.1-61

Cl_3CCHO
13.1-62

$AcOCH_2CH_2Cl$
13.1-66

$MeCCl_2C(O)O(CH_2)_4Me$
13.1-67

NCCl
13.1-68

Table 13.2. Olefinic Halogenated Organics

Species Number	Registry Number	Name	Emission Source	Emission Ref.	Detection Ref.	Code	Bioassay −MA	Bioassay +MA	Bioassay Ref.
13.2-1	116-14-3	**Perfluoroethylene**	volcano	234					
13.2-2S	75-01-4	**Vinyl chloride**	chemical mfr.	559,560	135,275	TRM	+		B21
			landfill	365		CCC	SP		B37
		plastics mfr. 58*,59*; polymer comb. 211,304; solvent 439;				STP35	NE,CE	+	B36
		tobacco smoke 298				STD00	+	+	B36
						STD35	+	+	B36
						MST	NEG		B5
						CYI	+		B8
						REC		+	B9
						YEA	NEG	+	B27
						SRL	+		B30
13.2-3	75-35-4	**1,1-Dichloroethylene**			797*,920*	TRM	+		B21
					952(I)	CCC	LP		B37
						STD00	+	+	B36
13.2-4	540-59-0	**1,2-Dichloroethylene**	biomass comb.	969*	333,429				
			building mtls.	856					
		chemical mfr. 462,564; petroleum mfr. 111,369; plastics comb. 354; refuse comb. 26(a); solvent 60,439							

Species Number	Registry Number	Name	Emission Source	Emission Ref.	Detection Ref.	Code	Bioassay −MA	Bioassay +MA	Bioassay Ref.
13.2-5	79-01-6	**Trichloroethylene**	chemical mfr.	559,1164	135*,163*	TRM	+		B21
			dry cleaning	426	973(a)	HMASC		+	B13
		landfill 365; solvent 7,324; turbine 414;			755*(r)	CCC	LP		B37
		volcano 234			713*(S)	STD00	NEG	−(D)	B36
					820(I),952(I)	MST	+		B5
						SRL	NEG		B30
						SPI	+		B25
13.2-6	127-18-4	**Perchloroethylene**	chemical mfr.	559,560	135*,233	CCC	LP		B37
			dry cleaning	426,560	713*(S)	STI00	NEG	NEG	B38
		landfill 365,628; solvent 50,60; tobacco smoke 421			820(I),952(I)	STI35	NEG	NEG	B38
						STI37	NEG	NEG	B38
						STI98	NEG	NEG	B38
						CTV	+,−		B24
13.2-7	25429-23-6	**Dibromoethylene**	auto	653*	653*,797*				
			solvent	60	1009*(S)				
13.2-8	79-38-9	**Chlorotrifluoroethylene**	volcano	234					
13.2-9	116-15-4	**Hexafluoropropene**	volcano	234					
13.2-10S	616-19-3	**1,3-Dichloro-2-methylenepropane**			628t				
13.2-11	513-37-1	**1-Chloro-2-methylpropene**			628				
13.2-12	107-05-1	**3-Chloropropene**			920*	REC	+		B9
13.2-13	563-47-3	**3-Chloro-2-methylpropene**			628	STI00	NEG	+/−	B38

Table 13.2. *(Continued)*

Species Number	Registry Number	Name	Emission Source	Emission Ref.	Detection Ref.	Bioassay Code	Bioassay −MA	Bioassay +MA	Bioassay Ref.
						STI35	NEG	NEG	B38
						STI37	NEG	NEG	B38
						STI38	NEG	NEG	B38
13.2-14	563-54-2	**1,2-Dichloropropene**	solvent	60					
			turbine	414					
13.2-15	542-75-6	**1,3-Dichloropropene**	solvent	60	755*(r)	STD00	+	+	B36
						STD35	+	+	B36
						STD98	+	−(C)	B36
						STP00	+	+	B36
						STP35	+	+	B36
						STP37	NEG	NEG	B36
						STP38	NEG	NEG	B36
						STI00	+	+	B38
						STI35	+	+	B38
						STI37	NEG	NEG	B38
						STI98	NEG	NEG	B38
13.2-16	20589-85-9	**1,2,3,3-Tetrachloropropene**			358				
13.2-17	39437-98-4	**Chlorobutene**	plastics comb.	354					
13.2-18	764-41-0	**1,4-Dichloro-2-butene**			597,628	SRL	+		B30

Species Number	Registry Number	Name	Emission Source	Emission Ref.	Detection Ref.	Code	Bioassay −MA	Bioassay +MA	Bioassay Ref.
13.2-19	126-99-8	**2-Chloro-1,3-butadiene**	adhesives	11	597	SRL	+		B30
			plastics comb.	354					
13.2-20	55880-77-8	**Pentachlorobutadiene**			1095(I)				
13.2-21	87-68-3	**Hexachloro-1,3-butadiene**			920*,1095*	STI00	NEG	NEG	B38
					1095*(I)	STI35	NEG	NEG	B38
						STI37	NEG	NEG	B38
						STI98	NEG	NEG	B38
13.2-22S	20395-23-7	**1-Chloro-1-buten-3-yne**			597				

$ClCH:CH_2$
13.2-2

$ClCH_2CHMeCH_2Cl$
13.2-10

$HC\vdots CCH:CHCl$
13.2-22

Table 13.3. Halogenated Cyclic and Aromatic Compounds

Species Number	Registry Number	Name	Emission Source	Emission Ref.	Detection Ref.	Code	Bioassay −MA	Bioassay +MA	Bioassay Ref.
13.3-1	115-25-3	**Perflourocyclobutane**	propellant	559,562		SRL	?		B30
13.3-2	372-46-3	**Fluorocyclohexane**	turbine	414					
13.3-3	29297-39-0	**Dichlorocyclohexane**	refuse comb.	840(a)					
13.3-4	392-56-3	**Hexafluorobenzene**	refuse comb.	790(a)					
13.3-5	108-90-7	**Chlorobenzene**	chemical mfr.	462*	358,597	STI00	NEG	NEG	B38
			landfill	365	876*(r)	STI35	NEG	NEG	B38
		plastics comb. 181,354; solvent 60,439; wood comb. 890			820(I),952(I)	STI37	NEG	NEG	B38
						STI98	NEG	NEG	B38
13.3-6	95-50-1	*o*-**Dichlorobenzene**			358,597	STI00	NEG	NEG	B38
					755*(r),876*(r)	STI35	NEG	NEG	B38
					952(I),1095(I)	STI37	NEG	NEG	B38
						STI98	NEG	NEG	B38
13.3-7	541-73-1	*m*-**Dichlorobenzene**	biomass comb.	597	358,591	STI00	NEG	NEG	B38
			refuse comb.	954*(a)	755*(r),876*(r)	STI35	NEG	NEG	B38
			tobacco smoke	421	820(I)t,952(I)t	STI37	NEG	NEG	B38
						STI98	NEG	NEG	B38

Species Number	Registry Number	Name	Emission Source	Emission Ref.	Detection Ref.	Code	Bioassay −MA	Bioassay +MA	Bioassay Ref.
13.3-8	106-46-7	*p*-**Dichlorobenzene**	industrial	60	86,334	STI00	NEG	NEG	B38
					755*(r),876*(r)	STI35	NEG	NEG	B38
					1095(I),1108(I)	STI37	NEG	NEG	B38
						STI98	NEG	NEG	B38
						ALC	+		B19
13.3-9	87-61-6	**1,2,3-Trichlorobenzene**			1095(I)	STI00	NEG	NEG	B38
						STI35	NEG	NEG	B38
						STI37	NEG	NEG	B38
						STI98	NEG	NEG	B38
13.3-10	120-82-1	**1,2,4-Trichlorobenzene**	refuse comb.	790(a),840(a)	358,633	STI00	NEG	NEG	B38
					633(a)	STI35	NEG	NEG	B38
					755*(r),876*(r)	STI37	NEG	NEG	B38
					1095(I)	STI98	NEG	NEG	B38
13.3-11	108-70-3	**1,3,5-Trichlorobenzene**			1095t	WPU	NEG		B1
					1095(I)	STI00	NEG	NEG	B38
						STI35	NEG	NEG	B38
						STI37	NEG	NEG	B38
						STI98	NEG	NEG	B38
13.3-12	634-66-2	**1,2,3,4-Tetrachlorobenzene**			1095(I)	STI00	NEG	NEG	B38
						STI35	NEG	NEG	B38
						STI37	NEG	NEG	B38
						STI98	NEG	NEG	B38

Table 13.3. *(Continued)*

Species Number	Registry Number	Name	Emission Source	Emission Ref.	Detection Ref.	Code	Bioassay −MA	Bioassay +MA	Bioassay Ref.
13.3-13	634-90-2	**1,2,3,5-Tetrachlorobenzene**	refuse comb.	788(a),790(a)	1095t,1125t	STI00	NEG	NEG	B38
					1095(I)	STI35	NEG	NEG	B38
						STI37	NEG	NEG	B38
						STI98	NEG	NEG	B38
13.3-14	608-93-5	**Pentachlorobenzene**	refuse comb.	788(a),790(a)	1095	STI00	NEG	NEG	B38
					1095(I)	STI35	NEG	NEG	B38
						STI37	NEG	NEG	B38
						STI98	NEG	NEG	B38
13.3-15	608-73-1	**Hexachlorobenzene**	pesticide	239,245	577,1028*				
			refuse comb.	788(a),790(a)	230(a),942(a)				
					755*(r),847*(r)				
					819*(s),864*(s)				
13.3-16	75181-94-1	**Chlorotrifluorobenzene**			1095*				
13.3-17	29758-89-2	**Dichlorotrifluorobenzene**			1095*				
13.3-18	90077-78-4	**Bromotetrachlorobenzene**	refuse comb.	878(a)					
13.3-19S	434-64-0	**Perfluorotoluene**			597				
13.3-20	95-49-8	***o*-Chlorotoluene**			797*,920*				
					1095(I)t				

Species Number	Registry Number	Name	Emission Source	Emission Ref.	Detection Ref.	Code	Bioassay −MA	Bioassay +MA	Bioassay Ref.
13.3-21	106-43-4	***p*-Chlorotoluene**			797*,1095t	STD00	NEG	NEG	B36
					1095(I)t	STD35	NEG	NEG	B36
						STD37	NEG	NEG	B36
						STD38	NEG	NEG	B36
						STD98	NEG	NEG	B36
						STP00	NEG	NEG	B36
						STP35	NEG	NEG	B36
						STP37	NEG	NEG	B36
						STP38	NEG	NEG	B36
						STP98	NEG	NEG	B36
13.3-22	29797-40-8	**Dichlorotoluene**			597t,628				
					1095(I)				
13.3-23	61878-57-7	**Trichlorotoluene**			1095*				
					1095(I)				
13.3-24	29733-70-8	**Tetrachlorotoluene**			1095*(I)				
13.3-25	28807-97-8	**Bromotoluene**			1095*(I)				
13.3-26S	100-44-7	**Benzyl chloride**			628	HMAST		NEG	B13
						CCC	LP		B37
						STP00	?(E)		B36
						STP35	−(C)		B36
						STP38	−(C)		B36
						STP98	−(C)		B36
						STI00	+(E)		B36
						REC	+		B9

Table 13.3. *(Continued)*

Species Number	Registry Number	Name	Emission Source	Emission Ref.	Detection Ref.	Code	Bioassay −MA	Bioassay +MA	Bioassay Ref.
13.3-27	28258-59-5	**Bromoxylene**	vitamin mfr.	274					
13.3-28	615-87-2	**4,6-Dibromoxylene**	vitamin mfr.	274					
13.3-29	27323-18-8	**Chlorobiphenyl**	refuse comb.	790(a)					
13.3-30	25323-68-6	**Trichlorobiphenyl**	refuse comb.	954*(a)					
13.3-31	26914-33-0	**Tetrachlorobiphenyl**	refuse comb.	26(a),878(a)					
13.3-32	25429-29-2	**Pentachlorobiphenyl**	biomass comb. refuse comb.	1089(a) 26(a),840(a)					
13.3-33	26601-64-9	**Hexachlorobiphenyl**	refuse comb.	26(a),954*(a)					
13.3-34	53742-07-7	**Nonachlorobiphenyl**	refuse comb.	878(a)					
13.3-35	2051-24-3	**Decachlorobiphenyl**	refuse comb.	878(a)	1003(a)				
13.3-36S	24942-77-6	**Chlorophenylethynylbenzene**	refuse comb.	878(a)					
13.3-37	25586-43-0	**Chloronaphthalene**	refuse comb.	790(a)	1095*(I)				
13.3-38	1321-64-8	**Pentachloronaphthalene**	refuse comb.	790(a),878(a)					
13.3-39	1335-87-1	**Hexachloronaphthalene**	refuse comb.	790(a),878(a)					
13.3-40	90077-79-5	**Tetrachloroacenaphthylene**	refuse comb.	878(a)					

Species Number	Registry Number	Name	Emission Source	Emission Ref.	Detection Ref.	Code	Bioassay −MA	Bioassay +MA	Bioassay Ref.
13.3-41	several	**Fluoroanisole**			597t				
13.3-42	53452-81-6	**Tetrachloroanisole**			1125t				
13.3-43	1825-21-4	**Pentachloroanisole**			1125				
13.3-44	several	**Dibromoanisole**			1125t				
13.3-45	607-99-8	**2,4,6-Tribromoanisole**			1125				
13.3-46		**2,3,5,6-Tetrachloro-1,4-dimethoxybenzene**			1125				
13.3-47S	25167-80-0	**Chlorophenol**	sewage tmt.	1161					
13.3-48	120-83-2	**2,4-Dichlorophenol**			755*(r)	STP00	NEG	NEG	B36
						STP35	NEG	NEG	B36
						STP37	NEG	NEG	B36
						STP38	NEG	NEG	B36
						STP98	NEG	NEG	B36
						STI00	NEG	?	B38
						STI35	NEG	NEG	B38
						STI37	NEG	NEG	B38
						STI98	NEG	NEG	B38
13.3-49	25167-82-2	**Trichlorophenol**	refuse comb.	790(a),878(a)					
13.3-50	25167-83-3	**Tetrachlorophenol**	refuse comb.	788(a),790(a)	363(a),897(a)				

Table 13.3. *(Continued)*

Species Number	Registry Number	Name	Emission Source	Emission Ref.	Detection Ref.	Bioassay Code	−MA	+MA	Ref.
13.3-51	87-86-5	**Pentachlorophenol**	fertilizer mfr.	272*	443(a),633*(a)	HMASM		NEG	B13
			refuse comb.	788(a),790(a)	755*(r)	HMAST		NEG	B13
			wood preserv.	1048*	632*(I),846*(I)	STP00	NEG	NEG	B36
						STP35	NEG	NEG	B36
						STP37	NEG	NEG	B36
						STP38	NEG	NEG	B36
						STP98	NEG	NEG	B36
						MST	NEG		B5
						STI00	NEG	NEG	B38
						STI35	NEG	NEG	B38
						STI37	NEG	NEG	B38
						STI98	NEG	NEG	B38
13.3-52	86006-43-1	**Bromodichloromethylphenol**	refuse comb.	878(a)					
13.3-53		**Chloromethylenediphenol**			973(a)t				
13.3-54		**Bromomethylenediphenol**			973(a)t				
13.3-55S	1341-24-8	**Chloroacetophenone**	refuse comb.	790(a)					
13.3-56	several	**Chloromethylindanone**			399(a),458(a)				
13.3-57	26444-41-7	**Tetrachlorobiphenylene**	refuse comb.	790(a)					
13.3-58S	85897-29-6	**Chloro-9-fluorenone**	refuse comb.	878(a)					
13.3-59	several	**Dichloro-9-fluorenone**	refuse comb.	878(a)					

Species Number	Registry Number	Name	Emission Source	Emission Ref.	Detection Ref.	Bioassay Code	Bioassay −MA	Bioassay +MA	Bioassay Ref.
13.3-60	several	**Trichloro-9-fluorenone**	refuse comb.	878(a)					
13.3-61	several	**Tetrachloro-9-fluorenone**	refuse comb.	878(a)					
13.3-62	several	**Pentachloro-9-fluorenone**	refuse comb.	878(a)					
13.3-63S	82-44-0	**1-Chloro-9,10-anthraquinone**	refuse comb.	878(a)					
13.3-64	35913-09-8	**Chlorobenzaldehyde**			597 1095*(I)				
13.3-65S	104-83-6	***p*-Chlorobenzoylchloride**	refuse comb.	790(a)					
13.3-66	51-44-5	**3,4-Dichlorobenzoic acid**	biomass comb. coal comb. refuse comb.	1089(a) 825(a) 26(a)	570(r)				
13.3-67S	106-89-8	**3-Chloro-1,2-epoxypropane**	building mtls. industrial	9 60		CCC STU00 STP00 CYI REC MNT SRL NEU SPI	SP + + + + ?/− + + NEG		B37 B36 B36 B8 B9 B29 B30 B32 B25
13.3-68	71926-11-9	**Tetrachlorobenzofuran**	refuse comb.	790(a),959(a)					

Table 13.3. *(Continued)*

Species Number	Registry Number	Name	Emission Source	Emission Ref.	Detection Ref.	Code	Bioassay −MA	Bioassay +MA	Bioassay Ref.
13.3-69S	43047-99-0	**Dichlorodibenzofuran**	refuse comb.	878(a)					
13.3-70	43048-00-6	**Trichlorodibenzofuran**	refuse comb.	878(a),959(a)					
13.3-71	30402-14-3	**Tetrachlorodibenzofuran**	coal comb. refuse comb. wood comb.	953*(a) 790(a),878(a) 954*(a)					
13.3-72	30402-15-4	**Pentachlorodibenzofuran**	coal comb. refuse comb. wood comb.	953*(a) 790(a),878(a) 954*(a)					
13.3-73	55684-94-1	**Hexachlorodibenzofuran**	coal comb. refuse comb. wood comb.	953*(a) 790(a),878(a) 954*(a)					
13.3-74	38998-75-3	**Heptachlorodibenzofuran**	coal comb. refuse comb. wood comb.	953*(a) 790(a),878(a) 954*(a)	1002*(a),1003(a)				
13.3-75	39001-02-0	**Octachlorodibenzofuran**	refuse comb. wood comb.	790(a),959*(a) 953*(a),954*(a)	1002*(a),1003*(a)				
13.3-76S	29446-15-9	**2,3-Dichlorodibenzodioxin**	refuse comb.	878(a)					
13.3-77	several	**Trichlorodibenzodioxin**	refuse comb.	959(a)					

Species Number	Registry Number	Name	Emission Source	Emission Ref.	Detection Ref.	Code	Bioassay −MA	Bioassay +MA	Bioassay Ref.
13.3-78	several	**Tetrachlorodibenzodioxin**	auto	1003*(a)	1002*(a),1003*(a)				
			coal comb.	953*(a)					
		diesel 1003*(a); refuse comb. 788*(a),790(a); wood comb. 953*(a),954*(a)							
13.3-79	several	**Pentachlorodibenzodioxin**	coal comb.	953*(a)					
			refuse comb.	788*(a),790(a)					
			wood comb.	953*(a),954*(a)					
13.3-80	several	**Hexachlorodibenzodioxin**	coal comb.	953*(a)	1002*(a),1003*(a)				
			diesel	1003*(a)					
		refuse comb. 788*(a),790(a); tobacco smoke 1003*(a); wood comb. 953*(a),954*(a)							
13.3-81	several	**Heptachlorodibenzodioxin**	auto	1003*(a)	1002*(a),1003*(a)				
			coal comb.	953*(a)					
		diesel 1003*(a); refuse comb. 788*(a),790(a); tobacco smoke 1003*(a); wood comb. 953*(a),954*(a)							
13.3-82	3268-87-9	**Octachlorodibenzodioxin**	auto	1003*(a)	1002*(a),1003*(a)				
			diesel	1003*(a)					
		refuse comb. 788*(a),790(a); tobacco smoke 1003*(a); wood comb. 953*,953*(a),954*(a)							
13.3-83S	106-47-8	*p*-**Chloroaniline**	industrial	250		REC	+/?		B9
13.3-84	106-40-1	*p*-**Bromoaniline**	industrial	250		SPI	NEG		B25
13.3-85	540-37-4	*p*-**Iodoaniline**	industrial	250		SPI	NEG		B25

Table 13.3. *(Continued)*

Species Number	Registry Number	Name	Emission Source	Emission Ref.	Detection Ref.	Code	Bioassay −MA	Bioassay +MA	Bioassay Ref.
13.3-86S	several	**Tetrachlorobenzenedi-carbonitrile**	refuse comb.	878(a)					
13.3-87S	54886-36-1	**Chlorocarbazole**	aluminum mfr.	850(a)					
13.3-88	several	**Chlorobenzo[*c*]carbazole**	aluminum mfr.	850(a)					
13.3-89S	several	**Chloromethylthiotriazine**	refuse comb.	840(a)					

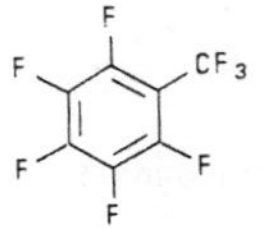

F, F, F, F, F, CF_3

13.3-19

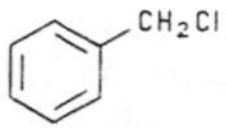

CH_2Cl

13.3-26

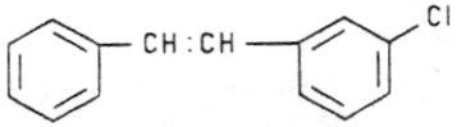

CH : CH, Cl

13.3-36

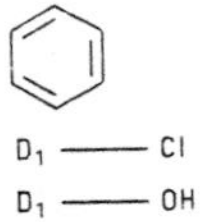

D_1 —— Cl

D_1 —— OH

13.3-47

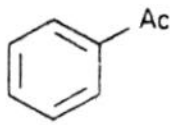

Ac

D_1 —— Cl

13.3-55

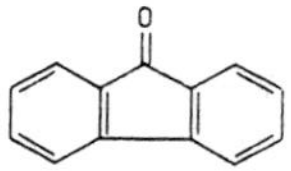

O

D_1 —— Cl

13.3-58

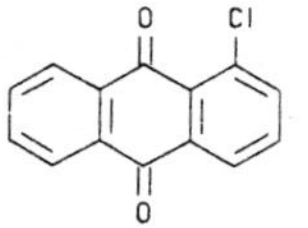

O Cl

O

13.3-63

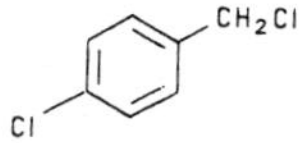

CH_2Cl

Cl

13.3-65

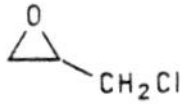

O

CH_2Cl

13.3-67

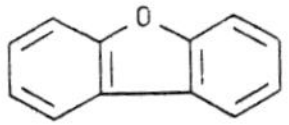

O

2 (D_1 —— Cl)

13.3-69

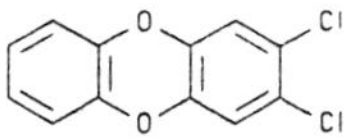

O Cl

O Cl

13.3-76

NH_2

Cl

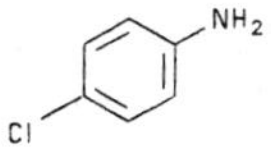

13.3-83

2 (D_1 — CN)
4 (D_2 — Cℓ)

13.3-86

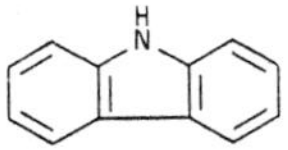

H
N

D_1 —— Cl

13.3-87

N N
N

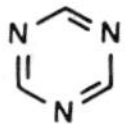

D_1 — SCCℓ$_3$

13.3-89

Table 13.4. Halogenated Pesticides

Species Number	Registry Number	Name	Emission Source	Emission Ref.	Detection Ref.	Bioassay Code	Bioassay −MA	Bioassay +MA	Bioassay Ref.
13.4-1S	8001-35-2	**Toxaphene**	pesticide	1118,126(a)	1026*	CCC	SP		B37
					21*(a),230*(a)				
					866*(r),1131*(r)				
13.4-2S	several	**Polychloropinene**	pesticide	126(a)					
13.4-3S	319-84-6	**α-Hexachlorocyclohexane**	pesticide	846,860	1026*,1028*				
					958*(a),977*(a)				
					864*(r),876*(r)				
					864*(s)				
13.4-4S	319-85-7	**β-Hexachlorocyclohexane**			1029*				
13.4-5S	58-89-9	**γ-Hexachlorocyclohexane**	pesticide	1118	1028*,1029*	HMASM		+/−	B13
					958*(a),977*(a)	HMAST		NEG	B13
					1026*(r),1028*(r)	CCC	LP		B37
					1080*(s)	ALC	+		B19
					846*(I),860*(I)	STI00	NEG	NEG	B38
						STI35	NEG	NEG	B38
						STI37	NEG	NEG	B38
						STI98	NEG	NEG	B38
						SRL	?		B30

Species Number	Registry Number	Name	Emission Source	Emission Ref.	Detection Ref.	Bioassay Code	Bioassay −MA	Bioassay +MA	Bioassay Ref.
13.4-6S	94-75-7	**2,4-D**	pesticide	194*	230*(a)	V79A	+		B7
						WPU	NEG		B1
						CCC	I		B37
						STP00	NEG	NEG	B36
						STP35	NEG	NEG	B36
						STP37	NEG	NEG	B36
						STP38	NEG	NEG	B36
						STP98	NEG	NEG	B36
						CYB	+		B8
						MDR	+		B12
						CYI	+		B8
						REC	+		B9
						VIC	+		B18
						ALC	+		B19
						MNT	?/−		B29
						SRL	NEG		B30
13.4-7S	94-11-1	**2,4-D, isopropyl ester**	pesticide	63,303*	303*				
					303*(a)				

Table 13.4. *(Continued)*

Species Number	Registry Number	Name	Emission Source	Emission Ref.	Detection Ref.	Code	Bioassay −MA	Bioassay +MA	Bioassay Ref.
13.4-8	94-80-4	**2,4-D, butyl ester**	pesticide	63,303*	63*,303* 194*(a),303*(a)				
13.4-9	1713-15-1	**2,4-D, isobutyl ester**	pesticide	63,303*	303*				
13.4-10	1929-73-3	**2,4-D, butoxyethanol ester**	pesticide	63,303*	303* 303*(a)				
13.4-11	25168-26-7	**2,4-D, isooctyl ester**	pesticide	63,303*	303* 303*(a)				
13.4-12	1320-18-9	**2,4-D propylene glycol butyl ether ester**	pesticide	63,303*	303* 303*(a)				
13.4-13S	1861-32-1	**Dacthal**	pesticide	567	1030				
13.4-14S	101-21-3	**Chlorpropham**	pesticide	575,1031*		MNT	?/−		B29
						SPI	NEG		B25
13.4-15S	1582-09-8	**Trifluralin**	pesticide	567,1031	1030*	WPU	NEG		B1
						CCC	LP		B37
						STP38	−(C)	−(C)	B36
						STP98	NEG	NEG	B36
						STP00	NEG	NEG	B36
						STP35	NEG	NEG	B36
						STP37	NEG	NEG	B36
						VIC	+		B18
						SRL	NEG		B30
						NEU	+		B32

Species Number	Registry Number	Name	Emission Source	Emission Ref.	Detection Ref.	Code	Bioassay −MA	Bioassay +MA	Ref.
13.4-16S	1912-24-9	**Atrazine**	herbicide	1032	1032*(r)	ASPH	NEG	+	B11
						SRL	?		B30
						NEU	NEG		B32
						ZMS	+		B22
13.4-17	several	**Polychlorobiphenyl**	electronics mfr.	1004*	20*,91*				
			fluorescent lamps	229*,1025	54(a),569*(a)				
		paint plasticizer 120; pesticide 20,91; refuse comb.		237*,557*	1015*(i)				
					847*(r),864*(r)				
					864*(s),1080*(s)				
					1025*(I)				
13.4-18	several	**Polychloroterphenyl**	refuse comb.	237*,1003(a)	1012				
			wood comb.	1003(a)	1003(a)				
					1012*(r)				
13.4-19S	50-29-3	***p,p′*-DDT**	pesticide	69*,265*	20*,1026*	HMASM		NEG	B13
					216(a),230*(a)	HMAST		NEG	B13
					847*(r),864*(r)	V79A	+		B7
					864*(s)	CCC	SP		B37
						STP35	−(CE)	−(CE)	B36
						STP37	−(CE)	−(CE)	B36
						STP38	−(CE)	−(CE)	B36
						STP00	NEG	NEG	B36
						STP98	NEG	NEG	B36
						CYC	+		B8
						MNT	−?		B29
						SRL	NEG		B30

Table 13.4. *(Continued)*

Species Number	Registry Number	Name	Emission Source	Emission Ref.	Detection Ref.	Code	Bioassay −MA	Bioassay +MA	Ref.
						SPI	NEG		B25
13.4-20S	789-02-6	**o,p′-DDT**	pesticide	20,568	20*,1026*	CYC	+		B8
					54*(a),230*(a)				
					1080*(r),1139*(r)				
					1080*(s)				
13.4-21S	72-54-8	**p,p′-TDE**	pesticide	265*,572	577	CYC	+		B8
			tobacco smoke	130*	1080*(r)	SRL	?		B30
					1080*(s)				
13.4-22S	53-19-0	**o,p′-TDE**	pesticide	130		CYC	+		B8
			tobacco smoke	130*					
13.4-23S	72-43-5	**Methoxychlor**	pesticide	1080	1080*(r)	WPU	NEG		B1
					1080*(s)	CCC	I,N		B37
						STP00	NEG	NEG	B36
						STP35	NEG	NEG	B36
						STP37	NEG	NEG	B36
						STP98		−(CE)	B36
						STP38	−(C)	−(C)	B36
						CTV	NEG		B24
						CTL	+		B24
						CTP	NEG		B24
						SRL	NEG		B30

Species Number	Registry Number	Name	Emission Source	Emission Ref.	Detection Ref.	Code	Bioassay −MA	Bioassay +MA	Bioassay Ref.
13.4-24S	72-55-9	***p,p'*-DDE**	pesticide	572,596	1027*,1028*	V79A	+		B7
					216(a),230*(a)	STP00	−(C)	−(C)	B36
					1080*(r)	STP35	−(C)	−(C)	B36
					1080*(s)	STP37	−(C)	−(C)	B36
						STP38	−(C)	−(C)	B36
						STP98	−(C)	−(C)	B36
						CYC	+		B8
						L5	+	NEG	B28
						SRL	?		B30
13.4-25S	3424-82-6	***o,p'*-DDE**	pesticide	230,596	230*(a)	CYC	NEG		B8
					866*(r)				
13.4-26S	1022-22-6	***p,p'*-TDEE**	pesticide	130					
			tobacco smoke	130*					
13.4-27	52645-53-1	**Permethrin**			943*(a)				
13.4-28S	76-44-8	**Heptachlor**	pesticide	574,1031*	1030*	CCC	LP		B37
					230*(a)	STP35	−(CE)	−(CE)	B36
						STP37	−(CE)	−(CE)	B36
						STP38	−(CE)	−(CE)	B36
						SRL	?		B30
						ZMS	+		B22
13.4-29S	1024-57-3	**Heptachlor epoxide**	pesticide	577,1118	577	STP35	−(C)	−(C)	B36
						STP37	−(C)	−(C)	B36
						STP38	−(C)	−(C)	B36
						SRL	?		B30

Table 13.4. *(Continued)*

Species Number	Registry Number	Name	Emission Source	Emission Ref.	Detection Ref.	Code	Bioassay −MA	Bioassay +MA	Ref.
13.4-30S	12789-03-6	**Chlordane**	pesticide	20,567	20*,1026*	V790	+		B7
					977*(a)	CCC	LP		B37
					847*(r),1080(r)	MDR	+		B12
					632*(I)				
13.4-31S	3734-49-4	**Nonachlor**			958(a),977(a)				
13.4-32S	309-00-2	**Aldrin**	pesticide	230,572	230*(a)	CCC	LP		B37
					819*(s)	STP00	NEG	NEG	B36
						STP35	NEG	NEG	B36
						STP37	NEG	NEG	B36
						STP38	NEG	NEG	B36
						STP98	NEG	NEG	B36
						SRL	?		B30
13.4-33S	72-20-8	**Endrin**	pesticide	230,1034	230*(a)	WPU	NEG		B1
					847*(r),1080(r)	CCC	I		B37
						STP00	NEG	NEG	B36
						STP35	NEG	NEG	B36
						STP37	NEG	NEG	B36
						STP38	NEG	NEG	B36
						STP98	NEG	NEG	B36
						HOC	+		B15
						VIC	+		B18
						SRL	?		B30

Species Number	Registry Number	Name	Emission Source	Emission Ref.	Detection Ref.	Code	Bioassay −MA	Bioassay +MA	Bioassay Ref.
13.4-34S	60-57-1	**Diedrin**	pesticide	212,265*	577,1026*	HMASC		NEG	B13
					230*(a),958*(a)	V790	+		B7
					847*(r),864*(r)	STP38	−(C)	−(C)	B36
					819*(s),1080*(s)	STP00	NEG	NEG	B36
					846*(I)	STP35	NEG	NEG	B36
						STP37	NEG	NEG	B36
						STP98	−(BC)	+(B)	B36
						MDR	+		B12
						STI00	NEG	NEG	B38
						STI35	NEG	NEG	B38
						STI37	NEG	NEG	B38
						STI98	NEG	NEG	B38
						SRL	?		B30
13.4-35S	13366-73-9	**Photodieldrin**			1030				
13.4-36S	36734-19-7	**Rovral**			943*(a)				
13.4-37S	115-29-7	**Endosulfan**			1026*				
					958*(a)				
					1080*(r)				

Table 13.4. *(Continued)*

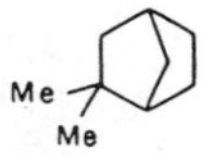

Me Me n(D$_1$—Cℓ)
13.4-1

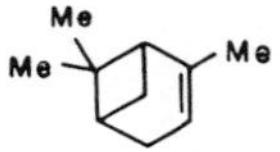

Me Me Me n(D$_1$—Cℓ)
13.4-2

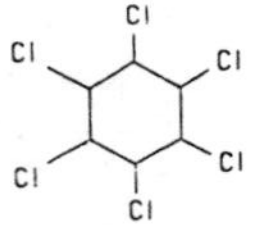

Cl Cl Cl Cl Cl Cl
13.4-3

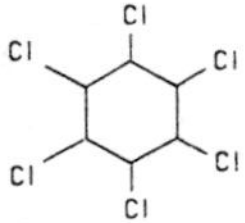

Cl Cl Cl Cl Cl Cl
13.4-4

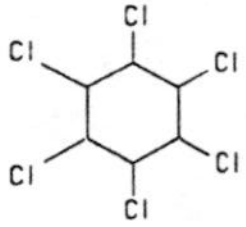

Cl Cl Cl Cl Cl Cl
13.4-5

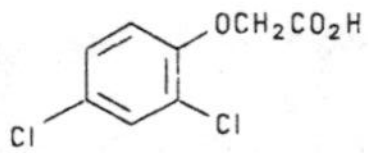

OCH_2CO_2H Cl Cl
13.4-6

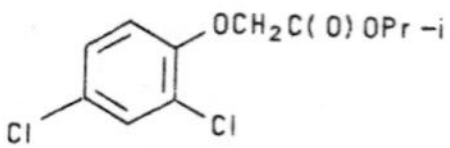

$OCH_2C(O)OPr$-i Cl Cl
13.4-7

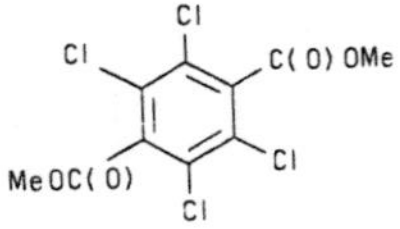

Cl Cl C(O)OMe MeOC(O) Cl Cl
13.4-13

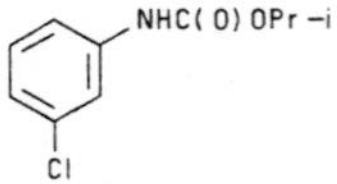

NHC(O)OPr-i Cl
13.4-14

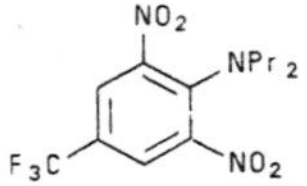

NO_2 NPr_2 F_3C NO_2
13.4-15

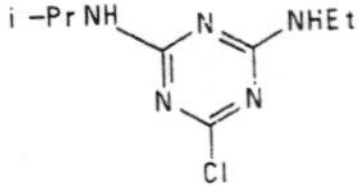

i-PrNH N N NHEt N Cl
13.4-16

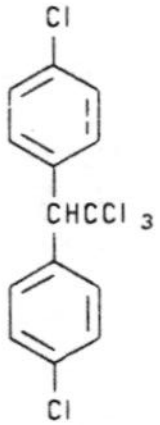

Cl $CHCCl_3$ Cl
13.4-19

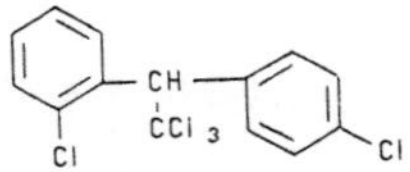

CH CCl_3 Cl Cl
13.4-20

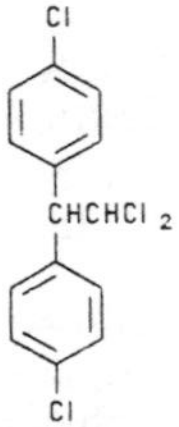

Cl $CHCHCl_2$ Cl
13.4-21

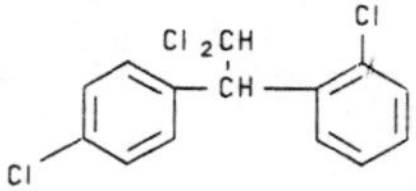

Cl_2CH CH Cl Cl
13.4-22

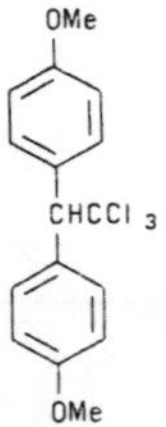

OMe $CHCCl_3$ OMe
13.4-23

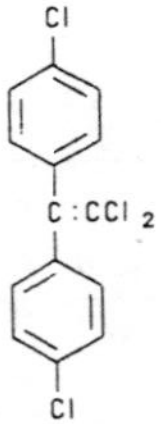

Cl $C:CCl_2$ Cl
13.4-24

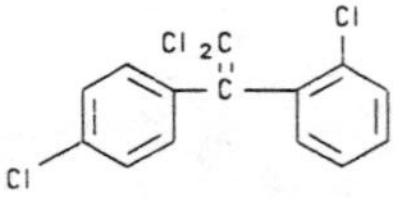

Cl_2C C Cl Cl
13.4-25

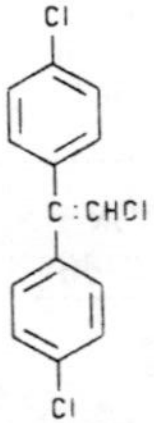

Cl C:CHCl Cl
13.4-26

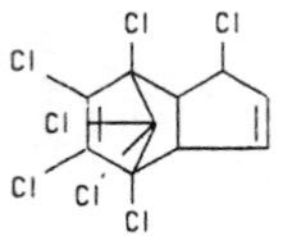

13.4-28

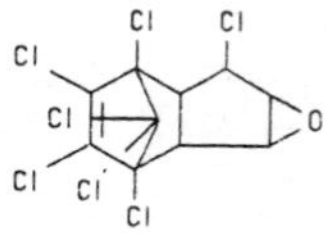

13.4-29

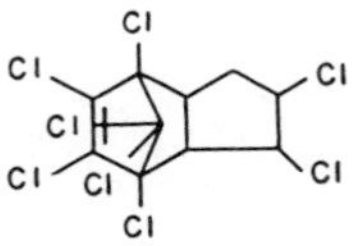

13.4-30

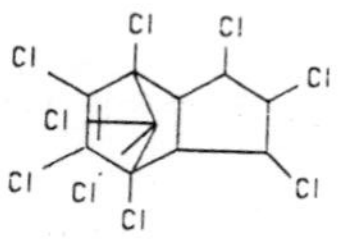

13.4-31

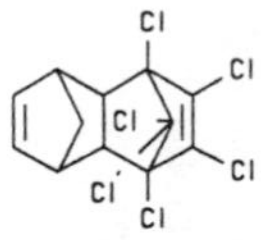

13.4-32

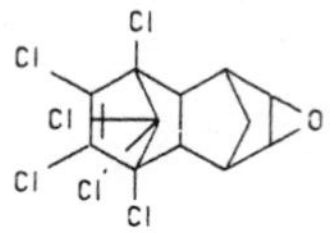

13.4-33

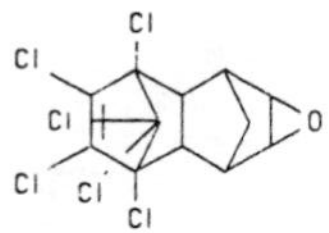

13.4-34

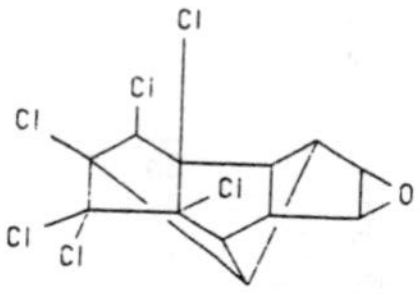

13.4-35

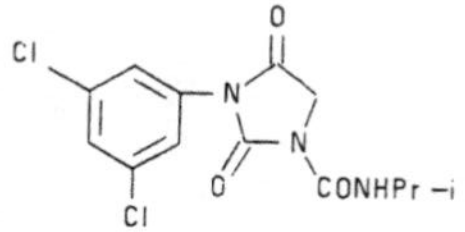

13.4-36

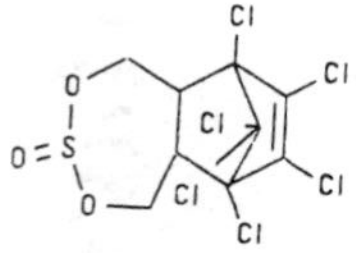

13.4-37

CHAPTER 14

Organometallic Compounds

14.0 Introduction

Organometallic compounds are those in which organic groups are linked through carbon to atoms other than carbon, hydrogen, nitrogen, halogens, and chalcogens.* Even though boron, silicon, and phosphorus do not behave much like metals, compounds containing C-B, C-Si, and C-P bonds are commonly designated organometallics.

Although organometallic chemistry is an extremely diverse and complex field, the organometallic compounds found in the atmosphere are restricted to three groups: (1) aliphatic derivatives of metal hydrides, (2) silicon-containing compounds, and (3) organophosphorus pesticides.

* The term *chalcogen* is a collective name for the atoms oxygen, sulfur, selenium, tellurium, and polonium.

14.1 Hydride Derivatives

Two groups of compounds are included in this table. The first is primarily made up of volatile alkanic derivatives of arsenic, mercury, and selenium. Microbial sources, for the most part, produce the small amounts of these compounds known to occur. The second group consists of volatile lead alkyls. These compounds are extensively used as gasoline additives, and are ubiquitous in both outdoor and indoor air in urban areas.

The atmospheric chemistry of the hydride derivatives has not received study, but hydroxyl radical attack on C-H bonds will certainly proceed along the lines of alkanes of similar structure. In the case of the lead compounds, both tetramethyl and tetraethyl lead are reported to react rapidly with OH· by H abstraction;[1284] the complete reaction sequences and products have not been specified.

14.2 Silicon Compounds

The nine compounds in this table are present as a result of volatilization from polymers to which they have been added as plasticizers. They are extremely long lived in the atmosphere and are thus useful as indicators of air transport from urban areas. In view of their lifetimes their atmospheric chemistry cannot be vigorous; its processes and products remain unknown at present.

14.3 Organophosphorus Pesticides

Halogenated organic compounds (Table 13.4) represent one of the main structural groups useful as pesticides. The present group, the organophosphorus compounds, represents the other. The compounds are characterized by a phosphorus atom with alkoxy ligands, often with other organic and organosulfur groups attached as well. As a result of the application of the pesticides to crops and soil, the compounds are found on aerosols in both outdoor and indoor air.

Although the soil and crop chemistry of these compounds has received extensive study, the atmospheric chemistry has not. There is no current evidence to suggest the direct involvement of any of these compounds in any of the principal atmospheric chemical cycles.

Table 14.1. Organometallics: Hydride Derivatives

Species Number	Registry Number	Name	Emission Source	Emission Ref.	Detection Ref.	Code	Bioassay −MA	Bioassay +MA	Bioassay Ref.
14.1-1S	107-44-8	**Fluoroisopropoxymethyloxophosphorus**	nerve gas mfr.	441					
14.1-2S	115-86-6	**Triphenylphosphate**	auto	311(a)	882(a)t				
			refuse comb.	26(a)					
14.1-3S	593-57-7	**Dimethylarsine**	microbes	34,302	112*				
14.1-4	593-88-4	**Trimethylarsine**	microbes	34,302	112*,367				
14.1-5S	593-69-1	**Ethylselenomercaptan**	chemical mfr.	1164					
14.1-6	593-79-3	**Dimethylselenide**	coal comb.	854	854*				
			microbes	393					
			sewage tmt.	854					
14.1-7	7101-31-7	**Dimethyldiselenide**	microbes	393	854*				
			vegetation	236,587					
14.1-8S	22089-69-6	**Dimethylselenone**	coal comb.	854	854*				
			sewage tmt.	854					
14.1-9	16056-34-1	**Methylmercury**	auto	397	388				
14.1-10S	115-09-3	**Chloromethyl mercury**	sewage tmt.	529*t	529t	REC	+		B9

Species Number	Registry Number	Name	Emission Source	Emission Ref.	Detection Ref.	Code	Bioassay −MA	Bioassay +MA	Ref.
14.1-11	593-74-8	**Dimethylmercury**	sewage tmt.	529t	215*,529	MDR	+		B12
			vegetation	615		CYO	NEG		B8
14.1-12	75-74-1	**Tetramethyl lead**	auto	988*	988*,989*	STI00	NEG	NEG	B38
			gasoline vapor	346,370	988*(I),989*(I)	STI35	NEG	NEG	B38
						STI37	NEG	NEG	B38
						STI98	NEG	NEG	B38
14.1-13	1762-26-1	**Ethyltrimethyl lead**	auto	988*	988*,989*				
			gasoline vapor	346,370	988*(I),989*(I)				
14.1-14	1762-27-2	**Diethyldimethyl lead**	auto	988*	988*,989*				
			gasoline vapor	346,370	988*(I),989*(I)				
14.1-15	1762-28-3	**Triethylmethyl lead**	auto	988*	988*,989*				
			gasoline vapor	346,370	988*(I),989*(I)				
14.1-16	78-00-2	**Tetraethyl lead**	auto	988*	988*,989*				
			gasoline vapor	346,370	988*(I),989*(I)				

O:PF(Me)OPr-i
14.1-1

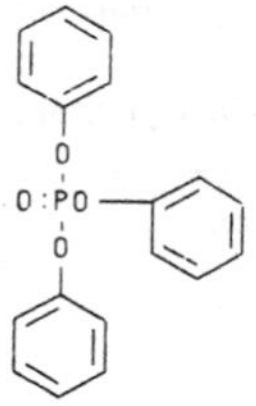

O:PO(OPh)OPh—Ph
14.1-2

$AsHMe_2$
14.1-3

EtSeH
14.1-5

$O:SeMe_2:O$
14.1-8

ClHgMe
14.1-10

Table 14.2. Organometallics: Silicon Compounds

Species Number	Registry Number	Name	Emission Source	Emission Ref.	Detection Ref.	Code	Bioassay −MA	Bioassay +MA	Bioassay Ref.
14.2-1S	992-94-9	**Methylsilane**	turbine	414	597				
14.2-2	75-76-3	**Tetramethylsilane**			358t				
14.2-3	1066-40-6	**Trimethylsilanol**			628				
14.2-4S	1529-17-5	**Trimethyl (phenoxy) silane**			628				
14.2-5S	107-46-0	**Hexamethyldisiloxane**			628				
14.2-6S	107-51-7	**Octamethyltrisiloxane**			1086(a)				
14.2-7S	541-05-9	**Hexamethylcyclotri-siloxane**			358,597t 1109*(I)				
14.2-8	556-67-2	**Octamethylcyclotetra-siloxane**			358t,597t				
14.2-9	541-02-6	**Decamethylcyclopenta-siloxane**			1086(a)				
14.2-10	107-52-8	**Tetradecamethylhexa-siloxane**			358t				

$MeSiH_3$
14.2-1

OSiMe$_3$
14.2-4

$Me_3SiOSiMe_3$
14.2-5

$Me_3SiOSiMe_2OSMe_3$
14.2-6

Me Me Si O O Si Si Me Me O Me Me
14.2-7

Table 14.3. Organophosphorus Pesticides

Species Number	Registry Number	Name	Emission Source	Emission Ref.	Detection Ref.	Code	Bioassay −MA	Bioassay +MA	Bioassay Ref.
14.3-1S	814-29-9	**Butyphos**	pesticide	228*,126(a)	888(a),1086(a)				
					570(r)				
					570(s)				
14.3-2S	78-51-3	**tris (2-Butoxyethyl) phosphate**			916(I),1109*(I)	STI00	NEG	NEG	B38
						STI35	NEG	NEG	B38
						STI37	NEG	NEG	B38
						STI98	NEG	NEG	B38
14.3-3	78-42-2	**tris (2-Ethylhexoxy) phosphate**			916(I),1109*(I)				
14.3-4S	62-73-7	**Dichlorvos**	pesticide	578,632	632*(I)	HMASM		NEG	B13
						HMAST		NEG	B13
						HMASC		NEG	B13
						WPU	+		B1
						WP2	+		B1
						CCC	I		B37
						STP35	+(EF)		B36
						SCE	NEG		B6
						CYB	NEG		B8
						CYS	NEG		B8
						CYT	NEG		B8
						REC	+		B9
						ALC	+		B19
						MNT	NEG		B29

Table 14.3. *(Continued)*

Species Number	Registry Number	Name	Emission Source	Emission Ref.	Detection Ref.	Code	Bioassay −MA	Bioassay +MA	Bioassay Ref.
						ASPD	+		B10
						SRL	NEG		B30
						SPI	?		B25
14.3-5S	52-68-6	**Trichlorfon**	pesticide	228*		HMAST		+	B13
						WPU	+		B1
						STP98	+	+	B36
						STP00	+	+	B36
						STP35	NEG	NEG	B36
						STP37	NEG	NEG	B36
						STP38	NEG	NEG	B36
						MNT	−?		B29
						SRL	NEG		B30
14.3-6S	78-48-8	**DEF**			230*(a)				
14.3-7S	8022-00-2	**Methylmercaptophos**	pesticide	228*,126(a)					
14.3-8S	60-51-5	**Dimethoate**	pesticide	228*,126(a)		CCC	I		B37
						VIC	+		B18
						ALC	+		B19
						STI00	+	+	B38
						STI35	NEG	NEG	B38
						STI37	NEG	NEG	B38
						STI98	NEG	NEG	B38
						SRL	NEG		B30
14.3-9S	121-75-5	**Malathion**	pesticide	228*,632	230*(a)	WPU	NEG		B1
					632*(I)	CCC	I,N		B37

Species Number	Registry Number	Name	Emission Source	Emission Ref.	Detection Ref.	Bioassay Code	Bioassay −MA	Bioassay +MA	Bioassay Ref.
						STP00	NEG	NEG	B36
						STP35	NEG	NEG	B36
						STP37	NEG	NEG	B36
						STP98	NEG	NEG	B36
						STP00	−(C)	−(C)	B36
						STP35	−(C)	−(C)	B36
						STP37	−(C)	−(C)	B36
						STP38	−(C)	−(C)	B36
						SCE	+		B6
						STI00	NEG	NEG	B38
						STI35	NEG	NEG	B38
						STI37	NEG	NEG	B38
						STI98	NEG	NEG	B38
						SRL	NEG		B30
14.3-10S	56-38-2	**Parathion**	pesticide	228*,1118	230*(a)	WPU	NEG		B1
						STP00	−(C)	−(C)	B36
						STP35	−(C)	−(C)	B36
						STP37	−(C)	−(C)	B36
						STP98	−(C)	−(C)	B36
						ALC	+		B19
						STI00	NEG	?	B38
						STI35	NEG	NEG	B38
						STI37	NEG	NEG	B38
						STI98	NEG	NEG	B38
						SRL	?		B30
14.3-11S	298-00-0	**Methylparathion**			230*(a)	WPU	NEG		B1
						STP00	−(C)	−(C)	B36

Table 14.3. *(Continued)*

Species Number	Registry Number	Name	Emission Source	Emission Ref.	Detection Ref.	Code	Bioassay −MA	Bioassay +MA	Bioassay Ref.
						STP35	−(C)	−(C)	B36
						STP37	−(C)	−(C)	B36
						STP38	−(C)	−(C)	B36
						STI00	+	+	B38
						STI35	NEG	+	B38
						STI37	NEG	NEG	B38
						STI98	?	+	B38
14.3-12S	122-14-5	**Fenitrothion**	pesticide	276		HMAST		NEG	B13
						SRL	?		B30
14.3-13S	299-84-3	**Ronnel**	pesticide	632	632*(I)				
14.3-14S	2921-88-2	**Dursban**	pesticide	632	632*(I)	WPU	NEG		B1
						STP00	−(C)	−(C)	B36
						STP35	−(C)	−(C)	B36
						STP37	−(C)	−(C)	B36
						STP38	−(C)	−(C)	B36
						REC	+		B9
						SRL	?		B30
14.3-15S	333-41-5	**Diazinon**	pesticide	578,632	632*(I)	WPU	NEG		B1
						STP00	NEG	NEG	B36
						STP98	NEG	NEG	B36
						STP35	NEG	NEG	B36
						STP37	NEG	NEG	B36
						STP38	NEG	NEG	B36

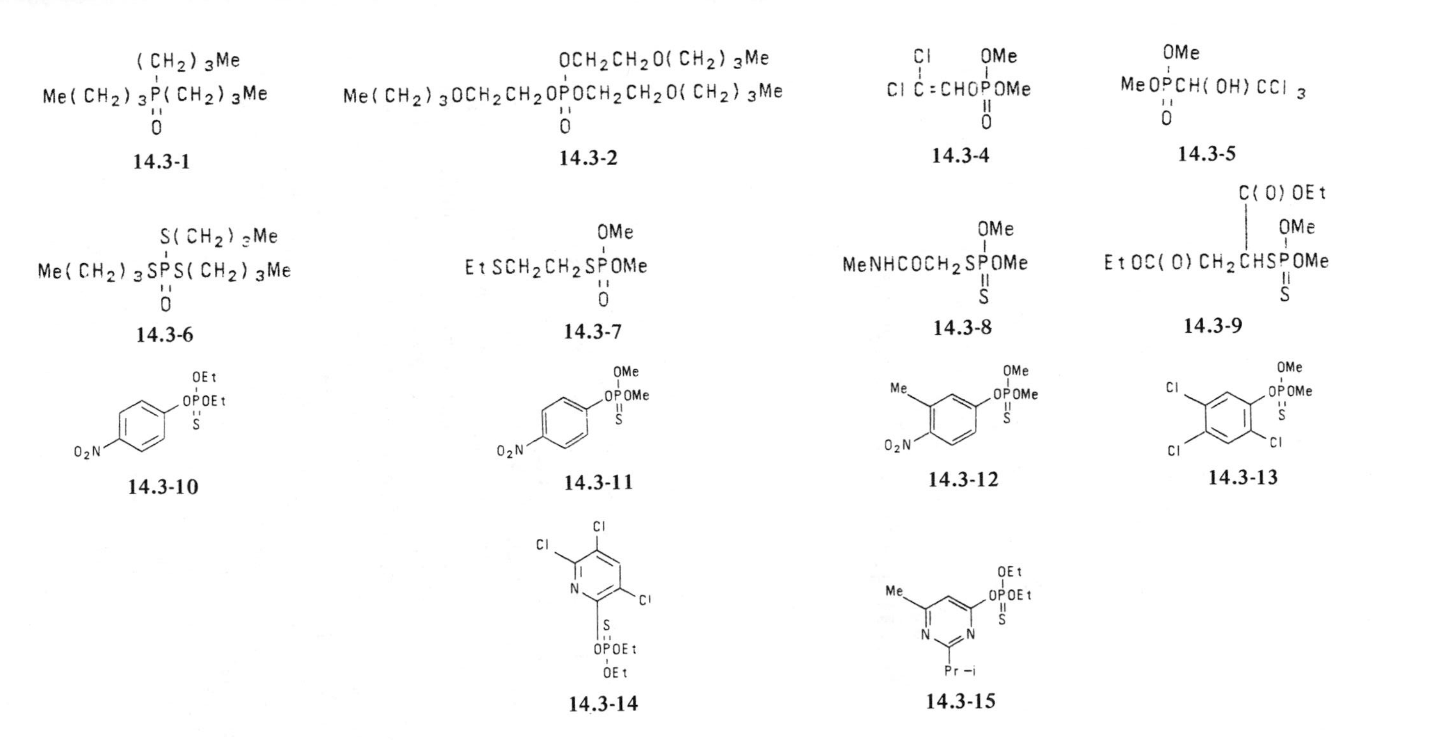
14.3-1
14.3-2
14.3-4
14.3-5
14.3-6
14.3-7
14.3-8
14.3-9
14.3-10
14.3-11
14.3-12
14.3-13
14.3-14
14.3-15

CHAPTER 15

Synthesis and Summary

15.0 Introduction

The primary purpose of this book has been to assemble from the diverse literature related in some way to atmospheric chemistry a substantial amount of material on the sources, occurrence, and concentrations of atmospheric compounds, their chemical reactions, and their genotoxicity. Extensive indexing and cross-referencing has been used to permit information on individual species or groups of chemically similar species to be located, and discussions of individual species or groups has been included. As with most taxonomic efforts, however, the whole is greater than the sum of its parts; that is, the broad perspective allowed by the ensemble of information provides insight on atmospheric processes which cannot be derived from data on individual species or groups. In this concluding chapter, we present and discuss topics which utilize the assembled information from the previous chapters.

It is reasonable to wonder whether the information in this chapter, and in the remainder of the book, for that matter, will rapidly become obsolete. Certainly many new atmospheric species will be discovered, their concentrations measured, and their chemical and biological consequences examined. In our view, such research will for the most part be evolutionary rather than revolutionary, however. We anticipate that at least the broad outlines that we draw in this work will endure for at least a decade or more.

15.1 Species Occurrence Summary

The tables in Chapters 2-14 include more than 2800 atmospheric species. The size of this data collection is such that attempts to summarize its characteristics are valuable. Accordingly, Table 15.1-1 presents statistical information for each individual chapter. In addition to listing the total number of species, the table shows the number found in each of the atmospheric regimes. It is worth noting here, as was done in the preface, that the numbers reflect not only which species are present, but also whether analytical chemists have tried to find them and whether techniques for finding them exist. For example, the earlier compendium on this topic[1271] listed 150 nitrogen-containing organic compounds, about nine percent of the total list of 1651 compounds. Table 15.1-1 shows 384 nitrogen-containing organic compounds in a total list of 2844, or about fourteen percent. This increased percentage does not reflect a change in atmospheric composition, but rather the increased attention given recently by atmospheric chemists to nitro compounds and nitrogen-containing heterocycles. Other groups of compounds with markedly increased numbers include ketones and carboxylic acids, as well as the completely new listing of minerals. It seems likely that alcohols, carboxylic acids, and nitrogen-containing organics are still underrepresented in the tables vis-a vis their true atmospheric occurrence.

Table 15.1-1 contains some surprises. One is that the inorganic compounds comprise less than ten percent of the total, although the atmospheric chemical literature devoted to inorganic species is probably at least ninety percent of the total literature. It is clear that organic atmospheric chemistry has been seriously slighted. Another interesting finding is that species detected in the aerosol phase in ambient air easily outnumber those found in the gas phase, yet the literature on gas phase atmospheric chemistry is much more extensive.

The distribution of compounds between gas and condensed phases in the atmosphere is obviously related to the volatility of the individual compounds. Junge[1311] has demonstrated that at equilibrium it will be approximately true that the phase distribution will be given by

Table 15.1-1. Total Species in Each Chemical Group and the Number Detected in Each Atmospheric Regime

Chapter	Chemical Group	Total† Species	Number of Species Detected* g	a	c	f	i	r	s	S	I
2	Inorganics	260	38	137	11	11	7	22	17	52	14
3	Hydrocarbons	729	310	314	0	0	1	52	24	4	110
4	Ethers	44	16	2	0	0	0	0	0	0	2
5	Alcohols	233	26	60	0	0	0	3	0	2	19
6	Ketones	227	23	51	1	0	1	2	1	2	10
7	Aldehydes	108	31	20	5	5	4	5	1	0	14
8	Acid derivatives	219	28	65	0	0	0	7	6	0	17
9	Carboxylic acids	174	32	118	1	7	0	69	35	0	19
10	O-Heterocycles	93	21	7	0	0	0	1	0	0	5
11	N-organics	384	30	92	0	0	0	9	0	2	10
12	S-organics	99	10	14	0	0	0	1	0	0	2
13	Halogenates	216	113	43	0	0	1	33	16	19	38
14	Organometallics	41	21	8	0	0	0	1	1	0	13
	Grand Totals	2827	699	931	18	23	14	205	101	81	273

† Includes compounds detected in emission as well as those detected in ambient air.

* g = gas phase, a = aerosol phase, c = cloud, f = fog, i = ice, r = rain, s = snow, S = stratosphere, I = indoors.

$$\Phi = \frac{c\Theta}{p_0 + c\Theta}$$

where Φ is the ratio of the amount of organic compound in the condensed phase to that in both condensed and vapor phases, p_0 is the saturation vapor pressure of the compound, Θ is the surface area of the condensed phase per unit volume of air, and c is a constant which depends on the molecular weight and the heat of condensation but does not vary greatly for the different surface-active molecules under discussion here. The consequences are illustrated by Figure 15.1-1, in which values of Φ for different values of aerosol Θ and p_0 are shown. For remote aerosol densities, p_0 values of $\leqslant 10^{-7}$ torr are required if substantial fractions of organic compounds are to be present upon the aerosol particles. For the urban aerosol, p_0 values of $\leqslant 10^{-6}$ torr are sufficient. There are at present too few measurements of gas to particle distribution ratios for individual compounds to confirm the correctness of this approach, but limited data[633,1312] is consistent with its predictions.

More than two hundred species (91 percent organic!) have been found in rain and more than one hundred (83 percent organic!) in snow. The chemistry within atmospheric droplets is clearly complicated and dominantly organic. As of this writing, that chemistry is relatively unexplored. The small number of compounds detected in clouds, fog, and ice reflect the very early state of knowledge of these systems rather than the presence of sparse numbers of compounds.

Stratospheric species occur almost entirely within either the inorganic group or that comprised of halogenated organics. This situation arises largely because stratospheric species must either be created there by chemical reactions or be transported from ground level, the latter process requiring at least a year. Only molecules with very long lifetimes will be thus transported, and only small molecules will be produced by stratospheric chemistry. The pattern of detection in Table 15.1-1 is in part due to the relative paucity of stratospheric sampling, however; it would not be surprising to find more stratospheric processes involving small organic molecules than are now known.

Table 15.1-1 notes the detection of nearly 300 different species in indoor air. The hydrocarbons are by far the largest group, but all chemical groups are reasonably well represented. Many more compounds will doubtless be found in indoor air as more investigations are undertaken.

Of the very large number of species in Table 15.1-1, many are uncommon or rare, but others are abundant. It seems useful to list by themselves those species which the data indicate are *always found* in the tropospheric gas, the tropospheric aerosol, or in precipitation (we combine all forms of precipitation to increase the data available). The result is presented in Table 15.1-2, which indicates that *any* sample of tropospheric air contains at least two dozen trace compounds. Many of these are unreactive or nearly so: all of the first row, plus N_2O, CO_2, and COS. Of the remaining compounds, CH_4 is by far the most abundant, at about 1.6 ppm. When all reactive molecules are considered, tropospheric air of minimum chemical complexity contains more than four times as many reactive organic gases as reactive inorganic gases.

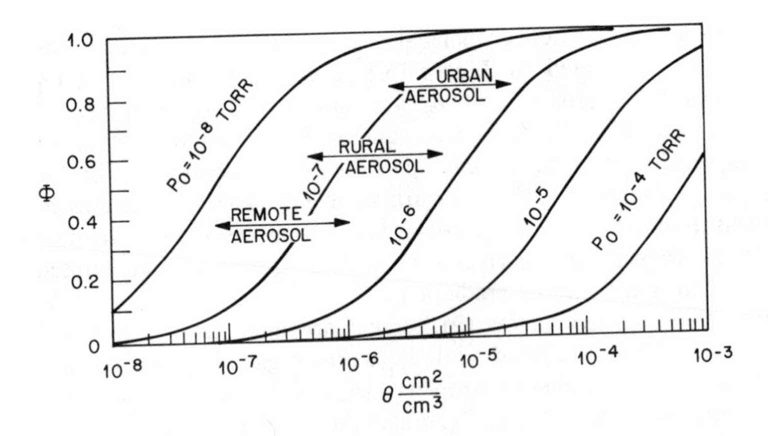

Fig. 15.1-1 The fraction Φ of atmospheric organic molecules present in the condensed phase at equilibrium, as a function of vapor pressure p_0 and surface density θ (square centimeters of condensed phase surface per cubic centimeter of air). Adapted from a diagram devised by C. Junge.[1311]

Tropospheric aerosols of minimum complexity contain about a dozen species, mostly the common inorganic ions and the common inorganic oxides found in the Earth's crust. Organic material is common on aerosol particles, but the present data are not sufficient to be sure that such material will always be found.

The ionic species always present in precipitation are similar to those found in aerosols, although the acidic inorganic anions nitrate and sulfate must be added to the list. It is likely that soil-derived oxides are universal in precipitation, but the data are still too sparse for certainty. Formaldehyde and formic acid appear always to be constituents of precipitation.

The search for abundant atmospheric species is carried one step farther in Table 15.1-3, which lists species *often* found in the tropospheric gas, the tropospheric aerosol, or in precipitation. These lists are longer then those of the previous table and include molecules of larger molecular weight. For the tropospheric gas, the list includes several reactive inorganics, sequences of alkanes and alkenes, a number of small oxygenates, and simple organic compounds of nitrogen, sulfur, and chlorine. Together with the list in Table 15.1-2, the total number of gas phase species commonly found in the troposphere is of the order of sixty. Reactive organic compounds are more abundant than are inorganics.

Table 15.1-2. Chemical Species Always Found in the Tropospheric Gas, Tropospheric Aerosols, or Precipitation

*Tropospheric Gas**
O_2, N_2, He, Ne, Ar, Kr, Xe, H_2O
H_2, N_2O, CO, CO_2, COS, SO_2
CH_4, C_2H_6, benzene, toluene
HCHO, HC(O)OH, HCN
CF_2Cl_2, $CFCl_3$, CH_3CCl_3

Tropospheric Aerosols†
H^+, Na^+, Mg^{2+}, K^+, Ca^{2+}, Cl^-
Na_2O, MgO, Al_2O_3, SiO_2, CaO, Fe_2O_3

Precipitation‡
H^+, NH_4^+, Na^+, Mg^{2+}, K^+, Ca^{2+}
Cl^-, NO_3^-, SO_4^{2-}
HCHO, HC(O)OH

* The identification numbers for these species are: 2.2-2, 2.3-1, 2.1-2, 2.1-5, 2.1-11, 2.1-19, 2.1-20, 2.2-4, 2.2-1, 2.3-12, 2.6-7, 2.6-8, 2.4-2, 2.4-10, 3.1-1, 3.1-2, 3.5-1, 3.5-2, 7.1-1, 9.1-1, 11.1-1, 13.1-14, 13.1-15, 13.1-28.

† The identification numbers for these species are: 2.1-1, 2.1-6, 2.1-7, 2.1-12, 2.1-13, 2.1-10, 2.6-12, 2.6-13, 2.6-14, 2.6-17, 2.6-21, 2.6-30.

‡ The identification numbers for these species are: 2.1-1, 2.3-3, 2.1-6, 2.1-7, 2.1-12, 2.1-13, 2.1-10, 2.3-11, 2.4-23, 7.1-1, 9.1-1

Nearly a hundred individual species are included in the tropospheric aerosol list in Table 15.1-3. This results largely from the presence of four groups of compounds: minerals, and the C_{16}-C_{30} alkanes, alcohols, and carboxylic acids. The minerals are, of course, from the earth's crust while the organic compounds are related to vegetative sources. Soot is also common, as are several of the polynuclear aromatic hydrocarbons (PNAH) and formaldehyde.

Table 15.1-3. Chemical Species Often Found in the Tropospheric Gas, Tropospheric Aerosols, or Precipitation

---

*Tropospheric Gas**

O_3, $OH\cdot$, $HO_2\cdot$, H_2O_2
NH_3, NO, NO_2, HNO_3, H_2S, HCl
C_3-C_8 alkanes, C_2-C_5 alkenes, C_2H_2
isoprene, cyclohexane, α-pinene, naphthalene
phenol, $CH_3C(O)CH_3$, CH_3CHO, $CH_2{=}CHCHO$
benzaldehyde, the methyl amines
$CH_3C(O)O_2NO_2$, CH_3SCH_3, CH_3Cl, $Cl_2C{=}CClH$

Tropospheric Aerosols†

NH_4^+, SO_4^{2-}, NO_3^-
NH_4NO_3, NH_4Cl, C_x(soot), minerals
C_{16}-C_{30} alkanes, C_{16}-C_{30} alcohols,C_{16}-C_{30} carboxylic acids
naphthalene, fluorene, phenanthrene, anthracene,
fluoranthene, pyrene, chrysene, HCHO

Precipitation‡

naphthalene, fluorene, phenanthrene, anthracene,
fluoranthene, pyrene, $CH_3C(O)OH$

---

* The identification numbers for these species are: 2.2-3, 2.2-6, 2.2-7, 2.2-8, 2.3-2, 2.3-7, 2.3-8, 2.3-15, 2.4-1, 2.5-6, 3.1-3, 3.1-4, 3.1-10, 3.1-30, 3.1-48, 3.1-66, 3.2-1, 3.2-3, 3.2-7, 3.2-26, 3.2-2, 3.2-19, 3.3-29, 3.4-1, 3.6-23, 5.4-1, 6.1-1, 7.1-3, 7.2-1, 7.3-1, 11.2-3, 11.2-4, 11.2-5, 11.3-6, 12.2-1, 13.1-2, 13.2-5. The simple alkyl derivatives of the organic compounds in this list are also common.

† The identification numbers for these species are: 2.3-3, 2.4-23, 2.3-11, 2.3-19, 2.5-13, 2.6-1, 2.7-1 to 2.7-31, 3.1-101, -104, -106, -108 to -115, -117, -119, -121, -123, 5.1-56 to -66, -68, -69, -71, -72, 9.1-77, -81, -83, -87, -88, -91, -92, -95, -96, -99 to -104, 3.6-23, 3.7-8, 3.7-40, 3.7-68, 3.8-3, 3.8-34, 3.8-60, 7.1-1. The simple alkyl and benzo derivatives of the organic compounds in this list are also common.

‡ The identification numbers for these species are: 3.6-23, 3.7-8, 3.7-40, 3.7-68, 3.8-3, 3.8-64, 9.1-3.

The precipitation list in Table 15.1-3 is short, doubtless because detailed chemical analyses of precipitation are uncommon. The list consists of acetic acid and several of the PNAH.

The inclusion of specific compounds in Tables 15.1-2 and 15.1-3 could be debated at length, but the key message of the tables is that some 170 distinct chemical species are common constituents of tropospheric air.

15.2 Sources of Atmospheric Compounds

The sources of atmospheric compounds are as diverse as are the compounds themselves. A number of sources, both anthropogenic and natural, emit only a few different compounds, some only one. At the opposite end of the spectrum are sources that produce hundreds of compounds. To assess this diversity, and to provide a useful reference tool, the source data contained in the tables of Chapters 2-14 have been cross-indexed by source name instead of by chemical species. This information is presented in index I4, where it is available to those concerned with emissions from a specific source.

As with most tabulations, it is of interest to select and examine those entries that stand out from the others. Thus, in Table 15.2-1 we list those sources emitting the largest numbers of different compounds. Heading the list is tobacco smoke, which we include in the book for two reasons: because it has a major influence on indoor air quality, and because it has received more analytical attention then a related source, biomass combustion (a source of obvious interest to atmospheric chemistry). Next in the table are emissions from gasoline and diesel engines. The analytical efforts devoted to these emissions demonstrate that incomplete combustion of fossil fuels can produce a very large number of compounds. Fuel fragment molecules and oxygenated and nitrated derivatives synthesized during the combustion process are included.

The fourth entry in Table 15.2-1 is vegetation. Despite the large number of compounds emitted by vegetation, our knowledge of the full range of compounds and especially the fluxes of vegetative emissions remains inadequate. The importance of natural organic emissions on atmospheric chemistry cannot be accurately assessed at present, although it seems likely to be substantial.

Sources five through nine in Table 15.2-1 are all combustion sources: biomass, coal, refuse, turbine, and wood. As with gasoline and diesel engines, these sources produce complex mixtures of saturated and unsaturated hydrocarbons, oxygenated compounds, and polycyclic aromatics. The exact compounds emitted reflect the feedstocks involved in the combustion processes.

Table 15.2-1. Sources That Emit Numerous Atmospheric Compounds

Ranking	Source	Number of Compounds
1	Tobacco smoke	461
2	Auto	459
3	Diesel	458
4	Vegetation	349
5	Biomass combustion	345
6	Coal combustion	283
7	Refuse combustion	260
8	Turbine	244
9	Wood combustion	151
10	Animal waste	117

The final source in the table is animal waste. The compounds consist mostly of small molecules produced by bacterial action, and include many sulfides, amines, and aldehydes.

It should be noted that the emission of large *numbers* of compounds does not necessarily imply the emission of large *fluxes* of compounds, nor does it indicate the impact of a source on atmospheric chemistry or atmospheric toxicity. Rather, the presence of a source in Table 15.2-1 indicates a combination of diverse emissions and extensive analytical study of those emissions, and serves as a guide to the probable molecular diversity of sources that have received less study.

Another topic of interest is the tabulation of the compounds which have the largest numbers of sources. Such compounds can be expected to be common atmospheric constituents as a result of this source diversity. The fifteen compounds with the most known sources are given in Table 15.2-2. Emission in condensed form is not precluded here, but it happens that all the compounds listed are emitted as gases. The compounds share several characteristics. First, they have relatively simple structures and low molecular weights (phenol is the highest, at 94; the average is 43). Second, they have both anthropogenic and natural sources. Third, they are emitted in reduced or intermediate oxidation states, so their

Table 15.2-2. Atmospheric Compounds with Multiple Sources

Rank	Identification Number	Name	Number of Sources	Typical Urban Conc. (ppb)	Atmospheric* Lifetime (days)
1	7.1-2	Acetaldehyde	34	5	1
2	3.5-2	Toluene	33	20	4
3	3.5-1	Benzene	32	25	20
4	2.3-2	Ammonia	30	5	150
4	2.4-1	Hydrogen sulfide	30	2	4
6	5.1-4	Ethanol	26	1(?)	8
6	6.1-1	Acetone	26	2	8
8	7.1-1	Formaldehyde	25	10	2
9	2.6-7	Carbon monoxide	24	10,000	70
9	5.4-1	Phenol	24	1	1
11	5.1-1	Methanol	23	4(?)	20
11	2.4-2	Carbonyl sulfide	23	0.5	$>10^3$
11	2.4-3	Carbon disulfide	23	0.2	$>10^4$
11	2.4-10	Sulfur dioxide	23	2	40
15	3.1-1	Methane	22	2,000	3,000
15	7.1-5	Propanal	22	2	1

* The lifetime estimates are based on the rate of reaction of the compounds with OH· and assume an OH· concentration of 5×10^5 radicals cm^{-3}.

emission implies atmospheric chemical reactions as well.

The fifth column in Table 15.2-2 lists typical urban concentrations for the compounds. The minimum concentrations are generally those of the more reactive compounds, and are a few parts per billion. In the last column we provide a rough estimate of the annually-averaged lifetimes of all of the compounds. The lifetimes are upper limits, since factors such as absorption onto droplets or surfaces and photolysis will in some cases also limit lifetimes to some degree. Compounds with long chemical lifetimes and low aqueous solubilities, i.e., carbon monoxide, methane, and carbon dioxide, and perhaps ethane and benzene, are most likely to have atmospheric concentrations that increase with time. Global concentrations of the first three of these compounds are in fact increasing; the concentrations of the latter two perhaps deserve to be monitored more closely.

Table 15.2-2 reiterates a point made several times elsewhere, that of the ubiquity and importance of the organic component of atmospheric chemistry: eleven of the sixteen compounds in the table are organic.

15.3 Sources of Indoor Compounds

In principle, the infiltration of outdoor air into buildings can result in the presence indoors of any of the outdoor atmospheric compounds. In practice, however, the trace constituents in indoor air are nearly all produced by very few sources. From the sources listed in Index I4, we have identified those present within residential and commercial buildings. The ten which produce the largest numbers of compounds are listed in Table 15.3-1.

Table 15.3-1. Indoor Sources That Emit Numerous Atmospheric Compounds

Ranking	Source	Number of Compounds
1	Tobacco smoke	461
2	Coal combustion	283
3	Wood combustion	151
4	Solvents	72
5	Building materials	58
6	Pesticides	44
7	Oil combustion	36

Tobacco smoke is by far the leading source in this list. Following it are sources related to the combustion of fuels for heating. A third grouping is that of emission from building materials, solvents, and pesticides. Emissions from building materials would be expected to decrease as the materials age; all the other sources reflect daily living practices and would thus be expected to repeat the same diurnal and seasonal patterns year after year.

15.4 Gas Phase Tropospheric Chemistry

Gas phase tropospheric chemistry involves surprisingly few of the elements. To illustrate this, we show in Figure 15.4-1 a "periodic table" for the tropospheric gas. This figure, and others like it on subsequent pages, was constructed from data in the tables of this book. It indicates that only sixteen elements participate in tropospheric gas phase chemistry (the noble gases being chemically inert in the atmosphere). Most of these elements are low molecular weight members of Groups IV-VII of the periodic table. Notable exceptions are hydrogen, mercury (which has a moderate vapor pressure), and lead (whose gas phase presence in several compounds is entirely anthropogenic in origin).

In the oxidizing atmosphere of the Earth, the virtually invariable thrust of the chemistry is in the direction of oxidizing any molecules that are emitted into the air. In the lower atmosphere, the principal oxidizing species are ozone (O_3) and the hydroxyl radical ($OH\cdot$). The former is produced in the troposphere by the photolysis of nitrogen dioxide

$$NO_2 \xrightarrow{h\nu} NO + O \quad (\lambda < 420 \text{ nm}) \tag{R15.4-1}$$

followed by

$$O + O_2 \xrightarrow{M} O_3 \, . \tag{R15.4-2}$$

Ozone is, however, also transported downward from the stratosphere following its formation from dissociated molecular oxygen.

Tropospheric destruction of ozone constitutes the primary source of hydroxyl radicals (Levy, 1971):

$$O_3 \xrightarrow{h\nu} O_2 + O(^1D) \; ; \; \lambda < 320 \text{ nm} \tag{R15.4-3}$$

$$O(^1D) + H_2O \rightarrow 2OH \, . \tag{R15.4-4}$$

TROPOSPHERIC GAS

1 H 291,294																	2 He 294,558
3	4											5 B 114,366	6 C 290,293	7 N 294,558	8 O 294,558	9 F 44,271	10 Ne 294,558
11	12											13	14 Si 597,628	15 P 147	16 S 23,100	17 Cℓ 388,429	18 Ar 294,558
19	20	21	22	23	24	25	26	27	28	29	30	31	32	33 As 112,367	34 Se 854	35 Br 373,462	36 Kr 294,558
37	38	39	40	41	42	43	44	45	46	47	48	49	50	51	52	53 I 135,322	54 Xe 294,558
55	56	57-71	72	73	74	75	76	77	78	79	80 Hg 388,529	81	82 Pb 988,989	83	84	85	86
87	88	89ff															

57	58	59	60	61	62	63	64	65	66	67	68	69	70	71
89	90	91	92	93	94	95								

Fig. 15.4-1 A periodic table for the tropospheric gas. The symbols of the elements found in the gas are displayed, together with reference numbers for publications reporting the detection.

A second $OH\cdot$ source is from oxygenated organic compounds. Formaldehyde (HCHO), the simplest example, photolyzes to produce hydroperoxyl radicals ($HO_2\cdot$):

$$HCHO \xrightarrow{h\nu} H\cdot + CHO\cdot \; ; \; \lambda < 335 \text{ nm} \qquad (R15.4\text{-}5)$$

$$H\cdot + O_2 + M \rightarrow HO_2\cdot + M \qquad (R15.4\text{-}6)$$

$$CHO\cdot + O_2 \rightarrow HO_2\cdot + CO \qquad (R15.4\text{-}7)$$

The $HO_2\cdot$ radicals are a source of $OH\cdot$ by disproportionation,

$$HO_2\cdot + HO_2\cdot \rightarrow H_2O_2 + O_2 \qquad (R15.4\text{-}8)$$

photodissociation of the resulting hydrogen peroxide,

$$H_2O_2 \xrightarrow{h\nu} OH\cdot + OH\cdot \; ; \; \lambda < 350 \text{ nm} \qquad (R15.4\text{-}9)$$

and by oxidation of nitric oxide:

$$NO + HO_2\cdot \rightarrow NO_2 + OH\cdot \; . \qquad (R15.4\text{-}10)$$

The alkylperoxyl radicals ($RO_2\cdot$) produced by similar chemical chains from the higher aldehydes and from ketones also generate alkoxyl ($RO\cdot$) radicals and oxidize NO. The presence of the hydroxyl radical in the atmosphere is central to virtually all gas phase tropospheric chemistry. As has been discussed in previous chapters, most molecules emitted into the atmosphere react with $OH\cdot$, the product being itself a reactive radical. The central role of $OH\cdot$ is evident from Figure 15.4-2. The interaction of the $OH\cdot$ oxidation chains with the variety of atmospheric species is most easily illustrated by examining the oxidation of methane, the simplest hydrocarbon. Its role is a crucial one, since methane oxidation is not only a potentially significant source of CO, but the atmospheric concentrations of $OH\cdot$ may largely be determined by the particular reaction paths that are followed during methane oxidation.[1278] Nitric oxide plays an important part in this process, as can be demonstrated by following simplified representations of the oxidation sequences. With enough NO present (> 10 ppt), the reaction path leading to formaldehyde (HCHO) is as follows:

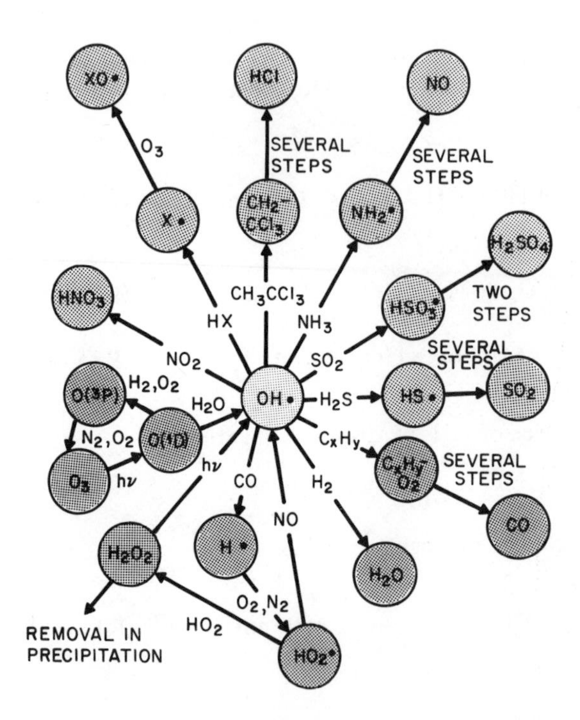

Fig. 15.4-2 The photochemistry of the free hydroxyl radical controls the rate at which many trace gases are oxidized and removed from the atmosphere. Processes that are of primary importance in controlling the concentration of OH· in the troposphere are indicated in dense hatching in the schematic diagram; those that have a negligible effect on OH· levels but are important because they control the concentrations of the associated reactants and products are indicated in lighter hatching. Circles indicate reservoirs of species in the atmosphere; arrows indicate reactions that convert one species to another, with the reactant or photon needed for each reaction indicated along each arrow. Multistep reactions actually consist of two or more sequential elementary reactions. HX = HCl, HBr, HI, or HF. C_xH_y denotes hydrocarbons. (Reproduced with permission.[1227])

$$
\begin{array}{lll}
CH_4 + OH\cdot & \rightarrow & CH_3\cdot + H_2O \\
CH_3\cdot + O_2 + M & \rightarrow & CH_3O_2\cdot + M \\
CH_3O_2\cdot + NO & \rightarrow & CH_3O\cdot + NO_2 \\
CH_3O\cdot + O_2 & \rightarrow & HCHO + HO_2\cdot \\
HO_2\cdot + NO & \rightarrow & OH\cdot + NO_2 \\
2(NO_2 + h\nu & \rightarrow & NO + O\ (< 420\ \text{nm})) \\
2(O + O_2 + M & \rightarrow & O_3 + M) \\
\hline
\text{net:}\quad CH_4 + 4O_2 & \rightarrow & HCHO + H_2O + 2O_3
\end{array}
$$

With little NO present, CH_4 oxidation may follow the pathway:

$$
\begin{array}{lll}
 & CH_4+OH\cdot & \rightarrow CH_3\cdot+H_2O \\
 & CH_3\cdot+O_2+M & \rightarrow CH_3O_2\cdot+M \\
 & CH_3O_2\cdot+HO_2\cdot & \rightarrow CH_3O_2H+O_2 \\
 & CH_3O_2H+h\nu & \rightarrow CH_3O\cdot+OH\cdot \\
 & CH_3O\cdot+O_2 & \rightarrow HCHO+HO_2\cdot \\
\hline
\text{net:} & CH_4+O_2 & \rightarrow HCHO+H_2O
\end{array}
$$

However, the photolysis of methyl hydroperoxide (CH_3O_2H) is slow, which results in a residence time of about one week for this compound. Thus, the methyl hydroperoxide may be rained out of the atmosphere or react with the earth's surface or aerosol particles. If this is the case, the oxidation of CH_4 will lead to a loss of two odd hydrogen radicals ($OH\cdot$ and $HO_2\cdot$) and no formation of CO would occur.

Depending on the concentrations of NO, the further oxidation of HCHO to CO and CO_2 may proceed largely by one of two routes, e.g.,

$$
\begin{array}{lll}
 & HCHO+h\nu & \rightarrow H\cdot+CHO\cdot \\
 & H\cdot+O_2+M & \rightarrow HO_2\cdot+M \\
 & CHO\cdot+O_2 & \rightarrow HO_2\cdot+CO \\
 & CO+OH\cdot & \rightarrow H\cdot+CO_2 \\
 & H\cdot+O_2+M & \rightarrow HO_2\cdot+M \\
3 & (HO_2\cdot+NO & \rightarrow OH\cdot+NO_2) \\
3 & (NO_2+h\nu & \rightarrow NO+O) \\
3 & (O+O_2+M & \rightarrow O_3+M) \\
\hline
\text{net:} & HCHO+6O_2 & \rightarrow CO_2+2OH\cdot+3O_3
\end{array}
$$

or

$$
\begin{array}{lll}
 & HCHO+h\nu & \rightarrow H\cdot+CHO\cdot \\
 & H\cdot+O_2+M & \rightarrow HO_2\cdot+M \\
 & CHO\cdot+O_2 & \rightarrow CO+HO_2\cdot \\
 & CO+OH\cdot & \rightarrow H\cdot+CO_2 \\
2 & (H\cdot+O_2+M & \rightarrow HO_2\cdot+M) \\
3 & (HO_2\cdot+O_3 & \rightarrow OH\cdot+2O_2) \\
\hline
\text{net:} & HCHO+3O_3 & \rightarrow CO_2+3O_2+2OH\cdot
\end{array}
$$

The important implications are that in the presence of enough NO there is a net production of five ozone molecules and two hydroxyl radicals per methane molecule oxidized. Otherwise, a net loss of three ozone molecules but production of two hydroxyl radicals may occur. The latter result will depend on the efficient removal of CH_3O_2H by rainfall or attachment to aerosol.

Atmospheric chemical reactions occur within a fluid in constant motion, as discussed in Chapter 1. An example of this property is provided by the $NO/NO_2/O_3$ "reaction triad," consisting of the following process:

$$NO_2 \xrightarrow{h\nu} NO + O \; ; \; (\lambda < 420 \text{ nm}) \qquad \text{(R15.4-1)}$$

$$O + O_2 \rightarrow O_3 \qquad \text{(R15.4-2)}$$

$$O_3 + NO \rightarrow NO_2 + O_2 \qquad \text{(R15.4-11)}$$

net: no reaction

A typical circumstance, especially near urban areas, is the emission of NO from vehicles with concomitant reduction of O_3 concentrations (R15.4-11), followed by reformation of O_3 downwind. A particularly vivid example is shown in Figure 15.4-3, where high ozone concentrations are seen to occur some one hundred miles downwind from the New York City metropolitan region.

Much of gas phase atmospheric chemistry can be summarized by schematically depicting the oxidation of hydrocarbons, as shown in Figure 15.4-4. In most cases the process is initiated by OH· attack, although ozone attack on alkenes and aromatics can sometimes be important as well. The product of the initial reaction is a free radical, which may (box I on the figure) or may not (II) contain an OH group. The next step is generally the addition of oxygen to form a peroxyl radical (III). If low concentrations of NO are present, the reaction of the $RO_2\cdot$ radical with $HO_2\cdot$ will produce an alkylhydroperoxide (IV). Alternatively, $RO_2\cdot$ radicals may react with NO to form alkoxy radicals (V). In either case, ketone (VI) and aldehyde (VII) products are generated. The final carbon-containing product of the oxidation sequence is CO_2, which is most efficiently produced from photosensitive aldehydes as shown. An alternative reaction sequence for HOR· radicals involves the production of organic nitro compounds (VIII); these eventually degrade in a manner similar to that of the simpler organic species. Nitrogen chemistry is thus crucial to tropospheric gas phase processes; in contrast, that of sulfur and

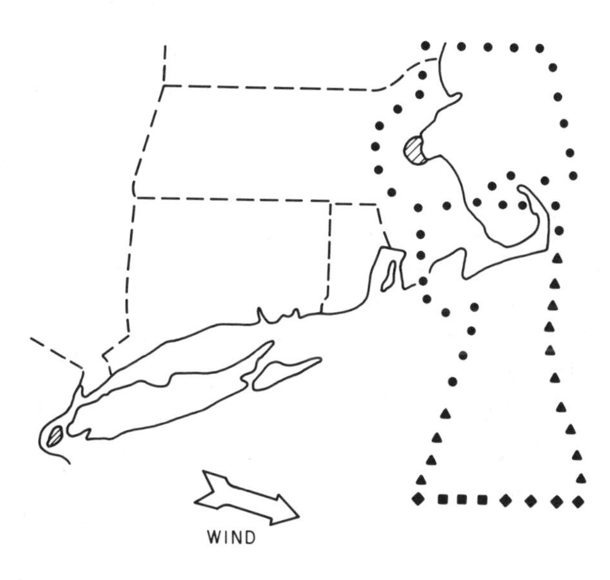

Fig. 15.4-3 Ozone concentrations (ppb) at 330 to 490 m altitude off the coast of the northeastern United States on the afternoon of 14 August 1975. The highest concentrations were seen about 250 km east of the New York City metropolitan complex, a region of high precursor emission fluxes. Trajectory analysis demonstrated that the high ozone air mass passed over the metropolitan area during the morning of the day on which measurements were made. (Adapted from Siple et al.[1331])

chlorine is almost completely independent of other chemical cycles except for interactions with $HO_x\cdot$ radicals.

On time scales of hours or days, atmospheric trace gas concentrations may vary because of fluctuations in emissions, local meteorological patterns, and concentrations of oxidizing species. On time scales of a year or longer, concentration changes indicate an imbalance between emission and removal rates for a molecule of interest. In several cases, anthropogenic emissions are known to be causing upward concentration trends. Examples of such behavior are shown in Figures 15.4-5, 6 and 7, for molecules whose tropospheric removal processes are slow or nonexistent. These gases and others absorb infrared radiation; their increasing concentrations have the potential to change the heat balance and hence the climate of the earth.

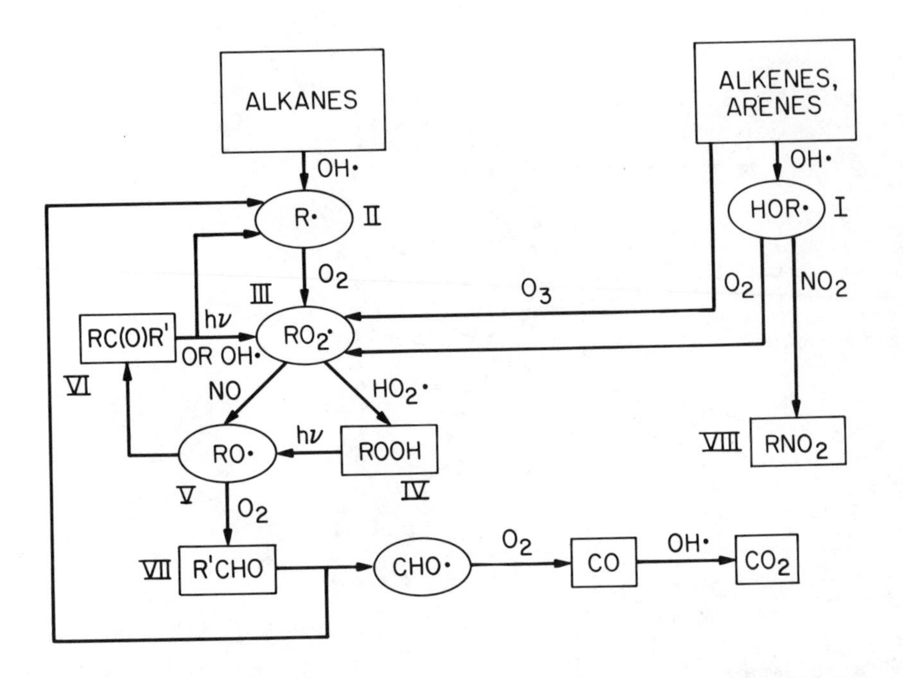

Fig. 15.4-4. The general sequence of gas-phase hydrocarbon oxidation in the troposphere. The symbol R refers to alkyl moieties, within which at certain stages of the sequence OH or other oxygenated ligands may be included. R′ refers to a similar but smaller sized group. Stable compounds appear in rectangles and transient intermediates in ovals. See text for discussion of this figure.

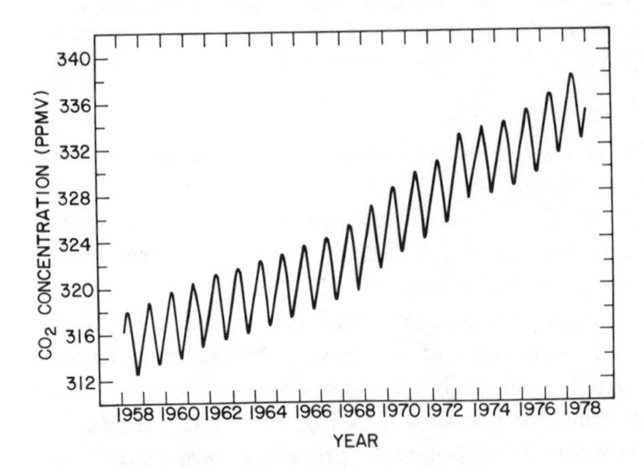

Fig. 15.4-5. The increasing atmospheric concentration of carbon dioxide, measured at Mauna Loa Observatory, Hawaii. (Reproduced with permission.[1316])

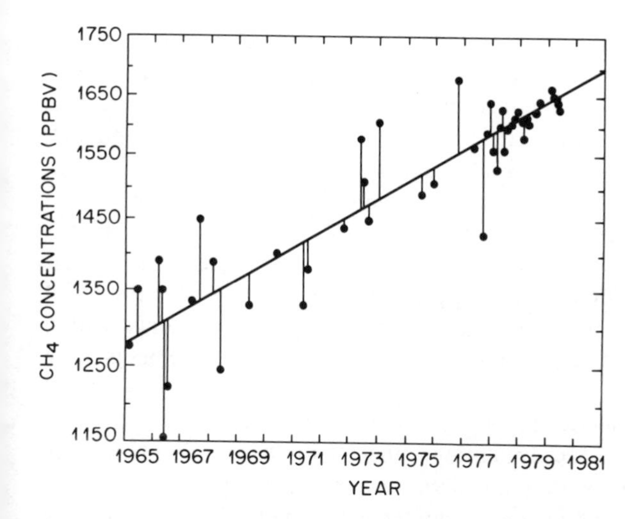

Fig. 15.4-6 The increasing atmospheric concentration of methane, measured at different sites in the northern hemisphere. (Reproduced with permission.[1317])

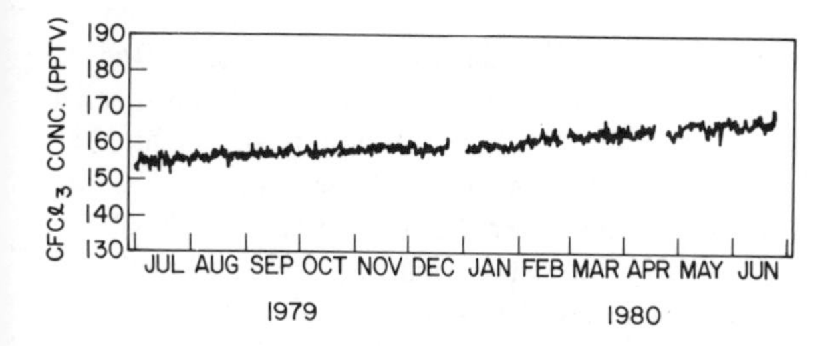

Fig. 15.4-7 The increasing atmospheric concentration of trichlorofluoromethane, measured at Cape Grim, Tasmania. (Reproduced with permission.[1318])

15.5 Aerosol Phase Tropospheric Chemistry

Atmospheric aerosol particles are the most easily noticed of the atmospheric trace constituents, since their ability to decrease visibility and their deposition on surfaces are widely appreciated. The principal sources of the particles were briefly summarized in the first chapter. The conversion of gaseous molecules to particle constituents is thought to be the largest (by mass) of the sources, followed fairly closely by sea salt generated by oceanic waves. The fluxes of particles from windblown dust, leaf litter, coal combustion, biomass combustion, and industrial operations are much smaller. They have little impact on the global aerosol budget, but may be important in regional or local situations.

Nearly all of the chemical elements are found in the earth's crust. As a result, nearly all are found in atmospheric aerosol particles as well. This is seen in Figure 15.5-1, where the only significant missing entries are the noble gases. Since particles frequently serve as nucleating sites for cloud droplets, one would expect to find the full range of elements within precipitation, but thus far only the more abundant of the elements have been reported there.

The sizes of atmospheric aerosol particles in the atmosphere are not static. Particles in the accumulation mode ($<0.05\ \mu m$ diameter, see Figure 1.1-9) coagulate with each other, growing rapidly in size until a diameter of several tenths of a micrometer is reached. At this size they are insufficiently mobile for further coagulation to be propitious, yet too small for gravitational setting to be significant.

During their lifetime in the atmosphere, several days at least, aerosol particles are exposed to atmospheric water vapor, which they readily adsorb. Several quite different studies[1293–95] have demonstrated that at relative humidities above 40 percent the aerosol water content can often be at least 30 percent of the total particle weight. Figure 15.5-2 shows such data over a fourteen hour period, the aqueous fraction being as high as 60 percent. The significance of this result can be appreciated by considering an individual particle of diameter $0.6\ \mu m$. If the weight of the water shell is thirty percent of the total weight, the shell will be approximately 150 monolayers thick. If the weight is only ten percent of the total, the shell will still be 50 monolayers thick. Such a shell is of ample thickness to be regarded as an aqueous solution in contact with an undissolved or partially dissolved core. As the ambient humidity changes, this shell will become thicker or thinner, but will always be present at moderate thicknesses under most atmospheric conditions. As well as providing a dynamic chemical environment at the aerosol surface, the varying water content affects aerosol microphysical properties such as the instantaneous settling velocity, the coagulation rate, etc.

In addition to its evolution in size, the aerosol particle evolves chemically. This is illustrated by Figure 15.5-3, which shows aerosol chemical analyses at different hours during the day. All of the measured components (ammonium, nitrate, and sulfate ions and total organics) show marked changes from hour to hour.

1 H 280, 476																	2
3 Li 280, 476	4 Be 280, 476											5 B 280,476	6 C 280, 476	7 N 280, 476	8 O 280, 476	9 F 280, 476	10
11 Na 280, 476	12 Mg 280, 476											13 Aℓ 280, 476	14 Si 280, 476	15 P 280, 476	16 S 280, 476	17 Cℓ 280, 476	18
19 K 280, 476	20 Ca 280, 476	21 Sc 280, 476	22 Ti 280, 476	23 V 280, 476	24 Cr 280, 476	25 Mn 280, 476	26 Fe 280, 476	27 Co 280, 476	28 Ni 280, 476	29 Cu 280, 476	30 Zn 280, 476	31 Ga 280, 476	32 Ge 280, 476	33 As 280, 476	34 Se 280, 476	35 Br 280, 476	36
37 Rb 280, 476	38 Sr 280, 476	39 Y 280, 476	40 Zr 280, 476	41 Nb 280, 476	42 Mo 280, 476	43 Te 482	44 Ru 476	45 Rh 476	46 Pd 476	47 Ag 280, 476	48 Cd 280, 476	49 In 476, 477	50 Sn 280, 476	51 Sb 280, 476	52 Te 280, 476	53 I 280, 476	54
55 Cs 280, 476	56 Ba 280, 476	57 - 71	72 Hf 280, 476	73 Ta 280, 476	74 W 280, 476	75 Re 280, 476	76 Os 476	77 Ir 476, 478	78 Pt 476, 544	79 Au 476, 478	80 Hg 280, 476	81 Tℓ 280, 476	82 Pb 280, 476	83 Bi 280, 476	84 Po 479	85	86 Ru 345, 483
87	88 Ra	89ff															

57 La 280, 476	58 Ce 280, 476	59 Pr 280, 476	60 Nd 280, 476	61 Pm 484	62 Sm 280, 476	63 Eu 280, 476	64 Gd 280, 476	65 Tb 280, 476	66 Dy 280, 476	67 Ho 280, 476	68 Er 280, 476	69 Tm 280, 476	70 Yb 280, 476	71 Lu 280, 476
89	90 Th 280, 483	91	92 U 280, 476	93 Np 1047	94 Pu 481, 485	95 Am 1047								

Fig. 15.5-1 A periodic table for tropospheric aerosol particles. The symbols of the elements found in particles are displayed, together with reference numbers for publications reporting the detection.

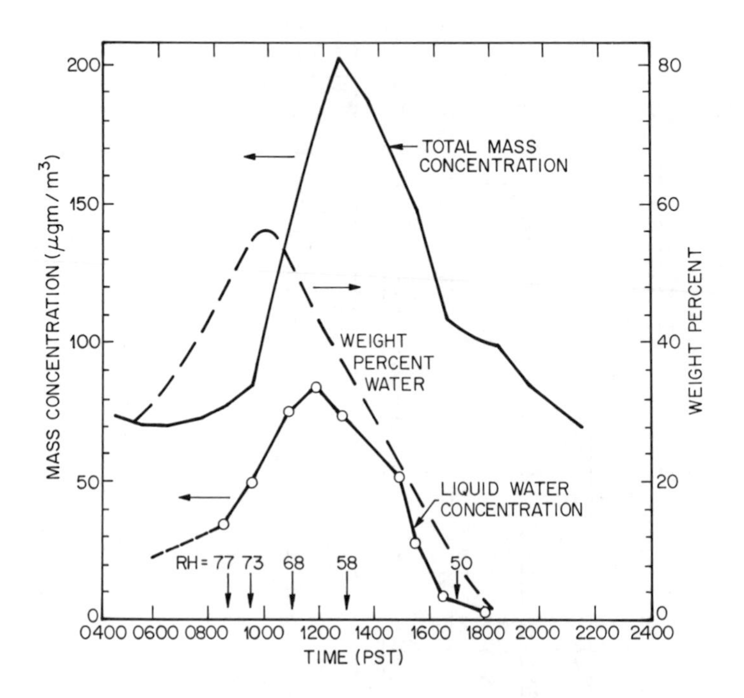

Fig. 15.5-2 The diurnal variation in liquid water concentration in atmospheric aerosols in Pasadena, CA, September 15, 1972. Note the dependence of the fractional water content on relative humidity, indicated at the bottom of the figure. (Adapted from Ho et al.[1295])

On the basis of information in this book and elsewhere, it is possible to say a good deal about the insoluble core of a typical aerosol particle. In urban areas the most common cores are primarily soot or fly ash. Fly ash is largely comprised of inorganic oxides; chemically labile transition metal atoms are present on the exterior of the fly ash particles.[1297] Soluble inorganic salts such as ammonium nitrate are also present. In less urban environments, soil constituents, both inorganic and organic, are common. The organic material contains a variety of lipids derived from vegetation. Also present are humic and fulvic acids, usually defined as the portion of soil organic matter soluble in dilute base but not in mineral acid or alcohol. Humic and fulvic acids are macromolecules with carboxyl, phenolic, and aliphatic hydroxyl groups; they are degradation products of plant tissue.[1298]

The diversity of oxidation states for elements present in aerosol particles is indicated by Figures 15.5-4 and 15.5-5, which show sulfur valence ranging from −2 to +6 and nitrogen valence ranging from −3 to +5. The

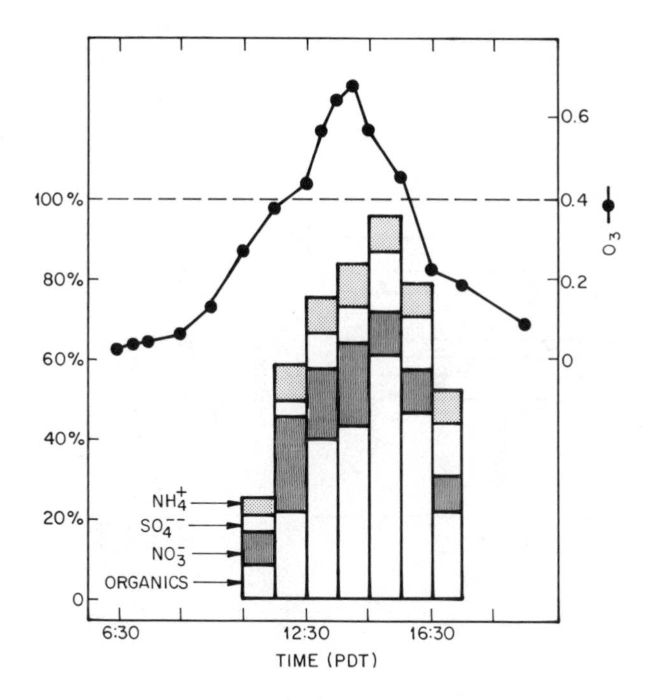

Fig. 15.5-3 Diurnal concentration patterns of nitrate, sulfate, and ammonium ions, total aerosol organics (as weight percent of the total dry aerosol), and ozone in Pasadena, CA on July 25, 1973. Concurrent ozone concentrations are also shown, since the generation of aerosol ions and molecules as well as the ozone concentrations are related to the vigor of the gas phase chemistry. (Reproduced with permission.[1296])

presence of inorganic, aliphatic, and heterocyclic compounds are all indicated.[1243]

The liquid water shell provides a concentrated chemical environment for reaction processes. Calculations suggest that ionic strengths are very high, perhaps of the order of 5 to 20.[1299] A few measurements of aerosol pH give values of 3.5-5.5,[1300] a further indication of a dynamic chemical environment. Finally, evidence suggests the frequent presence of organic surface films on aerosol particles.[1210] If such films occur, they may partially or totally inhibit exchange between atmospheric gas and condensed phases, thus creating small isolated reaction vessels in the atmosphere.

The evidence for chemical reactions within atmospheric aerosol particles is circumstantial but persuasive. One might first consider the case of sulfate ions. They are ubiquitous in aerosol particles, their concentration

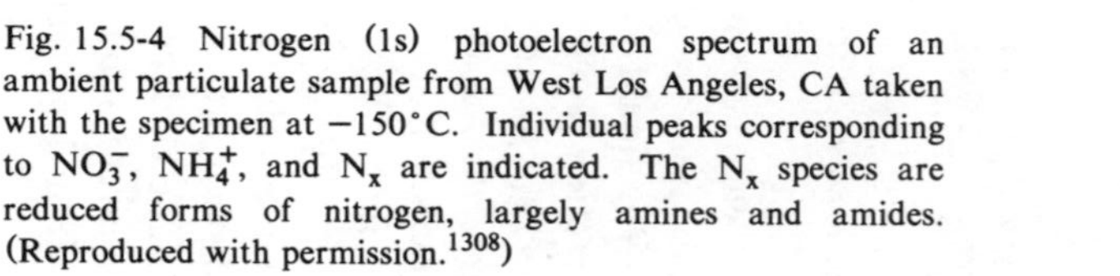

Fig. 15.5-4 Nitrogen (1s) photoelectron spectrum of an ambient particulate sample from West Los Angeles, CA taken with the specimen at $-150°C$. Individual peaks corresponding to NO_3^-, NH_4^+, and N_x are indicated. The N_x species are reduced forms of nitrogen, largely amines and amides. (Reproduced with permission.[1308])

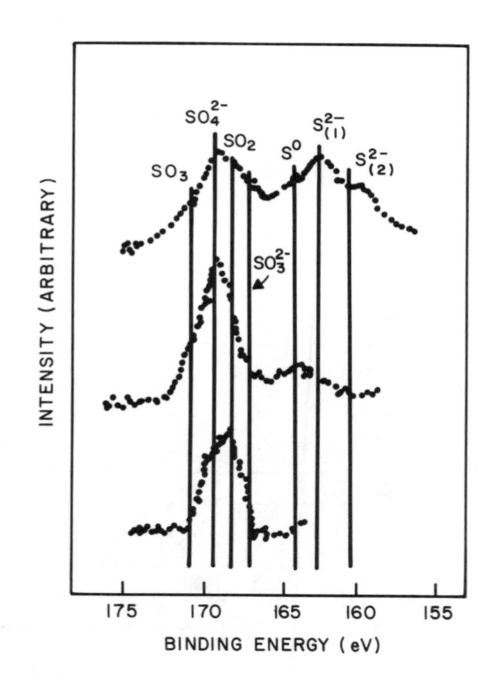

Fig. 15.5-5 Sulfur (2p) photoelectron spectra of ambient particulate samples from various locations in California. Individual peaks corresponding to SO_3, SO_4^{2-}, SO_3^{2-}, S^0, and two kinds of S^{2-} ions are indicated. (Reproduced with permission.[1309])

evolves with time, SO_2 (the precursor) is highly soluble in the water shell, and oxidation of dissolved SO_2 to sulfate is catalyzed by both soot and transition metals. These facts do not preclude formation of sulfuric acid in the gas phase followed by incorporation into particles, but suggest that oxidation within particles is likely to be very efficient and thus to play at least a minor role.

Oxidation of organic compounds in the condensed phase is also likely. In Figure 15.5-6 we show the evolution of aldehydes and carboxylic acids in the Los Angles basin in both gas and condensed phases. The time correlation between the phases is striking and suggests that the aldehydes are acid precursors. Such a chemical process is entirely consistent with the simultaneous presence of long-chain alkanes, alkanols, and carboxylic acids in the particles. The production of ketones, quinones, and nitro compounds from polynuclear aromatic hydrocarbons absorbed on atmospheric particles (see Chapter 3) is also evidence for hydrocarbon oxidation in the condensed phase.

With the evidence from this discussion, we can now proceed to sketch a typical physical and chemical cycle for atmospheric aerosol particles. The cycle is illustrated in Figure 15.5-7. Its first step is the emission of a condensable gas (SO_2 is the best example), a precursor to a condensable gas (e.g., H_2S), or a small particle. The precursor gas or condensible gas undergoes chemical reactions which result in condensation; we term the product the "sulfate aerosol." Sources for particles are of three types. The first is sea spray, which produces a droplet reminiscent of seawater, but often with enhanced trace metals and organics (such as fulvic or humic acids) from the sea surfaces microlayer. The resulting aerosol we term the "chloride aerosol." A second set of sources are natural ones which generate particles by mechanical processes on solid surfaces, leaf litter and windblown dust being good examples. Leaf litter is almost entirely organic, with C_{16}-C_{30} lipids predominating. Windblown dust is primarily composed of inorganic oxides, some humus or other organic material being common as well. We term these emittants the "organic aerosol." A third set of particle sources are those involving combustion of fossil fuels or biomass to generate soot, the "carbon aerosol."

Regardless of source, the aerosol particles agglomerate with each other and incorporate water vapor to arrive at a form typified by the diagram at the upper left of the figure. Here the particle has a solid core covered by an aqueous shell comprising perhaps thirty percent of the total particle mass. Atmospheric gases which are highly water soluble will then dissolve in this shell to form a complex and chemically reactive solution. Most atmospheric gases are water-soluble to some degree, but we have indicated six whose dissolution appears particularly likely.

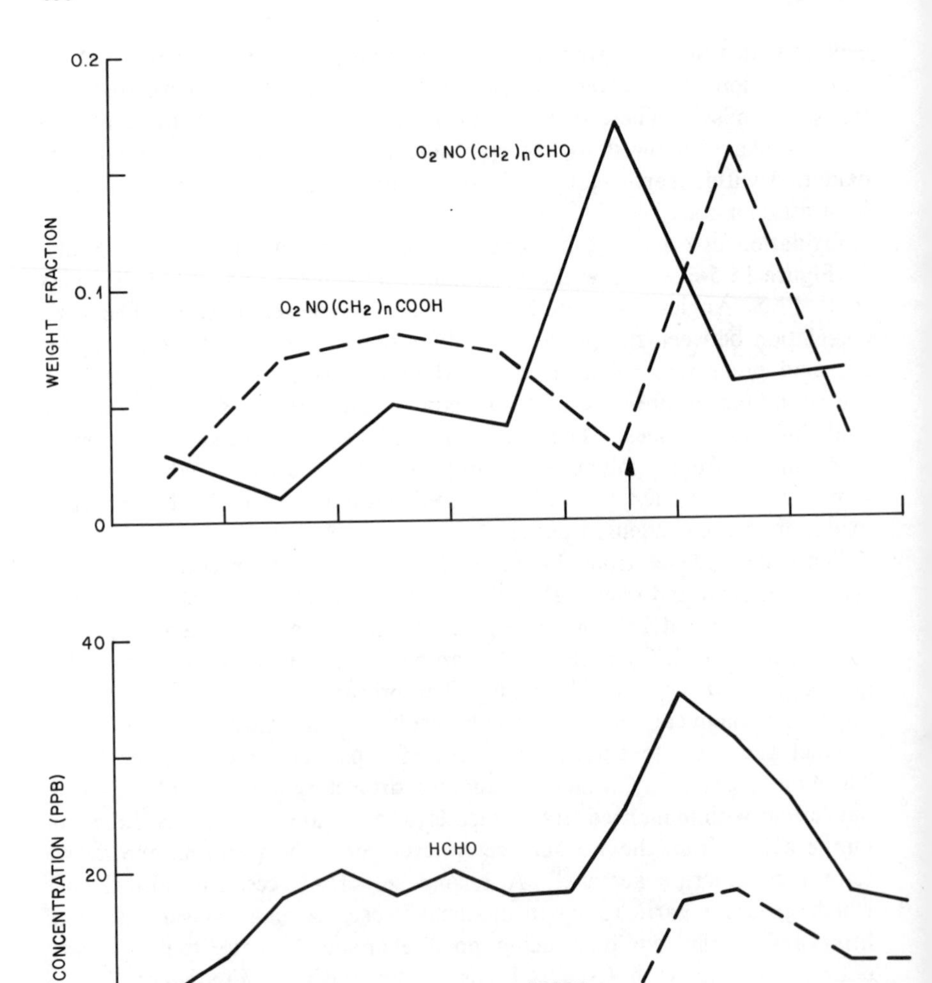

Fig. 15.5-6 The diurnal variation of airborne aldehydes and acids in the Los Angeles basin. The bottom diagram is for gaseous formaldehyde and formic acid in Claremont on Oct. 13, 1978.[1268] The top diagram is for summed aerosol aldehydic nitrate and acidic nitrate in West Covina on July 24, 1973.[1269] The ordinate on the latter plot is the weight fraction of the summed species relative to total alkanes and alkenes. The arrow on each diagram indicates the time of ozone maximum, a good measure of the peak time of smog chemistry. In each case the ozone maximum and aldehyde maximum coincide and precede the acid maximum by one to two hours.

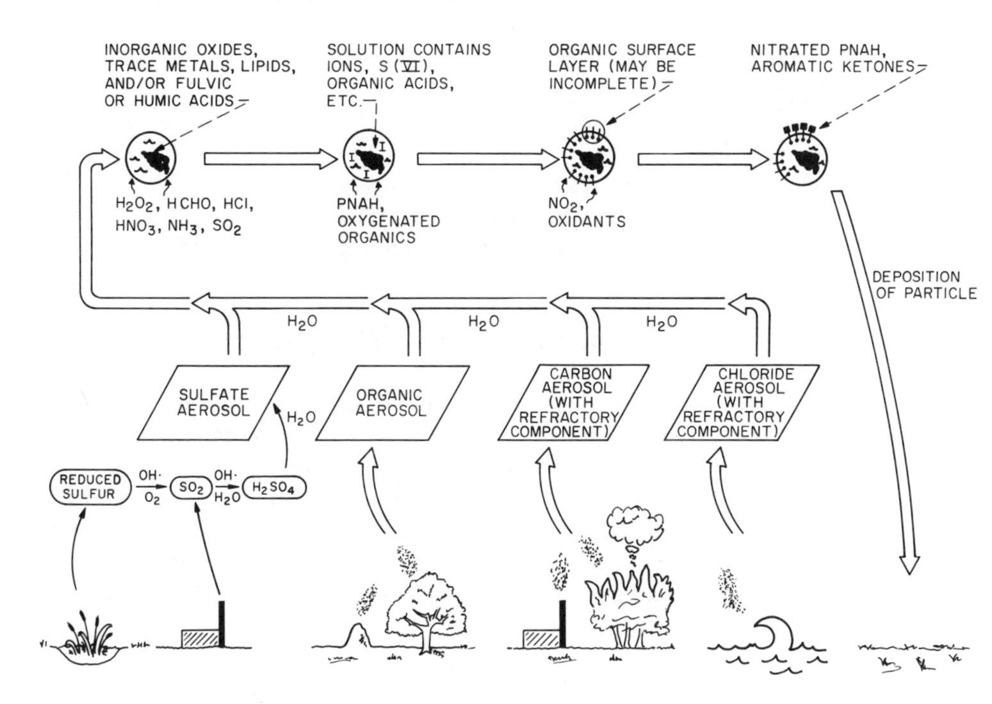

Fig. 15.5-7 A schematic cycle for atmospheric aerosol particles. The individual processes are discussed in the text. Particle deposition can occur at any stage but, for simplicity, is shown only for the last.

In the next stage of the particle cycle (top, second from left), species that are relatively insoluble begin to accumulate on the particle surface. Such a process is often favored when a relatively large molecule has both hydrophilic and hydrophobic parts; lipids such as decanol are good examples. An organic surface layer may thus form on the particle. Meanwhile, chemical reactions within the drop oxidize S(IV) to S(VI), aldehydes to organic acids, etc.

In the next stage (top, third from left), the organic layer rather than a water surface presents itself to gaseous molecules. If the organic layer does not react with the molecule (as with NH_3, say), the flux of the gaseous molecules into the particle will be depleted. If reactions do occur, however, some of the molecules in the organic layer are transformed. Two processes whose occurrence is reasonably well established are the formation of nitrated arenes by NO_2 reaction with PNAH and the formation of ketones from arenes. This produces a particle with the general characteristics shown in the top right of Figure 15.5-7.

The final step in the cycle is the deposition of the particle on a surface at the ground. (Deposition may, of course, occur at any of the stages). With only a few exceptions, it is these deposited particles that are available for analysis.

The complete cycle thus involves particle generation, growth by agglomeration and/or condensation, incorporation and chemical processing of atmospheric gases, and, finally, deposition to the earth's surface.

15.6 The Chemistry of Precipitation

The study of chemical reactions in cloud and fog droplets, raindrops, and snowflakes is in its infancy. As this is written, it appears that one can identify some of the major products of this chemistry, but the reaction paths that link those products to their precursors are known with considerably less confidence. It is clear, however, that chemical processes in liquid water droplets are governed in large part by the prevailing acidity. This is initially established by equilibration of the droplet with atmospheric carbon dioxide

$$CO_{2(g)} \rightleftarrows CO_{2(aq)}$$

$$CO_{2(aq)} + H_2O \rightleftarrows H^+ + HCO_3^- .$$

The addition of the strong mineral acids raises the acidity (i.e., lowers the pH), since the acid equilibria produce protons and the corresponding anions:

$$H_2SO_4 \rightleftarrows H^+ + HSO_4^-$$

$$HSO_4^- \rightleftarrows H^+ + SO_4^{2-}$$

$$HNO_3 \rightleftarrows H^+ + NO_3^-$$

$$HCl \rightleftarrows H^+ + Cl^-$$

Similar processes occur with the organic acids known to be present in the droplets

$$RC(O)OH \rightleftarrows H^+ + RC(O)O^- ,$$

but the equilibria are such that these acids are only partly ionized.

Once the electrolytic properties and the pH of the solution are fixed, the equilibria between the liquid and gas phases become established and further reactions will be governed by the oxidizers present. Hydrogen peroxide, which may be the most important, serves both as an oxidant and as a source of solution hydroxyl radicals when it is photolyzed

$$H_2O_2 \xrightarrow{h\nu} HO\cdot + HO\cdot \ ; \ \lambda < 380 \ \text{nm}$$

Ozone dissolved in water is a potential source of hydroxyl radicals as well:

$$O_3 + HO_2\cdot \rightarrow HO\cdot + 2O_2$$

$$O_3 \xrightarrow{h\nu} O_2 + O(^1D) \ ; \ \lambda < 320 \ \text{nm}$$

$$O(^1D) + H_2O \rightarrow H_2O_2$$

Ammonia [and the ammonium ion (NH_4^+)] and nitric acid [and the nitrate ion (NO_3^-)] are the most important inorganic nitrogen compounds in atmospheric water droplets. Ammonia is the principal species that neutralizes the strong acid anions, as evidenced by the large concentrations of ammonium salts found in aerosols. This leads to a lowering of the proton concentrations (acidity) of rainfall.

Sulfur compounds are also important constituents of atmospheric aqueous solutions. The most abundant are the sulfates, including sulfuric acid [H_2SO_4], ammonium hydrogen sulfate [NH_4HSO_4], and ammonium sulfate [$(NH_4)_2SO_4$]. In addition, SO_2 is soluble in water and chemically active. Its principal reactions (in the customary bisulfite form) are probably with H_2O_2 and O_3 to produce sulfate.

Transition metals are not abundant atmospheric species, but their ability to function as catalysts in the liquid phase oxidation of SO_2 to sulfuric acid makes them potentially important.[1274] They enter the atmosphere in particulate form, generally through emissions from smelters or coal combustion. In the latter case, at least, the transition metal atoms tend to be concentrated at the particle surfaces,[1275] so that their effects may be much more significant than their concentrations would indicate. The role of soot particles in SO_2 oxidation reactions is also potentially important,[1274] but the mechanism and magnitude of the effect are still not well enough known.

Oxidized organic compounds may turn out to be important participants in atmospheric solution chemistry, but have thus far received relatively little study. The research that has been done indicates that carboxylic acids are common. Organic hydroperoxides (ROOH) are readily formed in the gas phase, and may be precursors for carboxylic acids in atmospheric droplets. Precursors more likely to be important are the aldehydes (RCHO), which are hydrolyzed in solution to the glycol form ($RCH(OH)_2$). Oxidation of these precursors will lead to the carboxylic acids, but the precise mechanisms remain to be defined.

In solution, the equilibria are such that carboxylic acids are only partly ionized and are termed "weak". Nonetheless, they can be the major acid constituents of atmospheric water droplets if the precursors of the inorganic acids are present only at very low concentrations. Such conditions exist in more remote parts of the earth in biologically productive regions such as the tropical rainforests.

Present knowledge does not permit us to say definitively whether incorporation or solution oxidation is more important for the generation of acids found in clouds and rain. It is likely that both processes operate in all situations, with different ones becoming dominant under different conditions. Current evidence suggests that solution reactions are often important for the formation of sulfuric acid[1274] and less important for nitric acid.[1276] However, during the nighttime hours $NO_3\cdot$ and N_2O_5 are formed in the gas phase; their uptake by water droplets and wet aerosol surfaces will lead to the formation of nitric acid. The organic acids have received too little study for a dominant mechanism to be specified.

The discussion of precipitation chemistry presented above does not encompass very many of the chemical elements, but many are, in fact, found in precipitation phases. This is illustrated by Figure 15.6-1, which shows that more than thirty of the elements have been detected in rain. Virtually every element found in volatile atmospheric gases is included, clearly suggesting that gas to liquid transfer is an important process. Another group of interest is the first row transition metals. These are found as particle components in rain and presumably occur there as a consequence of droplet condensation around particles emitted by coal combustion, smelting, or other industrial processes. Mg(atomic number 12), Si(14), P(15), and Ti(21), among other elements, are common in aerosols and must also be present in raindrops, but their presence has not yet been reported.

More than forty of the elements have been detected in snow (Figure 15.6-2), perhaps because longer exposure to the atmosphere than

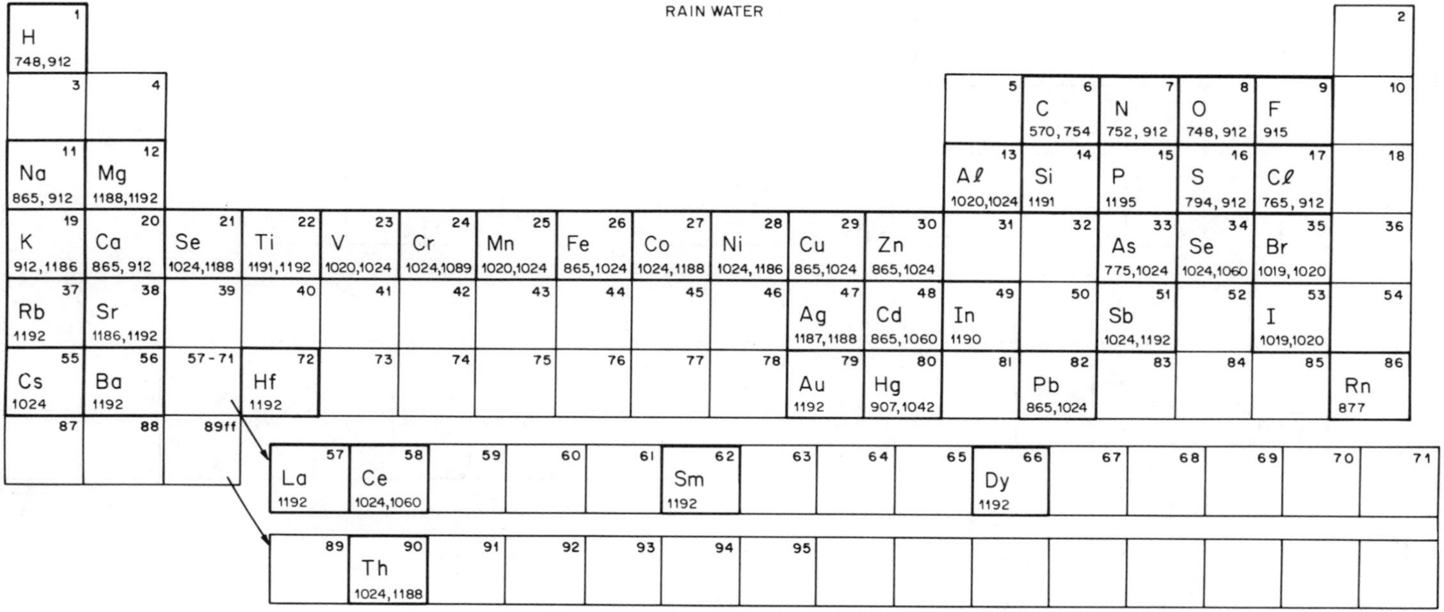

1 H 748,912																	2
3	4											5	6 C 570,754	7 N 752,912	8 O 748,912	9 F 915	10
11 Na 865,912	12 Mg 1188,1192											13 Aℓ 1020,1024	14 Si 1191	15 P 1195	16 S 794,912	17 Cℓ 765,912	18
19 K 912,1186	20 Ca 865,912	21 Se 1024,1188	22 Ti 1191,1192	23 V 1020,1024	24 Cr 1024,1089	25 Mn 1020,1024	26 Fe 865,1024	27 Co 1024,1188	28 Ni 1024,1186	29 Cu 865,1024	30 Zn 865,1024	31	32	33 As 775,1024	34 Se 1024,1060	35 Br 1019,1020	36
37 Rb 1192	38 Sr 1186,1192	39	40	41	42	43	44	45	46	47 Ag 1187,1188	48 Cd 865,1060	49 In 1190	50	51 Sb 1024,1192	52	53 I 1019,1020	54
55 Cs 1024	56 Ba 1192	57-71	72 Hf 1192	73	74	75	76	77	78	79 Au 1192	80 Hg 907,1042	81	82 Pb 865,1024	83	84	85	86 Rn 877
87	88	89ff															

57 La 1192	58 Ce 1024,1060	59	60	61	62 Sm 1192	63	64	65	66 Dy 1192	67	68	69	70	71
89	90 Th 1024,1188	91	92	93	94	95								

Fig. 15.6-1 A periodic table for raindrops. The symbols of elements included in dissolved species or in particles within raindrops are included, together with reference numbers for publications reporting their detection.

SNOW

1 H 912, 933																	2
3	4											5	6 C 819, 933	7 N 912, 933	8 O 912, 933	9 F 939	10
11 Na 852, 912	12 Mg 852, 923											13 Aℓ 852, 908	14	15 P 852	16 S 908, 912	17 Cℓ 852, 912	18
19 K 852, 912	20 Ca 852, 912	21 Se 1141	22 Ti 852	23 V 852, 1141	24 Cr 852, 1141	25 Mn 852, 908	26 Fe 852, 933	27 Co 852, 1141	28 Ni 852, 1193	29 Cu 819, 852	30 Zn 819, 852	31	32	33 As 852, 939	34 Se 852, 1140	35 Br 852	36
37 Rb 852	38 Sr 852	39	40	41	42 Mo 852, 1141	43	44	45	46	47 Ag 1187	48 Cd 819, 852	49	50	51 Sb 1141	52	53 I 983	54
55 Cs 1141	56	57-71	72 Hf 1141	73 Ta 1141	74	75	76	77 Ir 1141	78	79 Au 1141	80 Hg 907	81	82 Pb 819, 852	83	84	85	86
87	88	89ff															

57 La 1141	58 Ce 1141	59	60 Nd 1141	61	62 Sm 1141	63 Eu 1141	64	65	66	67	68	69	70	71
89	90	91	92 U 1141	93	94	95								

Fig. 15.6-2 A periodic table for snow. The symbols of the elements found in snow are displayed, together with reference numbers for publications reporting their detection.

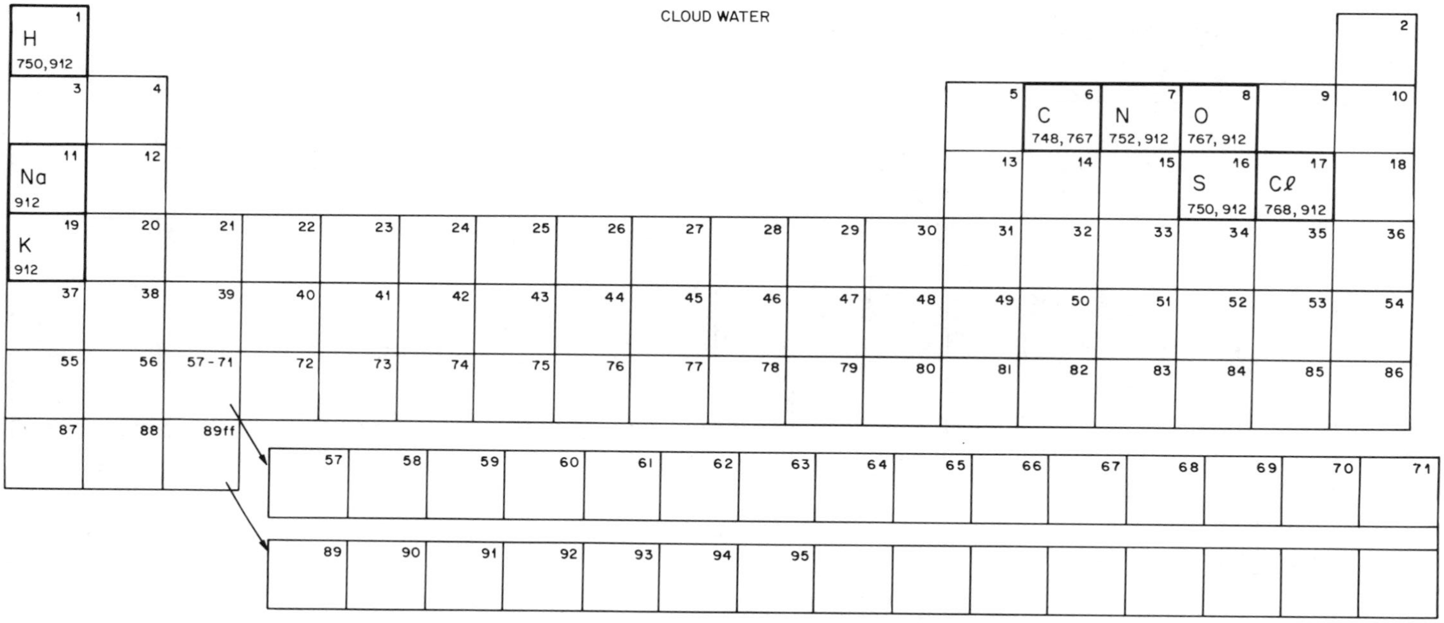

CLOUD WATER

1 H 750,912																	2
3	4											5	6 C 748,767	7 N 752,912	8 O 767,912	9	10
11 Na 912	12											13	14	15	16 S 750,912	17 Cℓ 768,912	18
19 K 912	20	21	22	23	24	25	26	27	28	29	30	31	32	33	34	35	36
37	38	39	40	41	42	43	44	45	46	47	48	49	50	51	52	53	54
55	56	57-71	72	73	74	75	76	77	78	79	80	81	82	83	84	85	86
87	88	89ff															

57	58	59	60	61	62	63	64	65	66	67	68	69	70	71
89	90	91	92	93	94	95								

Fig. 15.6-3 A periodic table for cloud water. The symbols of the elements found in cloud water are displayed, together with reference numbers for publications reporting their detection.

FOG WATER

1 H 796,1297																	2
3	4											5	6 C 746,796	7 N 796,1297	8 O 746,796	9 F 796	10
11 Na 796	12 Mg 796											13	14	15	16 S 796,1297	17 Cℓ 746,796	18
19 K 796	20 Ca 796	21	22	23	24	25	26 Fe 796	27	28 Ni 796	29	30	31	32	33	34	35	36
37	38	39	40	41	42	43	44	45	46	47	48	49	50	51	52	53	54
55	56	57 - 71	72	73	74	75	76	77	78	79	80	81	82 Pb 796	83	84	85	86
87	88	89ff															

57	58	59	60	61	62	63	64	65	66	67	68	69	70	71
89	90	91	92	93	94	95								

Fig. 15.6-4 A periodic table for fogwater. The symbols of the elements found in fogwater are displayed, together with reference numbers for publications reporting their detection.

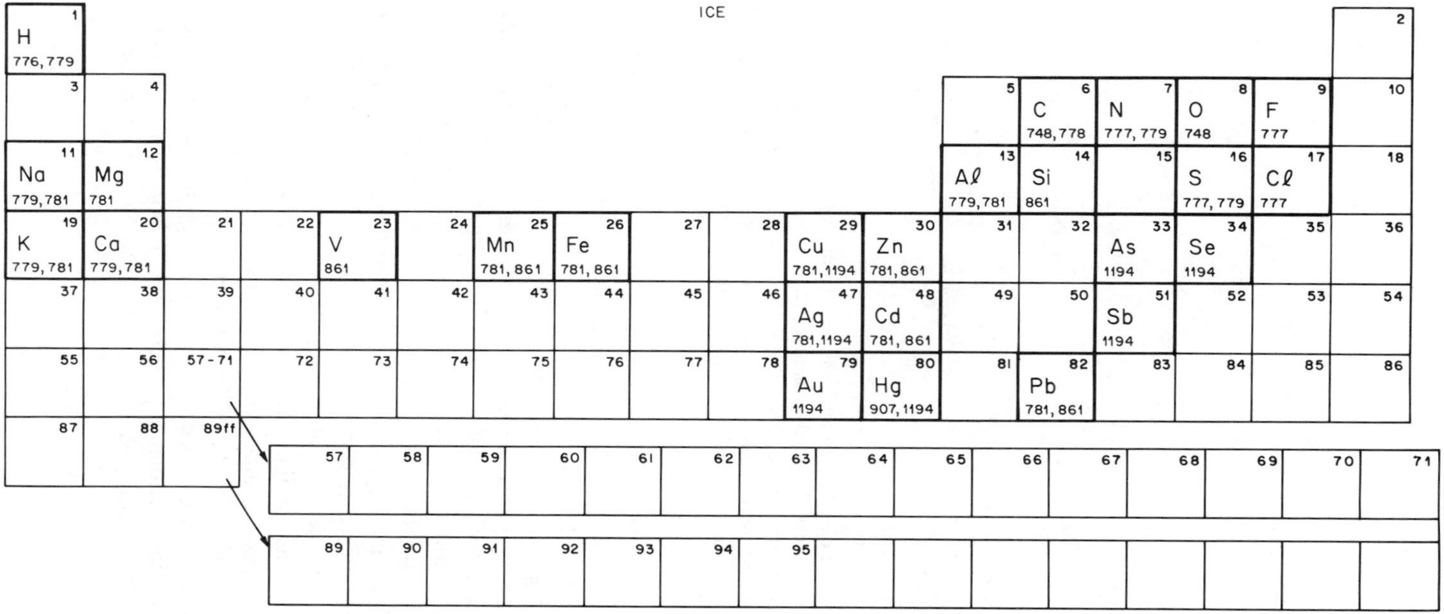

ICE

1 H 776, 779																	2
3	4											5	6 C 748, 778	7 N 777, 779	8 O 748	9 F 777	10
11 Na 779, 781	12 Mg 781											13 Aℓ 779, 781	14 Si 861	15	16 S 777, 779	17 Cℓ 777	18
19 K 779, 781	20 Ca 779, 781	21	22	23 V 861	24	25 Mn 781, 861	26 Fe 781, 861	27	28	29 Cu 781, 1194	30 Zn 781, 861	31	32	33 As 1194	34 Se 1194	35	36
37	38	39	40	41	42	43	44	45	46	47 Ag 781, 1194	48 Cd 781, 861	49	50	51 Sb 1194	52	53	54
55	56	57 - 71	72	73	74	75	76	77	78	79 Au 1194	80 Hg 907, 1194	81	82 Pb 781, 861	83	84	85	86
87	88	89ff															

57	58	59	60	61	62	63	64	65	66	67	68	69	70	71
89	90	91	92	93	94	95								

Fig. 15.6-5 A periodic table for atmospheric ice. The symbols of the elements found in atmospheric ice are displayed, together with reference numbers for publications reporting their detection.

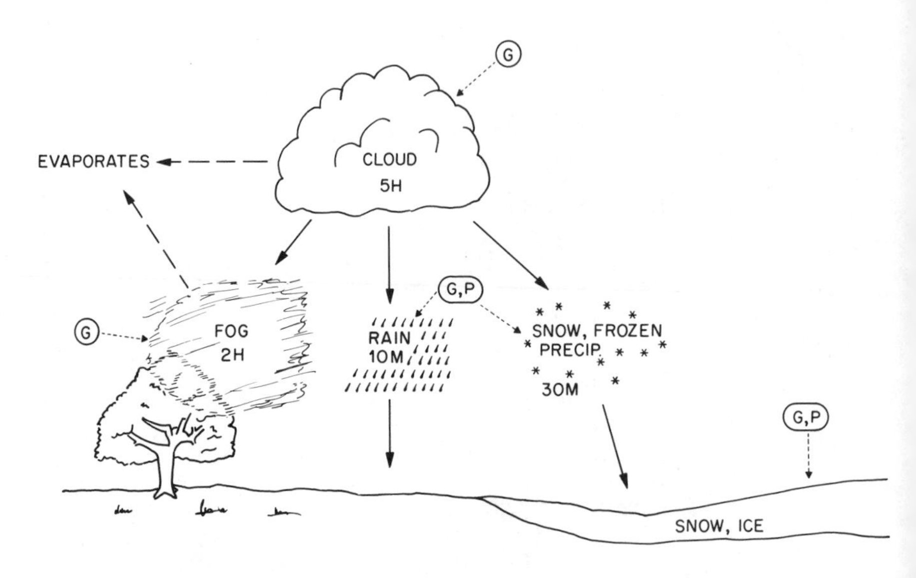

Fig. 15.6-6 A schematic diagram of transitions between atmospheric condensed water phases and inputs to them. Typical lifetimes in minutes (M) or hours (H) are indicated. Cloud droplets initially condense around particles (P). Particles are also scavenged by raindrops and snowflakes and deposited on snow and ice. Gaseous molecules (G) may be incorporated into any phase. If cloud and fog droplets evaporate, their constituents are returned to the background atmosphere.

for rain results in deposition of airborne particles on fallen snow. Silicon (14) is surely present as well, but its detection has not been announced. In general, the same patterns of occurrence are seen in snow as in rain.

Clouds and fog have been little studied to date and their periodic tables are as yet sparse (Figures 15.6-3,4). As more data are accumulated, one anticipates that the diagram for clouds will approach that for rain, since clouds should contain everything found in rain other than particles scavenged during rainfall.

Studies of the chemistry of ice are restricted to Antarctic and Greenland ice, where the material is comprised of frozen snow and deposited aerosol particles and gases. More than twenty elements have been identified. One anticipates patterns similar to those for snow but with a small anthropogenic aerosol component. In particular, P(15), Ti(22), Co(27), Ni(28), Br(35), and I(53) would be expected.

These periodic tables, though incomplete, suggest the interrelationships among precipitation elements. These are schematically illustrated in Figure 15.6-6. The sequence begins with the formation of small cloud droplets, always or nearly always around preexisting aerosol particles. Under some conditions, cloud droplets can grow in size and fall as rain, snow, or ice. Close to the ground under humid conditions fog may result. In all these cases, one would expect to find within the product phases the constituents of cloud droplets, plus additional material scavenged near the earth's surface. Snow, which may exist for months, and polar ice, which may exist for centuries, gather additional gaseous and particulate material by surface deposition processes.

Perhaps ninety percent of all clouds and fogs have a transitory existence followed by evaporation. This evaporation leaves behind the small particles around which the droplets nucleated; these particles are probably enriched chemically by the accretion of smaller particles and gases and by the chemical reactions that have occurred in the droplet phase. The particles, if not deposited on the ground, are then available to begin a new and more chemically active cycle of atmospheric droplet chemistry.

15.7 The Interactions of the Atmosphere With Surfaces

15.7.1 Atmospheric Effects on Vegetation

The atmosphere and the earth's biota interact in several different ways. The most universal and well recognized interaction is the photosynthetic cycle in which green plants absorb atmospheric carbon dioxide, convert it to sugars, and release molecular oxygen. A second interaction has been discussed throughout this book: the emission from vegetation of many organic compounds into the atmosphere. Here we briefly discuss a third interaction: damage to vegetation as a result of exposure to certain atmospheric chemical species.

It has been recognized for many years that vegetation is susceptible to damage in "smog" episodes. Extensive research has established that strong oxidants are generally the causitive species, with ozone, SO_2, peroxyacetyl nitrate (PAN) and peroxypropionyl nitrate (PPN) being identified as specific damage agents.[1332,1333] The damaging concentrations of these species are generally reached only near sources (SO_2) or during intense photochemical smog episodes (O_3, PAN, PPN). Precipitation chemistry or acidity has not been shown to have a direct impact on vegetation.

15.7.2 Effects on Materials - General

Most anthropogenically - produced materials degrade upon exposure to the atmosphere. The rate and form of this degradation is different for different materials, and the great range of atmospheric environments makes damage assessments quite complicated. The first step, however, is to assemble information on the interactions of materials with atmospheric constituents. This is done in Figure 15.7-1 for a number of different metals, for masonry and stone, and for art and artifacts (paints, paper, textiles, etc.). Ten different corrodent species or groups are included.

As shown in Figure 15.7-1, some atmospheric corrodents are known to degrade numerous materials, while others affect only a few. H_2O_2 and HCHO are fairly benign, for example, while H_2S interacts with many materials. It is important to realize that this figure indicates the *minimum level* of interactions, since (as the blanks indicate) a number of tests remain to be performed and the identification of other corrodent-material effects is likely.

The susceptibilities of the individual materials are of great interest, since they are used in diverse applications and environments. In the remainder of this section, we discuss each of the materials separately.

15.7.3 Atmospheric Effects on Aluminum

Aluminum is one of the most widely used metals because of its light weight, high strength, and good corrosion resistance. It is known to be susceptible to some degree to degradation by H_2O_2, NH_3/NH_4^+, H_2S, HCl/Cl^-, and HC(O)OH. In Figure 15.7-2, the degradation potential of aluminum in different atmospheric regimes is shown. To construct this (and subsequent) figures, the susceptibility data of Figure 15.7-1 is used to specify hatching, little, moderate, or high sensitivity being indicated by light, medium, and heavy hatching. The data from references cited in this book are then used to define the concentrations of corrodents in different atmospheric regimes. Each matrix element then contains symbols (or lack thereof) representing one of three possible conditions:

- If the matrix element contains hatching, the material is known to be degraded by the indicated corrodent. If the corrodent concentration in a particular regime is sufficiently high, degradation will occur. The conditions of most concern are those in which corrodent concentrations are high and the material is highly sensitive to that corrodent. Such conditions are indicated by a heavy box around the matrix element. This condition appears not to occur for aluminum, but does occur for iron.

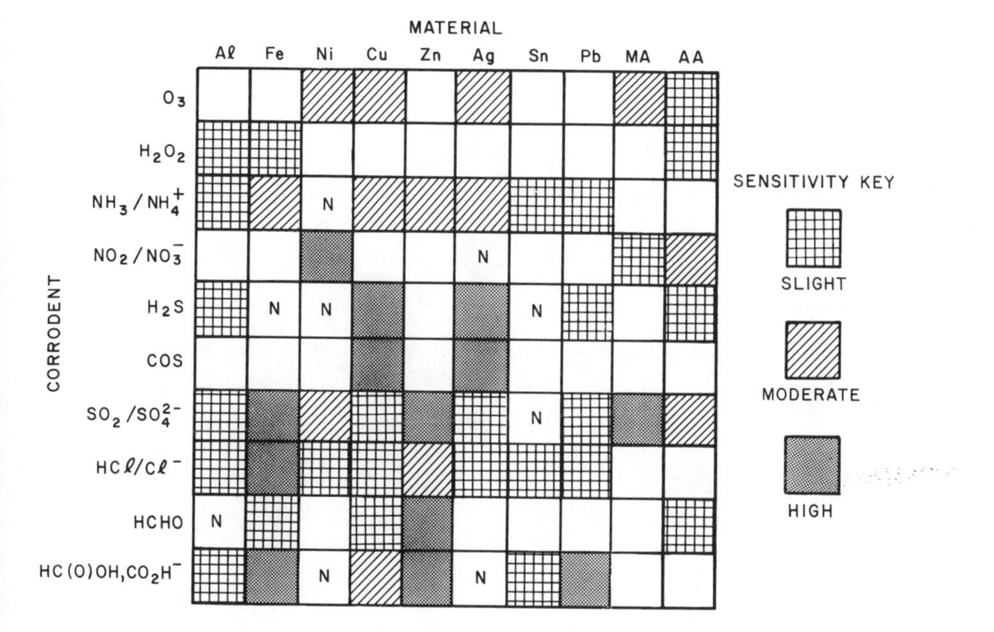

Fig. 15.7-1 The materials degradation potential of atmospheric species. The materials include eight metals (identified by their atomic symbols), masonry and stone (MA), and art and artifacts (AA). A filled matrix element indicates that the material in that column is known be degraded by the corrodent in that row, the type of hatching indicating the degree of sensitivity. N indicates that the material is unaffected by the corrodent; a blank indicates that degradation tests have not been conducted. Literature references for the degradation are given by Graedel[1279] for the metals, and by Yocum and Baer[1304] for masonry, stone, and art and artifacts.

- If the matrix element contains no hatching, the material is not susceptible to degradation by the corrodent under normal atmospheric conditions, and the presence or absence of the corrodent is of no interest. This is a condition of unconcern.
- If the matrix element is empty, the material has not been tested for susceptibility to the corrodent and the corrodent is not known to be present. This is a condition of possible concern.

Figure 15.7-3 indicates that aluminum is potentially susceptible to degradation in every atmospheric regime. This susceptibility is a consequence of aluminum's slight sensitivity to degradation by sulfate, chloride, ammonia, hydrogen sulfide, hydrogen peroxide, and formic acid. All of these species are at least sometimes present in the atmospheric gas

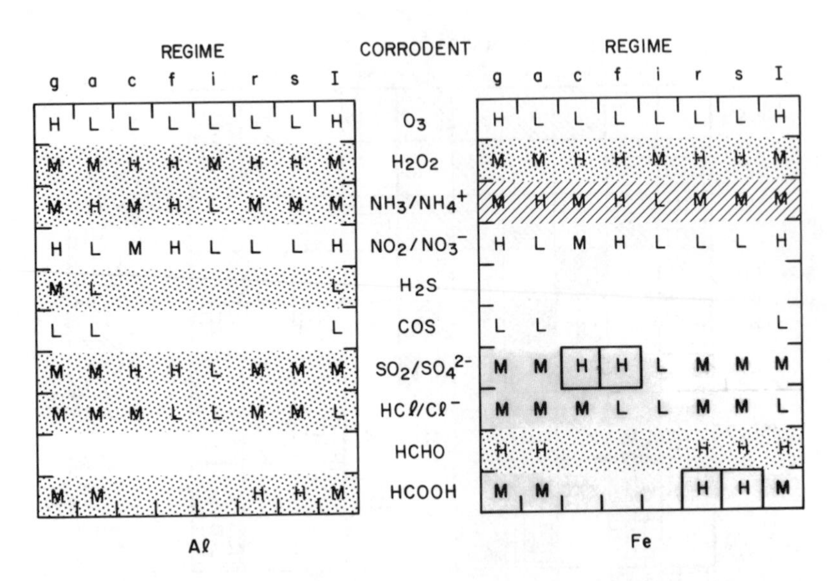

Fig. 15.7-2 The degradation potential of aluminum and iron in different atmospheric regimes. The regime codes are gas: g, aerosol: a, cloud: c, fog: f, ice: i, rain: r, snow: s, and indoor: I. A filled matrix element indicates that the metal is known to be degraded by the corrodent in that row, the type of hatching indicating the degree of sensitivity. The concentration of a particular corrodent in a particular regime is qualitatively indicated as high (H), moderate (M), or low (L). A blank row indicates that the material is insensitive to damage from the specific corrodent. A row with concentration overprints but no hatching indicates that the sensitivity of the material to the specific corrodent has not been established.

(both indoors and outdoors) and in aerosol particles. H_2S is not known to be contained in hydrometeors, but measurements may eventually indicate its presence in precipitation near H_2S sources. Formic acid is abundant in rain and snow, probably present in clouds and fog, and possibly present in ice, but measurements have not yet been made in the latter regimes. Aluminum's low sensitivity to corrodents suggests that its corrosion potential is slight, an assessment borne out by the good performance of aluminum in a wide range of outdoor and indoor environments.

15.7.4 Atmospheric Effects on Iron

Iron and its alloys are, by weight, the most widely used metals. Their environmental susceptibility is known to be a function of their composition, but only the pure metal has received detailed study. Its degradation potential is shown in Figure 15.7-2.

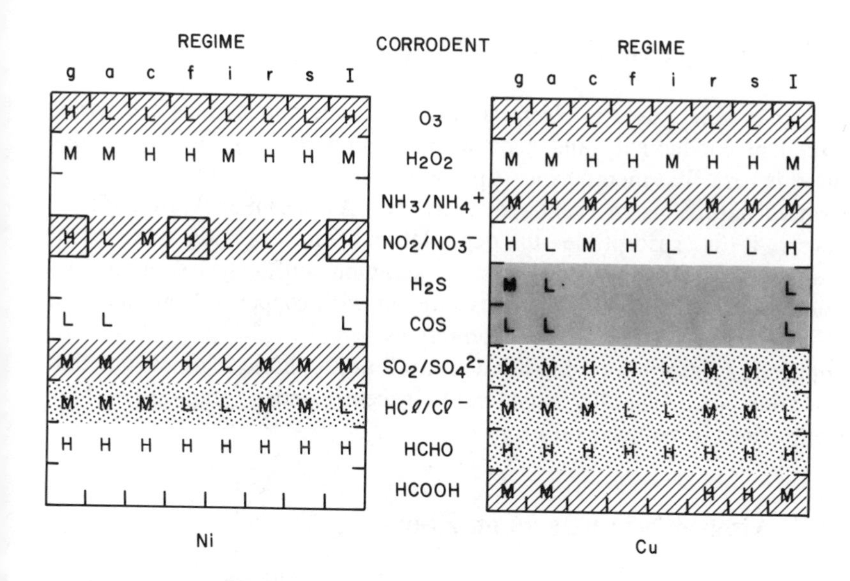

Fig. 15.7-3 The degradation potential of nickel and copper in different atmospheric regimes. The symbols are explained in the caption to Figure 15.7-2.

Iron degrades much more rapidly then aluminum, and does so under the attack of a broad spectrum of corrodents. Sulfur gases, nitrates, chlorides, and oxygenated organics are all of concern. Because of the presence of these common atmospheric constituents in the gas, aerosol, and precipitation regimes, iron and its alloys are subject to degradation under all typical atmospheric conditions. Four interactions of special concern are indicated by the boxed matrix elements: that with sulfate in clouds and fog and with formic acid in rain and snow.

15.7.5 Atmospheric Effects on Nickel

Nickel is rather stable under atmospheric exposure, as suggested by its frequent use in marine applications and as a plating material for more easily degraded metals. Nickel has some degree of susceptibility to NO_2, O_3, SO_2, and HCl, though generally only to rather high concentrations of those gases. Its degradation potential diagram is shown in Figure 15.7-3. The diagram shows some potential for deterioration in all regimes, with particular concern for interaction with NO_2 or nitrate in the indoor and outdoor atmosphere and in fog.

15.7.6 Atmospheric Effects on Copper

Copper is widely used in electrical and electronic applications because of its excellent conductivity and good working properties. It is also popular in uses where its metallurgical properties can be combined with its decorative nature, as in roofing or statuary. As shown in Figure 15.7-3, however, it is susceptible to degradation from a wide variety of atmospheric constituents. All of the common sulfur gases, ammonia, nitrates, chlorides, and organic acids interact with copper. Ozone tends to accelerate at least some of those interactions.

Copper is susceptible to degradation in all atmospheric regimes, but that degradation is generally moderate except under unusually high concentrations of reduced sulfur gases.

15.7.7 Atmospheric Effects on Zinc

Zinc is rarely used in pure form, but is often alloyed with copper to form brass or plated on iron (to give a "galvanized" product). It is quite resistant to some forms of atmospheric degradation, but shows sensitivity to NH_3, SO_2, HCl, HCHO, and HCOOH (Figure 15.7-4). As a result, it has some potential for degradation in each atmospheric regime. The relatively high concentrations of formaldehyde throughout the atmosphere and of sulfate and formic acid in hydrometeors provide a number of interaction possibilities in which significant zinc degradation is likely.

15.7.8 Atmospheric Effects on Silver

Silver is an excellent electrical conductor and one of the most intrinsically beautiful of metals. Its uses are many, but are limited by silver's sensitivity to several common atmospheric gases or ions. Figure 15.7-4 indicates that NH_3, H_2S, COS, NH_3, sulfates, and chlorides are of concern.

Because of its cost and its propensity for degradation, silver is rarely used outdoors. Thus its atmospheric interactions are limited, for all practical purposes, to the indoor environment. Within this environment, all of the potential problem species are known to be present. Most degradation takes the form of a black tarnish film of Ag_2S upon the silver with both H_2S and COS thought to be the most common causitive agents.

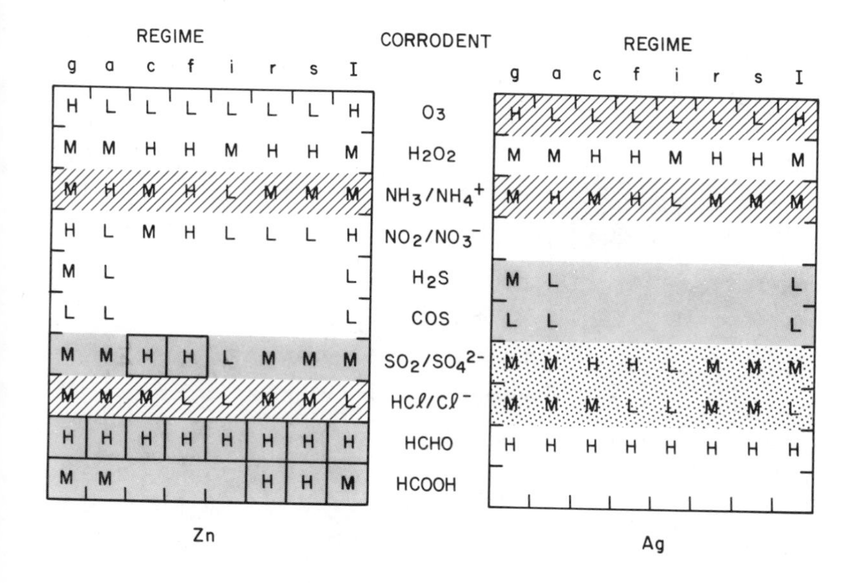

Fig. 15.7-4 The degradation potential of zinc and silver in different atmospheric regimes. The symbols are explained in the caption to Figure 15.7-2.

15.7.9 Atmospheric Effects on Tin

Tin is a soft metal rarely used in pure form, but often found in combination with other materials. Tin plating provides a highly corrosion resistant surface (the "tin can" is a steel can with tin plating). Common alloys are those with copper (termed "bronze") and with lead (termed "solder"). Pewter is a corrosion resistant alloy containing ninety percent or more tin, with the remainder being copper and arsenic. Tin's degradation potential diagram is given in Figure 15.7-5. Unlike many metals, tin is insensitive to sulfur compounds at atmospheric temperatures, but shows some sensitivity to NH_3, HCl, and HCOOH. High concentrations of these species have the potential to degrade tin in all atmospheric regimes.

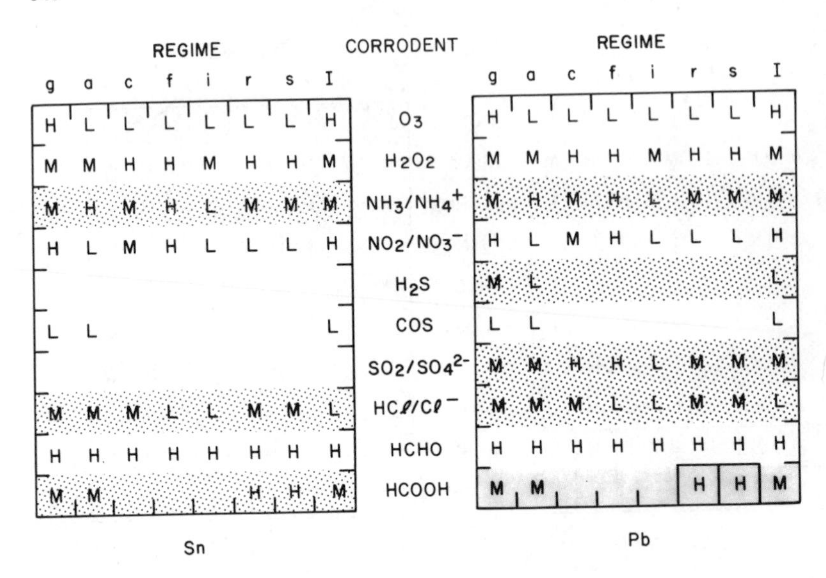

Fig. 15.7-5 The degradation of tin and lead in different atmospheric regimes. The symbols are explained in the caption to Figure 15.7-2.

15.7.10 Atmospheric Effects on Lead

Lead is not generally exposed to the atmosphere, since such uses as tank linings, battery components, etc. offer a degree of atmospheric protection. It is somewhat susceptible to degradation by NH_3, H_2S, SO_2, or chlorides, as seen in Figure 15.7-5. Sufficiently high concentrations of these species can degrade lead in all atmospheric regimes. Lead is very susceptible to degradation by formic acid, so its exposure to rain, snow, and perhaps other forms of atmospheric water should be viewed with concern.

15.7.11 Atmospheric Effects on Masonry

Masonry is relatively impervious to most atmospheric species, but is susceptible to degradation by NO_2 and SO_2 (Figure 15.7-6), especially the latter, perhaps augmented by ozone interaction. The degradation generally takes the form of oxidizing the calcite ($CaCO_3$) to soluble calcium sulfate.

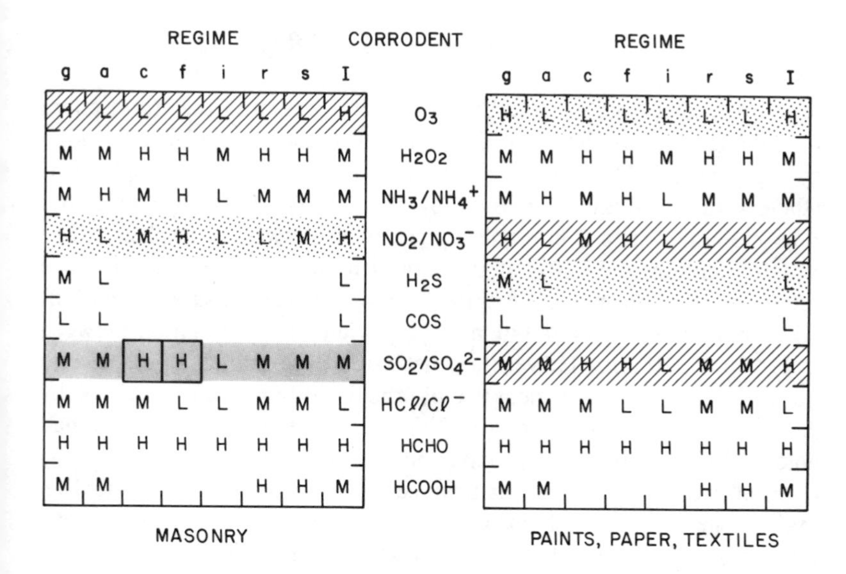

Fig. 15.7-6 The degradation of masonry and of art and artifacts in different atmospheric regimes. The symbols are explained in the caption to Figure 15.7-2.

$$CaCO_3 \xrightarrow[H_2O]{SO_2} CaSO_4 \qquad \text{(R15.7-1)}$$

The calcium sulfate that is produced can then be washed away and fresh calcite attacked. The mechanism by which reaction R15.7-1 occurs is not well understood, but is expected to be optimized in the interactions of masonry with clouds and fog.

15.7.12 Atmospheric Effects on Art and Artifacts

With the exception of metal or marble sculptures, art and artifacts are generally kept indoors. Thus, although potentially subject to degradation in all atmospheric regimes (Figure 15.7-6), our attention may be limited to the indoor environment. Potential problem constituents of the indoor atmosphere are O_3 (paints, textiles), NO_2 (textiles), H_2S (paints), and SO_2 (papers, photographic materials, leather). Since all of these species can have indoor sources (see section 15.9), indoor degradation of art and artifacts may be minimized or eliminated by a knowledge of the relationships between sources and indoor air species.

15.8 Stratospheric Chemistry

The temperature structure and the dynamic processes in the stratosphere are to a major degree determined by the absorption of solar ultraviolet energy by ozone. The total amount of ozone in the stratosphere is nevertheless rather small; if present by themselves, the ozone molecules could be contained in a 3 mm thick layer of the lower atmosphere. The troposphere contains only 10% of the total atmospheric ozone; most ozone is located in the stratosphere. Atmospheric ozone also plays an important ecological role, since it filters out most ultraviolet solar radiation between about 210 and 310 nm. However, penetration of ultraviolet radiation to ground level begins at wavelengths of about 300 nm and increases by orders of magnitude within the next 20 nm. It is this radiation that may be biologically harmful. Its penetration to the ground is enhanced by reduction of the ozone column: it has been estimated that a 1% reduction in total ozone would lead to a 2-4% increase in biological effects. Concomitantly, however, the penetration of ultraviolet radiation at wavelengths near 310 nm to the troposphere leads to the production of the OH· radical which initiates most atmospheric oxidation processes, leading to compounds which are effectively removed. This removal of atmospheric trace species, some of them biologically harmful, may balance to some extent the biological damage produced by increases in ultraviolet radiation.

Stratospheric ozone is particularly affected by compounds that are relatively inert in the troposphere. These compounds (nitrous oxide from biological and combustion sources and several chlorocarbon gases, such as natural CH_3Cl and industrially produced $CFCl_3$, CF_2Cl_2, CCl_4, and CH_3CCl_3) have a low solubility in water, a slow photolysis, and a very slow reaction with OH·. Stratospheric chemistry is also influenced by the direct injection of material into the upper troposphere and stratosphere by thunderstorms, volcanic eruptions, and emissions from jet aircraft.

An important initiator for research on the reactions involved in stratospheric gas phase chemistry was the proposal[1272] that NO_x $(NO+NO_2)$ would catalyze the destruction of ozone and limit its stratospheric abundance by a simple set of photochemical reactions:

$$O_3 + h\nu \rightarrow O + O_2 \; ; \; (\lambda < 150 \text{ nm}) \quad \text{(R15.8-1)}$$

$$O + NO_2 \rightarrow NO + O_2 \quad \text{(R15.8-2)}$$

$$NO + O_3 \rightarrow NO_2 + O_2 \quad \text{(R15.8-3)}$$

$$\text{net:} \quad 2O_3 \rightarrow 3O_2$$

Below 40 km this catalytic chain of reactions largely balances the formation of ozone in the stratosphere through the reaction sequence:

$$O_2 + h\nu \rightarrow 2O\ ;\ (\lambda < 240\ \text{nm}) \quad (R15.8\text{-}4)$$

$$2(O + O_2 + M \rightarrow O_3 + M) \quad (R15.8\text{-}5)$$

$$\text{net:}\quad 3O_2 \rightarrow 2O_3$$

The main source of NO_x in the stratosphere is probably the oxidation of nitrous oxide (N_2O) via

$$O_3 + h\nu \rightarrow O(^1D) + O_2\ ;\ \lambda < 320\ \text{nm} \quad (R15.8\text{-}5)$$

$$O(^1D) + N_2O \rightarrow 2NO \quad (R15.8\text{-}6)$$

The oxides of nitrogen play a remarkable catalytic role in the ozone balance of the atmosphere. Above about 25 km the net effect of NO_x additions will be a decrease of ozone concentrations by reactions R15.8-2 and R15.8-3. However, below about 25 km, NO_x protects ozone from destruction. This is because of another set of reactions:

$$HO_2\cdot + NO \rightarrow OH\cdot + NO_2 \quad (R15.8\text{-}7)$$

$$NO_2 + h\nu \rightarrow NO + O \quad (R15.8\text{-}8)$$

$$O + O_2 + M \rightarrow O_3 + M \quad (R15.8\text{-}5)$$

$$\text{net}\quad HO_2\cdot + O_2 \rightarrow OH\cdot + O_3$$

In the lower stratosphere, this chain of reactions tends to counteract the destruction of ozone by the catalytic reaction pair

$$OH\cdot + O_3 \rightarrow HO_2\cdot + O_2 \quad (R15.8\text{-}9)$$

$$HO_2\cdot + O_3 \rightarrow OH\cdot + 2O_2 \quad (R15.8\text{-}10)$$

$$\text{net:}\quad 2O_3 \rightarrow 3O_2$$

as a consequence of deflecting $HO_2\cdot$ into the chain

$$OH\cdot + O_3 \rightarrow HO_2\cdot + O_2 \quad (R15.8\text{-}9)$$
$$HO_2\cdot + NO \rightarrow OH\cdot + NO_2 \quad (R15.8\text{-}7)$$
$$NO_2 + h\nu \rightarrow NO + O \quad (R15.8\text{-}8)$$
$$O + O_2 + M \rightarrow O_3 + M \quad (R15.8\text{-}5)$$

no net chemical effect.

An additional role of NO_x in the stratosphere involves interactions with Cl· and ClO·. As with NO_x, chlorine atoms and chlorine monoxide radicals participate in an effective catalytic cycle that converts ozone back to molecular oxygen:

$$O_3 + h\nu \rightarrow O + O_2 \quad (R15.8\text{-}1)$$
$$O + ClO\cdot \rightarrow Cl\cdot + O_2 \quad (R15.8\text{-}11)$$
$$Cl\cdot + O_3 \rightarrow ClO\cdot + O_2 \quad (R15.8\text{-}12)$$

$$\text{net:} \quad 2O_3 \rightarrow 3O_2$$

whereby one molecule of ClO· is about three times more efficient than an NO_2 molecule in "odd oxygen" (O, O_3) conversion. The abundance of ClO· radicals in the stratosphere is increasing rapidly due to the release of chlorofluorocarbons. The presence of NO in the stratosphere makes the ClO· catalytic cycle less effective, however, as the reactions

$$NO + ClO\cdot \rightarrow Cl\cdot + NO_2 \quad (R15.8\text{-}13)$$
$$Cl\cdot + CH_4 \rightarrow HCl + CH_3\cdot \quad (R15.8\text{-}14)$$

transform the catalysts Cl· and ClO· into HCl, which does not react photochemically with ozone. Furthermore, since the reaction

$$ClO\cdot + NO_2 + M \rightarrow ClNO_3 + M \quad (R15.8\text{-}15)$$

ties up both some ClO· and some NO_2 as non-reactive $ClNO_3$, it is clear that ozone removal by additions of chlorine to the stratosphere is mitigated by the NO_x interference in the chlorine cycle. For additions of chlorine to the stratosphere which do not override those of NO_x, the reductions in total stratospheric ozone are calculated not to be very large because ozone reductions above 30 km are compensated by increases below.

The amount of water vapor present in the stratosphere is insufficient to form droplets, although ions with water molecule adducts appear common. This paucity of water restricts stratospheric condensed phase studies to those of particles. Three sources of particles are known: injection from outer space (meteorites, etc.), injection from the troposphere (especially as a consequence of volcanos or thunderstorms), and conversion from gas phase compounds. The latter is of the most interest, since sulfates or mixed compounds containing sulfur appear to be the primary constituents of stratospheric aerosol particles. These are formed from COS and SO_2, the principal stratospheric sulfur gases, by

$$COS \xrightarrow[\lambda<255\text{ nm}]{h\nu} CO + S \qquad (R15.8\text{-}16)$$

$$S + O_2 \rightarrow SO + O \qquad (R15.8\text{-}17)$$

$$SO + O_3 \rightarrow SO_2 + O_2 \qquad (R15.8\text{-}18)$$

$$SO_2 + OH\cdot \xrightarrow{M} HSO_3\cdot \qquad (R15.8\text{-}19)$$

$$HSO_3\cdot + O_2 \rightarrow SO_3 + HO_2\cdot \qquad (R15.8\text{-}20)$$

$$SO_3 + H_2O \rightarrow H_2SO_4 \qquad (R15.8\text{-}21)$$

$$H_2SO_4 + n H_2O \rightarrow H_2SO_4 \cdot n H_2O \qquad (R15.8\text{-}22)$$

The density of stratospheric aerosol particles is a rather complicated function of the latitude of particle or particle precursor injection. Contours of aerosol concentration (Figure 15.8-1) show that larger concentrations of particles occur at high altitudes in the tropics and lower altitudes in the polar region. Such a pattern is consistent with the tropical tropopause being the principal injection site of the sulfur gases.[1273] Further information is provided by typical size spectra of the stratospheric aerosol particles. As shown in Figure 15.8-2, many more small particles and fewer large particles occur in the stratosphere at low latitudes than at

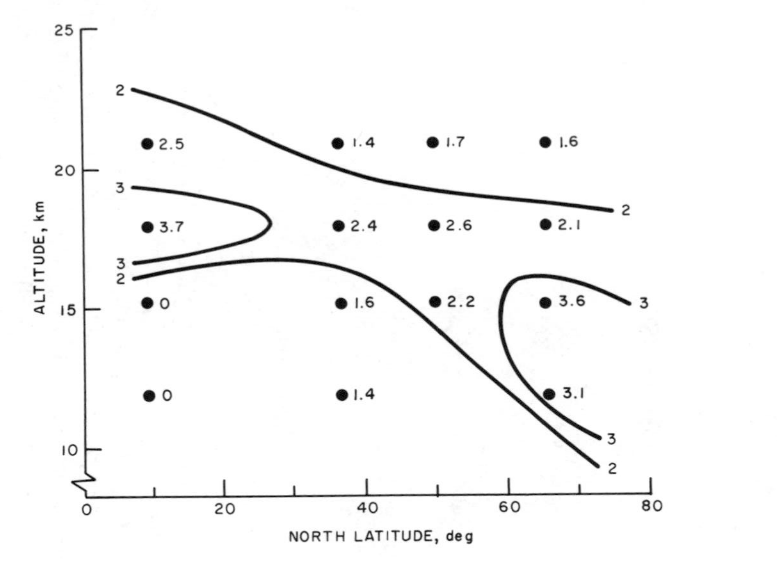

Fig. 15.8-1 Estimated contours of stratospheric aerosol particle concentrations at ambient conditions, as measured in 1976 and 1977 by particle collectors on high-flying aircraft. (Reproduced with permission.[1273])

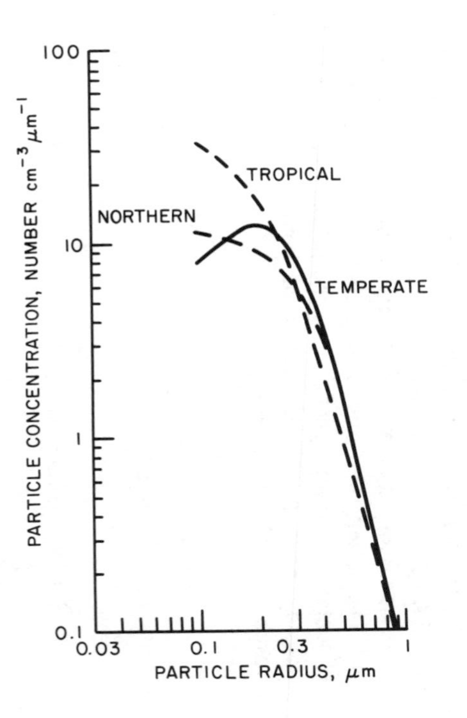

Fig. 15.8-2 Average size distributions for stratospheric aerosol particles: tropical latitudes (18-21 km), temperate latitudes (12-21 km), and north polar latitudes (12-21 km). (Adapted from Farlow et al.[1273])

1 H 725, 728																	2 He * 294, 558
3	4 Be † 1280											5	6 C 718, 724	7 N 702, 710	8 O 701, 720	9 F * 731, 737	10 Ne * 294, 558
11 Na 1022, 1039	12											13 Aℓ † 703	14 Si † 1022	15	16 S 731, 737	17 Cℓ 725, 727	18 Ar * 294, 558
19 K † 1022	20	21	22	23	24 Cr † 1039	25 Mn † 1022	26 Fe † 1039	27	28	29	30	31	32	33	34	35 Br 728, 1022	36
37	38 Sr † 1021, 1280	39	40 Zr † 1280	41	42	43	44	45	46	47	48	49	50	51	52	53	54
55 Cs † 1021, 1280	56	57-71	72	73	74	75	76	77	78	79	80	81	82 Pb † 1021, 1280	83 Bi † 1119	84	85	86
87	88 Ra † 1021	89ff															

57	58 Ce † 1280	59	60	61	62	63	64	65	66	67	68	69	70	71
89	90	91	92 U † 1196	93	94 Pu † 1280	95								

Fig. 15.8-3 A periodic table for the stratosphere. The symbols of elements detected in the gas phase or in particles are included, together with reference numbers for publications reporting their detection. An asterisk (*) indicates that the element has been found only in the gas phase and a dagger (†) indicates detection only in the aerosol phase. Elements whose symbols appear but which are unmarked have been found in both phases.

higher latitudes. One would anticipate just such a pattern if the low latitudes were the principal region in which new particles were forming and evolving.

The trace species composition of the stratosphere reflects the various inputs from below and above it. As seen in Figure 15.8-3, elements which comprise small, volatile molecules are present, as are a few from weapons testing (e.g., Sr) and (presumably) from meteorites (e.g., Fe). Most of the transition metals and rare earths found in the earth's crust have not (yet?) been found in the stratosphere.

15.9 Chemistry of Indoor Air

Nearly 300 different compounds have been identified in indoor air and appear in the tables throughout this book. That number is an underestimate, since gas chromatograms of indoor air include peaks for many unidentified compounds,[1113] and since cigarette smoke alone contains some 3000 compounds.[1292]

A small number of inorganic compounds have received virtually all of the analytical attention of indoor air quality analysts. As shown in Table 15.1-1, only fourteen inorganic compounds have been found in indoor air. The situation is vastly different for hydrocarbons, where more than a hundred have been identified. Other chemical groups well represented are alkanols, alkanic aldehydes, simple esters, carboxylic acids, and halogenated aromatic compounds.

It is of great interest to compare typical indoor concentrations of several well-measured species with those measured outdoors. This is done in Table 15.9-1 for gas phase compounds. If indoor sources of a compound are absent (as is often the case with ozone in the absence of gas stoves), outdoor concentrations will be much higher than indoor concentrations. The converse is true if strong indoor sources are present, such as occurs for formaldehyde in the presence of some building materials. A similar situation occurs in the case of compounds on indoor particles, as shown in Table 15.9-2. In general, if indoor sources of a gaseous or particulate species are present, the concentrations of that species are significantly higher indoors than out.

Since many chemical compounds are present in indoor air, it is of interest to ask whether they are chemically transformed in that environment or whether their life cycle is limited to emission followed by deposition to a surface or expulsion into outside air. This topic has yet to be explored in detail, but it is possible here to mention some possibilities that may exist. The first item to consider is the presence and extent of

Table 15.9-1. Typical Concentrations of Selected Gas Phase Species in Outdoor Urban Air and Indoor Air

Species	Outdoor conc. (ppb)	Ref.	Indoor conc. (ppb)	Ref.
O_3	0.1-500	Fig. 2.2-1	20-200	1290
H_2O_2	1-100	641	very low?	–
$OH\cdot$	10^{-5}-10^{-3}	259	very low?	–
CO	10^2-10^5	Fig. 2.6-2	10^3-10^5	1290
NO	.03-250	Fig. 2.3-2	50-110	1288
NO_2	1-80	1271	20-600	1290
SO_2	4-1500	Fig. 2.4-3	6-40	1290
HCHO	1-115	Fig. 7.1-1	80-4200	1290
Acetone	0.3-3	33	17-23	1290
Acrolein	1-15	5	1-20	1290
CH_3OH	5-100	90	26-84	1123
$CH_3C(O)OCH_3$	low?	-	6-11	1123
Phenol	0.05-2	652	3-7	1123
CH_3COOH	1-6	994	9-11	1123
α-Pinene	0.1-6	1320	.04-14	1112
Limonene	0.1-6	525	.01-29	1112
Camphene	.04-.05	886	.07-50	1112

short wavelength photons (i.e., those with wavelength less than about 420 nm). As was shown in the introductory chapter, incandescent lamps emit little or no radiation at these short wavelengths. Fluorescent lamps do emit short wavelength radiation, however. In Figure 15.9-1 we plot the spectrum of that emission together with those of three photochemically sensitive gases known to be found in indoor air. As can be seen, a significant overlap exists for NO_2, some for HCHO, and essentially none for O_3.

Consider now the possible chemical fate of these gases and related compounds. In the case of NO_2, emitted at relatively high flux from gas stoves, two fates seem possible. The first is deposition to surfaces. (Note that NO_2 is known to be damaging to textiles.) The second is the possibility of photodissociation:

Table 15.9-2. Typical Concentrations of Selected Particulate Species in Outdoor Urban Air and Indoor Air

Species	Outdoor con. ($\mu g/m^3$)	Ref.	Indoor conc. ($\mu g/m^3$)	Ref.
TPM*	10-550	1321	10-500	1290
SO_4^{2-}	1-20	Fig. 2.4-4	2-5	1290
NO_3^-	1-10	Fig. 2.3-5	0.4-0.7	806
Cl^-	0.02-3	Fig. 2.1-1	0.8-1.3	806
NH_4^+	0.1-2	Fig. 2.3-4	0.01-0.3	806
Na^+	0.3-2	911	0.2-1	806
Benzo[*a*]pyrene	2-80	113	0.01-5	1290
Pyrene	2-6	671	.001	857
Chrysene	0.5-18	671	.004	857
Chlordane	10^{-5}-10^{-3}	977	0.1-10	632
Dieldrin	10^{-5}-10^{-4}	958	0.1-0.3	846
PCB	10^{-5}-10^{-2}	569	0.04-12	1025

* Total particulate matter

$$NO_2 \xrightarrow[\lambda < 420\text{ nm}]{h\nu} NO + O$$

followed by

$$O + O_2 \rightarrow O_3 .$$

Ozone is also known to be damaging to rubber and textiles (perhaps the damage attributed to NO_2 is really due to O_3!), but in lieu of surface deposition it may react with unsaturated compounds such as terpenes or carbonyls. The radicals produced will generally be of the peroxyl ($RO_2\cdot$) type.

O_3 + [terpene] → [ozonide] → FRAGMENT MOLECULES AND RADICALS

O_3 + CHO → CHO → FRAGMENT MOLECULES AND RADICALS

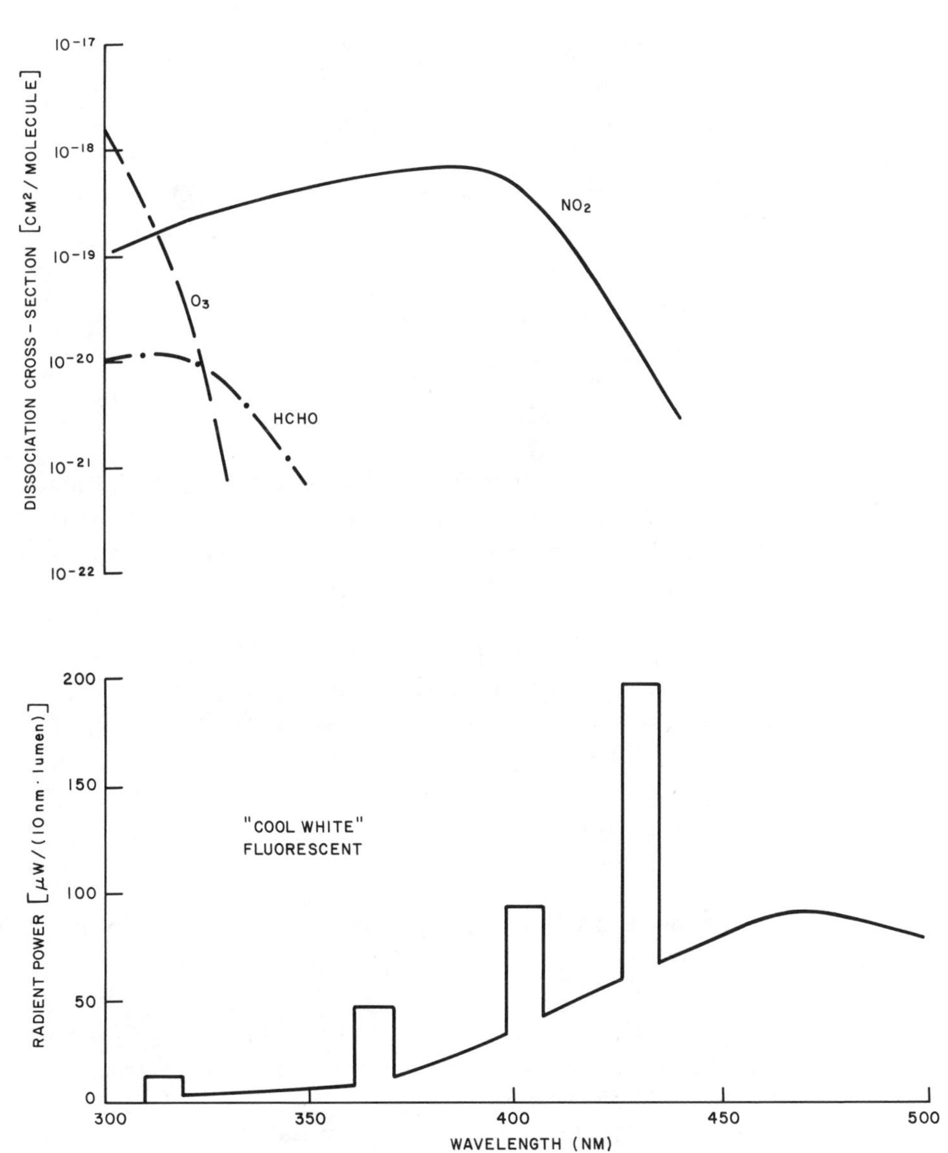

Fig. 15.9-1 The emission spectrum of a typical fluorescent lamp and the absorption spectra of three gases commonly present in indoor air.

In the case of formaldehyde, generated copiously from building materials, gas phase transformation probably depends on the proximity of ultraviolet light

$$HCHO \xrightarrow[\lambda < 335 \text{ nm}]{h\nu} H\cdot + HCO\cdot$$

$$H\cdot + O_2 \rightarrow HO_2\cdot$$

$$HCO\cdot + O_2 \rightarrow HO_2\cdot + CO$$

to give peroxyl radicals. (Alternatively, formaldehyde could readily attach itself to moist surfaces.) A possible fate for the gas phase radicals is disproportionation, as in the reaction

$$HO_2\cdot + HO_2\cdot \rightarrow H_2O_2 + O_2 \,.$$

One also wonders whether H_2O_2 or other peroxides might not be present in indoor air as a result of the use of industrial or household bleaches. In any event, if present, H_2O_2 or other peroxides could be lost to moist surfaces or might photodissociate:

$$H_2O_2 \xrightarrow[\lambda < 350 \text{ nm}]{h\nu} OH\cdot + OH\cdot \,.$$

A more likely source of $OH\cdot$ is reaction of the $HO_2\cdot$ radicals with NO:

$$NO + HO_2\cdot \rightarrow NO_2 + OH\cdot \,.$$

If the hydroxyl radical is indeed present in indoor air, as these equations suggest it might be, it could react with many different compounds to produce a variety of products.

Evidence suggesting the possibility of chemical processes on indoor aerosol particles is very sketchy. Note that several ingredients are present: sooty particles from tobacco smoking and home heating, moderate humidities to produce water shells on the particles, relatively short wavelength photons, and reactive species such as NO_2 from gas stoves.

In summary, little is known about indoor atmospheric chemistry but the potential for interesting processes is clearly present. Many compounds are found indoors; some are quite reactive and some are present at concentrations much higher than found outside. The light sources are generally of longer wavelength than outdoor sunlight but some action spectrum overlap exists. A building such as a mobile home, with fluorescent lights, a gas stove, and a high density of modern building materials clearly has the ingredients for extensive indoor atmospheric chemistry.

15.10 Summary of the Bioassay Information

Although chapters 2-14 list more than 2800 atmospheric compounds, only 303 or approximately 11% have been assayed for mutagenicity and/or carcinogenicity. This is an even more striking observation when one realizes that the references used for bioassay information contain results for 2346 compounds. The purpose of this section is to summarize the bioassay information contained within these thirteen chapters.

Since approximately half of the compounds were tested with more than one bioassay, the results of over 900 tests are provided within the tables. Please note, however, that although the result of each bacterial strain is listed separately within the table, the Salmonella results are grouped into plate and preincubation categories for the purpose of the tabulations. In other words, for any particular chemical compound a single overall result was listed within the summary tables of this section for the *Salmonella typhimurium* plate incorporation test even though all five tester strains (designated as STP00, STP98, STP38, STP37, and STP35 within the tables) may have been used. The same was done for the Salmonella preincubation assay. This was done with two reasons in mind. First, for purposes of determining whether or not a chemical is mutagenic in the Salmonella assay the results of multiple strains must be used. Second, this grouping of results prevented an artificial inflation of the bioassay result numbers. Another point worth considering is the fact that only 40 of approximately 100 existing bioassays are represented within this document. The other assays were not surveyed because they are not sufficiently evaluated, because they were used to screen relatively few compounds, or because they are used primarily for non-screening purposes (e.g., delineating genetic mechanisms). These other bioassays, then, would not significantly increase the number of identified atmospheric compounds that have been screened for genetic or carcinogenic activity.

Table 15.10-1 presents the degree of bioassay testing for each major category of chemical compounds listed within chapters 2 to 14. This table shows, as stated earlier, that 2827 atmospheric compounds are listed within the various tables and that 303 of these chemicals were bioassayed

Table 15.10-1. The Degree of Bioassay Testing by Chemical Category

•	Category	Number of Compounds	Number Tested	Total Number of Tests	Mean Number Test/Compound
2	INORGANICS	260	30	60	2.0
3	HYDROCARBONS	729	51	199	3.9
4	ETHERS	44	3	5	1.7
5	ALCOHOLS	233	28	65	2.3
6	KETONES	227	11	23	2.1
7	ALDEHYDES	108	6	13	2.2
8	CARBOXYLIC ACID DERIVATIVES	219	6	8	1.3
9	CARBOXYLIC ACIDS	174	5	5	1.0
10	HETEROCYCLIC OXYGEN COMPOUNDS	93	16	61	3.8
11	NITROGEN CONTAINING ORGANICS	384	59	214	3.6
12	SULFUR CONTAINING ORGANICS	99	4	15	3.8
13	HALOGEN CONTAINING ORGANICS	216	71	198	2.8
14	ORGANOMETALLIC COMPOUNDS	41	13	54	4.2
•	GRAND TOTALS	2827	303	920	3.0

in an average of three tests each. It is apparent that very few ethers, ketones, aldehydes, carboxylic acids and their derivatives, and sulfur compounds found in the ambient atmosphere have been tested. The two categories with the highest percentage of tested compounds are the halogen-containing compounds and the organometallic compounds. These categories have a high percentage of compounds that are of industrial and/or commercial use. For example, the tables in chapter 13 contain a number of halogenated compounds that are used in manufacturing, in refrigeration, as propellants, in building materials, in dry cleaning, and as pesticides. Most of the chemicals in chapter 14 that have associated bioassay data are organophosphorus pesticides. Of the 303 tested compounds, approximately half (153) of the compounds have been tested in only one bioassay. One quarter of the compounds (79) have been tested in either 2 or 3 bioassays. Twenty-five compounds have been tested in 4 or

Table 15.10-2. Compounds Tested in More Than Fifteen Bioassays

Compound	Registry Number	Species Number	Number of Tests
N-Nitrosodimethylamine	62-75-9	11.2-43	32
Benzo[*a*]pyrene	50-32-8	3.8-39	22
Caffeine	58-08-2	11.5-135	21
N-Nitrosodiethylamine	55-18-5	11.2-44	20
7,12-Dimethylbenz[*a*] anthracene	57-97-6	3.7-83	17
Dichlorvos	62-73-7	14.3-4	17
Aflatoxin B_1	1162-65-8	10.3-32	16

5 bioassays and 30 compounds were evaluated in 6 to 10 tests. Only 10 compounds have been tested with 11 to 15 bioassays, and only 7 have been tested with more than 15 bioassays. This is surprising since tier and phased approach testing schemes, in which a compound of interest would be tested in several assays, are often recommended.[B64,B65,B66] Table 15.10-2 lists the compounds tested in 15 or more bioassays. As one would expect, the bioassay of chemical compounds has been associated mainly with specific manufacturing processes and/or chemical compounds and with commercial products. Recently, efforts have begun to focus upon the identification of mutagens that are actually found in ambient air rather than those being emitted into ambient air. For compounds identified in this manner, there may not be any specifically known source since they may be generated by chemical reactions within the ambient air.[B67,B68,B69]

One of the main questions that we can address with this data base is how many genotoxic compounds have been found in the atmospheric or source air? This question has to be answered in light of the type of bioassay used, since data from one assay cannot be used in the same manner as information from another bioassay. For example, information from a bacterial bioassay cannot be applied for decision making in the same manner as information from a life-time rodent carcinogen bioassay. Table 15.10-3 shows, by test-type category, the number of compounds providing a positive response within each chemical class. Every chemical class, except carboxylic acids, contains some positively responding compounds. However, Table 15.10-1 shows that only 6 of the carboxylic

Table 15.10-3. Numbers of Positive Compounds by Bioassay and Chemical Category

•	Chemical Category	Type of Bioassay*							
		CCC	BAC	Y&F	INS	PLA	MC	MAM	HMA
2	INORGANICS	4	5	2	1	6	1	0	1
3	HYDROCARBONS	19	12	3	3	3	8	5	4
4	ETHERS	0	1	0	0	0	0	0	0
5	ALCOHOLS	0	1	0	0	8	0	0	0
6	KETONES	0	0	0	0	2	0	0	0
7	ALDEHYDES	1	4	1	1	0	2	0	0
8	CARBOXYLIC ACID DERIVATIVES	2	0	0	0	0	0	0	0
9	CARBOXYLIC ACIDS	0	0	0	0	0	0	0	0
10	HETEROCYCLIC OXYGEN COMPOUNDS	7	4	4	4	4	3	1	3
11	NITROGEN CONTAINING ORGANICS	12	22	2	5	11	11	6	5
12	SULFUR CONTAINING ORGANICS	1	1	1	2	1	1	0	0
13	HALOGEN CONTAINING ORGANICS	21	16	4	6	13	10	4	2
14	ORGANOMETALLIC COMPOUNDS	0	6	1	0	3	2	0	1
•	GRAND TOTALS	67	72	18	22	51	38	16	16

* See Table 1.4-1 for descriptions of the bioassay categories. Abbreviations are CCC, carcinogen bioassays; BAC, bacterial assays; Y&F, yeasts and fungi assays; INS, insect bioassays; PLA, plant bioassays; MC, mammalian cell (*in vitro*) bioassays; MAM, whole animal bioassays other than carcinogen bioassays; and HMA, host-mediated assays.

acids listed have been tested; therefore, it is not surprising that mutagens are not associated with this class. Three classes of compounds (hydrocarbons, nitrogen containing organics, and halogen containing organics) as a group contained well over one-half of the positive compounds for each category of bioassay. This, however, is not a true indicator of the types of compounds within the atmosphere which are responsible for any genotoxic activity seen. Instead, this is more an indicator of scientific interest among toxicologists in specific classes of compounds.

Table 15.10-4. Summary of Whole Animal Carcinogenicity Bioassays by Chemical Category

•	Category	Number Tested	Number of Compounds Positive	Indefinite	Negative
2	INORGANICS	7	4	2	1
3	HYDROCARBONS	23	19	4	0
4	ETHERS	0	0	0	0
5	ALCOHOLS	1	0	0	1
6	KETONES	4	0	3	1
7	ALDEHYDES	1	1	0	0
8	CARBOXYLIC ACID DERIVATIVES	3	2	1	0
9	CARBOXYLIC ACIDS	0	0	0	0
10	HETEROCYCLIC OXYGEN COMPOUNDS	8	7	1	0
11	NITROGEN CONTAINING ORGANICS	18	12	3	3
12	SULFUR CONTAINING ORGANICS	1	1	0	0
13	HALOGEN CONTAINING ORGANICS	25	21	4	0
14	ORGANOMETALLIC COMPOUNDS	3	0	3	0
•	GRAND TOTALS	94	67	21	6

Tables 15.10-4 through 15.10-11 provide a summary of tests performed within each category of bioassay. Table 15.10-4 summarizes the information for the combined list of potential rodent carcinogens; that is, for compounds that were tested in one or more rodent life-time carcinogen bioassays. Within the Gene-Tox literature, a compound is listed as a significant positive (SP) if the substance is positive in two or more species or in multiple separate experiments, or causes an unusually large increase in tumor incidence with regard to total incidence, site, or type of tumor, or causes a significant decrease in age of onset.[B37] A limited positive occurs when a clear positive response occurs; however, the constraints of the SP do not apply. For purposes of this overview, these two categories of positives have been combined into a single category. For purposes of this tabulation, the intermediate Gene-Tox categories between a positive and a negative response (e.g., inadequate positive, inadequate, inadequate

negative) have been replaced with a single category termed Indefinite (I). Likewise, the limited negative and significant negative categories have been combined for purposes of tabulation. Table 15.10-4 shows that 3% (94 of 2827) of the compounds were tested in a rodent carcinogen bioassay. These 94 compounds represent approximately 18% of the compounds for which there is adequate carcinogenicity bioassay data.[B37] Only 6 compounds are listed as negative. Again, most of the compounds and positive results are associated with three classes of chemicals: the hydrocarbons, the nitrogen containing organics, and the halogen containing organics. None of the atmospheric ether compounds or carboxylic acids have been tested. On examination of the less well tested classes in this and the other bioassays, it is apparent that other types of chemicals, e.g., the heterocyclic oxygen compounds, also may be important airborne carcinogens. Although these data are taken from a preliminary Gene-Tox registry, the percentages and numbers are not likely to change in a significant manner as more careful screening is performed.

The bacterial assays are the type of assay most commonly used to monitor or screen ambient air and source air mixtures for mutagenicity. Included in this class of assays are the DNA repair-deficient bacterial assays, the *E. coli* reverse mutation assays, and the *Salmonella typhimurium* assays. Since these assays, especially the Salmonella plate test, are used to test ambient particle-bound organics and mixtures of volatile organics collected from ambient air, it is important to understand how these assays respond to compounds within the various classes, and whether or not bacterial mutagenicity and carcinogenicity correlate. Of the 215 compounds within chapters 2 to 14 tested within bacterial bioassays, approximately one third (72) gave positive results (Table 15.10-5). These 215 compounds represent less than 15% of the compounds tested in these bacterial assays and published in the open literature.[B36] In contrast to the carcinogen bioassays, over half of the bacterial tests gave negative results. This finding appears to be due to at least two factors. First, researchers are more selective when choosing compounds for the carcinogen bioassays because of cost and time factors. Second, like each of the *in vitro* assays, bacterial assays do not detect the genotoxic activity of certain classes of compounds. For example, bacterial assays do not detect some of the halogenated organic carcinogens. This is due to a variety of factors such as differences in bacterial and mammalian cellular toxicity, differences in metabolism, and rates of exposure and elimination from the cellular environment. In spite of these differences, bacterial results correlate relatively well with carcinogenicity results. Within the publications and material available for this book, there were

Table 15.10-5. Summary of Bacterial Bioassays by Chemical Category

• Category	Number of Compound Tested	Positive Compound	Positive Tests	Indefinite Tests	Negative Tests
2 INORGANICS	13	5	5	4	7
3 HYDROCARBONS	42	12	14	16	23
4 ETHERS	3	1	1	0	3
5 ALCOHOLS	18	1	1	4	15
6 KETONES	7	0	0	3	4
7 ALDEHYDES	6	4	5	1	1
8 CARBOXYLIC ACID DERIVATIVES	3	0	0	1	2
9 CARBOXYLIC ACIDS	5	0	0	0	5
10 HETEROCYCLIC OXYGEN COMPOUNDS	13	4	6	5	6
11 NITROGEN CONTAINING ORGANICS	45	22	30	8	25
12 SULFUR CONTAINING ORGANICS	1	1	1	1	0
13 HALOGEN CONTAINING ORGANICS	52	16	21	16	33
14 ORGANOMETALLIC COMPOUNDS	11	6	9	6	9
• GRAND TOTALS	219	72	93	65	133

256 compounds (not all found in the atmosphere) for which there is both bacterial mutagenicity and whole animal carcinogenicity data available. Table 15.10-6 shows how the bacterial mutagenicity and animal carcinogenicity for these compounds correlate when all classes of chemicals are combined. Approximately, 70% of the carcinogenic significant positives (SP) are positive within these bacterial assays. Since there are only 2 significant negatives and 5 limited negatives within this data base, no meaningful correlation can be made between negative results. Within the indefinite carcinogens approximately 50% are also indefinite in bacterial assays, with the others being somewhat evenly split between positive and negative responses in the bacterial assays. If one examines specific chemical classes, this correlation changes. For example, there are 22 compounds in chapter 3 that have been tested for carcinogenicity and

Table 15.10-6. Correlation of Rodent Carcinogenicity Bioassays and *Salmonella typhimurium* Bioassays for 256 Compounds

Test	Carcinogen Bioassay Response*									
	SP**		LP		I		LN		SN	
Plate, +	75	(68.8)	28	(52.8)	17	(28.8)	2	(40.0)	0	(0.0)
Plate, I	20	(18.3)	17	(32.1)	27	(45.8)	0	(0.00)	1	(100.0)
Plate, Neg.	14	(12.8)	8	(15.1)	15	(25.4)	3	(60.0)	0	(0.0)
Plate, No. tested	109		53		59		5		1	
Plate, Not tested	16		5		7		0		1	
Preinc., +	29	(74.4)	2	(18.2)	4	(36.4)	0	(0.0)	0	(0.0)
Preinc., I	5	(12.8)	4	(36.4)	5	(45.5)	0	(0.0)	0	(0.0)
Preinc., Neg.	5	(12.8)	5	(45.5)	2	(18.2)	1	(100.0)	1	(100.0)
Preinc., No. Tested	39		11		11		1		1	
Preinc., Not Tested	86		47		55		4		1	
Both, +	86	(68.8)	29	(50.0)	21	(31.8)	2	(40.0)	0	(0.0)
Both, I	21	(16.8)	19	(32.8)	28	(42.4)	0	(0.0)	1	(50.0)
Both, Neg.	18	(14.4)	10	(17.2)	17	(25.8)	3	(60.0)	1	(50.0)
Both, Total	125	(100.0)	58	(100.0)	66	(100.0)	5	(100.0)	2	(100.0)

* Data is reported as number of compounds.

**Abbreviations: SP, significant positive; LP, limited positive; I, indefinite; LN, limited negative; SN, significant negative; +, positive; Neg., negative; No. tested, number tested; Both, when either or both assays are considered for each compound. Numbers in parentheses indicate per cent of agreement.

18 of these are positive, 4 are indefinite, and none are negative in bacterial assays. Half of the positive compounds were tested in a Salmonella bioassay and 8 of 9 (88.8%) are positive. The other compound is listed as giving an indefinite response. This result demonstrates that a researcher must be aware of the type of air shed under investigation or must know what types of compounds are being tested in order to take proper advantage of the less expensive and less time consuming bioassays. If one is monitoring a rural farming area that may have high concentrations of pesticides, a bacterial mutagenicity assay may not be the assay of choice. If one is monitoring an urban atmosphere impacted mainly by automotive exhaust and woodsmoke, however, bacterial assays are more likely to be the methods of choice. Researchers are continuing to develop new

Table 15.10-7. Summary of Yeast and Fungi Bioassays by Chemical Category

•	Category	Number of Cmpd. Tested	Positive Cmpd.	Positive Tests	Indefinite Tests	Negative Tests
2	INORGANICS	4	2	6	0	2
3	HYDROCARBONS	4	3	3	1	0
4	ETHERS	0	0	0	0	0
5	ALCOHOLS	8	0	0	0	8
6	KETONES	0	0	0	0	0
7	ALDEHYDES	1	1	2	0	0
8	CARBOXYLIC ACID DERIVATIVES	0	0	0	0	0
9	CARBOXYLIC ACIDS	0	0	0	0	0
10	HETEROCYCLIC OXYGEN COMPOUNDS	4	4	5	0	0
11	NITROGEN CONTAINING ORGANICS	7	2	3	2	4
12	SULFUR CONTAINING ORGANICS	2	1	3	0	1
13	HALOGEN CONTAINING ORGANICS	5	4	5	0	2
14	ORGANOMETALLIC COMPOUNDS	1	1	1	0	0
•	GRAND TOTALS	36	18	28	3	17

bacterial strains and methods for mutagenicity testing in order to be able to detect those carcinogens missed by current assays. With these efforts at improvement and with the already proven usefulness of these assays, bacterial assay methods will continue to be one of the main biological tools used in the study of atmospheric compounds and their toxicity.

Table 15.10-7 provides a summary of the yeasts and fungi bioassays by chemical category. Of the 36 atmospheric compounds tested with these organisms, 18 are positive. Although 4 classes of compounds (ethers, ketones, carboxylic acid derivatives, and carboxylic acids) have no associated bioassay results, no one class of compounds stands out among those that were tested.

Of the 62 atmospheric compounds tested in Drosophila, only 22, or about one third, were positive (Table 15.10-8). Half of the positive

Table 15.10-8. Summary of Insect (Drosophila) Bioassays by Chemical Category

•	Category	Number of				
		Cmpd. Tested	Positive Cmpd.	Positive Tests	Indefinite Tests	Negative Tests
2	INORGANICS	4	1	1	2	1
3	HYDROCARBONS	6	3	3	3	0
4	ETHERS	0	0	0	0	0
5	ALCOHOLS	1	0	0	1	0
6	KETONES	1	0	0	1	0
7	ALDEHYDES	1	1	1	0	0
8	CARBOXYLIC ACID DERIVATIVES	0	0	0	0	0
9	CARBOXYLIC ACIDS	0	0	0	0	0
10	HETEROCYCLIC OXYGEN COMPOUNDS	5	4	4	0	1
11	NITROGEN CONTAINING ORGANICS	10	5	5	3	2
12	SULFUR CONTAINING ORGANICS	3	2	2	1	0
13	HALOGEN CONTAINING ORGANICS	24	6	6	12	6
14	ORGANOMETALLIC COMPOUNDS	7	0	0	3	4
•	GRAND TOTALS	62	22	22	26	14

compounds were classified as either nitrogen containing organics or as halogen containing organics. The other eleven were divided among hydrocarbons, aldehydes, heterocyclic oxygen compounds, and sulfur containing organics. Over a third of the atmospheric compounds that were tested gave indefinite results. There are insect bioassays other than those used here; however, only the sex-linked recessive lethal assay has been used to any significant degree for the screening of chemical compounds.

Plant systems have been used for *in situ* screening of the ambient air and have been used to test volatile organics and gases that are difficult to test using other bioassays. Although many plant assays are ideal for screening volatile atmospheric compounds, only about 3% of the atmospheric compounds listed here were tested with plant assays. Of the

Table 15.10-9. Summary of Plant Bioassays by Chemical Category

•	Category	Number of				
		Cmpd. Tested	Positive Cmpd.	Positive Tests	Indefinite Tests	Negative Tests
2	INORGANICS	7	6	8	0	2
3	HYDROCARBONS	3	3	3	0	0
4	ETHERS	0	0	0	0	0
5	ALCOHOLS	17	8	9	3	7
6	KETONES	4	2	2	0	2
7	ALDEHYDES	0	0	0	0	0
8	CARBOXYLIC ACID DERIVATIVES	0	0	0	0	0
9	CARBOXYLIC ACIDS	0	0	0	0	0
10	HETEROCYCLIC OXYGEN COMPOUNDS	4	4	4	0	0
11	NITROGEN CONTAINING ORGANICS	14	11	13	2	5
12	SULFUR CONTAINING ORGANICS	2	1	1	0	1
13	HALOGEN CONTAINING ORGANICS	14	13	16	0	1
14	ORGANOMETALLIC COMPOUNDS	3	3	4	0	0
•	GRAND TOTALS	68	51	60	5	18

68 compounds tested (Table 15.10-9), 51 or 75% were positive. In contrast to the other bioassays, the plant assays have been used to test a number of alcohols, for which 8 of 9 were positive. The only other assay in which there is a positive response recorded for an alcohol is the bacterial assay. None of the atmospheric ethers, aldehydes, carboxylic acids, or carboxylic acid derivatives were tested in the plant systems reviewed. The available references listed 311 compounds that have been assayed with plant systems; therefore 21% of the screened compounds are known to be found in the atmosphere.

Mammalian cell assay systems allow the screening of compounds within eukaryotic cells that in many ways are similar to human cells. As pointed out in the first chapter, these systems, unfortunately, are more labor intensive, more costly, and sometimes more difficult to interpret than other

Table 15.10-10. Summary of *In Vitro* (Tissue Culture) Mammalian Cell Bioassay Systems by Chemical Category

•	Category	Number of				
		Cmpd. Tested	Positive Cmpd.	Positive Tests	Indefinite Tests	Negative Tests
2	INORGANICS	3	1	1	0	2
3	HYDROCARBONS	14	8	31	10	22
4	ETHERS	0	0	0	0	0
5	ALCOHOLS	2	0	0	0	5
6	KETONES	4	0	0	0	7
7	ALDEHYDES	2	2	2	0	0
8	CARBOXYLIC ACID DERIVATIVES	2	0	0	2	0
9	CARBOXYLIC ACIDS	0	0	0	0	0
10	HETEROCYCLIC OXYGEN COMPOUNDS	3	3	8	0	0
11	NITROGEN CONTAINING ORGANICS	13	11	30	3	6
12	SULFUR CONTAINING ORGANICS	1	1	3	0	0
13	HALOGEN CONTAINING ORGANICS	14	11	16	1	4
14	ORGANOMETALLIC COMPOUNDS	2	2	2	0	0
•	GRAND TOTALS	60	39	93	16	46

in vitro systems. These systems, however, can be used to screen for gene mutations, chromosomal aberrations, and cell transformation; therefore, the results can be more directly compared to whole animal and/or human results when the latter are available. Table 15.10-10 summarizes the overall response of these *in vitro* mammalian cell systems for the known atmospheric compounds. Approximately 2% of the atmospheric compounds listed were tested in these types of assays. Since this bioassay category is a grouping of 9 different types of tests, it is apparent that the screening of airborne compounds, to date, by mammalian cell systems is not extensive. It is also readily apparent that the emphasis for testing has been placed on compounds that fall into 3 chemical categories: hydrocarbons, nitrogen containing organics, and halogen containing organics.

Table 15.10-11. Summary of *In Vivo* Mammalian Bioassays by Chemical Category

•	Category	Number of				
		Cmpd. Tested	Positive Cmpd.	Positive Tests	Indefinite Tests	Negative Tests
2	INORGANICS	7	0	0	2	9
3	HYDROCARBONS	14	5	14	12	10
4	ETHERS	1	0	0	0	1
5	ALCOHOLS	7	0	0	5	5
6	KETONES	0	0	0	0	0
7	ALDEHYDES	0	0	0	0	0
8	CARBOXYLIC ACID DERIVATIVES	0	0	0	0	0
9	CARBOXYLIC ACIDS	0	0	0	0	0
10	HETEROCYCLIC OXYGEN COMPOUNDS	5	1	2	4	4
11	NITROGEN CONTAINING ORGANICS	13	6	6	13	17
12	SULFUR CONTAINING ORGANICS	0	0	0	0	0
13	HALOGEN CONTAINING ORGANICS	12	4	6	7	9
14	ORGANOMETALLIC COMPOUNDS	3	0	0	2	6
•	GRAND TOTALS	62	16	28	45	61

For the *in vivo* mammalian bioassay systems, the literature survey showed that approximately 350 compounds had been tested in at least one of several assays. These assays also screen a variety of endpoints, including gene mutation, chromosomal aberrations, and sperm morphology. Generally, these tests are more expensive and require more time than *in vitro* bioassays; however, they provide several advantages. For example, in many cases the animals can be exposed by the same route as humans are exposed. Also, there is obviously whole animal metabolism rather than metabolism of a single cell type. These and other advantages make the use of *in vivo* assays preferable when hazard and/or risk estimates are needed. As shown in Table 15.10-11, 62 compounds in chapters 2 to 14 have been tested in these types of assays and 16 of these compounds gave positive results. Of the 16 positive compounds, 15 are either hydrocarbons, nitrogen containing organics, or halogen containing organics. The only other positive compound is a heterocyclic oxygen compound. Two other classes, inorganics and alcohols, had a moderate number of compounds tested, but none were positive. Since so few compounds have been tested by a number of assay systems, it is difficult to know whether or not atmospheric chemicals are likely to have a significant impact upon these various endpoints.

The final bioassay category contains all of the host-mediated bioassays. In these bioassays, an indicator organism or tissue culture cell line is injected into a host animal along with the chemical under consideration. The primary idea is to provide whole animal metabolism while at the same time using a more rapid and less expensive indicator system. Only a small number of atmospheric compounds (34) have been tested using this approach (Table 15.10-12). Although approximately half of these gave positive results, no one of these systems has been used extensively enough to make any generalized observations valid.

It is accepted, but not proven, that carcinogens within the environment play a major role in the initiation and promotion of most cancers. While chemical compounds enter the body by absorption and ingestion, inhalation is the most common route of exposure in the environment. The recognition of the potentially important role of atmospheric genotoxicants to human health prompted the authors to identify which known atmospheric compounds are also known genotoxicants. According to Sawicki,[B70] "The main barriers to estimating human environmental risk stem from a lack of knowledge of the chemical composition of our environment, the failure to use the information we have, and our indecision as to what to measure." The information surveyed here is intended to help to overcome these barriers. The tabulations show that humans are exposed to a number of carcinogens and other genotoxicants. They also demonstrate, however, that many chemicals, some structurally very similar to the identified carcinogens, have not been bioassayed. The assessment of biological activity is necessary before either hazard assessment, risk assessment, and/or risk management can effectively occur. The present summary also shows that although continued interest should be placed on hydrocarbons, nitrogen containing organics, and halogen containing organics, greater emphasis should be placed on the identification and bioassay of other types of chemicals (e.g., heterocyclic oxygen compounds, aldehydes, carboxylic acids and their derivatives, and ketones).

Table 15.10-12. Summary of Host-Mediated Bioassays by Chemical Category

•	Category	Number of				
		Cmpd. Tested	Positive Cmpd.	Positive Tests	Indefinite Tests	Negative Tests
2	INORGANICS	1	1	1	0	0
3	HYDROCARBONS	9	4	4	1	7
4	ETHERS	0	0	0	0	0
5	ALCOHOLS	1	0	0	1	0
6	KETONES	0	0	0	0	0
7	ALDEHYDES	0	0	0	0	0
8	CARBOXYLIC ACID DERIVATIVES	0	0	0	0	0
9	CARBOXYLIC ACIDS	0	0	0	0	0
10	HETEROCYCLIC OXYGEN COMPOUNDS	4	3	3	0	1
11	NITROGEN CONTAINING ORGANICS	9	5	11	3	5
12	SULFUR CONTAINING ORGANICS	0	0	0	0	0
13	HALOGEN CONTAINING ORGANICS	7	2	2	1	8
14	ORGANOMETALLIC COMPOUNDS	3	1	1	0	4
•	GRAND TOTALS	34	16	22	6	25

15.11 Concluding Remarks

At the time of the second World War, barely two dozen different chemical species were known to be present in the Earth's atmosphere. A decade later, as scientists began investigating the chemical and biological implications of atmospheric composition, that number was still less than a hundred. Today nearly three thousand species are identified; many have been subject to detailed study. As is often the case in any immature, interdisciplinary field, however, the information now available is scattered very widely and is presented and discussed from different perspectives by the different scientific disciplines involved. To ameliorate this situation and to explore interrelationships and insights in this large amount of data, this book has presented, tabulated, and discussed atmospheric and bioassay information on 2827 chemical species. Separate compilations were provided for species present in the atmosphere in the gas phase, aerosol particles, cloud droplets, fog droplets, atmospheric ice, rain, snow, the stratosphere, and in indoor air.

The sources, concentrations, and genotoxicity of the atmospheric compounds are incredibly diverse, providing the wealth of detailed information that constitutes the heart of this book. Once one performs grouping and ordering of the trace atmospheric compounds, however, certain patterns and trends in the data become apparent and justify comment, as has been done in the final chapter. This final section is intended to briefly summarize the entire volume and to assess the status of knowledge of the chemistry and effects of atmospheric trace species.

Atmospheric chemistry is of interest and importance largely as a consequence of the fluxes and chemical natures of its "feedstocks": the inorganic compounds (Chapter 2), the hydrocarbons (Chapter 3), and light. Their interactions, often driven by sunlight, produce a very wide spectrum of oxygenated products. These products constitute substantial portions of Chapters 5-9. Other compounds containing oxygen, notably the ethers (Chapter 4) and the heterocycles (Chapter 10), are not abundant in the atmosphere and appear to play only minor roles in atmospheric chemistry.

The atmospheric roles of organic nitrogen and sulfur compounds (Chapters 11 and 12) are not yet fully explored, but it seems likely that their effects will be more important in condensed phases than in the gas phase. For example, the sulfur compounds are the precursors of sulfate and thus are involved in acidity and corrosion concerns. Nitro compounds, abundant in condensed phases, have significant biological impacts.

The halogenated species include some naturally-produced compounds, mostly from marine sources, and a large number of anthropogenically-produced compounds. The latter are of great utility in a variety of commercial and industrial applications. They are quite unlike most natural molecules, and neither atmospheric nor biological processes are efficient in reacting with them. Thus many are long lived and harmful in the environment.

A few organometallic compounds can be found in the atmosphere. Most are anthropogenic in origin and some of these are present in sufficiently high concentrations to be worthy of notice.

The individual compounds, nearby 3000 in all, may be present in the atmosphere in any of several phases: the gas phase, the aerosol phase, several aqueous phases, or indoors in both gas and aerosol phases. Of these, only the tropospheric and stratospheric gas phases are reasonably well understood. The aerosol phase has received extensive analytical attention, but little theoretical study. The aqueous phases require increased attention in both analysis and theory, as does the indoor atmospheric environment.

A mere handful of oxidizing species drive the vast majority of atmospheric chemical transitions. In the gas phase, the hydroxyl radical is by far the most important of these oxidizers. It reacts with virtually all atmospheric compounds, limiting the lifetime of most of them to a few hours or a few days. OH· is without doubt the dominant atmospheric vacuum cleaner, yet precious few measurements of its concentration are available.

In the liquid phase, hydrogen peroxide performs a number of oxidizing roles. Other species such as OH· or O_3 may also be chemically significant, however. In both the atmospheric liquid phase and aerosol phase, the major oxidation processes remain to be delineated.

It is common for people to regard the study of the chemistry and effects of atmospheric compounds as the study of anthropogenic molecules, and certainly anthropogenic molecules play important roles. Nonetheless, naturally emitted compounds are present in great abundance and diversity. Just as anthropogenic compounds dominate atmospheric chemistry in urban areas, so natural compounds have the potential to dominate atmospheric chemistry in marshland, in forests, and in the stratosphere following volcanic eruptions.

The occurrence and interactions of chemical compounds in the atmosphere would be of merely academic interest if it were not for the effects of these compounds on flora, fauna, man-made materials, and atmospheric stability itself. We have treated the biological impacts of these compounds in some detail, noting that polynuclear aromatic hydrocarbons, nitro compounds, and halogenated species are the potentially harmful groups. Most of these compounds are produced by man, either overtly or inadvertently. Although it is beyond the scope of this book to give similar treatment of the effects of atmospheric compounds on vegetation, materials, and atmospheric structure, we have addressed these concerns briefly. Continued research in all these impact areas is warranted.

This book has presented an array of information on atmospheric compounds, their chemistry, and their effects, reflecting the efforts of a large and dedicated group of scientists. It is apparent that there exists a moderate degree of understanding of many atmospheric chemical processes. Nonetheless, much remains to be done. Ambitious measurement programs, laboratory experiments, and computational studies are needed to assess a number of potential concerns. For few subjects will the results be as interesting or as potentially vital to mankind as will be those from the study of the sources, occurrence, chemistry, and bioassay of atmospheric chemical compounds.

References

The references are an integral part of this book since they provide the detailed information on detection, identification, site location, source type and proximity, and bioassay. Full reference information, including complete titles and inclusive pages, is given. In the initial iteration of this work, the references were alphabetized, a practice that was later abandoned. As a result, the first 278 references are alphabetically ordered. To permit the reader to search the entire list for references by a particular author, we provide an author index in an appendix.

The bioassay literature is almost completely distinct from the atmospheric chemical literature. Each was assembled independently for this compendium and we retain the division in our listings. References to the bioassay literature are indicated by a B preceding the reference number

Four abbreviations are used in the reference lists to aid the reader in obtaining access to specific documents. These are as follows:

APTIC Air Pollution Technical Information Center, United States Environmental Protection Agency, Research Triangle Park, NC 27711.

NTIS National Technical Information Service, US Department of Commerce, 5285 Port Royal Road, Springfield, VA 22151

SAHB Smoking and Health Bulletin, Technical Information Center, National Clearinghouse for Smoking and Health, Center for Disease Control, 1600 Clifton Road, N.E., Bldg. 14, Atlanta, GA 30333

CA Chemical Abstracts, Columbus, OH. The citation includes volume number, abstract number, and year.

[1] Aaronson, A. E. and R. A. Matula., Diesel odor and the formation of aromatic hydrocarbons during the heterogeneous combustion of pure cetane in a single-cylinder diesel engine, *Proc. 13th Int. Comb. Symp.*, Combustion Inst. Pittsburgh, Pa., p. 471-481, 1971.

[2] Abdoh, Y., N. Aghdaie, M. R. Darvich and M. H. Khorgami, Detection of some polynuclear aromatic hydrocarbons and determination of benzo(a) pyrene in Teheran atmosphere, *Atmos. Environ.*, **6**, 949-952, 1972.

[3] Abeles, F. B. and H. E. Heggested, Ethylene: an urban air pollutant, *J. Air Poll. Contr. Assoc.*, **23**, 517-521, 1973.

[4] Abelson, P. H., A damaging source of air pollution, *Science*, **158**, 1527, 1967.

[5] Altshuller, A. P. and S. P. McPherson, Spectrophotometric analysis of aldehydes in the Los Angeles atmosphere, *J. Air Poll. Contr. Assoc.*, **13**, 109-111, 1963.

[6] Altshuller, A. P., W. A. Lonneman, F. D. Sutterfield and S. L. Kopczynski, Hydrocarbon composition of the atmosphere in the Los Angeles basin-1967, *Environ. Sci. Technol.*, **5**, 1009-1016, 1971.

[7] Anonymous, Ford plant filters noxious vapors with activated carbon system, *Filtration Eng.* **1**, 4-5, 1969.

[8] Anonymous, Environmental protection and cultivation of the environment in lower Saxony, *Staedtehygiene (Uelzen/Hamburg)*, **22**, 266-267, 1971. (APTIC No. 37190).

[9] Antonyuk, O. K., Hygienic assessment of glycols present in polymer type building materials, *Gigena i Sanit.*, **9**, 106-107, 1974. (APTIC No. 71466).

[10] Augustine, F. E., *Airborne sampling of particles emitted to the atmosphere from Kraft paper mill processes and their characterization by electron microscopy*, Oregon State University (Corvallis, Oregon), Ph.D Dissertation, 1974,

[11] Babina, M. D., Determination of volatile substances emitted into the air by shoe industry plants, *Nov. Obl. Prom.-Sanit. Khim. (S.I. Muraveva, ed.)*, 227-232, 1969. (APTIC No. 33508).

[12] Balabaeva, L. and G. Petrova, Contamination of atmospheric air with fluoride compounds and their exeration in urine of persons, *Khig. Zdraueopazyane (Bulgaria)*, **15**, 162-168, 1972. (APTIC No. 47982).

[13] Ball, G. and E. A. Boettner, Combustion products of plastics and their contribution to incineration problems, *Am. Chem. Soc., Div. Water, Air, Waste Chem., Gen. Papers*, **10**, 236-239, 1970.

[14] Bandy, A. R., Studies of the importance of biogenic hydrocarbon emissions to the photochemical oxidant formation in Tidewater, Va., *Scientific Seminar on Automotive Pollutants*, EPA 600/9-75-003, Environmental Protection Agency, Washington, D.C., 1975.

[15] Barber, E. D. and J. P. Lodge, Jr., Paper chromatographic identification of carbonyl compounds as their 2,4- dinitrophenylhydrazones in automobile exhaust, *Anal. Chem.*, **35**, 348-350, 1963.

[16] Bentley, M. D., I. B. Douglass, J. A. Lacadie and D. R. Whittier, The photolysis of dimethyl sulfide in air, *J. Air Poll. Contr. Assoc.*, **22**, 359-363, 1972.

[17] Berglund, B., U. Berglund and T. Lindvall, Measurement of rapid changes of odor concentration by a signal detection approach, *J. Air Poll. Cont. Assoc.*, **24**, 162-164, 1974.

[18] Bethea, R. M. and R. S. Narayan, Identification of beef cattle feedlot odors, *Trans. Am. Soc. Agr. Engrs.*, **15**, 1135-1137, 1972.

[19] Bethge, P. O. and L. Ehrenborg, Identification of volatile compounds in Kraft mill emissions, *Svensk Papperstidning*, **70**, 347-350, 1967.

[20] Bidleman, T. F. and C. E. Olney, Chlorinated hydrocarbons in the Sargasso Sea atmosphere and surface water, *Science*, **183**, 516-518, 1974.
[21] Bidleman, T. F. and C. E. Olney, Long range transport of toxaphene insecticide in the atmosphere of the western North Atlantic, *Nature*, **257**, 475-477, 1975.
[22] Bigg, E. K., A. Ono and J. A. Williams, Chemical tests for individual submicron aerosol particles, *Atmos. Environ.*, **8**, 1-13, 1974.
[23] Breeding, R. J., J. P. Lodge, Jr., J. B. Pate, D. C. Sheesley, H. B. Klonis, B. Fogle, J. A. Anderson, T. R. Englert, P. L. Haagenson, R. B. McBeth, A. L. Morris, R. Pogue and A. F. Wartburg, Background trace gas concentrations in the Central United States, *J. Geophys. Res.*, **78**, 7057-7064, 1973.
[24] Brinkmann, W. L. F. and U. de M Santos, The emission of biogenic hydrogen sulfide from Amazonian floodplain lakes, *Tellus*, **26**, 261-267, 1974.
[25] Burnett, W. E., Air pollution from animal wastes: determination of maloders by gas chromatographic and organoleptic techniques, *Environ. Sci. Technol.*, **3**, 744-749, 1969.
[26] Busso, R. H., *Identification and determination of pollutants emitted by the principal types of urban waste incinerator plants (final report)*, Centre d Etudes Et Recherches des Charbonnages de France, Laboratoire du CERCHAR, Creil, France, Contract 69-01-758, A.R. 144, 1971.
[27] Caban, R. and T. W. Chapman, Losses of mercury from chlorine plants: a review of a pollution problem, *Am. Inst. Chem. Engrs. J.*, **18**, 892-903, 1972.
[28] Cadle, R. D., A. F. Wartburg, W. H. Pollock, B. W. Gandrud and J. P. Shedlovsky, Trace constituents emitted to the atmosphere by Hawaiian volcanos, *Chemosphere*, **2**, 231-234, 1973.
[29] Cadle, R. D., H. H. Wickman, C. B. Hall and K. M. Eberle, The reaction of atomic oxygen with formaldehyde, crotonaldehyde, and dimethyl sulfide, *Chemosphere*, **3**, 115-118, 1974.
[30] Cadle, R. D., Volcanic emissions of halides and sulfur compounds to the troposphere and stratosphere, *J. Geophys. Res.*, **80**, 1650-1652, 1975.
[31] Carotti, A. A. and E. R. Kaiser, Concentrations of 20 gaseous chemical species in the flue gas of a municipal incinerator, Paper presented at 64th Annual Meeting, Air Poll. Contr. Assoc. (Atlantic City, NJ), 1971.
[32] Carter, R. V. and B. Linsky, Gaseous emissions from whisky fermentation units, *Atmos. Environ.*, **8**, 57-62, 1974.
[33] Cavanagh, L. A., C. F. Schadt and E. Robinson, Atmospheric hydrocarbon and carbon monoxide measurements at Point Barrow, Alaska, *Environ. Sci. Technol.*, **3**, 251-257, 1969.
[34] Challenger, F., Biological methylation, *Chem. Rev.*, **36**, 315-361, 1936.
[35] Challis, E. J., The approach of industry to the assessment of environmental hazards, *Proc. Roy. Soc. B*, **185**, 183-197, 1974.
[36] Charlson, R. J., A. P. Waggoner, N. C. Ahlquist, D. S. Covert and R. Husar, Sulfates in lower tropospheric aerosols, Paper 76-20.2, 69th Annual Meeting, Air Poll. Contr. Assoc. (Portland, OR), 1976.
[37] Clemons, C. A., A. I. Coleman and B. E. Saltzman, Concentration and ultrasensitive chromatographic determination of sulfur hexafluoride for application to meteorological tracing, *Environ. Sci. Technol.*, **2**, 551-556, 1968.
[38] Collier, L., Determination of bis-chloromethyl ether at the ppb. level in air samples by high resolution mass spectroscopy, *Environ. Sci. Technol.*, **6**, 930-932, 1972.
[39] Collins-Williams, C., H. K. Kuo, E. A. Varga, S. Davidson, D. Collins-Williams and M. Fitch, Atmospheric pollen counts in Toronto, Canada, 1971, *Ann. Allergy*, **31**(2), 65-68, 1973.

[40] Collins-Williams, C., H. K. Kuo, D. N. Garey, S. Davidson, D. Collins-Williams, M. Fitch and J. B. Fischer, Atmospheric mold counts in Toronto, Canada, 1971, *Ann. Allergy*, **31**(2), 69-71, 1973.

[41] Cooper, R. L. and A. J. Lindsey, Atmospheric pollution by polycyclic hydrocarbons, *Chem. Ind. (London)*, 1177-1178, 1953.

[42] Cox, R. A. and F. J. Sandalls, The photo-oxidation of hydrogen sulphide and dimethyl sulphide in air, *Atmos. Environ.*, **8**, 1269-1281, 1974.

[43] Davis, D. D., G. Smith and G. Klauber, Trace gas analysis of power plant plumes via aircraft measurement: O_3, NO_x, SO_2 chemistry, *Science*, **186**, 733-736, 1974.

[44] Davison, A. W., A. W. Rand and W. E. Betts, Measurement of atmospheric fluoride concentrations in urban areas, *Environ. Pollut.*, **5**, 23-33, 1973.

[45] Dept. of Nature Conservation, *Air protection problems in the forestry industry*, Comm. on Air Protection in the Forestry Industry, (Stockholm), 1969. (APTIC No. 65119).

[46] Dept. of Trade and Industry, Great Britain, Programmer Analysis Unit, Appendix II: The contribution of plastics to air pollution in the United Kingdom, *An Economic and Technical Appraisal of Air Pollution in the United Kingdom*, (H.M. Stationery Office, London), p. 233-243, 1972.

[47] Dimitriades, B., B. H. Eccleston and R. W. Hurn, An evaluation of the fuel factor through direct measurement of photochemical reactivity of emissions, *J. Air Poll. Contr. Assoc.*, **20**, p. 150-160, 1970.

[48] Dimitriades, B. and T. C. Wesson, Reactivities of exhaust aldehydes, *J. Air Poll. Contr. Assoc.*, **22**, 33-38, 1972.

[49] Dishart, K. T., Exhaust HC composition: its relation to gasoline composition, Paper presented at 35th Midyear Meeting, Amer. Petroleum Inst. (Houston, Texas), 1970.

[50] Dow Chemical U.S.A., Dow chlorinated solvents and the clean air act, Inorganic Chemicals Dept, (Midland, Mich.), 1972.

[51] Dubois, L., T. Teichman and J. L. Monkman, The sulfuric acid content of soot, *Sci. Total Environ.*, **2**, 97-100, 1973.

[52] Dubrovskaya, F. I., *Gigiena i Sanit.*, **31**(1), 97-98, 1966.

[53] Dudley, H. C. and J. M. Dalla Valle, A study of the odors generated in the manufacture of Kraft paper, *Paper Trade J.*, **108**, 30-33, 1939.

[54] Ekstedt, J. and S. Oden, *Chlorinated hydrocarbons in the lower atmosphere in Sweden*, Forskningsavdelningen for Miljoward Inst. for Markvetenskap, Lantbrukshogskolan (Uppsala, Sweden), (APTIC No. 68784).

[55] Ellis, C. F., R. F. Kendall and B. H. Eccleston, Identification of some oxygenates in automobile exhausts by combined gas liquid chromatography and infrared techniques, *Anal. Chem.*, **37**, p. 511-516, 1965.

[56] Engdahl, R. B., Stationary combustion sources, in *Air Pollution, III*, (A. C. Stern, ed.) 2nd ed., New York: Academic Press, p. 3-54, 1968.

[57] Environmental Protection Agency, *Air quality criteria for hydrocarbons*, Report No. AP-64, Washington, D.C., 1970.

[58] Environmental Protection Agency, *Air Pollution Emission Factors*, (original plus supplements), Report No. AP-42, Research Triangle Park, (N.C.), 1973-75.

[59] Environmental Protection Agency, *Vinyl Chloride Monomer Investigation*, Tenneco Chemical, Inc., Flemington, New Jersey, Office of Enforcement and General Counsel, EPA, New Jersey, 1974.

[60] Environmental Protection Agency, *Guideline on use of reactivity criteria in control of organic emissions for reduction of atmospheric oxidants*, Washington, D.C., August 13, 1975.

[61] Eshleman, A., S. M. Siegel and B. Z. Siegel, Is mercury from Hawaiian volcanoes a natural source of pollution?, *Nature*, **233**, 471-472, 1971.

[62] Faith, W. L., Food and feed industries, in *Air Pollution, III*, (A. C. Stern, ed.) 2nd ed., New York: Academic Press, p. 269-288, 1968.

[63] Farwell, S. O., F. W. Bowes and D. F. Adams, Determination of chlorophenoxy herbicides in air by gas chromatography/mass spectrometry: selective ion monitoring, *Anal. Chem.*, **48**, 420-426, 1976.

[64] Fawcett, R. L., Air pollution potential of phthalic anhydride manufacture, *J. Air Poll. Contr. Assoc.*, **20**, 461-465, 1970.

[65] Finkelsteyn, D. H., A. P. Yelenevich and V. N. Dymchenko, Chemical composition of the dispersed phase of smokes generated in the process of manganese steel meeting, *Gigiena i Sanit.*, **11**(1), 25, 1953. (APTIC No. 40757).

[66] Foote, R. S., Mercury vapor concentrations inside buildings, *Science*, **177**, 513-514, 1972.

[67] Fracchia, M. F., F. J. Schuette and P. K. Mueller, A method for sampling and determination of organic carbonyl compounds in automobile exhaust, *Environ. Sci. Technol.*, **1**, 915-922, 1967.

[68] Fraser, H. S. and E. P. Swan, *Chemical analysis of veneer-dryer condensates*, Western Forest Products Lab. (Canadian Forestry Service, Vancouver, B.C.). (NTIS Document PB-212661).

[69] Freed, V. H., R. Haque and D. Schmedding, Vaporization and environmental contamination by DDT, *Chemosphere*, **1**, 61-66, 1972.

[70] Fujii, T., Offensive odors. 13. Constituents of, and countermeasures for, the odoriferous fumes from vinyl chloride 'leather' factories, *J. Japan Soc. Air Poll.*, **2**(1), 48-59, 1967.

[71] Fujii, T., N. Tajima, K. Yoshimura, K. Inoue, K. Taguchi, F. Kure, Y. Nishikawa and K. Oka, Studies on the air pollution by exhaust gases of aircraft(1), *J. Japan Soc. Air Poll.*, **8**, 515, 1973. (APTIC No. 58368).

[72] Gassmann, M., Freon 14 in purest krypton and in the atmosphere, *Naturwissenschaften*, **61**, 127, 1974.

[73] Gautier, A., Fluorine is an element always present in emissions from the Earth's core, *Compt. Rend.*, **157**, 820-825, 1913.

[74] Gerhold, H. D. and G. H. Plank, Monoterpene variations in vapor from white pines and hybrids, *Phytochemistry*, **9**, 1393-1398, 1970.

[75] Gerstle, R. W., *Atmospheric emissions from asphalt roofing processes*, PEDCo-Environmental, Inc. (Cincinnati, Ohio), 1974. (NTIS Document, PB-238445).

[76] Gerstle, R. W. and T. W. Devitt, Chlorine and hydrogen chloride emissions and their control, Paper 71-25, 64th Annual Meeting, Air Poll. Contr. Assoc. (Atlantic City, NJ), 1971.

[77] Giam, C. S., H. S. Chan and G. S. Neff, Rapid and inexpensive method for detection of polychlorinated biphenyls and phthalates in air, *Anal. Chem.*, **47**, 2319-2320, 1975.

[78] Gilbert, T. E., Rate of evaporation of liquids into air, *J. Paint Technol.*, **43**, 93-97, 1971.

[79] Gilbert, J. A. S. and A. J. Lindsey, Polycyclic hydrocarbons in tobacco smoke: pipe smoking experiments, *Brit. J. Cancer (London)*, **10**, 646-648, 1956.

[80] Goedseels, V., Evaluation of the odorous emissions in relation to the infrastructure of intense cattle raising, *Ingenieursblad (Holland)*, **42**, 557-564, 1973. (APTIC No. 57342).

[81] Gordon, R. J. and R. J. Bryan, Ammonium nitrate in airborne particles in Los Angeles, *Environ. Sci. Technol.*, **7**, 645-647, 1973.

[82] Gorodinskiy, S. M., M. I. Vakar, G. A. Gaziyev, Y. Y. Sotnikov and A. N. Mazin, Hygienic and chemical investigation of the hydrocarbons in the gas mixture in isolating pressure chambers, *Gigiena i Sanit.*, **7**, 106-109, 1974. (APTIC No. 66648).

[83] Graedel, T. E., B. Kleiner and C. C. Patterson, Measurements of extreme concentrations of tropospheric hydrogen sulfide, *J. Geophys. Res.*, **79**, 4467-4473, 1974.

[84] Gray, E. W., L. G. McKnight and J. M. Sawina, Identity and interactions of ions from relay break arcs, *J. Appl. Phys.*, **45**, 661-666, 1974.

[85] Grimsrud, E. P. and R. A. Rasmussen, Survey and analysis of halocarbons in the atmosphere by gas chromatography-mass spectroscopy, *Atmos. Environ.*, **9**, 1014-1017, 1975.

[86] Grob, K. and G. Grob, Gas-liquid chromatographic-mass spectrometric investigation of C_6-C_{20} organic compounds in an urban atmosphere, *J. Chromatogr.*, **62**, 1-13, 1971.

[87] Gross, G. P., *Gasoline composition and vehicle exhaust gas polynuclear aromatic content*, CRC-APRAC Project No. CAPE-6-68, Coordinating Research Council (New York, New York), 1973.

[88] Hansen, C. M., Solvents for coatings, *Chemtech*, 547-553, 1972.

[89] Hanst, P. L., L. L. Spiller, D. M. Watts, J. W. Spence and M. F. Miller, Infrared measurement of fluorocarbons, carbon tetrachloride, carbonyl sulfide, and other atmospheric trace gases, *J. Air Poll. Contr. Assoc.*, **25**, 1220-1226, 1975.

[90] Hanst, P. L., W. E. Wilson, R. K. Patterson, B. W. Gay, Jr., L. W. Chaney and C. S. Burton, *A spectroscopic study of California smog*, EPA 650/4-75-006, Environmental Protection Agency (Washington, D.C.), 1975. (NTIS Report PB 241022).

[91] Harvey, G. R. and W. G. Steinhauer, Atmospheric transport of polychlorobiphenyls to the North Atlantic, *Atmos. Environ.*, **8**, 777-782, 1974.

[92] Hayashi, M. and S. Hayashi, Environmental pollution caused by processing polyvinyl chloride resin, *Proc. Osaka Pub. Health Inst. (Ed. Ind. Health)*, **7**, 46-55, 1969. (APTIC No. 30687).

[93] Heller, A. N., S. T. Cuffe and D. R. Goodwin, Inorganic chemical industry, in *Air Pollution, III*, (A. C. Stern, ed.) 2nd ed., New York: Academic Press, p. 191-242, 1968.

[94] Hendrickson, E. R. and C. I. Harding, Air pollution problems associated with Kraft pulping, *Proc. Int. Clean Air Congress (London)*, p. 95-97, 1966.

[95] Hinkamp, J. B., M. E. Griffing and D. W. Zutaut, Aromatic aldehydes and phenols in the exhaust from leaded and unleaded fuels, *Am. Chem. Soc., Div. Petrol. Chem. (preprints)*, **16**(2), E5-E11, 1971.

[96] Hirschler, D. A., L. F. Gilbert, F. W. Lamb and L. M. Niebylski, Particulate lead compounds in automobile exhaust gas, *Ind. Eng. Chem.*, **49**, 1131-1142, 1957.

[97] Hoefig, R., Air pollution problems caused by the use of plastic binders, *Giessereitechnik*, **15**, 250-254, 1969. (APTIC No. 40595).

[98] Hoffmann, D. and E. L. Wynder, Chemical analysis and carcinogenic bioassays of organic particulate pollutants, *Air Pollution, II*, (A. C. Stern, ed.) 2nd ed., New York: Academic Press, p. 187-242, 1968.

[99] Hollingdale-Smith, P. A., *Gaseous atmospheric pollutants: a literature survey*, Chemical Defense Estab. (Porton Dowes, Salisbury, Witts), Tech. Note 144, Nov., 1972. (NTIS Document AD 905797).

[100] Sandalls, F. J., and S. A. Penkett, Measurements of carbonyl sulphide and carbon disulphide in the atmosphere, *Atmos. Environ.*, **11**, 197-199, 1977.

[101] Hori, M., N. Tanigawa and Y. Kobayashi, Investigation of lead compounds in exhaust gases from various kinds of sources, *J. Japan Soc Air Poll.*, **8**, 344, 1973. (APTIC No. 56970).

[102] Hoshika, Y., T. Ishiguro and Y. Shigeta, Analysis of odor components from feather rendering plants, *J. Japan Soc. Air Poll.*, **5**(1), 99, 1970.

[103] Hoshika, Y., T. Ishiguro, Y. Katori, S. Futaki and Y. Shigeta, An example of investigation methods for odor pollution, *J. Japan Soc. Air Poll.*, **6**, 227, 1971. (APTIC No. 36910).

[104] Hoshika, Y. and S. Kadowaki, Identification of evolved component from ABS copolymer by g.c., *J. Japan Soc. Air Poll.*, **8**, 271, 1973. (APTIC No. 58564).

[105] Hosono, Y., G. Kawaski, M. Yamazaki, S. Inada, M. Ishisaka, S. Maeda, S. Taka, R. Ishisaka, K. Senda, I. Yamazaki and I. Shinmura, On the hydrogen cyanide in exhaust gas of electric furnaces, *J. Japan Soc. Air Poll.*, **6**, 246, 1971. (APTIC No. 37322).

[106] Huntingdon, A. T., The collection and analysis of volcanic gases from Mount Etna, *Phil. Trans. R. Soc. (Lond.)*, **A274**, 119-128, 1973.

[107] Husar, J. D., R. B. Husar and P. K. Stubits, Determination of submicrogram amounts of atmospheric particulate sulfur, *Anal. Chem.*, **47**, 3062-3064, 1975.

[108] Ishiguro, T., K. Hishida and T. Yajima, Present state of public nuisance caused by offensive odors in Tokyo, *J. Water Waste (Japan)*, **13**(8), 972-978, 1971.

[109] Jacobson, A. R., Viable particles in the air, *Air Pollution, I*, (A. C. Stern, ed.) 2nd ed., New York: Academic Press, p. 95-119, 1968.

[110] Japan Environ. Health Center, *Reports on investigations of bad odors from animal offal processing and poultry farms in the Soka and Koshigaya areas, Saitama Prefect, and future countermeasures*, 1971. (APTIC No. 65669).

[111] Japan Environ. San. Center, *Report of survey of the specified poisonous substances and the prevention of offensive odor*, Report 4 (Tokyo), 67 p. ,1969. (APTIC No. 32475).

[112] Johnson, D. L. and R. S. Braman, Alkyl and inorganic arsenic in air samples, *Chemosphere*, **6**, 333-338, 1975.

[113] Just, J., S. Maziarka and H. Wyschinska, Benzpyrene and other aromatic hydrocarbons in the dust of Polish towns, *Wiss. Z. Humboldt Univ. (Berlin Math. Naturw. Raike)*, **19**, 513-515, 1970. (APTIC No. 29569).

[114] Kasparov, A. A. and V. G. Kiriy, *Gigiena i Sanit.*, **37**, 57, 1972. Cited by Saltzman and Cuddeback (206).

[115] Katari, V., G. Isaacs and T. W. Devitt, *Trace pollutant emissions from the processing of metallic ores*, EPA 650/2-74-115, Environmental Protection Agency (Washington, D.C.), 1974. (NTIS Document PB 238-655).

[116] Kipling, M. D. and R. Fothergill, *Br. J. Ind. Med.*, **21**, 74-77, 1964. Cited by Hollingdale-Smith (99).

[117] Kirsch, H. and K. Schwinkowski, Emission measurement of styrol and possibilities for its limitation in the waste gas, *Z. Hyg.*, **18**, 193-194, 1972. (APTIC No. 39340).

[118] Kitaoka, Y. and K. Murata, Experiments on the thermal degradation of ethylene low polymer, *J. Japan Fuel Soc. (Tokyo)*, **50**, 791-799, 1971. (APTIC No. 36274).

[119] Kitagawa, Y., A. Furukawa, A. Sueda, H. Ryoken and M. Ito, Investigation of odorous air pollution from factory of vulcanization accelerator, *J. Japan Soc. Air Poll.*, **8**, 379, 1973. (APTIC No. 58375).

[120] Klee, O., PCB in the wake of DDT, *Kosmos (Stuttgart)*, **2**, 65-66, 1972. (APTIC No. 46111).

[121] Kobayashi, S., H. Matsushita, H. Osaka and K. Morimura, Measurements of odorous pollutants dryer emissions and study on its deodorization, *J. Japan Soc. Air Poll.*, **8**, 374, 1973. (APTIC No. 58544).

[122] Kobayashi, Y., S. Tsukada, H. Hirobe, M. Takahashi and N. Ito, Report on the odor investigation in factory areas, *Mie Prefect. Pub. Nuisance Center Ann. Rpt.*, **2**, 102-107, 1974. (APTIC No. 69042).

[123] Kopczynski, S. L., W. A. Lonneman, F. D. Sutterfield and P. E. Darley, Photochemistry of atmospheric samples in Los Angeles, *Environ. Sci. Technol*, **6**, 342-347, 1972.

[124] Kremnera, S. N., et al., *Toksikol. Novykh. Press. Khim. Veshchestv.*, **5**, 123-135, 1963. Cited by Hollingdale-Smith (99).

[125] Krey, P. W. and R. J. Lagomarsino, Stratospheric concentrations of SF_6 and CCl_{3F}, *Environmental Quarterly (Health and Safety Laboratory, Atomic Energy Commission New York, New York)*, HASL-294, 1975.

[126] Kuchak, Y. A., Electro-aerosol pesticides and the hygienic aspects of their application, *Gigiena i Sanit.*, **1**, 10-13, 1974. (APTIC No. 62867).

[127] Kutuzova, L. N., A. F. Kononenko and G. P. Sokulskii, Discharges to the atmosphere from benzole plant, *Coke Chem. (USSR)*, **8**, 39-42, 1970. (APTIC No. 3508).

[128] Lang, O. and T. zur Muehlen, Air pollution by organic acids and esters and their analytical determination, *Zbl. Arbeitsmed.*, **2**, 39-45, Feb. 1971.

[129] Farmer, C. B., O. F. Raper and R. H. Norton, Spectroscopic detection and vertical distribution of HCl in the troposphere and stratosphere, *Geophys. Res. Lett.*, **3**, 13-16, 1976.

[130] Lee, M. L., M. Novotny and K. D. Bartle, Gas chromatography/mass spectrometric and nuclear magnetic resonance spectrometric studies of carcinogenic polynuclear aromatic hydrocarbons in tobacco and marijuana smoke condensates, *Anal. Chem.*, **48**, 405-416, 1976.

[131] Leonardos, G., D. Kendall and N. Barnard, Odor threshold determinations of 53 odorant chemicals, *J. Air Poll. Contr. Assoc.*, **19**, 91-95, 1969.

[132] Levaggi, D. A., Private communication, 1976.

[133] Levaggi, D. A. and M. Feldstein, The determination of formaldehyde, acrolein, and low molecular weight aldehydes in industrial emissions on a single collection sample, *J. Air Poll. Control Assoc.*, **20**, 312-314, 1970.

[134] Levy, A., S. E. Miller and F. Schofield, The photochemical smog reactivity of solvents, *Proc. 2nd. Int. Clean Air Congress*, (H. Englund, ed.) New York: Academic Press, p. 305-316, 1971.

[135] Lillian, D., H. B. Singh, A. Appleby, L. Lobban, R. Arnts, R. Gumpert, R. Hague, J. Toomey, J. Kazazis, M. Antell, D Hansen and B. Scott, Atmospheric fates of halogenated compounds, *Environ. Sci. Technol.*, **9**, 1042-1048, 1975.

[136] Linnell, R. H. and W. E. Scott, Diesel exhaust analysis, *Arch. Environ. Health*, **5**, 616-625, 1962.

[137] Little, A. D., Inc., *Chemical identification of the odor components in diesel engine exhaust*, Report C-71407/71475, HEW Contract No. CPA 22-69-63 (Cambridge, Mass), 1970.

[138] Little, A. D., Inc., *Chemical identification of the odor components in diesel engine exhaust*, Report ADL 62561-5, Cambridge, Mass., 1971.

[139] Lonneman, W. A., T. A. Bellar and A. P. Altshuller, Aromatic hydrocarbons in the atmosphere of the Los Angeles basin, *Environ. Sci. Technol.*, **2**, 1017-1020, 1968.

[140] Lovelock, J. E., R. J. Maggs and R. A. Rasmussen, Atmospheric dimethyl sulphide and the natural sulphur cycle, *Nature*, **237**, 452-453, 1972.

[141] Luebs, R. E., A. E. Laag and K. R. Davis, Ammonia and related gases emanating from a large dairy area, *Calif. Agr.*, **27**, (2), 10-12, 1973.

[142] Luzin, Yu.P. and V. V. TsarKov, Reduction of atmospheric discharges during the etching of steel, *Metallurg. (USSR)*, **16**, (12), 32, 1972. (APTIC No. 38011).

[143] Maarse, H. and M. C. Ten Noever de Brauw, Another catty odour compound causing air pollution, *Chem. Ind. (London)*, **1**, 36-37, 1974.

[144] Macfarlane, C., J. B. Lee and M. B. Evans, The qualitative composition of peat smoke, *J. Inst. Brewing*, **79**, 202-209, 1973.

[145] Masek, V., New findings concerning the properties of fly dust from coking plants. Part IV. Hard coal tar distillation plants, *Zbl. Arbeitsmed.*, **22**(11), 332-337, 1972. (APTIC No. 47099).

[146] Matheson Gas Products, *Ethylene FRP (Fruit Ripening Purity) in safe, easy-to-use Matheson lecture bottles*, Pamphlet 25M/11/72, East Rutherford, N.J., 1972.

[147] Mathu, M., S. K. Majumder and H. A. B. Parpia, *J. Agric. Food Chem.*, **21**, 184, 1973. Cited by Saltzman and Cuddeback (206).

[148] Matsuda, Y., On airborne true fungi, *J. Japan Air Cleaning Assoc. (Tokyo)*, **9**, (7), 42-58, 1972. (APTIC No. 50583).

[149] Matsumoto, H., Analysis of odor composition in starch lees, *J. Japan Soc. Air Poll.*, **9**, 461, 1974. (APTIC No. 70309).

[150] Matsuo, K., A study of air pollution at Wakayama City. Part 3. *J. Meteor. Res. (Tokyo)*, **25**, 481-486, 1973. (APTIC No. 61204).

[151] Matsuyama, T., Air pollution with dust, *J. Res. Assoc. Powder Technol. (Japan)*, **11**, 40-42, 1974. (APTIC No. 61569).

[152] Matteucci, M., Peculiarities of the eruption of Vesuvius, *C.R. Acad. Sci. (Paris)*, **129**, 65-67, 1899.

[153] McEwen, D. J., Automobile exhaust hydrocarbon analysis by gas chromatography, *Anal. Chem.*, **38**, 1047-1053, 1966.

[154] Miller, C. W. and T. M. Shafik, *Concentrations of OMS-33 in air following repeated indoor applications*, Dept. of Health, Educ. and Welfare (Atlanta, Ga.) and Environmental Protection Agency (Research Triangle Park, N.C.), 1973. (APTIC No. 67476).

[155] Mill, R. A., J. M. Robertson and B. Walker, A technique for airborne aerobiological sampling, *J. Environ. Health*, **35**, 51-53, 1972.

[156] Ministry of Labor, Health and Welfare of North Rhine-Westphalia (West Germany), Concept of clean air maintenance in North Rhine-Westphalia, Reine luft fuer morgen. Utopia odor Wirklichkeit?, Ein Konzept fuer das Nordrhein-Westfalen bis 1980, 13-14, 1972. (APTIC No. 59773).

[157] Mizutani, H., A. Kamiya, M. Kitasi, T. Aoyama and E. Ito, Odor pollution research in Nagoya City, *J. Japan Soc. Air Poll.*, **8**, 381, 1973. (APTIC No. 58583).

[158] Morie, G. P., Determination of hydrogen sulfide in cigarette smoke with a sulfide ion electrode, *Tobacco Sci.*, **15**, (29), 34, 1971.

[159] Morris, W. E. and K. T. Dishart, Influence of vehicle emission control systems on the relationship between gasoline and vehicle exhaust hydrocarbon composition, in *Effect of Automotive Emission Requirements on Gasoline Characteristics*, ASTM STP 487, Amer. Soc. for Testing and Materials, Philadelphia, Pa., p. 63-93, 1971.

[160] Mosier, A. R., C. E. Andre and F. G. Viets, Jr., Identification of aliphatic amines volatilized from cattle feedyard, *Environ. Sci. Technol.*, **7**, 642-644, 1973.

[161] Muhlrad, D., *The fight against air pollution caused by electro-metallurgic furnaces*, Eidgenoessische Kommission fuer Lufthygiene, Zurich, Probleme der Luftverunreinigung duroh die Industrie, 30-37, 1968. (APTIC No. 39309).

[162] Murata, M. Bad odor in Yokkaichi City, *Mie Prefect. (Japan) Pub. Nuisance Center Ann. Rep.*, **1**, 71-76, 1973. (APTIC No. 52659).

[163] Murray, A. J. and J. P. Riley, Occurrence of some chlorinated aliphatic hydrocarbons in the environment, *Nature*, **242**, 37-38, 1973.

[164] Naka, K., Y. Hasegawa and H. Hirobe, Studies on hydrocarbons present in the Yokkaichi City area, *Mie Prefect. (Japan) Pub. Nuisance Center Ann. Rep.*, **1**, 77-93, 1973. (APTIC No. 52669).

[165] Nakayama, H. *Report on the investigation on the offensive odor in Yokkaichi City*, Yokkaichi, Japan, 122 p., 1969. (APTIC No. 29349).

[166] Narayan, R. S., Identification and control of cattle feedlot odors, *Texas Tech. Univ. (Lubbock, Tex.)*, M.S. Thesis, 1971.

[167] National Air Pollution Control Administration, Publication AP-54, oo c Cited by Hollingdale-Smith (99).

[168] Naumann, R. J., Smoking and air pollution standards, *Science*, **182**, 334, 336, 1973.

[169] Nazyrov, G. N. and K. Y. Vengerskaya, *Gig. Tr. Prof. Zabol.*, **10**(11), 56-60, 1966.

[170] Neligan, R. E., Hydrocarbons in the Los Angeles atmosphere, *Arch. Environ. Health*, **5**, 581-591, 1962.

[171] Neligan, R. E., Paper presented at Princeton/BTL Conference on Air Pollution, Princeton, N.J., Dec. 10, 1974.

[172] Nichols, R., Ethylene production during senescence of flowers, *J. Hort. Sci.*, **41**, 279-290, 1966.

[173] Nieboer, H. and J. van Ham, Peroxyacetyl nitrate (PAN) in relation to ozone and some meteorological parameters at Delft in the Netherlands, *Atmos. Environ.*, **10**, 115-120, 1976.

[174] Nishida, K., T. Honda and T. Tsuji, The effect of ozone deodorization equipment for the sewage odor, *Akushu no Kenkya*, **4**(20), 24-30, 1975. (Ozone Chemistry and Technology Index No. 581, 1975).

[175] Obremski, R. J., Design notes-dioxane, *Poll. Eng.*, **7**(9), 22, 1975.

[176] O'Donnell, A. and A. Dravnieks, *Chemical species in engine exhaust and their contributions to exhaust odors*, Report No. IITRI. C6183-5, IIT Research Institute (Chicago, IL), 1970.

[177] Okada, S., Collected data on allowable and minimum perceptible concentrations of odor substances, *J. Water Waste (Japan)*, **13**, 1136-1142, 1971.

[178] Okita, T., M. Watanabe and S. Kiyono, Analysis of HC of low grade in atmosphere, Paper presented at 24th Ann. Mtg., *Japan Chem. Soc. (Tokyo)*, March, 1971. (APTIC No. 29482).

[179] Okuno, T., Measurement of polycyclic hydrocarbons in air, *J. Japan Soc. Air Poll.*, **5**, 74, 1970. (APTIC No. 32929).

[180] Okuno, T., M. Tsuji and K. Takada, *Problems of public nuisance caused by bad odors in Hyogo Prefecture*, Environmental Sci. Inst., Hyogo Prefacture, Kobe (Japan), Report 2, 18-24, 1971.

[181] Okuno, T., M. Tsuji, Y. Shintani and H. Watanabe, On the pyrolysis of PCB, *J. Japan Soc. Air Poll.*, **8**, 351, 1973. (APTIC No. 60236).

[182] Olsen, D. A. and J. L. Haynes, *Air Pollution aspects of Organic Carcinogens*, Litton Systems, Inc. (Bethesda, Maryland), 1969. (NTIS Document PB 188090).

[183] O'Mara, M. M., The combustion products from synthetic and natural products. Part 1. Wood, *J. Fire Flammability*, **5**, 34-53, 1974.

[184] O'Mara, M. M., L. B. Crider and R. L. Daniel, Combustion products from vinyl chloride monomer, *Am. Ind. Hyg. Assoc. J.*, **32**, 153-156, 1971.

[185] Pasco, L. I. and F.Kh. Yunusov, Air pollution in waste-water cleaning premises, *Bezop. Tr. Prom.*, **15**, 46-48, 1971. (APTIC No. 39402).

[186] Penkett, S. A., F. J. Sandalls and J. E. Lovelock, Observations of peroxyacetyl nitrate (PAN) in air in Southern England, *Atmos. Environ.*, **9**, 139-140, 1975.

[187] Pfaff, R. O., Swift Agricultural Chemicals, Inc., Birmingham, Alabama, Environ. Engg., Inc. (Gainesville, Fla., Office of Air Quality Planning and Standards), Contract Rpt. 73-FRT-6, 1973.

[188] Pierce, R. C. and M. Katz, Dependency of polynuclear aromatic hydrocarbon content on size distribution of atmospheric aerosols, *Environ. Sci. Technol.*, **9**, 347-353, 1975.

[189] Pierce, R. C. and M. Katz, Determination of atmospheric isomeric polycyclic arenes by thin-layer chromatography and fluorescence spectroscopy, *Anal. Chem.*, **47**, 1743-1748, 1975.

[190] Pierce, R. C. and M. Katz, Chromatographic isolation and spectral analysis of polycyclic quinones: application to air pollution analysis, *Environ. Sci. Technol.*, **10**, 45-51, 1976.

[191] Polgar, L. G., R. A. Duffee and L. J. Updyke, Odor characteristics of mixture of sulfur compounds emitted from the viscose process, Paper presented at 68th Annual Meeting, Air Pollution Control Assoc., Boston, Mass., 1975.

[192] Prakash, C. B. and F. E. Murray, Studies on air emissions from the combustion of wood-waste, *Combust. Sci. Technol.*, **6**, 81-88, 1972.

[193] Puchner, F., Air pollution problems of Dorog, Hungary, *Tatabanyai Szenbanyak Muszaki-Kozgazdasagi Kozl*, **2-3**, 74-76, 1974.

[194] Que Hee, S. S., R. G. Sutherland and M. Vetter, GLC analysis of 2,4-D concentrations in air samples from central Saskatchewan in 1972, *Environ. Sci. Technol.*, **9**, 62-66, 1975.

[195] Randall, C. W., Bacterial air pollution from activated sludge units-field and tracer studies, *College of Engg. (Texas Univ. Austin, Texas)*, Ph.D. Dissertation, 1966.

[196] Rasch, R., Lead in waste seen in connection with waste incineration, *Aufbereitungs Technik. (W. Ger.)*, **15**, 234-237, 1974. (APTIC No. 62515).

[197] Rasmussen, R. A., *Qualitative analysis of the hydrocarbon emission from veneer dryers*, Washington State Univ. (Pullman, Wash.), Research Grant AP 1232, 17, 1970. (APTIC No. 32180).

[198] Rasmussen, R. A., What do the hydrocarbons from trees contribute to air pollution?, *J. Air Poll. Contr. Assoc.*, **22**, 537-543, 1972.

[199] Rasmussen, R. A., Emission of biogenic hydrogen sulfide, *Tellus*, **26**, 254-260, 1974.

[200] Raymond, A. and G. Guiochon, Gas chromatographic analysis of C_8–C_{18} hydrocarbons in Paris air, *Environ. Sci. Technol.*, **8**, 143-148, 1974.

[201] Rondia, D., The solution of a hygienic problem in steel works. Exposure of workers to a fog containing 3,4-benzopyrene, *Arch. Maladies Profess. Med. Trav. Securite Sociale (Paris)*, **25**, 403-406, 1964. (APTIC No. 33279).

[202] Rosen, A. A., R. T. Skeel and M. B. Ettinger, Relationship of river water odor to specific organic contaminants, *J. Water Poll. Control Fed.*, **35**, 777-782, 1963. (APTIC No. 07089).

[203] Endo, R., Research on the source of odor development and the residential reaction in Hokkaido, *J. Pollution Control (Japan)*, **4**, (4), 209-220, 1968.

[204] Saijo, T., T. Tsujimoto and T. Takahashi, Odor pollution of Kashima District, *J. Japan Soc. Air Poll.*, **6**, 222, 1971. (APTIC No. 36845).

[205] Sakai, T. and E. Ito, Studies on vehicle exhaust(III) investigation on hydrogen cyanide and formaldehyde in the exhaust emissions, *Rep. Environ. Pollut. Res. Inst. (City of Nagoya)*, **2**, 39-42, 1973. (APTIC No. 69016).

[206] Saltzman, B. E. and J. E. Cuddeback, Air pollution, *Anal. Chem.*, **47**, 1R-15R, 1975.

[207] Sandberg, D. V., S. G. Pickford and E. F. Darley, Emissions from slash burning and the influence of flame retardant chemicals, *J. Air Poll. Contr. Assoc.*, **25**, 278-281, 1975.

[208] Saunders, R. A., J. R. Griffith and F. E. Saalfeld, Identification of some organic smog components based on rain water analysis, *J. Biomedical Mass Spectrometry*, **1**, 192-194, 1974.

[209] Sawicki, E., T. R. Hauser, W. C. Elbert, F. T. Fox and J. E. Meeker, Polynuclear aromatic hydrocarbon composition of the atmosphere in some large American cities, *Am. Ind. Hyg. Assoc. J.*, **23**, 137-144, 1962.

[210] Schlegel, H. G., Production, modification, and consumption of atmospheric trace gases by microorganisms, *Tellus*, **26**, 11-20, 1974.

[211] Schuler, M. and L. Borla, PVC and air hygiene, *Chem. Rundschau (Solothurn)*, **25**, (2), 17-18, 1972. (APTIC No. 44163).

[212] Schuphan, W. Potential and actual hazards of fertilizer and pesticide use for the environment, *Schriftenreihe Ver Wasser-Boden-Lufthyg (Berlin)*, No. 34, 35-50, 1971. (APTIC No. 57230).

[213] Schuetzle, D., Computer Controlled High Resolution Mass Spectrometric Analysis of Air Pollutants, *University of Washington*, Ph.D. Dissertation, 1972.

[214] Schuetzle, D., D. Cronn, A. L. Crittenden and R. J. Charlson, Molecular composition of secondary aerosol and its possible origin, *Environ. Sci. Technol.*, **9**, 838-845, 1975.

[215] Sciamanna, A. F. and A. S. Newton, *A survey of the occurrence of dimethyl mercury in the atmosphere*, Atomic Energy Comm. Contract W-7405-eng-48, Report TID-4500-R61, Lawrence Berkeley Lab., Berkeley, Cal., 410-413, 1974.

[216] Seba, D. B. and J. M. Prospero, Pesticides in the lower atmosphere of the northern equatorial Atlantic Ocean, *Atmos. Environ.*, **5**, 1043-1050, 1971.

[217] Seizinger, D. E. and B. Dimitriades, Oxygenates in exhaust from simple hydrocarbon fuels, *J. Air Pollut. Contr. Assoc.*, **22**, 47-51, 1972.

[218] Sforzolini, G., G. Scassellati and G. Saldi, Further research on the polycyclic hydrocarbons of cigarette smoke. Comparison of the inhaled smoke and that taken from the ambient atmosphere, *Bell. Soc. Ital. Biol. Sper. (Naples)*, **37**, 769-771, 1961. (APTIC No. 31630.

[219] Shaw, A. C. and K. T. Waldock, Vanillin analysis and odor evaluation by gas chromatography, *Pulp Paper Mag. Can.*, **68**(3), T-118 to T-122, 1967.

[220] Shendrikar, A. D. and P. W. West, Determination of selenium in the smoke from trash burning, *Environ. Lett.*, **5**, 35-39, 1973.

[221] Shigeta, Y., Examples of bad odor measurement designated by the bad odor control law, and sensitivity test, Preprint, Bad Odor Pollution Study Group (Japan), p. 46-54, 1974. (APTIC No. 69235).

[222] Sibbitt, D. J., R. H. Moyer and G. H. Milly, Emission of mercury from latex paints, *Am. Chem. Soc., Div. Water, Air, Waste Chem., Gen. Papers*, **12**(1), 20-26, 1972.

[223] Siegert, H., H. H. Oelert and J. Zajontz, Rapid gas chromatographic analysis for individual hydrocarbons in the exhaust of diesel engines, *Motortech. Z. (Stuttgart)*, **35**(4), 101-106, 1974. (APTIC No. 59888).

[224] Simmonds, P. G., S. L. Kerrin, J. E. Lovelock and F. H. Shair, Distribution of atmospheric halocarbons in the air over the Los Angeles basin, *Atmos. Environ.*, **8**, 209-216, 1974.

[225] Sohn, H. Y. and J. Szekely, On the oxidation of cyanides in the stack region of the blast furnace, Steel Ind. Environ, New York State Univ. (Buffalo), *2nd C.C. Furnas Mem. Conf.*, p. 249-264, 1973. (APTIC No. 69318).

[226] Spicer, C. L., Nonregulated photochemical pollutants derived from nitrogen oxides, in *Scientific Seminar on Automotive Pollutants*, EPA 600/9-75-003, Environ. Prot. Agency, Washington, D.C., 1975.

[227] Spindt, R. S., G. J. Barnes and J. H. Somers, The characterization of odor components in diesel exhaust gas, Paper presented at Midyear Meeting, Society of Auto. Engrs. Int. (Montreal, Can.), June 7, 1974.

[228] Spynu, Ye.I., L. N. Ivanova and A. V. Bolotnyy, Environmental pollution due to organophosphorus pesticides, *Gigiena i Sanit.*, **10**, 75-79, 1973. (APTIC No. 57347).

[229] Staiff, D. C., G. E. Quinby, D. L. Spencer and H. G. Starr, Jr., *Polychlorinated biphenyl emission from fluorescent lamp ballasts*, Environmental Protection Agency, Wenctchee, Wash., 1973. (APTIC No. 57527).

[230] Stanley, C. W., J. E. Barney, II, M. R. Helton and A. R. Yobs, Measurement of atmospheric levels of pesticides, *Environ. Sci. Technol.*, **5**, 430-435, 1971.

[231] Stephens, E. R., Chemistry of atmospheric oxidants, *J. Air Poll. Contr. Assoc.*, **19**, 181-185, 1969.

[232] Stephens, E. R., *Hydrocarbons in polluted air*, Coordinating Research Council (New York, New York), CRC Project CAPA 5-68, 1973.

[233] Stephens, E. R., Observations of peroxyacetyl nitrate (PAN) in air in Southern England, *Atmos. Environ.*, **9**, 461, 1975.

[234] Stoiber, R. E., D. C. Leggett, T. F. Jenkins, R. P. Murrmann and W. I. Rose, Organic compounds in volcanic gas from Santiaguito Volcano, Guatamala, *Geol. Soc. Amer. Bull.*, **82**, 2299-2302, 1971.

[235] Stoiber, R. E. and A. Jepsen, Sulfur dioxide contributions to the atmosphere by volcanoes, *Science*, **182**, 577-578, 1973.

[236] Lewis, B. G., C. M. Johnson, and C. C. Delwiche, Release of volatile selenium compounds by plants, *J. Agr. Food Chem.*, **14**, 638-640, 1966.

[237] Suzuki, F., An estimation of PCT's in incinerators, *J. Japan Soc. Air Pollution*, **9**, 440, 1974. (APTIC No. 70941).

[238] Suzuki, Y., K. Nishiyama, M. Oe and F. Kametani, Studies on the prevention of public nuisance by the exhaust gases from the Kraft pulp mill. Part 1. (Analysis of exhaust gases), *Tohoku J. Exp. Med. (Tokyo)*, **11**(2), 120-126, 1964. (APTIC No. 96240).

[239] Tachikawa, R., Mechanism of BHC pollution, *Kagaku Asahi (Japan)*, **30**(12), 45-51, 1970. (APTIC No. 30376).

[240] Takahashi, T., A few observations on the generation mechanism of obnoxious odor, *Odor Res. J. (Japan)*, **1**(4), 33-38, 1971.

[241] Takahashi, M., M. Yamazaki, Y. Sigeta, Y. Sakaida and H. Nagasawa, Studies on the offensive odor and odor substances originating from manufacture process fish-meal, bone-meal, and feather meal, *Human Hyg. (Japan)*, **38**(3), 122-136, 1973. (APTIC No. 53493).

[242] Tanaka, A., M. Hori and Y. Kobayashi, Preprint , *Japan Soc. for Safety Eng. (Tokyo)*, #28, 1973. Cited by Saltzman and Cuddeback(206).

[243] Tanimoto, M. and H. Uehara, Detection of acrolein in engine exhaust with microwave cavity spectrometer of Stark voltage sweep type, *Environ. Sci. Technol.*, **9**, 153-154, 1975.

[244] Tatsukawa, R., A new environmental pollutant-polychlorinated biphenyls (PCB), *J. Pollution Control (Japan)*, **7**, 419-425, 1971. (APTIC No. 29984).

[245] Tatsukawa, R. and W. Tadaaki, Pesticide residues in air. Part 2. Air pollution by BHC, *J. Japan Soc. Air Poll.*, **5**, 92, 1970. (APTIC No. 28776).

[246] Tazieff, H., Volcanism and atmospheric conditions, *Proc. Inst. Symp. Env. Meas.*, Bechman Instruments S.A. (Geneva), 131-132, 1973.

[247] Tebbens, B. D., Gaseous pollutants in the air, in *Air Pollution, I*, (A. C. Stern, ed.), New York: Academic Press, p. 23-46, 1968.

[248] Thomas, G. H., *Environ. Health Perspect.*, **3**, 23, 1973. Cited by Giam, et al.(77).

[249] Tichatschke, J., Studies of the emission from refuse incinerators, *Mitt. Ver. Grosskesselbesitzer*, **51**(3), 219-223, 1971.

[250] Trieff, N. M. and V. M. S. Ramanujam, Oxidation of aromatic amine air pollutants using chloramine-T and hypochlorous acid, Paper presented at 68th Annual Meeting, Air Poll. Cont. Assoc., Boston, MA, 1975. June 16, 1975.

[251] Tseudrovskaya, V. A., *Gigiena i Sanit.*, **1**, 62, 1973. Cited by Saltzman and Cuddeback(206).

[252] Tsifrinovich, A. N. and N. I. Lulova, Composition of organosulfur trace impurities in gases, *Neftepererab. Neftekhim.*, **6**, 33-35, 1972. (APTIC No. 54825).

[253] Tsuji, M. and T. Okuno, The GC analysis of carbonyl sulfide emitted by the viscose plant, *Proc. 13th Symp., Japan Soc. Air Poll.*, p. 119, 1972. (APTIC No. 49266).

[254] Tsuji, M., T. Okuno and N. Takada, On the concentration of amine and aldehyde compounds from fish-meal plants, *J. Japan Soc. Air Poll.*, **6**, 226, 1971. (APTIC No. 36850).

[255] Tsuji, M., T. Okuno and K. Takada, Investigation of the components of bad odor from fish meal factories, *Rpt. Public Nuisance Res. Inst. (Hyogo Prefect.)*, **3**, 18-20, 1972. (APTIC No. 48285).

[256] Vandegrift, A. E., L. J. Shannon, E. W. Lawless, P. G. Gorman, E. E. Sallee and M. Reichel, *Particulate pollutant system study. Volume III. Handbook of emission properties*, Midwest Research Institute (Kansas City, Mo.), Contract Report CPA 22-69-104, 1971.

[257] Vanderpol, A. H., F. D. Carsey, D. S. Covert, R. J. Charlson and A. P. Waggoner, Aerosol chemical parameters and air mass character in the St. Louis region, *Science*, **190**, 570, 1975.

[258] Vol'fson, U.Ya. and A. F. Sudak, Chromatographic determination of carbon disulfide, carbon dioxide, and sulfur dioxide micro- contaminants in air, *Ind. Lab (USSR)*, **36**, 1322-1323, 1970.

[259] Wang, C. C. , L. I. Davis, Jr., C. H. Wu, S. Japar, H. Niki, and B. Weinstock, Hydroxyl radical concentrations measured in ambient air, *Science*, **189**, 797-800, 1975.

[260] Watanabe, I. and T. Okita, Distribution and estimation of main sources of ambient light hydrocarbons in Tokyo, *J. Japan Soc. Air Poll.*, **8**, 710-728, 1973. (APTIC No. 65743).

[261] Wessler, M. A., Mass spectrographic analysis of exhaust products from an air aspirating diesel fuel burner, *Dept. of Mech. Eng. (Purdue Univ.)*, Ph.D. Thesis, 1968.

[262] Wilby, F. V., Variation in recognition odor threshold of a panel, *J. Air Poll. Contr. Assoc.*, **19**, 96-100, 1969.

[263] Williams, I. H., Gas chromatographic techniques for the identification of low concentrations of atmospheric pollutants, *Anal. Chem.*, **37**, 1723-1732, 1965.

[264] Williams, F. W. and J. E. Johnson, Atmospheric contamination with a cleaning solvent, *Chem. Res. in Nuclear Sub. Atmos. Purification*, Prog. Rpt. 7037, Naval Research Lab., Washington, D.C., 1970.

[265] Willis, G. H., J. F. Parr and S. Smith, Pesticides in air. Volatilizaion of soil-applied DDT and DDD from flooded plots, *Pesticides Monitoring J.*, **4**, 204-208, 1971.

[266] Willis, G. H., J. F. Parr, S. Smith and B. R. Carroll, Volatilization of dieldrin from fallow soil as affected by different soil water regimes, *J. Environ. Quality*, **1**, 193-196, 1972.

[267] Wilson, D. F. and B. F. Hrutfiord, Formation of volatile organic compounds in the Kraft pulping process, *TAPPI*, **54**, 1094-1098, 1971.

[268] Wilson, H. H. and L. D. Johnson, *Characterization of air pollutants emitted from brick plant kilns*, Environmental Protection Agency (Research Triangle Park, N.C.), 1973.
[269] Wilson, W. E., Jr., W. E. Schwartz and G. W. Kinzer, *Haze formation-its nature and origin*, Battelle Columbia Laboratories, Columbus, Ohio, 1972. (NTIS Document No. PB 212609).
[270] Winkler, H-D. and K. Welzel, Studies of the formaldehyde emission from the production of wooden boards, *Wasser Luft Betrieb.*, **16**, 213-215, 1972. (APTIC No. 43912).
[271] Wohlers, H. C. and G. B. Bell, Stanford Research Institute, Project No. SU-1816, 1956.
[272] Yamaguchi Prefecture (Japan) Research Inst. of Health, Result of PCP estimation in the air around a factory of agricultural chemicals in Ogori-town, *Ann. Rpt. Yamaguchi Prefect. Res. Inst. Health*, **13**, 77-78, 1971. (APTIC No. 37289).
[273] Yanagisawa, S. A report of investigation on actual conditions of air pollution in and around Osaka international airport, (Osaka kokusaikuko shuken ni okeru taiki osen jittai chosa hokokusho), 1971. (APTIC No. 61300).
[274] Yavorovskaya, S. F. and L. P. Anvayer, *Gig. Sanit.*, **8**, 64, 1973.
[275] Yokohama National Univ., Kanagawa-Ken Taiki Osen Chosa Kenkyu Hokuku, **14**, 152, 1972 (cited in Saltzman and Cuddeback (206)).
[276] Yule, W. N., A. F. W. Cole and I. Hoffman, A survey for atmospheric contamination following forest spraying with fenitrothion, *Bull. Environ. Contamination Toxicol.*, **6**, 289-296, 1971.
[277] Zdrazil, J. and F. Picha, Cancerogenic substances-3, 4-benzopyrene in moulding sand mixtures and foundry dust, *Prakovni Lekar. (Prague)*, **15**, 207-211, 1963. (APTIC No. 33297).
[278] Zielinski, M., M. Zamfirescu and M. Decusara, Air pollution by nitric oxide from fertilizer factories, *Wiss. Z. Humboldt Univ. Berlin Math. Naturw. Reihe*, **19**, 523-525, 1970.
[279] Bates, C. E. and L. D. Scheel, Processing emissions and occupational health in the ferrous foundry industry, *Am. Ind. Hyg. Assoc. J.*, **35**, 452-462, 1974.
[280] Blosser, E. R. and W. M. Henry, *Identification and estimation of ions, molecules, and compounds in particulate matter collected from ambient air*, Battelle Columbus Laboratories, Columbus, Ohio, 1971. (NTIS Document No. PB 201738).
[281] Environmental Protection Agency, *Air Quality Criteria for Carbon Monoxide*, Report No. AP-62, Washington, D.C., 1970.
[282] Environmental Protection Agency, *Air Quality Criteria for Nitrogen Oxides*, Report AP-84 (Washington, D.C.), 1971.
[283] Environmental Protection Agency, *Control of photochemical oxidants-technical basis and implications of recent findings*, Research Triangle Park (N.C.), 1975.
[284] Hafstad, L. R., Automobiles and air pollution, in *Universities, National Laboratories and Man's Environment*, (D. Jared, ed.), U.S. Atomic Energy Commission, Oak Ridge, Tenn., p. 122, 1969.
[285] Hangebrauck, R. P., D. J. von Lehmden and J. E. Meeker, *Sources of polynuclear hydrocarbons in the atmosphere*, U.S. Department of Health, Education, and Welfare, Cincinnati, Ohio, 1967. (NTIS Document No. PB 174706).
[286] Jepsen, A. F., Measurements of mercury vapor in the atmosphere, *ACS Adv. Chem. Ser.*, **123**, 81-95, 1973.
[287] Davis, D. D., W. Heaps, and T. McGee, Direct measurements of natural tropospheric levels of OH via an aircraft borne tunable dye laser, *Geophys. Res. Lett.*, **3**, 331-333, 1976.

[288] Popov, V. A., *Hyg. Sanit.*, **35**(5), 178-182, 1970.

[289] Environmental Protection Agency, *Air Quality Data-1973 Annual Statistics*, EPA 450/2-74-015, Research Triangle Park, N.C., 1974.

[290] Robinson, E. and R. C. Robbins, *Sources, Abundance and Fate of Gaseous Atmospheric Pollutants, with Supplement*, Stanford Research Institute, Project PR-6755, 1968, 1971. (NTIS Documents N71-25147 and ZZ18194).

[291] Schmidt, U., Molecular hydrogen in the atmosphere, *Tellus*, **26**, 78-89, 1974.

[292] Schulten, H. R. and U. Schurath, Aerosol analysis by field desorption mass spectrometry combined with a new sampling technique, *Atmos. Environ.*, **9**, 1107-1112, 1975.

[293] Seiler, W. and C. Junge, Carbon monoxide in the atmosphere, *J. Geophys. Res.*, **75**, 2217-2226, 1970.

[294] Valley, S. L., ed., *Handbook of Geophysics and Space Environments*, New York: McGraw-Hill, 1965.

[295] Watt, A. D., Placing atmospheric CO_2 in perspective, *IEEE Spectrum*, **8**(11), 59-72, 1971.

[296] Bates, D. R. and P. B. Hays, Atmospheric nitrous oxide, *Planet. Space Sci.*, **15**, 189-197, 1967.

[297] Hahn, J., The North Atlantic Ocean as a source of atmospheric N_2O, *Tellus*, **26**, 160-168, 1974.

[298] Hoffmann, D., C. Patrianakos, K. D. Brunnemann and G. B. Gori, Chromatographic determination of vinyl chloride in tobacco smoke, *Anal. Chem.*, **48**, 47-50, 1976.

[299] Lahue, M. D., J. B. Pate and J. P. Lodge, Jr., Atmospheric nitrous oxide concentrations in the humid tropics, *J. Geophys. Res.*, **75**, 2922-2926, 1970.

[300] Schutz, K., C. Junge, R. Beck and B. Albrecht, Studies of atmospheric N_2O, *J. Geophys. Res.*, **75**, 2230-2246, 1970.

[301] Koyama, T., Gaseous metobolism in lake sediments and paddy soils and the production of atmospheric methane and hydrogen, *J. Geophys. Res.*, **68**, 3971-3973, 1963.

[302] Babich, H. and G. Stotzky, Air pollution and microbial ecology, *CRC Critical Rev. Environ. Control*, **4**(3), 353-421, 1974.

[303] Farwell, S. O., E. Robinson, W. J. Powell and D. F. Adams, Survey of airborne 2, 4-D in South-Central Washington, *J. Air Poll. Contr. Assoc.*, **26**, 224-230, 1976.

[304] Woolley, W. D., Decomposition products of PVC for studies of fires, *British Polymer J.*, **3**, 186-193, 1971.

[305] Woodwell, G. M., R. A. Houghton, and N. R. Tempel, Atmospheric CO_2 at Brookhaven, Long Island, New York: Patterns of variation up to 125 meters, *J. Geophys. Res.*, **78**, 932-940, 1973.

[306] Robinson, E., R. A. Rasmussen, H. H. Westberg and M. W. Holdren, Nonurban nonmethane low molecular weight hydrocarbon concentrations related to air mass identification, *J. Geophys. Res.*, **78**, 5345-5351, 1973.

[307] Pitts, J. N., Jr., D. Grosjean, B. Shortridge, G. Doyle, J. Smith, T. Mischke and D. Fitz, The nature, concentration and size distribution of organic particulates on the eastern part of the Southern California air basin, Paper presented at Centennial Meeting, Amer. Chem. Soc., New York, New York, Apr. 5, 1976.

[308] Rappaport, S. M. and D. A. Fraser, Gas chromatographic-mass spectrometric identification of volatiles released from a rubber stock during simulated vulcanization, *Anal. Chem.*, **48**, 476-481, 1976.

[309] Smythe, R. J., The application of high resolution gas chromatography and mass spectrometry to analysis of engine exhaust emissions, *Univ. of Waterloo (Waterloo, Ontario)*, Ph.D Dissertation, 1973.

[310] Hauser, T. R. and J. N. Pattison, Analysis of aliphatic fraction of air particle matter, *Environ. Sci. Technol.*, **6**, 549-555, 1972.

[311] Boyer, K. W., Analysis of automobile exhaust particulates, *Univ. Of Illinois (Urbana-Champaign)*, Ph.D Dissertation, 1974.

[312] Hoffmann, D. and E. L. Wynder, Studies on gasoline engine exhaust, *J. Air Poll. Contr. Assoc.*, **13**, 322-327, 1963.

[313] Alperstein, M., and R. L. Bradow, Exhaust gas emissions related to engine combustion, *Soc. Automot. Eng. J.*, **4**, 85(8):52-53, 1968.

[314] Smythe, R. J. and F. W. Karasek, The analysis of diesel exhausts for low molecular weight carbonyl compounds, *J. Chromatog*, **86**, 228-231, 1973.

[315] Larsen, R. I., and W. C. Nelson, Preprint, 1967. 1967.

[316] Environmental Protection Agency, *Air Quality Criteria for Sulfur Dioxide*, Report AP-50, Washington, D.C., 1969.

[317] Chass, R. L. and R. E. George, Contaminant emissions from the combustion of fuels, *J. Air Poll. Contr. Assoc.*, **10**, 34-43, 1960.

[318] Chameides, W. L. Private communication, 1976.

[319] Gay, B. W., Jr. and J. J. Bufalini, Hydrogen peroxide in the urban atmosphere, *ACS Adv. Chem.*, **113**, 255-263, 1972.

[320] Nash, T. Nitrous acid in the atmosphere and laboratory experiments on its photolysis, *Tellus*, **26**, 175-179, 1974.

[321] Lovelock, J. E., Natural halocarbons in the air and in the sea, *Nature*, **256**, 193-194, 1975.

[322] Lovelock, J. E., R. J. Maggs and R. J. Wade, Halogenated hydrocarbons in and over the Atlantic, *Nature*, **241**, 194-196, 1973.

[323] Little, A. D., Inc., *Preliminary economic impact assessment of possible regulatory action to control atmospheric emissions of selected halocarbons*, EPA 450/3-75-073, Cambridge, Mass., 1975.

[324] Froelich, A. A., Private communication, 1970.

[325] Osag, T. R. and G. B. Crane, *Control of odors from inedible-rendering plants*, EPA450/1-74-006, Environmental Protection Agency, Research Triangle Park, N.C., 1974.

[326] Noller, C. R., *Textbook of Organic Chemistry*, Philadelphia, PA: W. B. Saunders Co., 2nd ed., 1958.

[327] White, R. K., Gas chromatogrylic analysis of dairy animal wastes, Ph.D. Dissertation, *Ohio State University*, (1970), *(Dissest. Abst. Int. B.)*, **31**(2), 643, 1970.

[328] Rosehart, R. G. and R. Chu., Methods for identification of arsenic compounds, *Water Air Soil Poll.*, **4**, 395-398, 1975.

[329] Lonneman, W. A., J. J. Bufalini and R. L. Seila, PAN and oxidant measurement in ambient atmospheres, *Environ. Sci. Technol.*, **10**, 374-380, 1976.

[330] Vogh, J. W., Nature of odor components in diesel exhaust, *J. Air Poll. Cont. Assoc.*, **19**, 773-777, 1969.

[331] Toyosawa, S., Y. Umezawa, T. Shirai, S. Yanagisawa, T. Yukawa and S. Nakamura, Analysis of gaseous products from rubber wear, *J. Japan Soc. Air Poll.*, **9**, 200, 1974. (APTIC No. 71608).

[332] Calcraft, A. M., R. J. S. Green and T. S. McRoberts, Burning of plastics. Part 1. Smoke formation, *J. Plastics Inst.*, **42**, 200-208, 1974.

[333] Sato, Y., I. Mizoguchi, K. Makino and H. Yagyu, On the correlation between light chlorinated hydrocarbons and other pollutants in Tokyo atmosphere, *Ann. Rpt. Tokyo Metro. Lab. Public Health*, **25**, 371-379, 1974. (APTIC No. 70587).

[334] Morita, M., H. Mimura and T. Nishizawa, Environmental pollution by p-Dichlorobenzene, *Japan. Soc. Public Health (Fukushima)*, p. 311, 1974. (APTIC No. 70517).

[335] Miller, C. W. and T. M. Shafik, Concentrations of propoxur in air following repeated indoor applications, *Bull. World Health Org. (Geneva)*, **51**(1), 41-44, 1974.

[336] Zeier, U. von and M. Semlitsch, Microstructure studies on magnetic spheroids found in Alpine snow and foundry dust samples in Switzerland, *Suizer Tech. Rev. (Switz.)*, **4**, 268-274, 1970.

[337] Egger, A. M. and E. Widmer, Environmental protection measures in petrochemical plant as exemplified by the gasoline cracking plant of Lonza AG, *Chimia*, **28**, 674-678, 1974.

[338] Numata, H. and S. Takano, A case of environmental pollution by coal-tar pitch, *J. Japan Soc. Air Poll.*, **9**, 540, 1974. (APTIC No. 72056).

[339] Schmeltz, I., J. Tosk and D. Hoffmann, Formation and determination of naphthalenes in cigarette smoke, *Anal. Chem.*, **48**, 645-650, 1976.

[340] Wallcave, L., D. L. Nagel, J. W. Smith and R. D. Waniska, Two pyrene derivatives of widespread environmental distribution: cyclopenta(cd) pyrene and acepyrene, *Environ. Sci. Technol.*, **9**, 143-145, 1975.

[341] Davies, I. W., R. M. Harrison, R. Perry, D. Ratnayaka and R. A. Wellings, Municipal incinerator as a source of polynuclear aromatic hydrocarbons in environment, *Environ. Sci. Technol.*, **10**, 451-453, 1976.

[342] Pitts, J. N., Jr., B. J. Finlayson-Pitts, and A. M. Winer, Optical systems unravel smog chemistry, *Environ. Sci. Technol.*, **11**, 568-573, 1977.

[343] King, R. B., J. S. Fordyce, A. C. Antoine, H. F. Leibecki, H. E. Neustadter and S. M. Sidik, *Extensive 1-year survey of trace elements and compounds in the airborne suspended particulate matter in Cleveland, Ohio*, National Aeronautics and Space Administration, Washington, D.C., NASA TN D-8110, 1976.

[344] Gibbard, S. and R. Schoental, Simple semi-quantitative estimation of sinapyl and certain related aldehydes in wood and in other materials, *J. Chromatog.*, **44**, 396-398, 1969.

[345] Moore, H. E. and S. E. Poet, Background levels of ^{226}Ra in the lower troposphere, *Atmos. Environ.*, **10**, 381-383, 1976.

[346] Chau, Y. K., P. T. S. Wong and H. Saitoh, Determination of tetraalkyl lead compounds in the atmosphere, *J. Chromat. Sci.*, **14**, 162-164, 1976.

[347] Bidleman, T. F. and C. P. Rice, TLC analysis of carbaryl insecticide on sprayed foliage, *J. Chem. Ed.*, **53**, 173, 1976.

[348] Temple, P. J. and S. N. Linzon, Boron as a phytotoxic air pollutant, *J. Air. Poll. Contr. Assoc.*, **26**, 498-499, 1976.

[349] Pauling, L., *General Chemistry*, 2nd ed., San Francisco: W.H. Freeman and Co., 1954.

[350] Anonymous, Zinc oxide emissions controlled by dry collection system, *Pollution Eng.*, **8**(5), 26, 1976.

[351] Anonymous, Planning permits economical air pollution control and product recovery, *Pollution Eng.*, **8**(5), 22, 1976.

[352] Finch, S. P., III, E. R. Stephens and M. A. Price, Paper presented at Pacific Conference on Chemistry and Spectroscopy, San Francisco, Cal., Oct. 16, 1974.

[353] Pierotti, D. and R. A. Rasmussen, Combustion as a source of nitrous oxide in the atmosphere, *Geophys. Res. Lett.*, **3**, 265-267, 1976.

[354] Hartstein, A. M. and D. R. Forshey, *Coal mine combustion products: neoprenes, polyvinyl chloride compositions, urethane foam, and wood*, Bureau of Mines Report of Investigations 7977, U.S. Dept. of the Interior, Washington, D.C., 1974.

[355] American Petroleum Institute, *Bulletin on photochemical evaporation loss from storage tanks*, API Bull. 2523, New York, NY, 1969.

[356] Sternling, C. V. and J. O. L. Wendt, *Kinetic mechanisms governing the fate of chemically bound sulfur and nitrogen in combustion*, EPA650/2-74-017, Environmental Protection Agency (Washington, D.C.), 1972. (NTIS Document No. PB 230895).

[357] Pearson, C. D., The determination of trace mercaptans and sulfides in natural gas by a gas chromatography-flame photometric detector technique, *J. Chromatog. Sci.*, **14**, 154-158, 1976.

[358] Pellizzari, E. D., J. E. Bunch, R. E. Berkley and J. McRae, Determination of trace hazardous organic vapor pollutants in ambient atmospheres by gas chromatography/mass spectrometry/computer, *Anal. Chem.*, **48**, 803-807, 1976.

[359] Katzman, H. and W. F. Libby, Hydrocarbon emissions from jet engines operated at simulated high-altitude supersonic flight conditions, *Atmos. Environ.*, **9**, 839-842, 1975.

[360] Candeli, A., G. Morozzi, A. Paolacci and L. Zoccolillo, Analysis using thin layer and gas-liquid chromatography of polycyclic aromatic hydrocarbons in the exhaust products from a European car running on fuels contaning a range of concentrations of these hydrocarbons, *Atmos. Environ.*, **9**, 843-849, 1975.

[361] Lao, R. C., R. S. Thomas and J. L. Monkman, Computerized gas chromatographic-mass spectrometric analysis of polycyclic aromatic hydrocarbons in environmental samples, *J. Chromatog.*, **112**, 681-700, 1975.

[362] Lao, R. C., R. S. Thomas, H. Oja and L. Dubois, Application of a gas chromatograph-mass spectrometer-data processor combination to the analysis of the polycyclic aromatic hydrocarbon content of airborne pollutants, *Anal. Chem.*, **45**, 908-915, 1973.

[363] Cautreels, W. and K. Van Cauwenberghe, Determination of organic compounds in airborne particulate matter by gas chromatography-mass spectrometry, *Atmos. Environ.*, **10**, 447-457, 1976.

[364] Kunen, S. M., M. F. Burke, E. L. Bandurskii and B. Nagy, Preliminary investigations of the pyrolysis products of insoluble polymer-like components of atmospheric particulates, *Atmos. Environ.*, **10**, 913-916, 1976.

[365] Arbesman, P., U.S. EPA monitoring results for hazardous substances in New Jersey, Memorandum, State of New Jersey Department of Environmental Protection (Trenton, New Jersey), July 16, 1976.

[366] Creac'h, P. V. and G. Point, Mise en evidence, dans l'atmosphere d'acide borique gazeux provement de l'evaporation de l'eau de mer, *C.R. Acad. Sci. (Paris)*, **Ser. B263**, 89-91, 1966.

[367] Johnson, D. L. and R. S. Braman, Distribution of atmospheric mercury species near the ground, *Environ. Sci. Technol.*, **8**, 1003-1009, 1974.

[368] Fine, D. H., D. P. Rounbehler, N. M Belcher and S. S. Epstein, N-Nitroso compounds: Detection in ambient air, *Science*, **192**, 1328-1330, 1976.

[369] Anonymous, *Investigation reports on the environmental atmosphere at a chemical plants region in the Kita Ward*, Kita Ward Office (Tokyo, Japan), Construction and Public Nuisance Div., April, 1975. (APTIC No. 78264).

[370] Katou, T. and R. Nakagawa, Determination of alkyl leads by combined system of gas chromatography and atomic absorption spectrophotometry, *Bull. Inst. Environ. Sci. Technol. (Yokohama Nat'l. Univ.)*, **1**, 19-24, 1975.

[371] Uchimura, R., Pollution caused by petrochemical complex: plants for butadiene and aromatic hydrocarbon, *Tech. Hum. Being (Japan)*, **4**, 112-117, 1975. (APTIC No. 78806).

[372] Axtmann, R. C., Emission control of gas effluents from geothermal power plants, *Environ. Lett.*, **8**, 135-146, 1975.

[373] Singh, H. B., L. J. Silas and L. A. Cavanagh, Distribution, sources and sinks of atmospheric halogenated compounds, *J. Air Poll. Contr. Assoc.*, **27**, 332-336, 1977.

[374] Gaddo, P. P., F. Corazzari and L. Giacomelli, Characterization of hydrocarbons pollution in the urban area of Turin, Paper presented at 69th Annual Meeting, Air Poll. Contr. Assoc., Portland, OR, June 28, 1976.

[375] Fox, M. A. and S. W. Staley, Determination of polycyclic aromatic hydrocarbons in atmospheric particulate matter by high pressure liquid chromatography coupled with fluorescence techniques, *Anal. Chem.*, **48**, 992-998, 1976.

[376] Bellar, T. A. and J. E. Sigsby, Jr., Direct gas chromatographic analysis of low molecular weight substituted organic compounds in emissions, *Environ. Sci. Technol.*, **4**, 150-156,1970.

[377] Pellizzari, E. D., J. E. Bunch, J. T. Bursey, R. E. Berkley, E. Sawicki and K. Krost, Estimation of N-nitrosodimethylamine levels in ambient air by capillary gas-liquid chromatography/mass spectrometry, *Anal. Lett.*, **9**, 579-594, 1976.

[378] Fine, D. H., D. P. Rounbehler, E. Sawicki, K. Krost and G. A. De Marrais, N-Nitroso compounds in the ambient community air of Baltimore, Maryland, *Anal. Lett.*, **9**, 595-604, 1976.

[379] Gardner, J., *Sulfuric Acid Emissions from ESB Battery Plant-Forming Room*, Report 77-BAT-5, York Research Co., Stamford, Conn., 1977. (APTIC No. 102953.)

[380] Giannovario, J. A., R. L. Grob and P. W. Rulon, Analysis of trace pollutants in the air by means of cryogenic gas chromatography, *J. Chromatog.*, **121**, 285-294, 1976.

[381] Murayama, T., Blue flame, white flame, and odor in compressed and ignited engine, *Internal Combust. Engine (Japan)*, **14**(2), 50-59, 1975. (ATPIC No. 73769).

[382] Nishida, K., T. Honda and Y. Miyake, On the exhausted odor component by spraying paints, *Environ. Conserv. Eng. (Japan)*, **4**(4), 240-247, 1975. (APTIC No. 73893).

[383] Gordon, S. J. and S. A. Meeks, A study of the gaseous pollutants in the Houston, Texas area, Paper presented at 79th National Meeting, Amer. Inst. of Chem. Engrs., Houston, Tex., Mar. 16-20, 1975.

[384] Saijo, T. and T. Nozaki, Fact-finding survey for offensive odor, Ibaraki-ken Research Cent. (Japan), *Ann. Rpt. on Environ. Pollution*, **6**, p. 83-93, 1973. (APTIC No. 74213).

[385] Hoshika, Y., S. Kadowaki, I. Kojima, K. Koike and K. Yoshimoto, The gas chromatographic analysis of sulfur compounds and aliphatic amines in the exhaust gas from a fishmeal dryer and cooker, *J. Japan Oil Chem. Soc.*, **24**(4), 27, 1975. (APTIC No. 77123).

[386] Uchiyama, M., Y. Shimada, I. Takahashi, Y. Saito and F. Hayashi, Studies on the photochemical air pollution: Halogen gas in the automobile exhaust gas, *Gunma Prefect. Res. Cent. Environ. Sci. (Japan)*, Ann. Rpt. No. 6, p. 134, 1974. (APTIC No. 76917).

[387] Neiser, J., M. Kaloc and V. Masek, Pyridine and its homologs in the atmosphere of coke oven plants, *Czech. Sh.Ved. Pr. Vys. Sk. Banske Ostrave. (Czechoslovakia)*, **19**(1), 287-309, 1973. (APTIC No. 79013).

[388] Braman, R. S. and D. L. Johnson, Ambient forms of mercury in air, *2nd Ann. Trace Contam. Conf. Proc.*, National Science Foundation., Washington., D.C., p. 75-78, 1974. (NTIS No. LBL-3217).

[389] Kagawa, F., K. Inazawa and H. Saruyama, Detection of cyanogen in atmosphere and rain, Paper presented at 1st Meeting, Soc. for General Research, Environ. Science (Japan), June, 1975. (APTIC No. 74803).
[390] Nagai, K., K. Naka, B. Matsuoayashi, M. Takatsuka and M. Murata, Methods of measurements and examples of measurement on odorous substances (lower aldehydes and styrene) in the petrochemical complex area, *Odor Research J. (Japan)*, **4**(19), 23-32, 1975. (APTIC No. 74561).
[391] Kunz, C. O. and C. J. Paperiello, Xenon-133: Ambient activity from nuclear power stations, *Science*, **192**, 1235-1237, 1976.
[392] Davis, D. D., R. T. Watson, T. McGee, W. Heaps, J. Chang and D. Wuebbles, Tropospheric residence times for several halocarbons based on chemical degradation via OH radicals, Paper ENVT-63 presented at Amer. Chem. Soc. Centennial Meeting, New York, New York, April 8, 1976.
[393] Chau, Y. K., P. T. S. Wong, B. A. Silverberg, P. L. Luxon and G. A. Bengert, Methylation of selenium in the aquatic environment, *Science*, **192**, 1130-1131, 1976.
[394] Singh, H. B., Phosgene in the ambient air, *Nature*, 264, 428-429, 1976.
[395] Franconeri, P. and L. Kaplan, Determination and evaluation of stack emissions from municipal incinerators, *J. Air Poll. Contr. Assoc.*, **26**, 887-888, 1976.
[396] Johnstone, R. A. W. and J. R. Plimmer, The chemical constituents of tobacco and tobacco smoke, *Chem. Rev.*, **59**, 885-936, 1959.
[397] Yamaguchi, M. and N. Shimojo, Methyl-mercury derivatives in engine exhaust and its synthesis, *J. Japan Soc. Air Poll.*, **10**, 413, 1975. (APTIC No. 79929).
[398] Suzuki, F., M. Mimakami, I. Ogawa, S. Oote, M. Hoshino, A. Matsura and K. Sato, Researches on iodine emitted from iodine manufacturing plants in Chiba, *J. Japan Soc. Air Poll.*, **9**, 439, 1974. (APTIC No. 73348).
[399] Kunen, S. M., *University of Utah*, Ph.D Dissertation, 1977.
[400] Crutzen, P. J., The possible importance of CSO for the sulfate layer of the stratosphere, *Geophys. Res. Lett.*, **3**, 73-76, 1976.
[401] Jones, J. H., J. T. Kummer, K. Otto, M. Shelef and E. E. Weaver, Selective catalytic reaction of hydrogen with nitric oxide in the presence of oxygen, *Environ. Sci. Technol.*, **5**, 790-798, 1971.
[402] Cavanagh, L. A. and R. E. Ruff, Sources and sampling of pollutants for geothermal steam areas, *1975 Environ. Sensing and Assessment*, Inst. of Elect. and Electronic Engrs., New York, NY, IEEE Pub. 75-CH 1004-1 ICESA, p. 18-5.1 to p. 18-5.5, 1976.
[403] Environmental Protection Agency, *Air Pollution Aspects of Sludge Incineration*, EPA 625/4-75-009, Washington, D.C., 1975.
[404] Council on Environmental Quality, *Carcinogens in the Environment*, Sixth Annual Report, Washington, D.C., 1975.
[405] Brunnemann, K. D., H. C. Lee and D. Hoffmann, Chemical studies on tobacco smoke. XLVII. On the quantitative analysis of catechols and their reduction, *Anal. Lett.*, **9**, 939-955, 1976.
[406] Severson, R. F., M. E. Snook, R. F. Arrendale and O. T. Chortyk, Gas chromatographic quantitation of polynuclear aromatic hydrocarbons in tobacco smoke, *Anal. Chem.*, **48**, 1866-1872, 1976.
[407] Dmitrieva, V. N., O. V. Meshkova and V. D. Bezuglyi, Determination of low contents of inorganic impurities in effluents from polymer production, *J. Anal. Chem. (USSR)*, **30**(7), Part 2, 1181-1183, 1975.
[408] Lee, M. L., M. Novotny and K. D. Bartle, Gas chromatography/mass spectrometric and nuclear magnetic resonance determination of polynuclear aromatic hydrocarbons in airborne particulates, *Anal. Chem.*, **48**, 1566-1572, 1976.

[409] Ishizaka, T. and K. Isono, Distribution of aerosols in urban atmospheres. I., Paper presented at Japan Met., Soc. Spring Meeting (Tokyo), May 21-23, 1975. (APTIC No. 74751).

[410] Dimitriades, B., *Photochemical Oxidants in the Ambient Air of the United States*, EPA-600/3-76-017, Environmental Protection Agency, Research Triangle Park, N.C., 1976.

[411] Kato, T. and Y. Hanai, GC and GC-MS environmental analysis of photochemical smog component, Paper presented at 24th Lecture Meeting, Japan Soc. for Anal. Chem., (Sapporo, Japan), Oct. 1-5, 1975. (APTIC No. 81207).

[412] Kozyr, N. P., A. Y. Chebanov and B. P. Titomer, Properties of gases and dust generated by titanium slag smelting in closed ore-smelting furnaces, *Tsvetn. Metal. (USSR)*, **11**, 52-53, 1974. (APTIC No. 71642).

[413] Kalpasanov, Y. and G. Kurchatova, A study of the statistical distribution of chemical pollutants in air, *J. Air Poll. Contr. Assoc.*, **26**, 981-985, 1976.

[414] Conkle, J. P., W. W. Lackey, C. L. Martin and R. L. Miller, Organic compounds in turbine combustor exhaust, *1975 Environ. Sensing and Assessment*, Inst. of Elect. and Electronic Engrs., New York, N.Y., IEEE Pub. 75-CH 1004-1 ICESA, p. 27-2.1 to 27-2.11, 1976.

[415] Hurn, R. W., F. W. Cox and J. R. Allsup, *Effects of Gasoline Additives on Gaseous Emissions. Part II.*, EPA 600/2-76-026, Energy Research and Development Admin., Bartlesville, Okla., 1976.

[416] Kaneko, M., Measurement of cyanide concentration in automobile exhaust gases, stack gases and environment, *Kankyo Sozo (Japan)*, **3**(7), 2-8, 1973. (APTIC No. 53231).

[417] Kanamaru, G., H. Shima, T. Matsui and T. Hashizume, An example of atmospheric pollution by fluorides, *Mie Prefect. Pub. Nuisance Center Ann. Rpt.*, **1**, 134-137, 1973. (APTIC No. 52664).

[418] Dietzmann, H. E., *Protocol to Characterize Gaseous Emissions as a Function of Fuel and Additive Composition*, EPA 600/2-75-048, Southwest Research Institute, San Antonio, Texas, 1975.

[419] Litton Systems, Inc., *Air Pollution Aspects of Ethylene*, Bethesda, Maryland, 1969. (NTIS Document PB-188069).

[420] Barger, W. R., and W. D. Garrett, Surface active organic material in air over the Mediterranean and over the eastern equatorial Pacific, *J. Geophys. Res.*, **81**, 3151-3157, 1976.

[421] Holzer, G., J. Oro, and W. Bertsch, Gas chromatographic-mass spectrometric evaluation of exhaled tobacco smoke, *J. Chromatog.*, **126**, 771-785, 1976.

[422] Rains, B. A., M. J. DePrimo, and I. L. Groseclose, *Odors Emitted from Raw and Digested Sewage Sludge*, EPA-670/2-73-098, Environmental Protection Agency, Washington, D.C., 1973.

[423] Mullins, B. J., Jr., D. E. Solomon, G. L. Austin, and L. M. Kacmarcik, *Atmospheric Emissions Survey of the Gas Processing Industry*, EPA-450/3-75-076, Environmental Protection Agency, Washington, D.C., 1975.

[424] Committee on Biologic Effects of Environmental Pollutants, *Particulate Polycyclic Organic Matter*, National Academy of Sciences, Washington, D.C., 1972.

[425] National Assn. of Secondary Material Industries, Inc., *Studies of dislocation factors: No. II. The secondary material industries and environmental problems*, New York, NY, 1968.

[426] Muenzer, M., and K. Heder, Results of the medical and technical inspection of chemical dry cleaning plants, *Zbl. Arbeitsmedizin.*, **22**(5), 133-138, 1972.

[427] Merz, O., Lacquers and surface coatings and their susceptible emissions, *Staub, Reinhaltung Luft*, **31**, 395-396, 1971.
[428] Guenther, R., *A study of the substances liberated from binding agents at the drying of lacquers with respect to air pollution*, Ph.D. Dissertation, Karlsruhe Univ. (West Germany), 1971.
[429] Hirose, K., S. Yamanaka, and S. Takada, Monthly variations of hydrocarbon concentrations in the urban atmosphere, *J. Japan Soc. Air. Poll.*, **10**(4), 586, 1975.
[430] Lambert, G., P. Bristeau, and G. Polian, Emission and enrichments of radon daughters from Etna volcano magma, *Geophys. Res. Lett.*, **3**, 724-726, 1976.
[431] Crittenden, B. D., and R. Long, Diphenylene oxide and cyclopentacenaphthylene(s) in flame soots, *Environ. Sci. Technol.*, **7**, 742-744, 1973.
[432] Scott Research Laboratories, *Diesel exhaust composition and odor: Progress report for year 1965*, CRC Project Rpt. CD-9-61, New York, NY, 1966.
[433] Scott Research Laboratories, *Diesel exhaust composition and odor: Progress report for year 1964*, CRC Project Rpt. No. CD-9-61, New York, NY, 1965.
[434] Walsh, W. H., *Analysis of diesel engine exhaust hydrocarbons*, M.S. Thesis, Penn. State Univ., 1963.
[435] Landen, E. W., and J. M. Perez, Some diesel exhaust reactivity information derived by gas chromatography, SAE Paper No. 740530, Soc. of Automotive Engineers, Inc., New York, NY, 1974.
[436] Smith, M. S., A. J. Francis, and J. M. Duxbury, Collection and analysis of organic gases from natural ecosystems: application to poultry manure, *Environ. Sci. Technol.*, **11**, 51-55, 1977.
[437] Banwart, W. L., and J. M. Bremner, Paper presented at Div. of Soil Microbiol. and Biochem., 66th Annual Meeting, ASA, Chicago, IL, 1974.
[438] Diebel, R. H., in *Agriculture and the quality of our environment*, N. C. Brady, Ed., Am. Assoc. for the Advancement of Science, Washington, D.C., 1967.
[439] Committee on Medical and Biologic Effects of Environ. Pollutants, *Vapor-Phase Organic Pollutants*, National Academy of Sciences, Washington, D.C., 1976.
[440] Rasmussen, R. A., *Terpenes: their analysis and fate in the atmosphere*, Ph.D. Dissertation, Washington Univ., St. Louis, Mo., 1964.
[441] Epstein, J., G. T. Davis, L. Eng, and M. M. Demek, Potential hazards associated with spray drying operations, *Environ. Sci.Technol.*, **11**, 70-75,1977.
[442] Suprenant, N. F., and M. I. Bornstein, *New source classification codes for processes which cause hydrocarbon and organic emissions*, EPA-450/3-75-067, GCA Corporation, Bedford, Mass., 1975.
[443] Cautreels, W., and K. Van Cauwenberghe, Extraction of organic compounds from airborne particulate matter, *Water, Air, Soil Poll.*, **6**, 103-110, 1976.
[444] Kircher, D. S., and D. P. Armstrong, *An Interim Report on Motor Vehicle Emission Estimation*, EPA-450/2-73-003, Environmental Protection Agency, Research Triangle Park, N.C., 1973.
[445] Baker, R. R., The formation of oxides of carbon by the pyrolysis of tobacco, *Beit. zur Tabak.*, **8**(1), 16-27, 1975 (SHB 75-0462).
[446] Ceschini, P., and D. Chem, Effect of sampling conditions on the composition of the volatile phase of cigarette smoke, *Beit. zur Tabak.*, **7**(5), 294-301, 1974 (SHB 75-0156).
[447] Klus, H., and H. Kuhn, The determination of nitrophenols in tobacco smoke condensate (preliminary results), *Fach. Mitt. der Austria Tabakwerke A.G.*, **15**, 275-288, 1974 (SHB 75-0164).
[448] Spears, A. W., Effect of manufacturing variables on cigarette smoke composition, *CORESTA Information Bulletin*, Special Issue: 65-78, 1974 (SHB 75-0175).

[449] Sander, J., F. Schweinsberg, J. LaBar, G. Burkle, and E. Schweinsberg, Nitrate and nitrosable amino compounds in carcinogenesis, *GANN Monograph on Cancer Research*, **17**, 145-160, 1975 (SHB 76-0016).

[450] Tso, T. C., J. L. Sims, and D. E. Johnson, Some agronomic factors affecting N-dimethylnitrosamine content in cigarette smoke, *Beir. zur Tabak.*, **8**(1), 34-38, 1975 (SHB 75-0483).

[451] Walters, D. B., W. J. Chamberlain, and O. T. Chortyk, *Cis* and *trans* fatty acids in cigarette smoke condensate, *Anal. Chim. Acta*, **77**, 309-311, 1975.

[452] Mestres, R., S. Illes, C. Espinoza, and C. Chevallier, Detection and assay of pesticide residues in tobaccos, *Trav. de la Soc. de Pharm. de Montpellier*, **34**(3), 255-266, 1974 (SHB 76-0313).

[453] Burton, H. R., and G. Childs, Jr., The thermal degradation of tobacco.VI. Influence of extraction on the formation of some major gas phase constituents, *Beit. zur Tabak.*, **8**(4), 174-180, 1975 (SHB 76-0473).

[454] Hoshika, Y., and Y. Iida, Gas chromatographic determination of sulphur compounds in town gas, *J. Chromatog.*, **134**, 423-432, 1977.

[455] Jones, P. W., Analysis of nonparticulate organic compounds in ambient atmospheres, Paper presented at 67th Annual Meeting, Air Poll. Cont. Assoc., Denver, CO, 1974.

[456] Moore, B. J., *Analyses of Natural Gases, 1975*, Bureau of Mines Inform. Circ. 8717, U.S. Dept. of the Interior, Washington, D.C., 1976.

[457] Environmental Health Research Center, *Proposed Standard and Health Effects of Ambient Chlorine*, Rept. IIEQ 76-07, Chicago, IL, 1976. (NTIS PB-258983).

[458] Kunen, S. M., K. J. Voorhees, A. C. Hill, F. D. Hileman, and D. N. Osborne, Chemical analysis of the insoluble carbonaceous components of atmospheric particulates with pyrolysis/gas chromatography/mass spectrometry techniques, Paper 77-36.4, 70th Annual Meeting, Air Poll. Cont. Assoc., Montreal, Ont., Can., June, 1977.

[459] Rupp, W. H., Air pollution sources and their control, in *Air Pollution Handbook*, ed. P. L. Magill, F. R. Holden, and C. Ackley, New York: McGraw-Hill, p. 1-7, 1956.

[460] Dillard, J. G., R. D. Seals, and J. P. Wightman, A study of aluminum-containing atmospheric particulate matter, paper presented at Spring Annual Meeting, American Chemical Society, New Orleans, LA, 1977.

[461] Gay, B. W., Jr., R. R. Arnts, and W. A. Lonneman, Naturally emitted hydrocarbons, their atmospheric chemistry and effects on rural and urban oxidant formation, paper presented at Spring Annual Meeting, American Chemical Society, New Orleans, LA, 1977.

[462] Berkley, R. E., E. D. Pellizzari, and J. T. Bursey, Analysis of trace levels of organic vapors in ambient air by glass capillary gc/ms: industrial pollutants in the Los Angeles basin, paper presented at Spring Annual Meeting, American Chemical Society, New Orleans, La., 1977.

[463] Maddox, W. L., and G. Mamantov, Analysis of cigarette smoke by fourier transform infrared spectrometry, *Anal. Chem.*, **49**, 331-336, 1977.

[464] Moore, R. E., Volatile compounds from marine algae, *Acc. Chem. Res.*, **10**, 40-47, 1977.

[465] Mayrsohn, H., J. H. Crabtree, M. Kuramoto, R. D. Sothern, and S. H. Mano, Source reconciliation of atmospheric hydrocarbons 1974, *Atmos. Environ.*, **11**, 189-192, 1977.

[466] Shultz, J. L., T. Kessler, R. A. Friedel, and A. G. Sharkey, Jr., High-resolution mass spectrometric investigation of heteroatom species in coal-carbonization products, *Fuel*, **51**, 242-246, 1972.

[467] Nozaki, Y., D. J. McMaster, L. K. Benninger, D. M. Lewis, W. C. Graustein, and K. K. Turekian, Atmospheric Pb-210 fluxes determined from soil profiles, *EOS-Trans. AGU*, **58**, 396, 1977.

[468] Birkenheuer, D. L., and B. L. Davis, X-ray quantitative analysis of hi-volume aerosol samples, *EOS-Trans. AGU*, **58**, 396, 1977.

[469] Stephens, E. R., and F. R. Burleson, Unpublished data cited by Winer, et al. (1033).

[470] Ryan, J. A., and N. R. Mukherjee, Sources of atmospheric gaseous chloring, *Rev. Geophys. Space Phys.*, **13**, 650-658, 1975.

[471] Morgan, E. D., and R. C. Tyler, Microchemical methods for the identification of volatile pheromones, *J. Chromatog.*, **134**, 174-177, 1977.

[472] Rasmussen, R. A., R. B. Chatfield, and M. W. Holdren, Hydrocarbon species in rural Missouri air, paper presented at symposium on "The Non-Urban Tropospheric Composition", Hollywood, FL, Nov. 11, 1976.

[473] Mason, B. *Principles of Geochemistry*, 2nd ed., New York: John Wiley, 1958.

[474] Murrell, J. T., and R. B. Channell, Fragrence analyses of *Trillium luteum* and *Trillium cuneatum*, J. Tenn. Acad. Sci., **48**, 101-103, 1973.

[475] Lodge, J. P., Jr., J. B. Pate, B. E. Ammons, and G. A. Swanson, The use of hypodermic needles as critical orifices in sampling, *J. Air Poll. Cont. Assoc.*, **16**, 197-200, 1966.

[476] Rahn, K. A., *The Chemical Composition of the Atmospheric Aerosol*, Technical Report, Graduate School of Oceanography, University of Rhode Island, Kingston, RI, July, 1976.

[477] Dams, R., J. A. Robbins, K. Rahn, and J. W. Winchester, Nondestructive neutron activation analysis of pollution particulates, *Anal. Chem.*, **42**, 861-867, 1970.

[478] King, R. B., J. S. Fordyce, A. C. Antoine, H. F. Leibecki, H. E. Neustadler, and S. M. Sidik, Elemental composition of airborne particulates and source identification: an extensive one-year survey, *J. Air Poll. Cont. Assoc.*, **26**, 1073-1084, 1976.

[479] Moore, H. E., E. A. Martell, and S. E. Poet, Sources of polonium-210 in the atmosphere, *Environ. Sci. Technol*, **10**, 586-591, 1976.

[480] Grosjean, D., K. Van Cauwenberghe, J. P. Schmid, P. E. Kelley, and J. N. Pitts, Jr., Identification of C_3-C_{10} aliphatic dicarboxylic acids in airborne particulate matter, *Environ. Sci. Technol.*, **12**, 313-317, 1978.

[481] Volchok, H. L., and B. Krajewski, Radionuclides and lead in surface air, *Health and Safety Laboratory Fallout Program*, Quarterly Summary Report, p. C-1 to C-9 and C-102 to C-107, New York, NY, June 1, 1972.

[482] Thomas, C. W., Preliminary finding of the concentration of ^{99}Tc in environmental samples, *Pacific Northwest Laboratory Annual Report for 1972*, BNWL-1951, Pt. 2, UC-48, 1973.

[483] Junge, C. *Air Chemistry and Radioactivity*, New York: Academic Press, 1963.

[484] Krey, P. W., private communication, 1976.

[485] Koide, M., J. J. Griffin, and E. D. Goldberg, Records of plutonium fallout in marine and terrestrial samples, *J. Geophys. Res.*, **80**, 4153-4162, 1975.

[486] Black, F. M., L. E. High, and A. Fontijn, Chemiluminescent measurement of reactivity weighted ethylene-equivalent hydrocarbons, *Environ. Sci. Technol.*, **11**, 597-601, 1977.

[487] Kaiser, K. L. E., and J. Lawrence, Polyelectrolytes: potential chloroform precursors, *Science*, **196**, 1205-1206, 1977.

[488] Dong, M. W., D. C. Locke, and D. Hoffmann, Characterization of aza-arenes in basic organic portion of suspended particulate matter, *Environ. Sci. Technol.*, **11**, 612-618, 1977.

[489] Robertson, D. E., E. A. Crecelius, J. S. Fruchter, and J. D. Ludwick, Mercury emissions from geothermal power plants, *Science*, **196**, 1094-1097, 1977.

[490] Billings, C. E., A. M. Sacco, W. R. Matson, R. M. Griffin, W. R. Coniglio, and R. A. Hartley, Mercury balance on a large pulverized coal-fired furnace, *J. Air Poll. Cont. Assoc.*, **23**, 773-777, 1973.

[491] Maroulis, P. J., and A. R. Bandy, Estimate of the contribution of biologically produced dimethyl sulfide to the global sulfur cycle, *Science*, **196**, 647-648, 1977.

[492] Iverson, R. E., Air pollution in the aluminum industry, *J. Metals*, **25**, 19-23, 1973.

[493] Stahly, E. E., Removal of metal carbonyls from tobacco smoke, Patent U.S. 3, 246, 654, *Chem. Abstr.*, **65**, 9358, 1966.

[494] Wilkniss, P. E., J. W. Swinnerton, D. J. Bressan, R. A. Lamontagne, and R. E. Larson, CO, CCl_4, Freon-11, CH_4 and Rn-222 concentrations at low altitude over the Arctic ocean in January, 1974, *J. Atmos. Sci.*, **32**, 158-162, 1975.

[495] Palmer, T. Y., Combustion sources of atmospheric chlorine, *Nature*, **263**, 44-46, 1976.

[496] Loper, G. M., and A. M. Lapidi, Photoperiodic effects on the emanation of volatiles from alfalfa *(Medicago sativa L.)* florets, *Plant. Physiol.*, **49**, 729-732, 1971.

[497] Kingston, B. H., Selection of "naturals" in modern perfumery, *Soap, Perfum., Cosmetics*, **44**, 553-558, 1971.

[498] Saxena, K. N., and S. Prabha, Relationship between the olfactory sensilla of *Papileo demoleus L.* larvae and their orientation response to different odours, *J. Entomol.*, **A50**, 119-126, 1975.

[499] Holman, R. T., and W. H. Heimermann, Identification of components of orchid fragrances by gas chromatography-mass spectrometry, *Bull. Amer. Orchid Soc.*, **42**, 678-682, 1973.

[500] Dodson, C. H., R. L. Dressler, H. G. Hills, R. M. Adams, and N. H. Williams, Biologically active compounds in orchid fragrances, *Science*, **164**, 1243-1249, 1969.

[501] Williams, N. H., and C. H. Dodson, Selective attraction of male euglossine bees to orchid floral fragrances and its importance in long distance pollen flow, *Evolution*, **26**, 84-95, 1972.

[502] Waller, G. D., G. M. Loper, and R. L. Berdel, A bioassay for determining honey bee response to flower volatiles, *Environ. Entomol.*, **2**, 255-259, 1973.

[503] Opdyke, D. L. J., Monographs on fragrance raw materials, *Food Cosmet. Toxicol.*, **11**, 95-115, 1973.

[504] Opdyke, D. L. J., Monographs on fragrance raw materials, *Food Cosmet. Toxicol.*, **11**, 477-495, 1973.

[505] Opdyke, D. L. J., Monographs on fragrance raw materials, *Food Cosmet. Toxicol.*, **11**, 855-876, 1973.

[506] Opdyke, D. L. J., Monographs on fragrance raw materials, *Food Cosmet. Toxicol.*, **11**, 1011-1081, 1973.

[507] Opdyke, D. L. J., Monographs on fragrance raw materials, *Food Cosmet. Toxicol.*, **12**, 385-405, 1974.

[508] Opdyke, D. L. J., Monographs on fragrance raw materials, *Food Cosmet. Toxicol.*, **12**, 517-537, 1974.

[509] Opdyke, D. L. J., Monographs on fragrance raw materials, *Food Cosmet. Toxicol.*, **12**, 703-736, 1974.

[510] Opdyke, D. L. J., Monographs on fragrance raw materials, *Food Cosmet. Toxicol.*, **12**, 807-1016, 1974.

[511] Opdyke, D. L. J., Monographs on fragrance raw materials, *Food Cosmet. Toxicol.*, 91-112, 1975.

[512] Opdyke, D. L. J., Monographs on fragrance raw materials, *Food Cosmet. Toxicol.*, **13**, 449-457, 1975.
[513] Opdyke, D. L. J., Monographs on fragrance raw materials, *Food Cosmet. Toxicol.*, **13**, 545-554, 1975.
[514] Opdyke, D. L. J., Monographs on fragrance raw materials, *Food Cosmet. Toxicol.*, **13**, 681-923, 1975.
[515] Opdyke, D. L. J., Monographs on fragrance raw materials, *Food Cosmet. Toxicol.*, **14**, 307-338, 1976.
[516] Opdyke, D. L. J., Monographs on fragrance raw materials, *Food Cosmet. Toxicol.*, **14**, 443-481, 1976.
[517] Opdyke, D. L. J., Monographs on fragrance raw materials, *Food Cosmet. Toxicol.*, **14**, 601-633, 1976.
[518] Hocking, G. M., Plant flavor and aromatic values in medicine and pharmacy, in *Current Topics in Plant Science*, New York: Academic Press, 1969.
[519] Hagen, D. F., and G. W. Holiday, The effects of engine operating and design variables on exhaust emissions, *SAE TP-6*, Society of Automotive Engineers, New York, NY, p. 206, 1964.
[520] Lee, R. E., Jr., and F. V. Duffield, EPA's catalyst research program: environmental impact of sulfuric acid emissions, *J. Air Poll. Cont. Assoc.*, **27**, 631-635, 1977.
[521] Akagi, H., and A. Kobayashi, Analysis of plasticizers and hydrogen chloride during processing of polyvinyl chloride compounds, *J. Japan Soc. Air Poll.*, **10**, 374, 1975, (APTIC No. 84207).
[522] Banwart, W. L., and J. M. Bremner, Identification of sulfur gases evolved from animal manures, *J. Environ. Qual.*, **4**, 363-366, 1975.
[523] Hoshika, Y., I. Kozima, K. Koike, and K. Yoshimoto, Analysis of the sulfur compounds in the exhaust gases from two corn starch factories, *J. Japan Oil Chem. Soc.*, **24**, 317-318, 1975.
[524] Stephens, E. R., and F. R. Burleson, Distribution of light hydrocarbons in ambient air, *J. Air Poll. Contr. Assoc.*, **19**, 929-936, 1969.
[525] Bertsch, W., R. C. Chang, and A. Zlatkis, The determination of organic volatiles in air pollution studies: characterization of profiles, *J. Chromatog. Sci.*, **12**, 175-182, 1974.
[526] Papa, L. J., D. L. Dinsel, and W. C. Harris, Gas chromatographic determination of C_1 to C_{12} hydrocarbons in automotive exhaust, *J. Gas Chromatog.*, **6**, 270-279, 1968.
[527] Jacobs, E. S., Rapid gas chromatographic determination of C_1 to C_{10} hydrocarbons in automotive exhaust gas, *Anal. Chem.*, **38**, 43-48, 1966.
[528] Ketseridis, G., J. Hahn, R. Jaenicke, and C. Junge, The organic constituents of atmospheric particulate matter, *Atmos. Environ.*, **10**, 603-610, 1976.
[529] Soldano, B. A., P. Bien, and P. Kwan, Air-borne organo-mercury and elemental mercury emissions with emphasis on central sewage facilities, *Atmos. Environ.*, **9**, 941-944, 1975.
[530] Shendrikar, A. D., and P. W. West, Air sampling methods for the determination of selenium, *Anal. Chim. Acta*, **89**, 403-406, 1977.
[531] Arctander, S., *Perfume and Flavor Materials of Natural Origin*, Privately published, Elizabeth, New Jersey, 1960.
[532] Cronn, D. R., R. J. Charlson, R. L. Knights, A. L. Crittenden, and B. R. Appel, A survey of the molecular nature of primary and secondary components of particles in urban air by high-resolution mass spectrometry, *Atmos. Environ.*, **11**, 929-937, 1977.

[533] Wakayama, S., S. Namba, and M. Ohno, Odorous constituents of lilac flower oil, *Nippon Kagaku Zasshi*, **92**, 256-259, 1971.
[534] Schmeltz, I., and D. Hoffmann, Nitrogen-containing compounds in tobacco and tobacco smoke, *Chem. Rev.*, **77**, 295-311, 1977.
[535] Levaggi, D. A., and M. Feldstein, The collection and analysis of low molecular weight carbonyl compounds from source effluents, *J. Air Poll. Contr. Assoc.*, **19**, 43-45, 1969.
[536] Hendel, F. J., Aerothermochemistry of the terrestrial atmosphere, *CRC Crit. Rev. Environ. Control*, **3**, 129-152, 1973.
[537] Kimura, K., H. Kaji, N. Nakano, and T. Ito, Measurement of trace hydrocarbons in automobile exhaust, *Proc. Lecture Meeting, Soc. Auto. Engr. Japan*, p. 609-614, 1975 (APTIC No. 79610).
[538] McLane, J. E., R. B. Finkelman, and R. R. Larson, *Minerological examination of particulate matter from the fumeroles of Sherman Crater, Mt. Baker, Washington (State)*, Preprint, Geological Survey, Reston, VA, 2 p., 1975 (APTIC No. 82574).
[539] Michalak, L., Air pollution by a cement plant at Chelm Lubelski, *Ochrona Powietrza (Warsaw)*, no. 3, 81-86, 1975 (APTIC No. 79697).
[540] Terajima, T., Offensive odor, *Mokuzai Koryo*, **30**, 508-510, 1975 (APTIC No. 82209).
[541] Blosser, E. R., L. J. Hillenbrand, J. Lathouse, W. R. Pierson, and J. W. Butler, Sampling and analysis for sulfur compounds in automobile exhaust, *Proc. IMR Symp.*, 389-400, (National Bureau of Standards Pub. 422), 1976.
[542] Chamberlain, W. J., Polar lipid materials in cigarette smoke condensate, *Tobacco*, **178**(#23), 51-53, 1976.
[543] Trijonis, J. C., and K. W. Arledge, *Utility of Reactivity Criteria in Organic Emission Control Strategies for Los Angeles*, Contract Report No. 68-02-1735, TRW Environ. Systems, Redondo Beach, CA, 1975.
[544] Carter, J. A., and W. R. Musick, Platinum metals in air particulates near a catalytic converter test site as measured by isotope dilution SSMS, in *Proc. Platinum Res. Rev. Conf.*, Catal. Res. Program, Rougemont, NC, 1975.
[545] Hrutfiord, B. F., L. N. Johanson, and J. L. McCarthy, *Steam Stripping Odorous Substances from Kraft Effluent Streams*, EPA-R2-73-196, Environmental Protection Agency, Washington, D. C., 1973. (NTIS PB-221335).
[546] Dong, M., D. Hoffmann, D. C. Locke, and E. Ferrand, The occurrence of caffeine in the air of New York City, *Atmos. Environ.*, **11**, 651-653, 1977.
[547] Rasmussen, R., private communication, 1977.
[548] Tabor, E. C., T. E. Hauser, J. P. Lodge, and R. H. Burttschell, Characteristics of the organic particulate matter in the atmosphere of certain American cities, *Arch. Ind. Health.*, **17**, 58-63, 1958.
[549] Meijer, G. M., and H. Nieboer, Determination of peroxybenzoyl nitrate (PBzN) in ambient air, *Ver. Deutsch. Ingen.*, **270**, 55-56, 1977.
[550] Hoshika, Y., Simple and rapid gas-liquid-solid chromatographic analysis of trace concentrations of acetaldelyde in urban air, *J. Chromatogr.*, **137**, 455-460, 1977.
[551] Tausch, H., and G. Stehlik, Bestimmung polycyclischer aromaten in russ mittels gas-chromatographie-massenspektrometerie, *Chromatographia*, **10**, 350-357, 1977.
[552] Nicholas, H. J., Miscellaneous volatile plant products, in *Phytochemistry*, (L. P. Miller, ed.), v. 2, p. 381-399, 1973
[553] Krstulovic, A. M., D. M. Rosie, and P. R. Brown, Distribution of some atmospheric polynuclear aromatic hydrocarbons, *Amer. Lab.*, **9**(#7), 11-18, 1977.
[554] Mansfield, C. T., B. T. Hodge, R. B. Hege, Jr., and W. C. Hamlin, Analysis of formaldehyde in tobacco smoke by high performance liquid chromatography, *J. Chromatog. Sci.*, **15**, 301-302, 1977.

[555] Ito, Y. Offensive smell in tap water, Paper presented at Chem. Soc. Japan 34th Ann. Meeting, Tokyo, 1976 (APTIC No. 101209).

[556] Grimmer, G., Analysis of automobile exhaust condensates, in *Air Pollut. Cancer Man*, Lyon, World Health Org., p. 29-39, 1977.

[557] Suzuki, R., M. Ito, M. Noma, I. Moritani, Y. Watanabe, T. Nakaya, and N. Saito, Determination of PCB in dust, ash and combustion gas from city wastes incinerators, *Rept. Environ. Pollut. Res. Cent. Aichi Pref.* (Japan), **2**, 43-49 (1974) (APTIC No. 77236).

[558] Nicolet, M., The properties and constitution of the upper atmosphere, in *Physics of the Upper Atmosphere*, J. A. Ratcliffe, ed., New York: Academic Press, p. 17, 1960.

[559] Little, A. D., Inc. *Preliminary Economic Input of Possible Regulatory Action to Control Atmosphere Emissions of Selected Halocarbons*, EPA Contract No. 68-02-1349, Task 8, EPA-450/3-75-073, Research Triangle Park, NC, 1975.

[560] Panel on Atmospheric Chemistry, *Halocarbons: Effects on Stratospheric Ozone*, National Academy of Sciences, Washington, D. C., 1976.

[561] Lapp, T. W., H. M. Gadberry, R. R. Wilkinson, and T. Weast, *Chemical Technology and Economics in Environmental Perspectives. Task III - Chlorofluorocarbon Emission Control in Selected End-Use Applications*, EPA-560/1-76-009, Environmental Protection Agency, Washington, D.C., 1976.

[562] Lapp, T. W., G. J. Hennon, H. M. Gadberry, I. C. Smith, and K. Lawrence, *Chemical Technology and Economics in Environmental Perspectives; Task I - Technical Alternatives to Selected Chlorofluorocarbon Uses*, EPA-560/1-76-002, Environmental Protection Agency, Washington, D.C., 1976.

[563] Federal Task Force on Inadvertent Modification of the Stratosphere, *Fluorocarbons and the Environment*, Council on Environmental Quality and Federal Council for Science and Technology, Washington, D.C., 1975.

[564] Amato, W. S., B. Bandyopadhyay, B. E. Kurtz, and R. H. Fitch, Development of a low emissions process for ethylene dichloride production, *Int. Conf. on Photochem. Oxidant*, EPA-600/3-77, 001b, Environmental Protection Agency, Research Triangle Park, NC, 1977.

[565] Bernstein, L. J., K. K. Kearby, A. K. S. Roman, J. Vardi, and E. E. Wigg, Application of catalysts to automotive NO_x emissions control, *Tech. Paper 710014*, Soc. Automot. Engineers, Warrenton, PA, 1971.

[566] Ciccioli, P., G. Bertoni, E. Brancaleoni, R. Fratarcangeli, and F. Bruner, Evaluation of organic pollutants in the open air and atmospheres in industrial sites using graphetized carbon black traps and gas chromatographic-mass spectrometric analysis with specific detectors, *J. Chromatogr.*, **126**, 757-770, 1976.

[567] Turner, B. C., and D. E. Glotfelty, Field air sampling of pesticide vapors with polyurethane foam, *Anal. Chem.*, **49**, 7-10, 1977.

[568] Plimmer, J. R., Photochemistry of organochlorine insecticides, *Proc. 2nd Int. Conf. on Pesticide Chem.*, **1**, 413-432, 1972.

[569] Arnts, R. R., A. Appleby, D. Lillian, and H. B. Singh, Vapor phase photodecomposition of ortho-chlorobiphenyl, Paper ENVT-68, 172nd Meeting, American Chemical Society, San Francisco, CA, 1976.

[570] Lunde, G., J. Gether, N. Gjos, and M.-B. S. Lande, Organic micropollutants in precipitation in Norway, *Atmos. Environ.*, **11**, 1007-1014, 1977.

[571] Rubel, F. N., Incinerator emissions, in *Incineration of Solid Wastes*, Noyes Data Corp., Park Ridge, NJ, p. 49-67, 1974.

[572] Singmaster, J. N. III, and D. G. Crosby, Volatilization of hydrophobic pesticides from water, Paper PEST-6, 173rd Meeting, American Chemical Society, New Orleans, LA, 1977.

[573] Miller, C. W., and M. T. Shafik, Occurrence of propoxur in air following intradomicilliary spraying, Paper PEST-26, 173rd Meeting, American Chemical Society, New Orleans, LA, 1977.

[574] Taylor, A. W., D. E. Glotfelty, B. C. Turner, R. E. Silver, H. P. Freeman, and A. Weiss, Volatilization of dieldrin and heptachlor residues from field vegetation, Paper PEST-7, 173rd Meeting, American Chemical Society, New Orleans, LA, 1977.

[575] Turner, B. C., D. E. Glotfelty, A. W. Taylor, and D. R. Watson, Volatilization of microencapsulated and conventionally applied CIPC in the field, Paper PEST-8, 1973rd Meeting, American Chemical Society, New Orleans, LA, 1977.

[576] Caro, J. H., B. A. Bierl, H. P. Freeman, D. E. Glotfelty, and B. C. Turner, Disparlure: volatilization rates of two microencapsulated formulations from a grass field, Paper PEST-9, 173rd Meeting, American Chemical Society, New Orleans, LA, 1977.

[577] Abbott, D. C., R. B. Harrison, J. O'G. Tatton, and J. Thomson, Organochlorine pesticides in the atmosphere, *Nature*, **211**, 259-261, 1966.

[578] Miles, J. W., L. E. Fetzer, and G. W. Pearce, Collection and determination of trace quantities of pesticides in air, *Environ. Sci. Technol.*, **4**, 420-425, 1970.

[579] Schmeltz, I., J. Tosk, G. Jacobs, and D. Hoffmann, Redox potential and quinone content of cigarette smoke, *Anal. Chem.*, **49**, 1924-1929, 1977.

[580] Cheney, J. L., and C. R. Fortune, The present relationships of sulfuric acid concentration to acid dew point for flue gases, *Anal. Lett.*, **10**, 797-816, 1977.

[581] Meguerian, G. H., and C. R. Lang, NO_x reduction catalysts for vehicle emission control, Tech. Paper 710291, Soc. Automot. Engineers, Warrenton, PA, 1971.

[582] Louw, C. W., J. F. Richards, and P. K. Faure, The determination of volatile organic compounds in city air by gas chromatography combined with standard addition, selective subtraction, infrared spectrometry and mass spectrometry, *Atmos. Environ.*, **11**, 703-717, 1977.

[583] Woodhead, S., and E. Bernays, Changes in release rates of cyanide in relation to palatability of Sorghum to insects, *Nature*, **270**, 235-236, 1977.

[584] Oaks, D. M., H. Hartmann, and K. P. Dimick, Analysis of sulfur compounds with electron capture/hydrogen flame dual channel gas chromatography, *Anal. Chem.*, **36**, 1560-1565, 1964.

[585] Duce, R. A., W. H. Zoller, and J. L. Moyers, Particulate and gaseous halogens in the Antarctic atmosphere, *J. Geophys. Res.*, **78**, 7802-7811, 1973.

[586] Savenko, V. S., Is boric acid evaporation from the sea water the major boron source in the atmosphere?, *Okeanolognya*, **17**, 445-448, 1977.

[587] Evans, C. S., C. J. Asher, and C. M. Johnson, Isolation of dimethyl diselenide and other volatile selenium compounds from *Astragalus racemosus* (pursh.), *Aust. J. Biol. Sci.*, **21**, 13-20, 1968.

[588] Winer, A. M., A. C. Lloyd, K. R. Darnall, R. Atkinson, and J. N. Pitts, Jr., Rate constants for the reaction of OH radicals with *n*-propyl acetate, *sec*-butyl acetate, tetrahydrofuran, and peroxyacetyl nitrate, *Chem. Phys. Lett.*, **51**, 221-226, 1977.

[589] Schuetzle, D., A. L. Crittenden, and R. J. Charlson, Application of computer controlled high resolution mass spectrometry to the analysis of air pollutants, *J. Air Poll. Control. Assoc.*, **23**, 704-709, 1973.

[590] Drivas, P. J., and F. H. Shair, Dispersion of a crosswind line source of tracer released from an urban highway, *Atmos. Environ.*, **8**, 1155-1164, 1974.

[591] Holzer, G., H. Shanfield, A. Zlatkis, W. Bertsch, P. Juarez, H. Mayfield, and H. M. Liebich, Collection and analysis of trace organic emissions from natural sources, *J. Chromatogr.*, **142** 755-764, 1977.

[592] Hurn, R. W., Mobile Combustion sources, in *Air Pollution, III*, (A. C. Stern, ed.) 2nd Ed., New York: Academic Press, p. 55-95, 1968.

[593] Krey, P. W., R. J. Lagomarsino, and L. E. Toonkel, Gaseous halogens in the atmosphere in 1975, *J. Geophys. Res.*, **82**, 1753-1766, 1977.

[594] Fujii, M., M. Matsuoka, K. Nakamura, K. Oikawa, K. Honma, and Y. Hashimoto, Air pollutants in exhaust gas of jet-aircraft engine, *Japan J. Public Health*, **23**(4), 299-300, 1976. (APTIC No. 103711.)

[595] Wofsy, S. C., M. B. McElroy, and Y. C. Yung, The chemistry of atmospheric bromine, *Geophys. Res. Lett.*, **2**, 215-218, 1975.

[596] Fishbein, J., Chromatographic and biological aspects of DDT and its metabolites, *J. Chromatog.*, **98**, 177-251, 1974.

[597] Pellizzari, E. D., *The Measurement of Carcinogenic Vapors in Ambient Atmospheres*, EPA-600/7-77-055, Environmental Protection Agency, Research Triangle Park, NC, 1977.

[598] Summer, W., *Odour Pollution of Air*, Cleveland: CRC Press, 1971.

[599] Campbell, K., and P. L. Dartnell, *Air Pollution Control in Transport Engines*, London: Inst. of Mech. Engrs., 1972.

[600] Klippel, W., and P. Warneck, Formaldehyde in rain water and on the atmospheric aerosol, *Geophys. Res. Lett.*, **5**, 177-179, 1978.

[601] Ioffe, B. V., V. A. Isidorov, and I. G. Zenkevich, Gas chromatographic - mass spectrometric determination of volatile organic compounds in an urban atmosphere, *J. Chromatogr.*, **142**, 787-795, 1977.

[602] Vanhaelen, M., R. Vanhaelen-Fastré, and J. Geeraerts, Isolation and characterization of trace amounts of volatile compounds affecting insect chemosensory behavior by combined pre-concentration on Tenax GC and gas chromatography, *J. Chromatogr.*, **144**, 108-112, 1977.

[603] Harsch, D. E., and R. A. Rasmussen, Identification of methyl bromide in urban air, *Anal. Lett.*, **10**, 1041-1047, 1977.

[604] Giggenbach, W. F., and F. LeGuern, The chemistry of magmatic gases from Erta'Ale, Ethiopia, *Geochim. Cosmochim. Acta*, **40**, 25-30, 1976.

[605] Biggins, P. D. E., and R. M. Harrison, Identification of lead compounds in urban air, *Nature*, **272**, 531-532, 1978.

[606] Cadle, S. H., and R. L. Williams, Gas and particle emissions from automobile tires in laboratory and field studies, *J. Air Poll. Contr. Assoc.*, **28**, 502-507, 1978.

[607] Schulman, B. L., and P. A. White, Pyrolysis of scrap tires using the Tosco II process - a progress report, Paper presented at 175th Annual Meeting, Amer. Chem. Soc., Anaheim, CA, 1978.

[608] Huebert, B. J., and A. L. Lazrus, Global tropospheric measurement of nitric acid vapor and particulate nitrate, *Geophys. Res. Lett.*, **5**, 577-580, 1978.

[609] Lewis, F. M., and P. W. Chartrand, A scrap tire-fired boiler, *Proc. Natl. Waste Process. Conf.*, New York: Amer. Soc. Mech. Engrs., p. 301-311, 1976.

[610] Bleibtreu, J. N., *The Parable of the Beast*, New York: Collier, 1968.

[611] Morimoto, T., K.-I. Takeyama, and F. Konishi, Composition of gaseous combustion products of polymers, *J. Appl. Polymer Sci.*, **20**, 1967-1976, 1976.

[612] Terrill, J. B., R. R. Montgometry, and C. F. Reinhardt, Toxic gases from fires, *Science*, **200**, 1343-1347, 1978.

[613] Zeldes, S. G., and A. D. Horton, Trapping and determination of labile compounds in the gas phase of cigarette smoke, *Anal. Chem.*, **50**, 779-782, 1978.

[614] Karasek, F. W., D. W. Denney, K. W. Chan, and R. E. Clement, Analysis of complex organic mixtures on airborne particulate matter, *Anal. Chem.*, **50**, 82-87, 1978.

[615] Gay, D. D., L. C. Fortmann, and K. O. Wirtz, Dimethylmercury: volatilization from plants, Paper presented at 4th Conf. Sensing Environ. Pollut., New Orleans, LA, 1977.

[616] Harshbarger, W., private communication, 1978.

[617] Etz, E. S., G. J. Rosasco, and W. C. Cunningham, The chemical identification of airborne particles by laser Raman spectroscopy, in *Environmental Analysis*, ed. G. W. Ewing, New York: Academic Press, 1977.

[618] Chatfield, H. E., *et al.*, Chemical processing equipment, in *Air Pollut. Eng. Manual*, Environ. Prot. Agency, Research Triangle Park, NC, p. 699-851, 1973.

[619] Matta, J. E., J. C. LaScola, and F. N. Kissell, *Methane Emissions from Gassy Coals in Storage Silos*, Report RI-8269, Bureau of Mines, Pittsburgh, PA, 1978 (APTIC No. 104670).

[620] Ishiguro, T., Y. Nagata, T. Hasegawa, N. Takeuchi, O. Furukawa, and Y. Shigeta, Analysis of odors emitting from metal printing factory, *Odor Control Assoc. J. (Japan)*, **50**, 5 (25), 8-14, 1976 (APTIC No. 104761).

[621] McDermott, H. J., and S. E. Killiany, Jr., Quest for a gasoline TLV, *Amer. Ind. Hyg. Assoc. J.*, **39**, 110-117, 1978.

[622] Committee on Medical and Biologic Effects of Environment Pollutants, *Chlorine and Hydrogen Chloride*, Washington, D.C.: National Academy of Sciences, 1976.

[623] Driscoll, J. N., *Flue Gas Monitoring Techniques*, Ann Arbor, MI: Ann Arbor Science Pub., 1974.

[624] Herlan, A., On the formation of polycyclic aromatics: investigation of fuel oil and emissions by high-resolution mass spectrometry, *Comb. Flame*, **31**, 297-307, 1978.

[625] Giger, W., and C. Schaffner, Determination of polycyclic aromatic hydrocarbons in the environment by glass capillary gas chromatography, *Anal. Chem.*, **50**, 243-249, 1978

[626] Hughes, T. W., and D. A. Horn, *Source Assessment: Acrylonitrile Manufacture (Air Emissions)*, Monsanto Research Corp., Dayton, OH, 1977 (NTIS Document PB 271969).

[627] Crittenden, A. L., *Analysis of Atmospheric Organic Aerosols by Mass Spectrometry*, Dept. of Chem., Univ. of Washington, Seattle, WA, 1976 (NTIS Document PB 258822).

[628] Pellizzari, E. D., *Analysis of Organic Air Pollutants by Gas Chromatography and Mass Spectrometry*, Research Triangle Institute, Research Triangle Park, NC, 1977 (NTIS Document PB 269654).

[629] Chi, C. T., and T. W. Hughes, *Phthalic Anhydride Plant Air Pollution Control*, Monsanto Research Corp., Dayton, OH, 1977 (NTIS Document PB 272102).

[630] Homolya, J. B., and J. L. Cheney, A study of the emissions of SO_2, H_2SO_4, and sulfate from a coal-fired boiler incorporating a wet-limestone scrubber, Paper ENVT-67, 176th Meeting, American Chemical Society, Miami, FL, 1978.

[631] Joselow, M. M., E. Tobias, R. Koehler, S. Coleman, J. Bogden, and D. Gause, Manganese pollution in the city environment and its relationship to traffic density, *Amer. J. Pub. Health*, **68**, 557-560, 1978.

[632] Beall, J. R., and A. G. Ulsamer, Toxicity of volatile organic compounds present indoors, *Bull. NY Acad. Med.*, **57**, 978-996, 1981.

[633] Cautreels, W., and K. Van Cauwenberghe, Experiments on the distribution of organic pollutants between airborne particulate matter and the corresponding gas phase, *Atmos. Environ.*, **12**, 1133-1141, 1978.

[634] Stedman, R. L., The chemical composition of tobacco and tobacco smoke, *Chem. Rev.*, **68**, 153-207, 1968.

[635] Ballance, P. E., Production of volatile compounds related to the flavor of foods from the Strecker degradation of DL-methionine, *J. Sci. Food Agric.*, **12**, 532-536, 1961.

[636] Carson, J. F., and F. F. Wong, The volatile flavor components of onions, *J. Agric. Food Chem.*, **9**, 140-143, 1961.

[637] Miller, T. L., W. T. Davis, and M. P. Barkdoll, Control of odors from rendering plants, in *Proc. 15th Purdue Air Qual. Conf.*, Indianapolis, IN, p. 284-305, 1976.

[638] Giam, C. S., H. S. Chan, G. S. Neff, and E. L. Atlas, Phthalate ester plasticizers: a new class of marine pollutant, *Science*, **199**, 419-421, 1978.

[639] Rasmussen, R. A., S. A. Penkett, and N. Prosser, Measurement of carbon tetrafluoride in the atmosphere, *Nature*, **277**, 549-551, 1978.

[640] Inaba, H., and T. Kobayasi, Laser-raman radar - Laser-raman scattering methods for remote detection and analysis of atmospheric pollution, *Opto-electronics*, **4**, 101-123, 1972.

[641] Kok, G. L., K. R. Darnall, A. M. Winer, J. N. Pitts, Jr., and B. W. Gay, Ambient air measurement of hydrogen peroxide in the California South Coast Air Basin, *Environ. Sci. Technol.*, **12**, 1077-1080, 1978.

[642] Moore, H., and K. R. Williams, Emissions into the atmosphere due to the ceramic salt glazing process, *J. Air Poll. Contr. Assoc.*, **28**, 1043-1045, 1978.

[643] Kuwata, K., M. Uebori, and Y. Yamasaki, Determination of aliphatic and aromatic aldehydes in polluted airs as their 2,4-dinitrophenylhydrazones by high performance liquid chromatography, *J. Chromatogr. Sci.*, **17**, 264-268, 1979.

[644] Levine, J. S., R. E. Hughes, W. L. Chameides, and W. E. Howell, N_2O and CO production by electric discharge: atmospheric implications, *Geophys. Res. Lett.*, **6**, 557-559, 1979.

[645] Crescentini, G., and F. Bruner, Evidence for the presence of Freon 21 in the atmosphere, *Nature*, **279**, 311-312, 1979.

[646] Farwell, S. O., S. J. Gluck, W. L. Bamesberger, T. M. Schutte, and D. F. Adams, Determination of sulfur-containing gases by a deactivated cryogenic enrichment and capillary gas chromatographic system, *Anal. Chem.*, **51**, 609-615, 1979.

[647] Brunnemann, K. D., G. Stahnke, and D. Hoffmann, Chemical studies on tobacco smoke. LXI. Volatile pyridines: quantitative analysis in mainstream and sidestream smoke of cigarettes and cigars, *Anal. Lett.*, **A11**, 545-560, 1978.

[648] Eatough, D. J., N. L. Eatough, M. W. Hill, N. F. Mangelson, J. Ryder, L. D. Hansen, R. G. Meisenheimer, and J. W. Fischer, The chemical composition of smelter flue dusts, *Atmos. Environ.*, **13**, 489-506, 1979.

[649] Klepper, L., Nitric oxide (NO) and nitrogen dioxide (NO_2) emissions from herbicide-treated soybean plants, *Atmos. Environ.*, **13**, 537-542, 1979.

[650] Liao, J. C., and R. F. Browner, Determination of polynuclear aromatic hydrocarbons in poly(vinyl chloride) smoke particulates by high pressure liquid chromatography and gas chromatography-mass spectrometry, *Anal. Chem.*, **50**, 1683-1686, 1978.

[651] Elliott, L. F., and T. A. Travis, Detection of carbonyl sulfide and other gases emanating from beef cattle manure, *Soil Sci. Soc. Amer. Proc.*, **37**, 700-702, 1973.

[652] Hoshika, Y., and G. Muto, Gas-liquid-solid chromatographic determination of phenols in air using Tenax-GC and alkaline precolumns, *J. Chromatogr.*, **157**, 277-284, 1978.

[653] Leinster, P., R. Perry, and R. J. Young, Ethylene dibromide in urban air, *Atmos. Environ.*, **12**, 2383-2387, 1978.

[654] Appel, B. R., E. M. Hoffer, E. L. Kothny, S. M. Wall, M. Haik, and R. L. Knights, Analysis of carbonaceous material in Southern California atmospheric aerosols. 2., *Environ. Sci. Technol.*, **13**, 98-104, 1979.

[655] Crutzen, P. J., L. E. Heidt, J. P. Krasnec, W. H. Pollock, and W. Seiler, Biomass burning as a source of the atmospheric gases CO, H_2, N_2O, NO, CH_3Cl, and COS, *Nature*, **282**, 253-256, 1979.

[656] Weiss, R. F., and H. Craig, Production of atmospheric N_2O by combustion, *Geophys. Res. Lett.*, **3**, 751-753, 1976.

[657] Wouterlood, H. J., and K. McG. Bowling, Removal and recovery of arseneous oxide from flue gases, *Environ. Sci. Technol.*, **13**, 93-97, 1979.

[658] Bruner, F., G. Bertoni, and G. Crescentini, Critical evaluation of sampling and gas chromatographic analysis of halocarbons and other organic air pollutants, *J. Chromatogr.*, **167**, 399-407, 1978.

[659] Colmsjö, A., and U. Stenberg, Identification of polynuclear aromatic hydrocarbons by Shpol'skii low temperature fluorescence, *Anal. Chem.*, **51**, 145-150, 1979.

[660] Daisey, J. M., and M. A. Leyko, Thin-layer gas chromatographic method for the determination of polycyclic aromatic and aliphatic hydrocarbons in airborne particulate matter, *Anal. Chem.*, **51**, 24-26, 1979.

[661] Gerstle, R. W., and D. A. Kemnitz, Atmospheric emissions from open burning, *J. Air Poll. Contr. Assoc.*, **17**, 324-327, 1967

[662] Darley, E. F., F. R. Burleson, E. H. Mateer, J. T. Middleton, and V. P. Osterli, Contribution of burning of agricultural wastes to photochemical air pollution, *J. Air Poll. Contr. Assoc.*, **11**, 685-690, 1966.

[663] Tyson, B. J., W. A. Dement, and H. A. Mooney, Volatilisation of terpenes from *Salvia mellifera, Nature*, **252**, 119-120, 1974.

[664] Weete, J. D., W. Y. Huang, and J. L. Laseter, *Streptomyces* sp: a source of odorous substances in potable water, *Water, Air, Soil Poll.*, **11**, 217-223, 1979.

[665] Gold, A., C. E. Dubé, and R. B. Perni, Solid sorbent for sampling acrolein in air, *Anal. Chem.*, **50**, 1839-1841, 1978.

[666] Robertson, D. J., R. H. Groth, and A. G. Glastris, HCN content of turbine engine exhaust, *J. Air Poll. Contr. Assoc.*, **29**, 50-51, 1979.

[667] Kelly, T. J., D. H. Stedman, and G. L. Kok, Measurements of H_2O_2 and HNO_3 in rural air, *Geophys. Res. Lett.*, **6**, 375-378, 1979.

[668] Biggins, P. D. E., and R. M. Harrison, The identification of specific chemical compounds in size-fractionated atmospheric particulates collected at roadside sites, *Atmos. Environ.*, **13**, 1213-1216, 1979.

[669] Harrison, R. M., and H. A. McCartney, Some measurements of ambient air pollution arising from the manufacture of nitric acid and ammonium nitrate fertiliser, *Atmos. Environ.*, **13**, 1105-1120, 1979.

[670] Aneja, V. P., J. H. Overton, Jr., L. T. Cupitt, J. L. Durham, and W. E. Wilson, Direct measurements of emission rates of some atmospheric biogenic sulfur compounds, *Tellus*, **31**, 174-178, 1979.

[671] Rietz, E. B., A study on air pollution by asphalt fumes, *Anal. Lett.*, **12**, 143-153, 1979.

[672] Snyder, C. A., and D. A. Isola, Assay for arsenic trioxide in air, *Anal. Chem.*, **51**, 1478-1480, 1979.

[673] Hinkle, M. E., and T. F. Harms, CS_2 and COS in soil gases of the Roosevelt Hot Springs known geothermal resource area, Beaver County, Utah, *J. Res. US Geol. Survey*, **6**, 571-578, 1978.

[674] Ioffe, B. V., V. A. Isadorov, and I. G. Zenkevich, Certain regularities in the composition of volatile organic pollutants in the urban atmosphere, *Environ. Sci. Technol.*, **13**, 864-868, 1979.

[675] Flyckt, D. L., *Seasonal Variation in the Volatile Hydrocarbon Emissions from Ponderosa Pine and Red Oak*, M. S. Thesis, Washington State University, 1979.

[676] Björseth, A., G. Lunde, and A. Lindskog, Long-range transport of polycyclic aromatic hydrocarbons, *Atmos. Environ.*, **13**, 45-53, 1979.

[677] Eichmann, R., P. Neuling, G. Ketseridis, J. Hahn, R. Jaenicke, and C. Junge, *n*-Alkane studies in the troposphere-I. Gas and particulate concentrations in North Atlantic air, *Atmos. Environ.*, **13**, 587-599, 1979.

[678] Monroe, F. L., R. A. Rasmussen, W. L. Bamesberger, and D. F. Adams, *Investigation of Emissions from Plywood Veneer Dryers*, Contract Report CPA-70-138, Coll. of Engg., Wash. State Univ., 1972.

[679] Peyton, T. O., R. V. Steele, and W. R. Mabey, *Carbon Disulfide, Carbonyl Sulfide - Literature Review and Environmental Assessment*, Stanford Research Inst., Contract Rpt., EPA-600/9-78-009, 1978. (NTIS Document PB-257947).

[680] Biggins, P. D. E., and R. M. Harrison, Characterization and classification of atmospheric sulfates, *J. Air Poll. Contr. Assoc.*, **29**, 838-840, 1979.

[681] Grosjean, D., Formaldehyde and other carbonyls in Los Angeles ambient air, *Environ. Sci. Technol.*, **16**, 254-262, 1982.

[682] Hampton, C. V., W. R. Pierson, T. M. Harvey, W. S. Updegrove, and R. S. Marano, Hydrocarbon gases emitted from vehicles on the road. I. A qualitative gas chromatography/mass spectrometry survey, *Environ. Sci. Technol.*, **16**, 287-298, 1982.

[683] Platt, U., D. Perner, A. M. Winer, G. W. Harris, and J. N. Pitts, Jr., Detection of NO_3 in the polluted atmosphere by differential optical absorption, *Geophys. Res. Lett.*, **7**, 89-92, 1980.

[684] Newton, D. L., M. D. Erickson, K. B. Tomer, E. D. Pellizzari, P. Gentry, and R. B. Zweidinger, Identification of nitroaromatics in diesel exhaust particulate using gas chromatography/negative ion chemical ionization mass spectrometry and other techniques, *Environ. Sci. Technol.*, **16**, 206-213, 1982.

[685] Yergey, J. A., T. H. Risby, and S. S. Lestz, Chemical characterization of organic adsorbates on diesel particulate matter, *Anal. Chem.*, **54**, 354-357, 1982.

[686] Creech, G., R. T. Johnson, and J. O. Stoffer, Part I. A comparison of three different high pressure liquid chromatography systems for the determination of aldehydes and ketones in diesel exhaust, *J. Chromatogr. Sci.*, **20**, 67-72, 1982.

[687] Zeman, A., and K. Koch, Determination of odorous volatiles in air using chromatographic profiles, *J. Chromatogr.*, **216**, 199-207, 1981.

[688] Schuetzle, D., T. L. Riley, T. J. Prater, T. M. Harvey, and D. F. Hunt, Analysis of nitrated polycyclic aromatic hydrocarbons in diesel particulates, *Anal. Chem.*, **54**, 265-271, 1982.

[689] Rappaport, S. M., Y. Y. Wang, E. T. Wei, R. Sawyer, B. E. Watkins, and H. Rapoport, Isolation and identification of a direct-acting mutagen in diesel-exhaust particulates, *Environ. Sci. Technol.*, **14**, 1505-1509, 1980.

[690] Schuetzle, D., F.S.-C. Lee, T. J. Prater, and S. B. Tejada, The identification of polynuclear aromatic hydrocarbon (PAH) derivatives in mutagenic fractions of diesel particulate extracts, *Int. J. Environ. Anal. Chem.*, **9**, 93-144, 1981.

[691] Penkett, S. A., N. J. D. Prosser, R. A. Rasmussen, and M. A. K. Khalil, Atmospheric measurements of CF_4 and other fluorocarbons containing the CF_3 grouping, *J. Geophys. Res.*, **86**, 5172-5178, 1981.

[692] Jonsson, A., and B. M. Bertilsson, Formation of methyl nitrite in engines fueled with gasoline/methanol and methanol/diesel, *Environ. Sci. Technol.*, **16**, 106-110, 1982.

[693] Choudhury, D. R., Characterization of polycyclic ketones and quinones in diesel emission particulates by gas chromatography/mass spectrometry, *Environ. Sci. Technol.*, **16**, 102-106, 1982.

[694] Eatough, D. J., M. L. Lee, D. W. Later, B. E. Richter, N. L. Eatough, and L. D. Hansen, Dimethyl sulfate in particulate matter from coal- and oil-fired power plants, *Environ. Sci. Technol.*, **15**, 1502-1506, 1981.

[695] Hoshika, Y., Y. Nihei, and G. Muto, Simple circular odor chart for characterization of trace amounts of odorants discharged from thirteen odor sources, *J. Chromatogr. Sci.*, **19**, 200-215, 1981.

[696] Platt, U., D. Perner, G. W. Harris, A. M. Winer, and J. N. Pitts, Jr., Observation of nitrous acid in an urban atmosphere by differential optical absorption, *Nature*, **285**, 312-314, 1980.

[697] Wauters, E., F. Vangaever, P. Sandra, and M. Verzele, Polar organic fraction of air particulate matter, *J. Chromatogr.*, **170**, 133-138, 1979.

[698] Schuetzle, D., L. M. Skewes, G. E. Fisher, S. P. Levine, and R. A. Gorse, Jr., Determination of sulfates in diesel particulates, *Anal. Chem.*, **53**, 837-840, 1981.

[699] Jäger, J., Detection and characterization of nitro derivatives of some polycyctic aromatic hydrocarbons by fluorescence quenching after thin-layer chromatography: application to air pollution analysis, *J. Chromatogr.*, **152**, 575-578, 1978

[700] Maier, E. J., A. C. Aikin, and J. E. Ainsworth, Stratospheric nitric oxide and ozone measurements using photoionization mass spectrometry and UV absorption, *Geophys. Res. Lett.*, **5**, 37-40, 1978.

[701] Anderson, J. G., J. J. Margitan, and D. H. Stedman, Atomic chlorine and the chlorine monoxide radical in the stratosphere: three in situ observations, *Science*, **198**, 501-503, 1977.

[702] Murcray, D. G., D. B. Barker, J. N. Brooks, A. Goldman, and W. J. Williams, Seasonal and latitudinal variation of the stratospheric concentration of HNO_3, *Geophys. Res. Lett.*, **2**, 223-225, 1975.

[703] Brownlee, D. E., G. V. Ferry, and D. Tomandi, Stratospheric aluminum oxide, *Science*, **191**, 1270-1271, 1976.

[704] Anderson, J. G., The absolute concentration of $O(^3P)$ in the earth's stratosphere, *Geophys. Res. Lett.*, **2**, 231-234, 1975.

[705] Anderson, J. G., The absolute concentration of $OH(X^2\pi)$ in the earth's stratosphere, *Geophys. Res. Lett.*, **3**, 165-168, 1976.

[706] Ehhalt, D. H., and L. E. Heidt, Vertical profiles of CH_4 in the troposphere and stratosphere, *J. Geophys. Res.*, **78**, 5265-5271, 1973.

[707] Cumming, C., and R. P. Lowe, Balloon-borne spectroscopic measurment of stratospheric methane, *J. Geophys. Res.*, **78**, 5259-5264, 1973.

[708] Noxon, J. F., R. B. Norton, and W. R. Henderson, Observation of atmospheric NO_3, *Geophys. Res. Lett.*, **5**, 675-678, 1978.

[709] Ackerman, M., J. C. Fontanella, D. Frimout, A. Girard, N. Louisnard, and C. Muller, Simultaneous measurements of NO and NO_2 in the stratosphere, *Planet. Space Sci.*, **23**, 651-660, 1975.

[710] Lazrus, A. L., and B. W. Gandrud, Distribution of stratospheric nitric acid vapor, *J. Atmos. Sci.*, **31**, 1102-1108, 1974.

[711] Jaeschke, W., R. Schmitt, and H.-W. Georgii, Preliminary results of stratospheric SO_2 measurements, *Geophys. Res. Lett.*, **3**, 517-519, 1976.

[712] Rudolph, J., D. H. Ehhalt, and A. Tonnissen, Vertical profiles of ethane and propane in the stratosphere, *J. Geophys. Res.*, **86**, 7267-7272, 1981.

[713] Cronn, D. R., R. A. Rasmussen, E. Robinson, and D. E. Harsch, Halogenated compound identification and measurement in the troposphere and lower stratosphere, *J. Geophys. Res.*, **82**, 5935-5944, 1977.

[714] Fabian, P., R. Borchers, S. A. Penkett, and N. J. D. Prosser, Halocarbons in the stratosphere, *Nature*, **29**,4, 733-735, 1981.

[715] Inn, E. C. Y., J. F. Vedder, and D. O'Hara, Measurement of stratospheric sulfur constituents, *Geophys. Res. Lett.*, **8**, 5-8, 1981.

[716] Mankin, W. G., M. T. Coffey, D. W. T. Griffith, and S. R. Drayson, Spectoscopic measurement of carbonyl sulfide (OCS) in the stratosphere, *Geophys. Res. Lett.*, **6**, 853-856, 1979.

[717] Cronn, D., and E. Robinson, Tropospheric and lower stratospheric vertical profiles of ethane and acetylene, *Geophys. Res. Lett.*, **6**, 641-644, 1979.

[718] Fabian, P., R. Borchers, K. H. Weiler, U. Schmidt, A. Volz, D. H. Ehhalt, W. Seiler, and F. Müller, Simultaneously measured vertical profiles of H_2, CH_4, CO, N_2O, $CFCl_3$, and CF_2Cl_2 in the mid-latitude stratosphere and troposphere *J. Geophys. Res.*, **84**, 3149-3154, 1979.

[719] Arijs, E., D. Nevejans, P. Frederick, and J. Ingels, Negative ion composition measurements in the stratosphere, *Geophys. Res. Lett.*, **8**, 121-124, 1981.

[720] Parrish, A., R. L. deZafra, P. M. Solomon, J. W. Barrett, and E. R. Carlson, Chlorine oxide in the stratospheric ozone layer: Ground-based detection and measurement, *Science*, **211**, 1158-1161, 1981.

[721] Viggiano, A. A., and F. Arnold, Stratospheric sulfuric acid vapor: new and updated measurements, *J. Geophys. Res.*, **88**, 1457-1462, 1983.

[722] Penkett, S. A., R. G. Derwent, P. Fabian, R. Borchers, and U. Schmidt, Methyl chloride in the stratosphere, *Nature*, **283**, 58-60, 1980.

[723] Arijs, E., D. Nevejans, and J. Ingels, Unambiguous mass determination of major stratospheric positive ions, *Nature*, **288**, 684-686, 1980.

[724] Zander, R., H. Leclercq, and L. D. Kaplan, Concentration of carbon monoxide in the upper stratosphere, *Geophys. Res. Lett.*, **8**, 365-368, 1981.

[725] Zander, R., Recent observations of HF and HCl in the upper stratosphere, *Geophys. Res. Lett.*, **8**, 413-416, 1981.

[726] Coffey, M. T., W. G. Mankin, and R. J. Cicerone, Spectroscopic detection of stratospheric hydrogen cyanide, *Science*, **214**, 333-335, 1981.

[727] Ackerman, M., D. Frimout, A. Girard, M. Gottignies, and C. Muller, Stratospheric HCl from infrared spectra, *Geophys. Res. Lett.*, **3**, 81-84, 1976.

[728] Lazrus, A. L., B. W. Gandrud, R. N. Woodard, and W. A. Sedlacek, Direct measurement of stratospheric chlorine and bromine, *J. Geophys. Res.*, **81**, 1067-1070, 1976.

[729] Leifer, R., L. Toonkel, R. Larsen, and R. Lagomarsino, Trace gas concentrations in the stratosphere of the northern hemisphere during 1976, *J. Geophys. Res.*, **85**, 1069-1072, 1980.

[730] Henschen, G., and F. Arnold, Extended positive ion composition measurements in the stratosphere-implications for neutral trace gases, *Geophys. Res. Lett.*, **8**, 999-1001, 1981.

[731] Cronn, D. R., D. E. Harsch, and E. Robinson, Tropospheric and lower stratospheric profiles of halocarbons and related chemical species, Abstract. ENVR- 132, 176th National Meeting, American Chemical Society, Miami, FL, 1978.

[732] Zander, R., G. Roland, and L. Delbouille, Confirming the presence of hydrofluoric acid in the upper stratosphere, *Geophys. Res. Lett.*, **4**, 117-120, 1977.

[733] Arnold, F., and T. Bührke, New H_2SO_4 and HSO_3 vapour measurements in the stratosphere-evidence for a volcanic influence, *Nature*, **301**, 293-295, 1983.

[734] Wilcox, R. W., G. D. Nastrom, and A. D. Belmont, Periodic variations of total ozone and of its vertical distribution, *J. Appl. Met.*, **16**, 290-298, 1977.

[735] Kendall, D. J. W., and H. L. Buijs, Stratospheric NO_2 and upper limits of CH_3Cl and C_2H_6 from measurements at 3.4 μm, *Nature*, **303**, 221-222, 1983.

[736] Bischof, W., P. Fabian, and R. Borchers, Decrease in CO_2 mixing ratio observed in the stratosphere, *Nature*, **288**, 347-348, 1980.

[737] Leifer, R., R. Larsen, and L. Toonkel, Trends in stratospheric concentrations of trace gases in the northern hemisphere during the years 1974-1979, *Geophys. Res. Lett.*, **9**, 755-758, 1982.

[738] de Jonckheere, C. G., A measurement of the mixing ratio of water vapor from 15 to 45 km, *Quart. J. Roy. Met. Soc.*, **101**, 217-226, 1975.

[739] Bertaux, J.-L., and A. Delannoy, Vertical distribution of H_2O in the stratosphere as determined by UV fluorescence in-situ measurements, *Geophys. Res. Lett.*, **5**, 1017-1020, 1978.

[740] Levine, J. S., and E. F. Shaw, Jr., *In situ* aircraft measurements of enhanced levels of N_2O associated with thunderstorm lightning, *Nature*, **303**, 312-314, 1983.

[741] Farlow, N. H., K. G. Snetsinger, D. M. Hayes, H. Y. Lem, and B. M. Tooper, Nitrogen-sulfur compounds in stratospheric aerosols, *J. Geophys. Res.*, **83**, 6207-6211, 1978.

[742] Heaps, W. S., T. J. McGee, R. D. Hudson, and L. O. Caudill, Stratospheric ozone and hydroxyl radical measurements by balloon-borne lidar, *Appl. Optics*, **21**, 2265-2274, 1982.

[743] Mihelcic, D., D. H. Ehhalt, J. Klomfalb, G. F. Kulessa, U. Schmidt, and M. Trainer, Measurements of free radicals in the atmosphere by matrix isolation and electron paramagnetic resonance, *Ber. Bunsenges. Phys. Chem.*, **82**, 16-19, 1978.

[744] Hayes, D., K. Snetsinger, G. Ferry, V. Oberbeck, and N. Farlow, Reactivity of stratospheric aerosols to small amounts of ammonia in the laboratory environment, *Geophys. Res. Lett.*, **7**, 974-976, 1980.

[745] Arnold, F., and G. Henschen, First mass analysis of stratospheric negative ions, *Nature*, **275**, 521-522, 1978.

[746] Waldman, J. M., J. W. Munger, D. J. Jacob, R. C. Flagan, J. J. Morgan, and M. R. Hoffmann, Chemical composition of acid fog, *Science*, **218**, 677-680, 1982.

[747] Norton, R. B., J. M. Roberts, and B. J. Huebert, Tropospheric oxalate, *Geophys. Res. Lett.*, **10**, 517-520, 1983.

[748] Grosjean, D., and B. Wright, Carbonyls in urban fog, ice fog, cloudwater, and rainwater, *Atmos. Environ.*, **17**, 2093-2096, 1983.

[749] Ramdahl, T., Polycyclic aromatic ketones in environmental samples, *Environ. Sci. Technol.*, **17**, 666-670, 1983.

[750] Daum, P. H., S. E. Schwartz, and L. Newman, Studies of the gas- and aqueous-phase composition of stratiform clouds, in *Precipitation Scavenging, Dry Deposition, and Resuspension, I*, H. R. Pruppacher, R. G. Semonin, and W. G. N. Slinn, Eds., New York: Elsevier, P. 31-44, 1983.

[751] Kadlacek, J., S. McLaren, N. Camerota, V. Mohnen, and J. Wilson, Cloud water chemistry at Whiteface Mountain, in *Precipitation Scavenging, Dry Deposition, and Resuspension, I*, H. R. Pruppacher, R. G. Semonin, and W. G. N. Slinn, Eds., New York: Elsevier, p. 103-112, 1983.

[752] Raynor, G. S., and J. V. Hayes, Differential rain and snow scavenging efficiency implied by ionic concentration differences in winter precipitation, in *Precipitation Scavenging, Dry Deposition, and Resuspension, I*, H. R. Pruppacher, R. G. Semonin, and W. G. N. Slinn, Eds., New York: Elsevier, p. 249-262, 1983.

[753] Huebert, B. J., F. C. Fehsenfeld, R. B. Norton, and D. Albritton, The scavenging of nitric acid vapor by snow, in *Precipitation Scavenging, Dry Deposition, and Resuspension, I*, H. R. Pruppacher, R. G. Semonin, and W. G. N. Slinn, Eds., New York: Elsevier, p. 293-300, 1983.

[754] Georgii, H.-W., and G. Schmitt, Distribution of polycyclic aromatic hydrocarbons in precipitation, in *Precipitation Scavenging, Dry Deposition, and Resuspension, I*, H. R. Pruppacher, R. G. Semonin, and W. G. N. Slinn, Eds., New York: Elsevier, p. 395-402, 1983.

[755] Pankow, J. F., L. M. Isabelle, W. E. Asher, T. J. Kristensen and M. E. Peterson, Organic compounds in Los Angeles and Portland rain: identities, concentrations, and operative scavenging mechanisms, in *Precipitation Scavenging, Dry Deposition, and Resuspension, I*, H. R. Pruppacher, R. G. Semonin, and W. G. N. Slinn, Eds., New York: Elsevier, p. 403-415, 1983.

[756] Kapoor, R. K., and S. K. Paul, A study of the chemical components of aerosols and snow in the Kashmir region, *Tellus*, **32**, 33-41, 1980.

[757] Sidle, A. B., Amino acid content of atmospheric precipitation, *Tellus*, **19**, 128-135, 1967.

[758] Hoffman, W. A., Jr., S. E. Lindberg, and R. R. Turner, Some observations of organic constituents in rain above and below a forest canopy, *Environ. Sci. Technol.*, **14**, 999-1002, 1980.

[759] Kok, G. L., Measurements of hydrogen peroxide in rainwater, *Atmos. Environ.*, **14**, 653-656, 1980.

[760] Galloway, J. N., G. E. Likens, W. C. Keene, and J. M. Miller, The composition of precipitation in remote area of the world, *J. Geophys. Res.*, **87**, 8771-8786, 1982.

[761] Likens, G. E., E. S. Edgerton, and J. N. Galloway, The composition and deposition of organic carbon in precipitation, *Tellus*, **35B**, 16-24, 1983.

[762] Meyers, P. A., and R. A. Hites, Extractable organic compounds in midwest rain and snow, *Atmos. Environ.*, **16**, 2169-2175, 1982.

[763] Rasmussen, R. A., M. A. K. Khalil, and S. D. Hoyt, Methane and carbon monoxide in snow, *J. Air Poll. Contr. Assoc.*, **32**, 176-178, 1982.

[764] Galloway, J. N., G. E. Likens, and E. S. Edgerton, Acid precipitation in the northeastern United States: *p*H and acidity, *Science*, **194**, 722-724, 1976.

[765] Church, T. M., J. N. Galloway, T. D. Jickells, and A. H. Knap, The chemistry of western Atlantic precipitation at the mid-Atlantic coast and on Bermuda, *J. Geophys. Res.*, **87**, 11013-11018, 1982.

[766] Zika, R., E. Saltzman, W. L. Chameides, and D. D. Davis, H_2O_2 levels in rainwater collected in south Florida and the Bahama Islands, *J. Geophys. Res.*, **87**, 5015-5017, 1982.

[767] Richards, L. W., J. A. Anderson, D. L. Blumenthal, J. A. McDonald, G. L. Kok, and A. L. Lazrus, Hydrogen peroxide and sulfur (IV) in Los Angles cloud water, *Atmos. Environ.*, **17**, 911-914, 1983.

[768] Parungo, F., C. Nagamoto, I. Nolt, M. Dias, and E. Nickerson, Chemical analysis of cloud water collected over Hawaii, *J. Geophys. Res.*, **87**, 8805-8810, 1982.

[769] Castillo, R. A., J. E. Jiusto, and E. McLaren, The *p*H and ionic composition of stratiform cloud water, *Atmos. Environ.*, **17**, 1497-1505, 1983.

[770] Van Valin, C. C., L. P. Stearns, C. T. Nagamoto, and R. F. Pueschel, Analysis results of rainwater and cloudwater samples, *EOS-Trans. AGU*, **63**, 894, 1982.

[771] Cadle, S. H., and P. A. Mulawa, Sulfide emission from catalyst-equipped cars, *SAE Tech. Paper Ser. No. 780200*, 1978.

[772] Braddock, J. N., and P. A. Gabele, Emission patterns of diesel-powered passenger cars - Part II, *SAE Tech. Paper Ser. No. 770168*, 1977.

[773] Yokouchi, Y., T. Fujii, Y. Ambe, and K. Fuwa, Determination of monoterpene hydrocarbons in the atmosphere, *J. Chromatogr.*, **209**, 293-298, 1981.

[774] Okamoto, W. K., R. A. Gorse, Jr., and W. R. Pierson, Nitric acid in diesel exhaust, *J. Air Poll. Contr. Assoc.*, **33**, 1098-1100, 1983.

[775] Andreae, M. O., Arsenic in rain and the atmospheric mass balance of arsenic, *J. Geophys. Res.*, **85**, 4512-4518, 1980.

[776] Hammer, C. U., H. B. Clausen, and W. Dansgaard, Greenland ice sheet evidence of post-glacial volcanism and its climatic impact, *Nature*, **288**, 230-235, 1980.

[777] Herron, M. M., Impurity sources of F^-, Cl^-, NO_3^-, and SO_4^{2-} in Greenland and Antarctic precipitation, *J. Geophys. Res.*, **87**, 3052-3060, 1982.

[778] Craig, H., and C. C. Chou, Methane: the record in polar ice cores, *Geophys. Res. Lett.*, **9**, 1221-1224, 1982.

[779] Aristarain, A. J., R. J. Delmas, and M. Briat, Snow chemistry on James Ross Island (Antarctic Peninsula), *J. Geophys. Res.*, **87**, 11004-11012, 1982.

[780] Neftel, A., H. Oeschger, J. Schwander, and B. Stauffer, Carbon dioxide concentration in bubbles of natural cold ice, *J. Phys. Chem.*, **87**, 4116-4120, 1983.

[781] Boutron, C., Respective influence of global pollution and volcanic eruptions on the past variations of the trace metals content of Antarctic snows since 1880's, *J. Geophys. Res.*, **85**, 7426-7432, 1980.

[782] Farmer, J. C., and G. A. Dawson, Condensation sampling of soluble atmospheric trace gases, *J. Geophys. Res.*, **87**, 8931-8942, 1982.

[783] Cronn, D. R., and W. Nutmagul, Characterization of trace gases in 1980 volcanic plumes of Mt. St. Helens, *J. Geophys. Res.*, **87**, 11153-11160, 1982.

[784] Rasmussen, R. A., and M. A. K. Khalil, Global production of methane by termites, *Nature*, **301**, 700-702, 1983.

[785] Zimmerman, P. R., J. P. Greenberg, S. O. Wandiga, and P. J. Crutzen, Termites: a potentially large source of atmospheric methane, carbon dioxide, and molecular hydrogen, *Science*, **218**, 563-565, 1982.

[786] Robertson, D. J., R. H. Groth, and T. J. Blasko, Organic content of particulate matter in turbine engine exhaust, *J. Air Poll. Contr. Assoc.*, **30**, 261-266, 1980.

[787] Tsuchiya, Y., and J. G. Boulanger, Carbonyl sulphide in fire gases, *Fire and Materials*, **3**, 154-155, 1979.

[788] Eiceman, G. A., R. E. Clement, and F. W. Karasek, Variations in concentrations of organic compounds including polychlorinated dibenzo-*p*-dioxins and polynuclear aromatic hydrocarbons in fly ash from a municipal incinerator, *Anal. Chem.*, **53**, 955-959, 1981.

[789] Kuwata, K., M. Uebori, and Y. Yamazaki, Reversed-phase liquid chromatographic determination of phenols in auto exhaust and cigarette smoke as *p*-nitrobenzeneazophenol derivatives, *Anal. Chem.*, **53**, 1531-1534, 1981.

[790] Eiceman, G. A., R. E. Clement, and F. W. Karasek, Analysis of fly ash from municipal incinerators for trace organic compounds, *Anal. Chem.*, **51**, 2343-2350, 1979.

[791] Cronn, D. R., S. G. Truitt, and M. J. Campbell, Chemical characterization of plywood veneer dryer emissions, *Atmos. Environ.*, **17**, 201-211, 1983.

[792] Cadle, S. H., and P. A. Mulawa, Low molecular weight aliphatic amines in exhaust from catalyst-equipped cars, *Environ. Sci. Technol.*, **14**, 718-723, 1980.

[793] Panter, R., and R.-D. Penzhorn, Alkyl sulfonic acids in the atmosphere, *Atmos. Environ.*, **14**, 149-151, 1980.

[794] Saltzman, E. S., D. L. Savoie, R. G. Zika, and J. M. Prospero, Methane sulfonic acid in the marine atmosphere, *J. Geophys. Res.*, **88**, 10897-10902, 1983.

[795] Pierson, W. R., and W. W. Brachaczek, Emission of ammonia and amines from vehicles on the road, *Environ. Sci. Technol.*, **17**, 757-760, 1983.

[796] Brewer, R. L., R. J. Gordon, L. S. Shepard, and E. C. Ellis, Chemistry of mist and fog from the Los Angeles urban area, *Atmos. Environ*, **17**, 2267-2270, 1983.

[797] Harkov, R., B. Kebbekus, J. W. Bozzelli, and P. J. Lioy, Measurement of selected volatile organic compounds at three locations in New Jersey during the summer season, *J. Air Poll. Contr. Assoc.*, **33**, 1177-1183, 1983.

[798] Heitmann, H., and F. Arnold, Composition measurements of tropospheric ions, *Nature*, **306**, 747-751, 1983.

[799] Erickson, M. D., D. L. Newton, E. D. Pellizzari, K. B. Tomer, and D. Dropkin, Identification of alkyl-9-fluorenones in diesel exhaust particulate, *J. Chromatog. Sci.*, **17**, 449-454, 1979.

[800] Ramdahl, T., G. Becher, and A. Bjφrseth, Nitrated polycyclic aromatic hydrocarbons in urban air particles, *Environ. Sci. Technol.*, **16**, 861-865, 1982.

[801] Paputa-Peck, M. C., R. S. Marano, D. Schuetzle, T. L. Riley, C. V. Hampton, T. J. Prater, L. M. Skewes, T. E. Jensen, P. H. Ruehle, L. C. Bosch, and W. P. Duncan, Determination of nitrated polynuclear aromatic hydrocarbons in particulate extracts by capillary column gas chromatography with nitrogen selective detection, *Anal. Chem.*, **55**, 1946-1954, 1983.

[802] Jensen, T. E., and R. A. Hites, Aromatic diesel emissions as a function of engine conditions, *Anal. Chem.*, **55**, 594-599, 1983.

[803] König, J., E. Balfanz, W. Funcke, and T. Romanowski, Determination of oxygenated polycyclic aromatic hydrocarbons in airborne particulate matter by capillary gas chromatography and gas chromatography/mass spectrometry, *Anal. Chem.*, **55**, 599-603, 1983.

[804] Yu, M.-L., and R. A. Hites, Identification of organic compounds on diesel engine soot, *Anal. Chem.*, **53**, 951-954, 1981.

[805] Hov, Ø., J. Schjoldager, and B. M. Wathne, Measurement and modeling of the concentrations of terpenes in coniferous forest air, *J. Geophys. Res.*, **88**, 10679-10688, 1983.

[806] Walker, M. V., and C. J. Weschler, Water-soluble components of size-fractionated aerosols collected after hours in a modern office building, *Environ. Sci. Technol.*, **14**, 594-597, 1980.

[807] Urban, C. M., and R. J. Garbe, Regulated and unregulated exhaust emissions from malfunctioning automobiles, *SAE Tech. Paper Ser.*, **790696**, 1979.

[808] Hare, C. T., and R. L. Bradow, Characterization of heavy-duty diesel gaseous and particulate emissions, and effects of fuel composition, *SAE Tech. Paper Ser.*, **790490**, 1979.

[809] Hare, C. T., and T. M. Baines, Characterization of particulate and gaseous emissions from two diesel automobiles as functions of fuel and driving cycle, *SAE Tech. Paper Ser.*, **790424**, 1979.

[810] Papa, L. J., Gas chromatography — measuring exhaust hydrocarbons down to parts per billion, *SAE Tech. Paper Ser.*, **670494**, 1967.

[811] Spindt, R. S., G. J. Barnes, and J. H. Somers, The characterization of odor components in diesel exhaust gas, *SAE Tech Paper Ser.*, **710605**, 1971.

[812] Landen, E. W., and J. M. Perez, Some diesel exhaust reactivity information derived by gas chromatography, *SAE Tech. Paper Ser.*, **740530**, 1974.

[813] Spitzer, T., and W. Dannecker, Glass capillary gas chromatography of polynuclear aromatic hydrocarbons in aircraft turbine particulate emissions using stationary phases of varying polarity, *J. Chromatogr.*, **267**, 167-174, 1983.

[814] Ramdahl, T., Retene — a molecular marker of wood combustion in ambient air, *Nature*, **306**, 580-582, 1983.

[815] Becker, K. H., and A. Ionescu, Acetonitrile in the lower troposphere, *Geophys. Res. Lett.*, **9**, 1349-1351, 1982.

[816] Snider, J. R., and G. A. Dawson, Surface acetonitrile near Tucson, Arizona, *Geophys. Res. Lett.*, **11**, 241-242, 1984.

[817] Brasseur, G., E. Arijs, A. DeRudder, D. Nevejans, and J. Ingels, Acetonitrile in the atmosphere, *Geophys. Res. Lett.*, **10**, 725-728, 1983.

[818] Fogg, T. R., R. A. Duce, and J. L. Fasching, Sampling and determination of boron in the atmosphere, *Anal. Chem.*, **55**, 2179-2184, 1983.

[819] Schrimpff, E., W. Thomas, and R. Herrmann, Regional patterns of contaminants (PAH, pesticides and trace metals) in snow of northeast Bavaria and their relationship to human influence and orographic effects, *Water, Air, Soil Poll.*, **11**, 481-497, 1979.

[820] Jarke, F. H., A. Dravnieks, and S. M. Gordon, Organic contaminants in indoor air and their relation to outdoor contaminants, *ASHRAE Trans.*, **87**, Part 1, 153-165, 1981.

[821] Greenberg, J. P., P. R. Zimmerman, L. Heidt, and W. Pollock, Hydrocarbon and carbon monoxide emissions from biomass burning in Brazil, *J. Geophys. Res.*, **89**, 1350-1354, 1984.

[822] Lee, F. S.-C., T. J. Prater, and F. Ferris, PAH emissions from a stratified-charge vehicle with and without oxidation catalyst: sampling and analysis evaluation, in *Polynuclear Aromatic Hydrocarbons*, P. W. Jones and P. Leber, Eds., Ann Arbor: Ann Arbor Science Pubs., pp. 83-110, 1979.

[823] Colmsjö, A., and U. Stenberg, The identification of polynuclear aromatic hydrocarbon mixtures in high-performance liquid chromatography fractions utilizing the Shpol'skii effect, in *Polynuclear Aromatic Hydrocarbons*, P. W. Jones and P. Leber, Eds., Ann Arbor: Ann Arbor Science Pubs., pp. 121-139, 1979.

[824] Stenberg, U., T. Alsberg, L. Blomberg, and T. Wännman, Gas chromatographic separation of high-molecular polynuclear aromatic hydrocarbons in samples from different sources, using temperature-stable glass capillary columns, in *Polynuclear Aromatic Hydrocarbons*, P. W. Jones and P. Leber, Eds., Ann Arbor: Ann Arbor Science Pubs., pp. 313-326, 1979.

[825] Bennett, R. L., K. T. Knapp, P. W. Jones, J. E. Wilkerson, and P. E. Strup, Measurement of polynuclear aromatic hydrocarbons and other hazardous organic compounds in stack gases, in *Polynuclear Aromatic Hydrocarbons*, P. W. Jones and P. Leber, Eds., Ann Arbor: Ann Arbor Science Pubs., pp. 419-428, 1979.

[826] Lao, R. C., and R. S. Thomas, The gas chromatographic separation and determination of PAH from industrial processes using glass capillary and packed columns, in *Polynuclear Aromatic Hydrocarbons*, P. W. Jones and P. Leber, Eds., Ann Arbor: Ann Arbor Science Pubs., pp. 429-452, 1979.

[827] Zelenski, S. G., G. T. Hunt, and N. Pangaro, Comparison of SIM GC/MS and HPLC for the detection of polynuclear aromatic hydrocarbons in fly ash collected from stationary combustion sources, in *Polynuclear Aromatic Hydrocarbons: Chemistry and Biological Effects*, A. Bjørseth and A. J. Dennis, Eds., Columbus, OH: Battelle Press, pp. 589-597, 1980.

[828] Hanson, R. L., R. L. Carpenter, and G. J. Newton, Chemical characterization of polynuclear aromatic hydrocarbons in airborne effluents from an experimental fluidized bed combustor, in *Polynuclear Aromatic Hydrocarbons: Chemistry and Biological Effects*, A. Bjørseth and A. J. Dennis, Eds., Columbus, OH: Battelle Press, pp. 599-616, 1980.

[829] Hites, R. A., M.-L. Yu, and W. G. Thilly, Compounds associated with diesel exhaust particulates, in *Chemical Analysis and Biological Fate: Polynuclear Aromatic Hydrocarbons*, M. Cooke and A. J. Dennis, Eds., Columbus, OH: Battelle Press, pp. 455-466, 1981.

[830] Peters, J. A., D. G. Deangelis, and T. W. Hughes, An environmental assessment of POM emissions from residential wood-fired stoves and fireplaces, in *Chemical Analysis and Biological Fate: Polynuclear Aromatic Hydrocarbons*, M. Cooke and A. J. Dennis, Eds., Columbus, OH: Battelle Press, pp. 571-581, 1981.

[831] Gibson, T. L., Nitro derivatives of polynuclear aromatic hydrocarbons in airborne and source particulate matter, *Atmos. Environ.*, **16**, 2037-2040, 1982.

[832] Choudhury, D. R., and B. Bush, Chromatographic-spectrometric identification of airborne polynuclear aromatic hydrocarbons, *Anal. Chem.*, **53**, 1351-1356, 1981.

[833] Van Langenhove, H. R., F. A. Van Wassenhove, J. K. Koppin, M. R. Van Acker, and N. M. Schamp, Gas chromatography/mass spectrometry identification of organic volatiles contributing to rendering odors, *Environ. Sci. Technol.*, **16**, 883-886, 1982.

[834] Neuling, P., R. Neeb, R. Eichmann, and C. Junge, Qualitative and quantitative analysis of the n-alkanes C_9-C_{17} and pristane in clean air masses, *Fresenius Z. Anal. Chem.*, **302**, 375-381, 1980.

[835] McGregor, W. K., B. L. Seiber, and J. D. Few, Concentration of OH and NO in YJ93-GE-3 engine exhausts measured in situ by narrow-line UV absorption, *Second Conf. CIAP*, Washington: U.S. Dept. of Transportation, pp. 214-229, 1972.

[836] Hurn, R. W., J. W. Vogh, F. W. Cox, and D. E. Seizinger, Trace components of aviation turbine fuels and exhausts, *Second Conf. CIAP*, Washington: U.S. Dept. of Transportation, pp. 194-198, 1972.

[837] Barnes, R. D., and A. J. MacLeod, Analysis of the composition of the volatile malodour emissions from six animal rendering factories, *Analyst*, **107**, 711-715, 1982.

[838] Ramdahl, T., and G. Becher, Characterization of polynuclear aromatic hydrocarbon derivatives in emissions from wood and cereal straw combustion, *Anal. Chim. Acta*, **144**, 83-91, 1982.

[839] Post, J. E., and P. R. Buseck, Characterization of individual particles in the Phoenix urban aerosol using electron-beam instruments, *Environ. Sci. Technol.*, **18**, 35-42, 1984.

[840] Karasek, F. W., and A. C. Vian, Gas chromatographic-mass spectrometric analysis of polychlorinated dibenzo-*p*-dioxins and organic compounds in high-temperature fly ash from municipal waste incineration, *J. Chromatogr.*, **265**, 79-88, 1983.

[841] Grosjean, D., Chemical ionization mass spectra of 2,4-dinitrophenylhydrazones of carbonyl and hydroxycarbonyl atmospheric pollutants, *Anal. Chem.*, **55**, 2436-2439, 1983.

[842] Barber, E. D., F. T. Fox, J. P. Lodge, and L. M. Marshall, Organic acids in a selected dialysate of air particulate matter, *J. Chromatogr.*, **2**, 615-619, 1959.

[843] Hoshika, Y., Gas chromatographic determination of lower fatty acids in air at part-per-trillion levels, *Anal. Chem.*, **54**, 2433-2437, 1982.

[844] Roberts, J. M., F. C. Fehsenfeld, D. L. Albritton, and R. E. Sievers, Measurement of monoterpene hydrocarbons at Niwot Ridge, Colorado, *J. Geophys. Res.*, **88**, 10667-10678, 1983.

[845] Ramdahl, T., and K. Urdal, Determination of nitrated polycyclic aromatic hydrocarbons by fused silica capillary gas chromatography/negative ion chemical ionization mass spectrometry, *Anal. Chem.*, **54**, 2256-2260, 1982.

[846] Dobbs, A. J. and N. Williams, Indoor air pollution from pesticides used in wood remedial treatments, *Environ. Pollut.*, **B6**, 271-296, 1983.

[847] Strachan, W. M. J., and H. Huneault, Automated rain sampler for trace organic substances, *Environ. Sci. Technol.*, **18**, 127-130, 1984.

[848] Matsumoto, G., Alcohols in atmospheric fallout in the Tokyo area, *Atmos. Environ.*, **17**, 83-85, 1983.

[849] Spitzer, T., and W. Dannecker, Profile of aromatic hydrocarbons in aircraft turbine particulate emissions, *J. High Resol. Chromatogr. & Chromatogr. Comm.*, **5**, 98-99, 1982.

[850] Oehme, M., S. Mano, and H. Stray, Determination of nitrated polycyclic hydrocarbons in aerosols using capillary gas chromatography combined with different electron capture detection methods, *J. High Resol. Chromatogr. & Chromatogr. Comm.*, **5**, 417-423, 1982.

[851] Tusseau, D., M. Barbier, J.-C. Marty, and A. Saliot, Les stérols de l'atmosphère marine, *C. R. Acad. Sc. Paris*, **290C**, 109-111, 1980

[852] Jervis, R. E., S. Landsberger, R. Lecomte, P. Paradis, and S. Monaro, Determination of trace pollutants in urban snow using PIXE techniques, *Nucl. Inst. Meth.*, **193**, 323-329, 1982.

[853] Woolley, W. D., Smoke and toxic gas production from burning polymers, *J. Macromol. Sci. – Chem.*, **A17**, 1-33, 1982.

[854] Jiang, S., H. Robberecht, and F. Adams, Identification and determination of alkylselenide compounds in environmental air, *Atmos. Environ.*, **17**, 111-114, 1983.

[855] Yasuhara, A., and K. Fuwa, Isolation and analysis of odorous components in swine manure, *J. Chromatogr.*, **281**, 225-236, 1983.

[856] Mølhave, L., Indoor air pollution due to organic gases and vapors of solvents in building materials, *Environ. Internat.*, **8**, 117-127, 1982.

[857] Butler, J. D., and P. Crossley, An appraisal of relative airborne sub-urban concentrations of polycyclic aromatic hydrocarbons monitored indoors and outdoors, *Sci. Total Environ.*, **11**, 53-58, 1979.

[858] Behymer, T. D., and R. A. Hites, Similarity of some organic compounds in spark-ignition and diesel engine particulate extracts, *Environ. Sci. Technol.*, **18**, 203-206, 1984.

[859] Hansen, L. D., D. Silberman, G. L. Fisher, and D. J. Eatough, Chemical speciation of elements in stack-collected, respirable-size, coal fly ash, *Environ. Sci. Technol.*, **18**, 181-186, 1984.

[860] Zimmerli, B., and H. Zimmermann, Einfaches verfahren zur schätzung von schadstoffkonzentrationen in der luft von innenräumen, *Mitt. Gebiete Lebensm. Hyg.*, **70**, 429-442, 1979.

[861] Herron, M. M., C. C. Langway, Jr., H. V. Weiss, and J. H. Cragin, Atmospheric trace metals and sulfate in the Greenland Ice Sheet, *Geochim. Cosmochim. Acta.*, **41**, 915-920, 1977.

[862] Penzhorn, R. D., and W. G. Filby, Eine methode zur spezifischen bestimmung von schwefelhaltigen säuren in atmosphärischen aerosol, *Staub-Reinhalt. Luft*, **36**, 205-207, 1976.

[863] Bodek, I., and K. T. Menzies, Ion chromatographic analysis of organic acids in diesel exhaust and mine air, in *Proc. Symp. Process Meas. Environ.*, EPA Report 600/9-81-018, pp. 155-168, Environ. Protect Agency, Research Triangle Park, NC, 1981.

[864] Strachan, W. M. J., and H. Huneault, Polychlorinated biphenyls and organochlorine pesticides in Great Lakes precipitation, *J. Great Lakes Res.*, **5**, 61-68, 1979.

[865] Slanina, J., J. G. Van Raaphorst, W. L. Zijp, A. J. Vermeulen, and C. A. Roet, An evaluation of the chemical composition of precipitation sampled with 21 identical collectors on a limited area, *Intern. J. Environ. Anal. Chem.*, **6**, 67-81, 1979.

[866] Harder, H. W., E. C. Christensen, J. R. Matthews, and T. F. Bidleman, Rainfall input of toxaphene to a South Carolina estuary, *Estuaries*, **3**, 142-147, 1980.

[867] Tsani-Bazaca, E., A. E. McIntyre, J. N. Lester, and R. Perry, Concentrations and correlations of 1,2-dibromoethane, 1,2-dichloroethane, benzene, and toluene in vehicle exhaust and ambient air, *Environ. Technol. Lett.*, **2**, 303-316, 1981.

[868] Jaeger, R. J., Carbon monoxide in houses and vehicles, *Bull. NY Acad. Med.*, **57**, 860-872, 1981.

[869] Johnson, L., B. Josefsson, P. Marstorp, and G. Eklund, Determination of carbonyl compounds in automobile exhausts and atmospheric samples, *Intern. J. Environ. Anal. Chem.*, **9**, 7-26, 1981.

[870] Tenner, R. L., and R. Fajer, Determination of nitro-polynuclear aromatics in ambient aerosol samples, *Intern. J. Environ. Anal. Chem.*, **14**, 231-241, 1983.

[871] Burg, W. R., O. L. Shotwell, and B. E. Saltzman, Measurements of airborne aflatoxins during the handling of 1979 contaminated corn, *Am. Ind. Hyg. Assoc. J.*, **43**, 580-586, 1982.

[872] Goff, E. U., J. R. Coombs, D. H. Fine, and T. M. Baines, Nitrosamine emissions from diesel engine crankcases, *SAE Tech. Pap. Ser.*, **801374**, 1980.

[873] Saltzman, E. S., D. L. Savoie, R. G. Zika, and J. M. Prospero, Methane sulfonic acid in the marine atmosphere, *J. Geophys. Res.*, **88**, 10897-10902, 1983.

[874] Neftel, A., S. Breitenbach, W. Elbert, and J. Hahn, Critical evaluation of the Br-Mmc fluorescent labeling technique for the determination of organic acids in precipitation, paper presented at Conf. on Gas-Liquid Chemistry of Natural Waters, Brookhaven National Lab., April, 1984.

[875] Lipari, F., J. M. Dasch, and W. F. Scruggs, Aldehyde emissions from wood-burning fireplaces, *Environ. Sci. Technol.*, **18**, 326-330, 1984.

[876] Pankow, J. F., L. M. Isabelle, and W. E. Asher, Trace organic compounds in rain. 1. Sampler design and analysis by adsorption/thermal desorption (ATD), *Environ. Sci. Technol.*, **18**, 310-318, 1984.

[877] Minato, S., Estimate of radon-222 concentrations in rainclouds from radioactivity of rainwater observed at ground level, *J. Radioanal. Chem.*, **78**, 199-207, 1983.

[878] Tong, H. Y., D. L. Shore, F. W. Karasek, P. Helland, and E. Jellum, Identification of organic compounds obtained from incineration of muncipal waste by high-performance liquid chromatographic fractionation and gas chromatography – mass spectrometry, *J. Chromatogr.*, **285**, 423-441, 1984.

[879] Barnes, R. D., L. M. Law, and A. J. MacLeod, Comparison of some porous polymers as absorbents for collection of odour samples and the application of the technique to an environmental malodour, *Analyst*, **106**, 412-418, 1981.

[880] Mendenhall, G. D., P. W. Jones, P. E. Strup, and W. L. Margard, *Organic Characterization of Aerosols and Vapor Phase Compounds in Urban Atmospheres*, EPA-600/3-78-031, Environmental Protection Agency, Washington, D.C., 1978 (NTIS Document PB-280050).

[881] Smith, L. R., *Unregulated Emissions for Vehicles Operated Under Low Speed Conditions*, EPA-460/3-83-006, Environmental Protection Agency, Washington, D.C., 1983 (NTIS Document PB 83-216366).

[882] Brenner, S., R. L. Brewer, I. R. Kaplan, and W. W. Wong, *Chemical Measurements in the Los Angeles Atmosphere*, EPRI EA-1466, Project 1315-2, Electric Power Research Institute, Palo Alto, CA, 1980.

[883] Hunter, S. C., Formation of SO_3 in gas turbines, *J. Eng. Power*, **104**, 44-51, 1982.

[884] Dong, M., I. Schmeltz, E. LaVoie, and D. Hoffmann, Aza-arenes in the respiratory environment, in *Carcinogenesis, Vol. 3: Polynuclear Aromatic Hydrocarbons*, P. W Jones and R. I. Freudenthal, Eds., New York: Raven Press, 1978.

[885] Pitts, J. N., Jr., H. W. Biermann, A. M. Winer, and E. C. Tuazon, Spectroscopic identification and measurement of gaseous nitrous acid in dilute auto exhaust, *Atmos. Environ.*, **18**, 847-854, 1984.

[886] Greenberg, J. P., and P. R. Zimmerman, Nonmethane hydrocarbons in remote tropical, continental, and marine atmospheres, *J. Geophys. Res.*, **89**, 4767-4778, 1984.

[887] Koenig, J., Bestimmung organischer Inhaltsstoffe von Stäuben der Aufenluft, *VDI-Ber.*, **429**, 325-242, 1982.

[888] Knoeppel, H., B. Versino, H. Schlitt, A. Peil, H. Schauenberg, and E. Vissers, Organics in air. Sampling and identification, in *Proc. 1st Eur. Symp. Phys. – Chem. Behav. Atmos. Pollut.*, Comm. Eur. Commun. – ISPRA, pp. 25-40, 1980.

[889] Richter, B. E., *S(IV) and Alkylating Agents in Airborne Particulate Matter*, Ph.D. Dissertation, Brigham Young University, 1981.

[890] Rudling, L., and B. Ahling, Chemical and biological characterization of emissions from combustion of wood and wood-chips in small furnaces and stoves, *Proc. Intl. Conf. Resid. Solid Fuels*, pp. 34-53, 1982.

[891] Mast, T. J., D. P. H. Hsieh, and J. N. Seiber, Mutagenicity and chemical characterization of organic constituents in rice straw smoke particulate matter, *Environ. Sci. Technol.*, **18**, 338-348, 1984.

[892] Hammond, E. G., C. Fedler, and R. J. Smith, Analysis of particle-borne swine house odors, *Agric. Environ.*, **6**, 395-401, 1981.

[893] Hollowell, C. D., J. V. Berk, and G. W. Traynor, *Indoor Air Quality Measurements in Energy Efficient Buildings*, Paper 78-60.6, 71st meeting, Air Pollution Control Assoc., Houston, TX, June 25-29, 1978.

[894] Giam, C. S., E. Atlas, and K. Sullivan, Vapor-phase n-aldehydes in high volume air samples from New Zealand, *SEAREX Newsletter*, **7(2)**, 13, 1984.

[895] Matthews, R. D., Estimated permissible levels, ambient concentrations, and adverse effects of the nitrogenous products of combustion: the cyanides, nitro-olefins, and nitroparaffins, *J. Comb. Toxicol.*, **7**, 157-172, 1980.

[896] Hagemann, R., D. Gaudin, and H. Virelizier, Analysis of oxygenated organic compounds in combustion and urban atmospheres, in *Adv. Mass. Spectrom.*, **7B**, N. R. Daly, Ed., London: Heyden and Son, pp. 1691-1696, 1978.

[897] Cautreels, W., G. Broddin, and K. Van Cauwenberghe, Fast quantitative analysis of organic compounds in airborne particulate matter by mass chromatography, in *Adv. Mass. Spectrom.*, **7B**, N. R. Daly, Ed., London: Heyden and Son, pp. 1674-1686, 1978.

[898] Fosnaugh, J., and E. R. Stephens, *Identification of Feedlot Odors*, Final Report, Dept. HEW, PHS Grant No. UI 00531-02, Statewide Air Pollution Research Center, Univ. Cal. – Riverside, 1969.

[899] Merkel, J. A., T. E. Hazen, and J. R. Miner, Identification of gases in a confinement swine building atmosphere, *Trans. Am. Soc. Agric. Eng.*, **12**, 310, 1969.

[900] Miner, J. R., and T. E. Hazen, Ammonia and amines: components of swine building odor, *Trans. Am. Soc. Agric. Eng.*, **12**, 772-774, 1969.

[901] Wils, E. R. J., A. G. Hulst, and J. C. den Hartog, The occurrence of plant wax constituents in airborne particulate matter in an urbanized area, *Chemosphere*, **11**, 1087-1096, 1982.

[902] Lahmann, E., The pollution of ambient air and rainwater by organic components of motor vehicle exhaust gases, *Proc. 4th Intl. Clean Air Cong.*, 595-597, 1977.

[903] Hites, R. A., M.-L. Yu, and W. G. Thilly, Compounds associated with diesel exhaust particulates, in *5th Intl. Symp. on Polynuclear Aromatic Hydrocarbons*, Columbus, OH: Battelle Columbus Laboratories, pp. 455-466, 1981.

[904] Cotton, M. L., N. D. Johnson, and K. G. Wheeland, Removal of arsine from process emissions, *Can. Metall. Q.*, **16**, Report 1-4, Metall. Soc. Cim., 205-209, 1977.

[905] Simoneit, B. R. T., M. A. Mazurek, and W. E. Reed, Characterization of organic matter in aerosols over rural sites: phytosterols, *Advances in Organic Geochemistry*, New York: John Wiley, pp. 355-361, 1983.

[906] Cox, R. E., M. A. Mazurek, and B. R. T. Simoneit, Lipids in Harmattan aerosols of Nigeria, *Nature*, **296**, 848-849, 1982.

[907] Airey, D., Contributions from coal and industrial materials to mercury in air, rainwater and snow, *Sci. Total Environ.*, **25**, 19-40, 1982.

[908] Landsberger, S., R. E. Jervis, G. Kajrys, S. Monaro, and R. Lecomte, Total soluble and insoluble sulfur concentrations in urban snow, *Environ. Sci. Technol.*, **17**, 542-546, 1983.

[909] Willison, M. J., and A. G. Clarke, Analysis of atmospheric aerosols by nonsuppressed ion chromatography, *Anal. Chem.*, **56**, 1037-1039, 1984.

[910] Benner, W. H., P. M. McKinney, and T. Novakov, Evidence for primary oxidants of SO_2, *Atmospheric Aerosol Research – FY 1982 Annual Report*, LBL-15298, Lawrence Berkeley Laboratory, Berkeley, CA, pp. 4-3 to 4-5, 1983.

[911] Harrison, R. M., and C. A. Pio, Major ion composition and chemical associations of inorganic atmospheric aerosols, *Environ. Sci. Technol.*, **17**, 169-174, 1983.

[912] Nagamoto, C. T., F. Parungo, R. Reinking, R. Pueschel, and T. Gerish, Acid clouds and precipitation in eastern Colorado, *Atmos. Environ.*, **17**, 1073-1082, 1983.

[913] MacDonald, G. A., *Volcanoes*, Englewood Cliffs, NJ: Prentice-Hall, 1972.

[914] Siegel, B. Z., and S. M. Siegel, Mercury emission in Hawaii: aerometric study of the Kalalua eruption of 1977, *Environ. Sci. Technol.*, **12**, 1036-1039, 1978.

[915] Barnard, W. R., and D. K. Nordstrom, Fluoride in precipitation – II. Implications for the geochemical cycling of fluorine, *Atmos. Environ.*, **16**, 105-111, 1982.

[916] Weschler, C. J., Characterization of selected organics in size-fractionated indoor aerosols, *Environ. Sci. Technol.*, **14**, 428-431, 1980.

[917] Wilson, M. J. G., Indoor air pollution, *Proc. Roy. Soc.*, **A300**, 215-221, 1968.

[918] Lunde, G., and A. Bjorseth, Polycyclic aromatic hydrocarbons in long-range transported aerosols, *Nature*, **268**, 518-519, 1977.

[919] Singh, H. B., L. J. Salas, A. J. Smith, and H. Shigeishi, Measurements of some potentially hazardous organic chemicals in urban environments, *Atmos. Environ.*, **15**, 601-612, 1981.

[920] Singh, H. B., L. J. Salas, and R. E. Stiles, Distribution of selected gaseous organic mutagens and suspect carcinogens in ambient air, *Environ. Sci. Technol.*, **16**, 872-880, 1982.

[921] Goff, E. U., J. R. Coombs, D. H. Fine, and T. M. Baines, Determination of N-nitrosamines from diesel engine crankcase emissions, *Anal. Chem.*, **52**, 1833-1836, 1980.

[922] Radtke, L. F., P. V. Hobbs, and D. A. Hegg, Aerosols and trace gases in the effluents produced by the launch of large liquid- and solid-fueled rockets, *J. Appl. Met.*, **21**, 1332-1345, 1982.

[923] Peel, D. A., and E. W. Wolff, Recent variations in heavy metal concentrations in firn and air from the Antarctic Peninsula, *Ann. Glaciol.*, **3**, 255-259, 1982.

[924] Cucco, J. A., and P. R. Brown, Confirming the presence of N-nitrosamines in ambient air and cigarette smoke by converting to and photochemically altering their corresponding N-nitroamines, *J. Chromatogr.*, **213**, 253-263, 1981.

[925] Matsushita, H., T. Shiozaki, M. Fujiwara, S. Goto, and T. Handa, Determination of nitrated polynuclear aromatic hydrocarbons by capillary column gas chromatography, *Taiki Osen Gakkaishi*, **18**, 241-249, 1983 (CA, **99**, 145214b, 1983).

[926] Tokiwa, H., S. Kitamori, R. Nakagawa, K. Horikawa, and L. Matamala, Demonstration of a powerful mutagenic dinitropyrene in airborne particulate matter, *Mutation Res.*, **121**, 107-116, 1983.

[927] Nojima, K., A. Kawaguchi, T. Ohya, S. Kanno, and M. Hirobe, Studies on photochemical reaction of air pollutants. X. Identification of nitrophenols in suspended particulates, *Chem. Pharm. Bull.*, **31**, 1047-1051, 1983.

[928] Tokiwa, H., S. Kitamori, R. Nakagawa, and Y. Ohnishi, Mutagens in airborne particulate pollutants and nitro derivatives produced by exposure of aromatic compounds to gaseous pollutants, *Environ. Sci. Res.*, **27** (Short-Term Bioassays Anal. Complex Environ. Mixtures 3), 555-567, 1983 (CA, **98**, 193139g, 1983).

[929] Gaiduk, M. I., Comparative hygienic evaluation of air pollution in gas-heated apartments, *Gig. Sanit.*, #11, 68-70, 1981 (CA, **96**, 90823e, 1982).

[930] Burckle, J. O., G. H. Marchant, and R. L. Meek, *Arsenic Emissions and Control Technology Gold Roasting Operations*, EPA-600/9-80-039C, Environ. Protect. Agency, Cincinnati, OH, 1980.

[931] Uemura, T., I. Kibune, and H. Murayama, Odors from livestock farms. I., *Niigata-ken Kogai Kenkyusho Kenkyu Hokoku*, **5**, 51-54, 1980 (CA, **95**, 85452h, 1981).

[932] Trout, D. A., and R. S. Reimers, Air pollutants associated with wastewater treatment, *Proc. Ind. Waste Conf.*, **30**, 65-77, 1977.

[933] Herrmann, A., Chemical composition of a polluted snow cover, *IAHS-AISH Publ. 118*, 121-126, 1977 (CA, **88**, 176859t, 1978).

[934] Butler, F. E., R. H. Jungers, L. F. Porter, A. E. Riley, and F. J. Toth, Analysis of air particulates by ion chromatography: comparison with accepted methods, in *Ion Chromatogr. Anal. Environ. Pollut.*, E. Sawicki, J. D. Mulik, and E. Wittgenstein, Eds., Ann Arbor, MI: Ann Arbor Sci. Pub., pp. 65-76, 1978.

[935] Woods, D. C., R. J. Bendura, and D. E. Wornom, *Launch Vehicle Effluent Measurements During the August 20, 1977 Titan 3 Launch at Air Force Eastern Test Range*, NASA-TM-78778, L-12551, National Aeronautics and Space Adm., Washington, D.C., 1979.

[936] Lubyanskii, M. L., M. M. Levina, V. M. Blagodatin, G. V. Sudakova, G. I. Kuz'mina, and Yu. P. Tikhomorov, Hygienic assessment of the thermal detoxication of acrylate production wastes, *Gig. Sanit.*, #9, 43-46, 1977 (CA, **88**, 54417b, 1978).

[937] Izatt, R. M., D. J. Eatough, M. L. Lee, T. Major, B. E. Richter, L. D. Hansen, R. G. Meisenheimer, and J. W. Fischer, The formation of inorganic and organic sulfur (IV) species in aerosols, *Proc. 4th Jt. Conf. Sens. Environ. Pollut.*, Am. Chem. Soc., Washington, D.C., pp. 821-824, 1978.

[938] Morikawa, T., Evolution of soot and polycyclic aromatic hydrocarbons in combusion, *Shobo Kenkyusho Hokoku*, **45**, 13-24, 1978 (CA, **89**, 151718a, 1978).

[939] Elpat'evskii, P. V., Chemical composition of snow-melt water and its alteration by technogenic factors, *Geokhim. Zony Gipergeneza Tekh. Deyat. Chel.*, Yu. P. Badenkov, Ed., Akad. Nauk SSSR, Vladivostok, pp. 48-56, 1976 (CA, **91**, 42408y, 1979).

[940] Dmitriev, M. T., and V. A. Mishchikhin, Determination of toxic substances given off by polymeric materials under experimental conditions, *Gig. Sanit*, #6, 45-48, 1979 (CA, **91**, 95916u, 1979).

[941] Akhmadulina, L. A., and A. I. Lesnyak, Chemical composition of rain water and snow in the Samarkand water reservoir region, *Teor. Osn. Pererab. Miner. Org. Syr'ya*, **3**, 43-54, 1976 (CA, **91**, 111835z, 1979).

[942] Lindskog, A., Atmospheric transport of PCB and HCB, Publ. IVL B-527, Inst. Vatten Luftvardsforsk, Stockholm, 19 pp., 1980 (CA, **93**, 30992n, 1980).

[943] Wils, E. R. J., A. G. Hulst, and J. C. den Hartog, The occurrence of Rovral and Permethrin in airborne particulate matter, *Chemosphere*, **11**, 585-589, 1982.

[944] Koenig, J., W. Funcke, E. Balfanz, B. Grosch, T. Romanowski, and F. Pott, Untersuchung von 135 polyzyklischen aromatischen Kohlenwasserstoffen in atmosphärischen Schwebstoffen aus 5 Städten der Bundesrepublik Deutschland, *Staub-Reinhalt. Luft*, **41**, 73-78, 1981.

[945] Morlin, Z., M. Kertesz, A. Kiss, and J. Szeili, Presence of 20-methylcholanthrene in the atmosphere, *Zentr. Bakteriol. Parasitenkd. Infekt. Hyg., Abt. 1: Orig., Reihe B*, **169**, 446-455, 1979 (CA, **93**, 30990k, 1980).

[946] Iida, H., T. Tokunaga, and K. Miwa, Protective methods for occurrence of bad odors and the removal of odors in fish processing factories. IV. Deodorization of volatile compounds by the combination method of chlorine solution- and water-washings, *Tokai-ku Suisan Kenkyusho Kenkyu Hokoku*, **87**, 15-23, 1976. (CA, **88**, 41061f, 1978).

[947] Appel, B. R., E. M. Hoffer, E. L. Kothny, S. M. Wall, M. Haik, and R. L. Knights, Analysis of carbonaceous material in Southern California atmospheric aerosols. 2, *Environ. Sci. Technol.*, **13**, 98-104, 1979.

[948] Rosen, H., A. D. A. Hansen, R. L. Dod, and T. Novakov, Soot in urban atmospheres: determination by an optical absorption technique, *Science*, **208**, 741-744, 1980.

[949] Ogren, J. A., R. J. Charlson, and P. J. Groblicki, Determination of elemental carbon in rainwater, *Anal. Chem.*, *55*, 1569-1572, 1983.

[950] Ogren, J. A., P. J. Groblicki, and R. J. Charlson, Measurement of the removal rate of elemental carbon from the atmosphere, *Sci. Total Environ.*, **36**, 329-338, 1984.

[951] Van Vaeck, L., K. Van Cauwenberghe, and J. Janssens, The gas-particle distribution of organic aerosol constitutents: measurement of the volatilisation artefact in hi-vol cascade impactor sampling, *Atmos. Environ.*, **18**, 417-430, 1984.

[952] Wallace, L. A., E. D. Pellizzari, T. D. Hartwell, C. Sparacino, and H. Zelon, *Personal Exposure to Volatile Organics and Other Compounds Indoors and Outdoors - The TEAM Study*, EPA-600/D-83-082, Research Triangle Inst., Research Triangle Park, NC, 1983.

[953] Chiw, C., R. S. Thomas, J. Lockwood, K. Li, R. Halman, and R. C. C. Lao, Polychlorinated hydrocarbons from power plants, wood burning, and municipal incinerators, *Chemosphere*, **12**, 607-616, 1983.

[954] Tiernan, T. O., M. L. Taylor, J. H. Garrett, G. F. Van Ness, J. G. Solch, D. A. Deis, and D. J. Wagel, Chlorodibenzodioxins, chlorodibenzofurans and related compounds in the effluents from combustion processes, *Chemosphere*, **12**, 595-606, 1983.

[955] Simoneit, B. R. T., and M. A. Mazurek, Organic matter of the troposphere-II. Natural background of biogenic lipid matter in aerosols over the rural western United States, *Atmos. Environ.*, **16**, 2139-2159, 1982.

[956] Simoneit, B. R. T., Organic matter of the troposphere-III. Characterization and sources of petroleum and pyrogenic residues in aerosols over the western United States, *Atmos. Environ.*, **18**, 51-67, 1984.

[957] Harker, A. B., P. J. Pagni, T. Novakov, and L. Hughes, Manganese emissions from combustors, *Chemosphere*, **6**, 339-347, 1975.

[958] Bidleman, T. F., E. J. Christensen, W. N. Billings, and R. Leonard, Atmospheric transport of organochlorines in the North Atlantic gyre, *J. Marine Res.*, **39**, 443-464, 1981.

[959] Liberti, A., G. Goretti, and M. V. Russo, PCDD and PCDF in the combusion of vegetable wastes, *Chemosphere*, **12**, 661-663, 1983.

[960] Hryhorczuk, D. O., W. A. Withrow, C. S. Hesse, and V. R. Beasley, A wire reclamation incinerator as a source of environmental contamination with tetrachlorodibenzo-p-dioxins and tetrachlorodibenzofurans, *Arch. Environ. Health*, **36**, 228-234, 1981.

[961] Herget, W. F., and J. D. Brasher, Remote measurement of gaseous pollutant concentrations using a mobile Fourier transform interferometer system, *Appl. Optics*, **18**, 3604-3420, 1979.

[962] Fulford, J. E., T. Sakuma, and D. A. Lane, Real-time analysis of exhaust gases using triple quadrupole mass spectrometry, in *Polynucl. Aromat. Hydrocarbons: Phys. Biol. Chem.*, M. Cooke, A. J. Dennis, and G. L. Fisher, Eds., Columbus, OH: Battelle Press, pp. 297-303, 1982.

[963] Thompson, C. R., E. G. Hensel, and G. Kats, Outdoor-indoor levels of six air pollutants, *J. Air Poll. Contr. Assoc.*, **23**, 881-886, 1973.

[964] Derham, R. L., G. Peterson, R. H. Sabersky, and F. H. Shair, On the relation between the indoor and outdoor concentration of nitrogen oxides, *J. Air Poll. Contr. Assoc.*, **24**, 158-161, 1974.

[965] Mueller, F. X., L. Loeb, and W. H. Mapes, Decomposition rates of ozone in living areas, *Environ. Sci. Technol.*, **7**, 342-346, 1973.

[966] Derouane, A., and G. Verduyr, Comparaison des concentrations en SO_2 a l'interier et a l'exterieur de deux batiments, *Atmos. Environ.*, **7**, 891-899, 1973.

[967] Grosjean, D., and G. L. Kok, *Interlaboratory Comparison Study of Methods for Measuring Formaldehyde and Other Aldehydes in Ambient Air*, Report for CAPA-17-80, Coordinating Research Council, Claremont, CA: Harvey Mudd College, 1981.

[968] Gerstle, R. W., and D. A. Kemnitz, Atmospheric emissions from open burning, *J. Air Poll. Contr. Assoc.*, **17**, 324-327, 1967.

[969] Fritschen, L. J., H. Bovee, C. Buettner, R. J. Charlson, L. Monteith, S. Pickford, J. Murphy, and E. F. Darley, *Slash Fire Atmospheric Pollution*, Forest Service Research Paper PNW-97, U.S. Dept. of Agriculture, Washington, D.C., 1970.

[970] Altshuller, A. P., and C. A. Clemons, Gas chromatographic analysis of aromatic hydrocarbons at atmospheric concentrations using flame ionization detection, *Anal. Chem.*, **34**, 466-472, 1962.

[971] Mulawa, P. A., and S. H. Cadle, Measurement of phenols in automobile exhaust, *Anal. Lett.*, **14**, 671-687, 1981.

[972] McEwan, D. J., Automobile exhaust hydrocarbon analysis by gas chromatography, *Anal. Chem.*, **38**, 1047-1053, 1966.

[973] Bunn, W. W., E. R. Deane, D. W. Klein, and R. D. Kleopfer, Sampling and characterization of air for organic compounds, *Water, Air, Soil Poll.*, **4**, 367-380, 1975.

[974] Schuchmann, H. P., and K. J. Laidler, Nitrogen compounds other than NO in automobile exhaust gas, *J. Air Poll. Contr. Assoc.*, **22**, 52-53, 1972.

[975] Slemr, F., and W. Seiler, Field measurements of NO and NO_2 emissions from fertilized and unfertilized soils, *J. Atm. Chem.*, **2**, 1-24, 1984.

[976] Kim, C. M., Influence of vegetation types on the intensity of ammonia and nitrogen dioxide liberation from soil, *Soil Biol. Biochem.*, **5**, 163-166, 1973.

[977] Billings, W. N., and T. F. Bidleman, High volume collection of chlorinated hydrocarbons in urban air using three solid adsorbents, *Atmos. Environ.*, **17**, 383-391, 1983.

[978] Goldman, A., D. G. Murcray, F. J. Murcray, G. R. Cook, J. W. Van Allen, F. S. Bonomo, and R. D. Blatherwick, Identification of the ν_3 vibration-rotation band of CF_4 in balloon-borne infrared solar spectra, *Geophys. Res. Lett*, **6**, 609-612, 1979.

[979] Shaver, C. L., G. R. Cass, and J. R. Druzik, Ozone and the deterioration of works of art, *Environ. Sci. Technol.*, **17**, 748-752, 1983.

[980] Moschandreas, D. J., D. J. Pelton, and D. R. Berg, The effects of woodburning on indoor pollutant concentrations, Paper 81-22.2, 74th Ann. Mtg., Air Pollution Control Assoc., Philadelphia, PA, June 23, 1981.

[981] Rudling, L., B. Ahling, and G. Löfroth, *Emissions from Combusion of Wood-Chips in a Small Central Heating Furnace and from Combustion of Wood in Closed Fireplace Stoves*, Report SNV-PM-1331 (DE82-900626), Inst. för Vatten-och Luftvardsforskning, Stockholm, 1980. (CA, **97**, 97529, 1982).

[982] Ramdahl, T., I. Alfheim, S. Rustad, and T. Olsen, Chemical and biological characterization of emission from small residential stoves burning wood and charcoal, *Chemosphere*, **11**, 601-611, 1982.

[983] Rasmussen, R. A., M. A. K. Khalil, and S. D. Hoyt, Trace gases in snow and rain, in *Precipitation Scavenging, Dry Deposition, and Resuspension, II*, H. R. Pruppacher, R. G. Semonin, and W. G. N. Slinn, Eds., New York: Elsevier, p. 1301-1314, 1983.

[984] Bondarev, V. B., N. V. Porshnev, and D. F. Nenarakov, Gas chromatographic analysis of the vapours and gases discharged from the thermal fields of Kamchatka, *J. Chromatogr.*, **247**, 347-351, 1982.

[985] Schuetzle, D., and J. M. Perez, Factors influencing the emissions of nitrated-polynuclear aromatic hydrocarbons (nitro-PAH) from diesel engines, *J. Air Poll. Contr. Assoc.*, **33**, 751-755, 1983.

[986] Rinsland, C. P., M. A. H. Smith, P. L. Rinsland, A. Goldman, J. W. Brault, and G. M. Stokes, Ground-based infrared spectroscopic measurements of atmospheric hydrogen cyanide, *J. Geophys. Res.*, **87**, 11119-11125, 1982.

[987] Pearson, C. D., and W. J. Hines, Determination of hydrogen sulfide, carbonyl sulfide, carbon disulfide, and sulfur dioxide in gases and hydrocarbon streams by gas chromatography/flame photometric detection, *Anal. Chem.*, **49**, 123-126, 1977.

[988] Reamer, D. C., W. H. Zoller, and T. C. O'Haver, Gas chromatograph - microwave plasma detector for the determination of tetraalkyllead species in the atmosphere, *Anal. Chem.*, **50**, 1449-1452, 1978.

[989] De Jonghe, W. R. A., D. Chakraborti, and F. C. Adams, Sampling of tetraalkyllead compounds in air for determination by gas chromatography - atomic absorption spectrometry, *Anal. Chem.*, **52**, 1974-1977, 1980.

[990] Yasuhara, A., M. Morita, and K. Fuwa, Determination of naphtho[2,1,8-*qra*] naphthacene in soots, *Environ. Sci. Technol*, **16**, 805-808, 1982.

[991] Yasuhara, A., and K. Fuwa, Determination of fatty acids in airborne particulate matter, dust and soot by mass chromatography, *J. Chromatogr.*, **240**, 369-376, 1982.

[992] Pope, D., B. J. Davis, and R. L. Moss, Multi-stage absorption of rendering plant odours using sodium hypochlorite and other reagents, *Atmos. Environ.*, **15**, 251-262, 1981.

[993] Harsch, D. E., R. A. Rasmussen, and D. Pierotti, Identification of a potential source of chloroform in urban air, *Chemosphere*, **11**, 769-775, 1977.

[994] Dawson, G. A., J. C. Farmer, and J. L. Moyers, Formic and acetic acids in the atmosphere of the southwest U.S.A., *Geophys. Res. Lett.*, **7**, 725-728, 1980.

[995] Anderson, J. G., H. J. Grassl, R. E. Shetter, and J. J. Margitan, HO_2 in the stratosphere: three *in situ* measurements, *Geophys. Res. Lett.*, **8**, 289-292, 1981.

[996] Noxon, J. F., R. B. Norton, and E. Marovich, NO_3 in the troposphere, *Geophys. Res. Lett.*, **7**, 125-128, 1980.

[997] Hartung, L. D., E. G. Hammond, and J. R. Miner, Identification of carbonyl compounds in a swine-building atmosphere, *Livest. Waste Manage. Pollut. Abatement, Proc. Int. Symp.*, Amer. Soc. Agric. Eng., St. Joseph, Mich., pp. 105-106, 1971 (CA, **80**, 99679q, 1974).

[998] Ludwick, J. D., D. E. Robertson, J. S. Fruchter, and C. L. Wilkerson, Analysis of well gases from areas of geothermal power potential, *Atmos. Environ.*, **16**, 1053-1059, 1982.

[999] Barber, E. D., E. Sawicki, and S. P. McPherson, Separation and identification of phenols in automobile exhaust by paper and gas liquid chromatography, *Anal. Chem.*, **36**, 2442-2445, 1964.

[1000] Sexton, K., and H. Westberg, Photochemical ozone formation from petroleum refinery emissions, *Atmos. Environ.*, **17**, 467-475, 1983.

[1001] Ketseridis, G., and R. Jaenicke, Organische Beimengungen in atmosphärisher Reinluft: Ein Beitrag zur Budget-Abschatzung, in *Organische Verunreinigungen in der Umwelt-Erkennen, Berwerben, Vermindern*, K. Aurand, et al., Ed., pp. 379-390, Berlin: E. Schmidt, 1978.

[1002] Lamparski, L. L., and T. J. Nestrick, Determination of tetra-, hexa-, hepta-, and octachlorodibenzo-*p*-dioxin isomers in particulate samples at parts per trillion levels, *Anal. Chem.* **52**, 2045-2054, 1980.

[1003] Bumb, R. R., W. B. Crummett, S. S. Cutie, J. R. Gledhill, R. H. Hummell, R. O. Kagel, L. L. Lamparski, E. V. Luoma, D. L. Miller, T. J. Nestrick, L. A. Shadoff, R. H. Stehl, and J. S. Woods, Trace chemistries of fire: a source of chlorinated dioxins, *Science,* **210**, 385-390, 1980.

[1004] Rosenblatt, G., D. Mozzon, N. G. H. Guilford, and G. H. S. Thomas, Polychlorinated biphenyl source emission strengths from municipal incineration and electrical component filling systems, *Proc. 4th Intl. Clean Air Congr.*, S. Kasuga, N. Suzuki, and T. Yamada, Eds., Tokyo: Japan Union Air Pollut. Prev. Assoc., pp. 637-640, 1977.

[1005] Müller, J., and E. Rohbock, Method for measurement of polycyclic aromatic hydrocarbons in particulate matter in ambient air, *Talanta,* **27**, 673-675, 1980.

[1006] Nojima, K., K. Fukaya, S. Fukui, S. Kanno, S. Nishiyama, and Y. Wada, Studies on photochemistry of aromatic hydrocarbons, *Chemosphere*, **1**, 25-30, 1976.

[1007] Mukhin, L. M., V. B. Bondarev, and E. N. Safonova, The role of volcanic processes in the evolution of organic compounds on the primitive earth, *Mod. Geol.*, **6**, 119-122, 1978.

[1008] Chuan, R. L., and D. C. Woods, The appearance of carbon aerosol particles in the lower stratosphere, *Geophys. Res. Lett.*, **11**, 553-556, 1984.

[1009] Berg, W. W., L. E. Heidt, W. Pollock, P. D. Sperry, and R. J. Cicerone, Brominated organic species in the arctic atmosphere, *Geophys. Res. Lett.*, **11**, 429-432, 1984.

[1010] Rasmussen, R. A., and M. A. K. Khalil, Gaseous bromine in the Arctic and Arctic haze, *Geophys. Res. Lett.*, **11**, 433-436, 1984.

[1011] Khalil, M. A. K., and R. A. Rasmussen, Statistical analysis of trace gases in Arctic haze, *Geophys. Res. Lett.*, **11**, 437-440, 1984.

[1012] Wingender, R. J., and R. M. Williams, Evidence for the long-distance atmospheric transport of polychlorinated terphenyl, *Environ. Sci. Technol.*, **18**, 625-628, 1984.

[1013] Majima, T., K. Tadao, M. Naruse, and M. Hiraoka, Studies on pyrolysis process of sewage sludge, *Prog. Water Technol.*, **9**, 381-396, 1977.

[1014] Heintzenberg, J., H.-C. Hansson, and H. Lannefors, The chemical composition of Arctic haze at Ny-Ålesund, Spitsbergen, *Tellus*, **33**, 162-171, 1981.

[1015] Murphy, T. J., and A. W. Schinsky, Net atmospheric inputs of PCBs to the ice cover on Lake Huron, *J. Great Lakes Res.*, **9**, 92-96, 1983.

[1016] Bartle, K. D., M. L. Lee, and M. Novotny, Identification of environmental polynuclear aromatic hydrocarbons by pulse fourier-transform ^1H nuclear magnetic resonance spectroscopy, *Analyst*, **102**, 731-738, 1977.

[1017] Sexton, K., and H. Westberg, Ambient hydrocarbon and ozone measurements downwind of a large automotive painting plant, *Environ. Sci. Technol.*, **14**, 329-332, 1980.

[1018] Westberg, H., K. Sexton, and D. Flyckt, Hydrocarbon production and photochemical ozone formation in forest burn plumes, *J. Air Poll. Contr. Assoc.*, **31**, 661-664, 1981.

[1019] Sadasivan, S., and S. J. S. Anand, Chlorine, bromine and iodine in monsoon rains in India, *Tellus*, **31**, 290-294, 1979.

[1020] Martens, C. S., and R. C. Harriss, Chemistry of aerosols, cloud droplets, and rain in the Puerto Rican marine atmosphere, *J. Geophys. Res.*, **78**, 949-957, 1973.

[1021] Jaworowski, Z., and L. Kownacka, Lead and radium in the lower stratosphere, *Nature*, **263**, 303-304, 1976.

[1022] Delany, A. C., J. P. Shedlovsky, and W. H. Pollock, Stratospheric aerosol: the contribution from the troposphere, *J. Geophys. Res.*, **79**, 5646-5650, 1974.

[1023] Matsumoto, G., and T. Hanya, Organic constituents in atmospheric fallout in the Tokyo area, *Atmos. Environ.*, **14**, 1409-1419, 1980.

[1024] Peirson, D. H., P. A. Cawse, and R. S. Cambray, Chemical uniformity of airborne particulate material, and a maritime effect, *Nature*, **251**, 675-679, 1974.

[1025] MacLeod, K. E., Polychlorinated biphenyls in indoor air, *Environ. Sci. Technol.*, **15**, 926-928, 1981.

[1026] Bidleman, T. F., and R. Leonard, Aerial transport of pesticides over the Northern Indian Ocean and adjacent seas, *Atmos. Environ.*, **16**, 1099-1107, 1982.

[1027] Giam, C. S., E. Atlas, H. S. Chan, and G. S. Neff, Phthalate esters, PCB and DDT residues in the Gulf of Mexico atmosphere, *Atmos. Environ.*, **14**, 65-69, 1980.

[1028] Atlas, E., and C. S. Giam, Global transport of organic pollutants: ambient concentrations in remote marine atmospheres, *Science*, **211**, 163-165, 1981.

[1029] Tanabe S., R. Tatsukawa, M. Kawano, and H. Hidaka, Global distribution and atmospheric transport of chlorinated hydrocarbons: HCH(BCH) isomers and DDT compounds in the Western Pacific, Eastern Indian and Antarctic Oceans, *J. Ocean. Soc. Japan*, **38**, 137-148, 1982.

[1030] Glotfelty, D. E., A. W. Taylor, and W. H. Zoller, Atmospheric dispersion of vapors: are molecular properties unimportant?, *Science*, **219**, 843-845, 1983.

[1031] Taylor, A. W., Post-application volatilization of pesticides under field conditions, *J. Air Poll. Contr. Assoc.*, **28**, 922-927, 1978.

[1032] Wu, T. L., Atrazine residues in estuarine water and in rainwater, Paper ENVT-40, 178th Nat. Mtg., Amer. Chem. Soc., Washington, D.C., Sept., 1979.

[1033] Lazrus, A. L., B. Gandrud, and R. D. Cadle, Chemical composition of air filtration samples of the stratospheric sulfate layer, *J. Geophys. Res.*, **76**, 8083-8088, 1971.

[1034] Spencer, W. F., W. J. Farmer, and M. M. Cliath, Pesticide volatilization, *Pesticide Rev.*, **49**, 1-47, 1973.

[1035] Yocom, J. E., Indoor-outdoor air quality relationships: a critical review, *J. Air Poll. Contr. Assoc.*, **32**, 500-520, 1982.

[1036] Grob, K., and G. Grob, Trace analysis on capillary columns. Selected practical applications: insecticides in raw butter extract; aroma head space from liquors; auto exhaust gas, *J. Chromatogr. Sci.*, **8**, 635-639, 1970.

[1037] Noxon, J. F., Atmospheric nitrogen fixation by lightning, *Geophys. Res. Lett.*, **3**, 463-465, 1976.

[1038] Fruchter, J. S., D. E. Robertson, J. C. Evans, K. B. Olsen, E. A. Lepel, J. C. Laul, K. H. Abel, R. W. Sanders, P. O. Jackson, N. S. Wogman, R. W. Perkins, H. H. VanTuyl, R. H. Beauchamp, J. W. Shade, J. L. Daniel, R. L. Erikson, G. A. Sehmel, R. N. Lee, A. V. Robinson, O. R. Moss, J. K. Briant, and W. C. Cannon, Mount St. Helens ash from the 18 May 1980 eruption: chemical, physical, mineralogical, and biological properties, *Science*, **209**, 1116-1125, 1980.

[1039] Shedlovsky, J. P., and S. Paisley, On the meteoritic component of stratospheric aerosols, *Tellus*, **18**, 499-503, 1966.

[1040] Andersson, B., K. Andersson, and C.-A. Nilsson, Mass spectrometric identification of 2-ethylhexanol in indoor air: recovery studies by charcoal sampling and gas chromatographic analysis at the micrograms per cubic metre level, *J. Chromatogr.*, **291**, 257-263, 1984.

[1041] Cantrell, C. A., D. H. Stedman, and G. J. Wendel, Measurement of atmospheric peroxy radicals by chemical amplification, *Anal. Chem.*, **56**, 1496-1502, 1984.

[1042] Högström, U., L. Enger, and I. Svedung, A study of atmospheric mercury dispersion, *Atmos. Environ.*, **13**, 465-476, 1979.

[1043] Kozuchowski, J., and D. L. Johnson, Gaseous emissions of mercury from an aquatic vascular plant, *Nature*, **274**, 468-469, 1978.

[1044] Cho, P., and J. H. Chambers, Municipal refuse: an alternate energy resource in power plants, *Proc. 4th Natl. Conf. Energy Environ.*, Air Pollut. Contr. Assoc., Pittsburgh, pp. 204-211, 1976.

[1045] Zoller, W. H., E. A. Lepel, E. J. Mroz, and K. M. Stefanssen, Trace elements from volcanoes: Augustine, 1976, *Air Pollution Measurement Techniques*, Spec. Environ. Rpt. 10, Geneva: World Meteorol. Org., pp. 155-164, 1977.

[1046] Lane, D. A., B. A. Thomson, A. M. Lovett, and N. M. Reid, Real-time tracking of industrial emissions through propulated areas using a mobile APCI mass spectrometer system, Paper presented at 8th Intl. Mass Spectrom. Conf., Oslo, 1979.

[1047] Thomas, C. W., and R. W. Perkins, Transuranium elements in the atmosphere, Report BNWL-1881, UC-48, Biol. Sci. Dept., Battelle Pacific Northwest Labs., Richland, WA, 1974.

[1048] DaRos, B., R. Merrill, H. K. Willard, and C. D. Wolbach, *Emission and Residue Values from Waste Disposal During Wood Preserving*, EPA-600/2-82-062, Environ. Prot. Agency, Cincinnati, OH, 1982.

[1049] Adams, D. F., S. O. Farwell, M. R. Pack, and E. Robinson, Biogenic sulfur gas emissions from soils in Eastern and Southeastern United States, *J. Air Poll. Contr. Assoc.*, **31**, 1083-1089, 1981.

[1050] Knapp, K. T., and J. L. Cheney, *Measurement of Sulfuric Acid and HCl in Stationary Source Emissions*, EPA-600/D-83-069, Environ. Prot. Agency, Research Triangle Park, NC, 1983.

[1051] Henry, W. M., and K. T. Knapp, Compound forms of fossil fuel fly ash emissions, *Environ. Sci. Technol.*, **14**, 450-456, 1980.

[1052] Ruud, C. O., and P. A. Russell, X-ray diffraction, in *Analysis of Airborne Particles by Physical Methods*, H. Malissa, Ed., West Palm Beach, FL: CRC Press, pp. 179-189, 1978.

[1053] Wostradowski, R. A., S. P. Bhatia, and S. Prahacs, The study of the air pollution significance of carbonyl sulfide (COS) emissions from Kraft recovery furnaces, *Trans. Tech. Soc., Can. Pulp Paper Assoc.*, **2**, B173-B180, 1976.

[1054] Hinkle, M. E., and T. F. Harms, CS_2 and COS in soil gases of the Roosevelt Hot Springs known geothermal resource area, Beaver County, Utah, *J. Res. US Geol. Survey*, **6**, 571-578, 1978.

[1055] Rasmussen, R. A., M. A. K. Khalil, and S. D. Hoyt, The oceanic source of carbonyl sulfide (OCS), *Atmos. Environ.*, **16**, 1591-1594, 1982.

[1056] Aneja, V. P., J. H. Overton, Jr., L. T. Cupitt, J. L. Durham, and W. E. Wilson, Direct measurements of emission rates of some atmospheric biogenic sulfur compounds, *Tellus*, **31**, 174-178, 1979.

[1057] Aneja, V. P., J. H. Overton, L. T. Cupitt, J. L. Durham, and W. E. Wilson, Carbon disulphide and carbonyl sulphide from biogenic sources and their contributions to the global sulphur cycle, *Nature*, **282**, 493-496, 1979.

[1058] Murayama, H., N. Moriyama, and I. Kifune, Determination of condensed aromatic and aliphatic hydrocarbons in airborne dust samples taken around the aluminum smelting plant and particulate matters emitted from the aluminum smelter, *Niigata-Ken Kogai Kenkyusho Kenkyu Hokoku*, **5**, 1-10, 1981 (CA, **95**, 48260y, 1981).

[1059] Bradley, J. P., P. Goodman, I. Y. T. Chan, and P. R. Buseck, Structure and evolution of fugitive particles from a copper smelter, *Environ. Sci. Technol.*, **15**, 1208-1212, 1981.

[1060] Ronneau, C., and J. P. Hallet, Heavy elements in acid rain, Rpt. EUR 8307, Acid Deposition, pp. 149-154, Comm. Eur. Communities, 1983.

[1061] Gendreau, R. M., R. J. Jakobsen, W. M. Henry, and K. T. Knapp, Fourier transform infrared spectroscopy for inorganic compound speciation, *Environ. Sci. Technol.*, **14**, 990-995, 1980.

[1062] Bloch, P., F. Adams, J. Van Landuyt, and L. Van Goethem, Morphological and chemical characterization of individual aerosol particles in the atmosphere, Rpt. EUR 6621, Proc. 1st Eur. Symp. Phys.-Chem. Behav. Atmos. Pollut., Comm. Eur. Communities, pp. 307-321, 1980.

[1063] Lee, M. L., D. W. Later, D. K. Rollins, D. J. Eatough, and L. D. Hansen, Dimethyl and monomethyl sulfate: presence in coal fly ash and airborne particulate matter, *Science*, **207**, 186-188, 1980.

[1064] Muhlbaier, J. L., and R. L. Williams, Fireplaces, furnaces and vehicles as emission sources of particulate carbon, in *Particulate Carbon: Atmospheric Life Cycle*, G. T. Wolff and R. L. Klimisch, Eds., New York: Plenum, pp. 185-198, 1982.

[1065] Grimmer, G., J. Jacob, K.-W. Naujack, and G. Dettbarn, Determination of polycyclic aromatic compounds emitted from brown-coal-fired residential stoves by gas chromatography/mass spectrometry, *Anal. Chem.*, **15**, 892-900, 1983.

[1066] Renwick, J. A. A., and J. Potter, Effects of sulfur dioxide on volatile terpene emission from balsam fir, *J. Air Poll. Contr. Assoc.*, **31**, 65-66, 1981.

[1067] Winner, W. E., C. L. Smith, G. W. Koch, H. A. Mooney, J. D. Bewley, and H. R. Krouse, Rates of emission of H_2S from plants and patterns of stable sulphur isotope fractionation, *Nature*, **289**, 672-673, 1981.

[1068] Farquhar, G. D., R. Wetselaar, and P. M. Firth, Ammonia volatilization from senescing leaves of maize, *Science*, **203**, 1257-1258, 1979.

[1069] Post, J. E., Characterization of particles in the Phoenix aerosol, and structure refinements of Hollandite minerals, PhD dissertation, Arizona State Univ., 1981.

[1070] Bain, D. C., and J. M. Tait, Mineralogy and origin of dust fall on Skye, *Clay Minerals*, **12**, 353-355, 1977.

[1071] Hoshika, Y., Y. Nihei, and G. Muto, Pattern display for characterization of trace amounts of odorants discharged from nine odour sources, *Analyst*, **106**, 1187-1202, 1981.

[1072] Broddin, G., L. Van Vaeck, and K. Van Cauwenberghe, On the size distribution of polycyclic aromatic hydrocarbon containing particles from a coke oven emission source, *Atmos. Environ.*, **11**, 1061-1064, 1977.

[1073] Gagosian, R. B., O. C. Zafiriou, E. T. Peltzer, and J. B. Alford, Lipids in aerosols from the tropical North Pacific: temporal variability, *J. Geophys. Res.*, **87**, 11133-11144, 1982.

[1074] Kamiya, A., and Y. Ose, Study of odorous compounds produced by putrefaction of foods. V. Fatty acids, sulphur compounds, and amines, *J. Chromatogr.*, **292**, 383-391, 1984.

[1075] Miner, J. R., *Odors from Confined Livestock Production*, Report EPA-660/2-74-023, Washington, D.C.: US Environ. Prot. Agency, 1974.

[1076] Committee on Odors, *Odors from Stationary and Mobile Sources*, Washington, D.C.: National Academy of Sciences, 1979.

[1077] Doty, D. M., R. H. Snow, and H. G. Reilich, *Investigation of Odor Control in the Rendering Industry*, Contract No. 68-02-0260, Washington, D.C.: US Environ. Prot. Agency, 1972.

[1078] Osag, T. R., and G. B. Crane, *Control of Odors from Inedibles-Rendering Plants*, Report EPA-450/1-74-006, Research Triangle Park, NC: US Environ. Prot. Agency, 1974.

[1079] Cooper, J. A., and R. W. Perkins, Versatile Ge(Li)-NaI(Ti) coincidence-anticoincidence gamma-ray spectrometer for environment and biological problems, *Nucl. Instrum. Methods*, **99**, 125-132, 1972.

[1080] Strachan, W. M. J., H. Huneault, W. M. Schertzer, and F. C. Elder, Organochlorines in precipitation in the Great Lakes Region, *Environ. Sci. Res.*, **16**, 387-396, 1980.

[1081] Perkins, M. D., and F. L. Eisele, First mass spectrometric measurements of atmospheric ions at ground level, *J. Geophys. Res.*, **89**, 9649-9657, 1984.

[1082] Ip, W. M., R. J. Gordon, and E. C. Ellis, Characterization of organics in aerosol samples from a Los Angeles receptor site, paper presented at 2nd Intl. Conf. Carbonaceous Particles in the Atmosphere, Linz, Austria, August, 1983.

[1083] Eatough, D. J., and L. D. Hansen, Bis-hydroxymethyl sulfone: a major product of atmospheric reactions of SO_2(g), *Sci. Total Environ.*, **36**, 319-328, 1984.

[1084] Karasek, F. W., and M. C. Parsons, *Analysis of Houston Aerosol Samples by GC/MS Methods*, Rpt. EPA-600/2-80-071, Research Triangle Park, NC: US Environ. Prot. Agency, 1980 (NTIS Document PB 80-196785).

[1085] Guiang, S. F. III, S. V. Krupa, and G. C. Pratt, Measurements of S(IV) and organic anions in Minnesota rain, *Atmos. Environ.*, **18**, 1677-1682, 1984.

[1086] Sparacino, C. M., *GC/MS Analysis of Ambient Air Aerosols in the Houston, Texas Area*, Rpt. EPA-600/2-80-194, Research Triangle Park, NC: US Environ. Prot. Agency, 1980 (NTIS Document PB 81-126377).

[1087] Ter Haar, G. L., M. E. Griffing, M. Brandt, D. G. Oberding, and M. Kapron, Methylcyclopentadienyl manganese tricarbonyl as an antiknock: composition and fate of manganese exhaust products, *J. Air. Poll. Contr. Assoc.*, **25**, 858-860, 1975.

[1088] Price, J. A., and K. J. Saunders, Determination of airborne methyl *tert*-butyl ether in gasoline atmospheres, *Analyst*, **109**, 829-834, 1984.

[1089] Pereira, W. E., C. E. Rostad, H. E. Taylor, and J. M. Klein, Characterization of organic contaminants in environmental samples associated with Mount St. Helens 1980 volcanic eruption, *Environ. Sci. Technol*, **16**, 387-396, 1982.

[1090] Stedman, D. H., M. Z. Creech, P. L. Cloke, S. E. Kealer, and M. Gardner, Formation of CS_2 and OCS from decomposition of metal sulfides, *Geophys. Res. Lett.*, **11**, 858-860, 1984.

[1091] Rasmussen, R. A., and L. E. Rasmussen, Trace gases of volcanic origin, Paper presented at 26th Ann. Mtg., Pacific Northwest Section, Amer. Geophys. Union, Bend, Oregon, Sept. 17, 1979.

[1092] Thompson, C. R., G. Kats, and R. W. Lennox, Phytotoxicity of air pollutants formed by high explosive production, *Environ. Sci. Technol.*, **13**, 1263-1268, 1979.

[1093] Carpenter, B. H., R. Liepins, J. Sickles, II, H. L. Hamilton, D. W. Van Osdell, G. F. Weant, III, and L. M. Worsham, *Specific Air Pollutants from Munitions Processing and Their Atmospheric Behavior*, Contract Rept. DAMD17-76-C-6067, Research Triangle Inst., Research Triangle Park, NC, 1978 (NTIS Document AD-A060155).

[1094] Miner, S., *Air Pollution Aspects of Ammonia*, Litton System, Inc., Bethesda, MD, 1969 (NTIS Document PB 188082).

[1095] Pellizzari, E. D., Analysis for organic vapor emissions near industrial and chemical waste disposal sites, *Environ. Sci. Technol.*, **16**, 781-785, 1982.

[1096] Davis, B. L., Quantitative analysis of crystalline and amorphous airborne particulates in the Provo-Orem vicinity, Utah, *Atmos. Environ.*, **15**, 613-618, 1981.

[1097] Wood, M. B., An application of gas chromatography to measure concentrations of ethane, propane, and ethylene found in interstitial soil gases, *J. Chromatogr. Sci.*, **18**, 307-310, 1980.

[1098] Lynch, J. M., Ethylene in soil, *Nature*, **256**, 576-577, 1975.

[1099] Sexton, K., and H. Westberg, Nonmethane hydrocarbon composition of urban and rural atmospheres, Paper 81-47.3, Annual Mtg., Air Pollut. Contr. Assoc., Philadelphia, PA, June, 1981.

[1100] Schuetzle, D., T. L. Riley, T. J. Prater, I. Salmeen, and T. M. Harvey, The identification of mutagenic chemical species in air particulate samples, in *Environ. Sci.* **7**; *Anal. Tech. Environ. Chem.* **2**, 259-280, New York: Pergamon, 1982.

[1101] Dana, M. T., Dissolved sulfur dioxide in Eastern US precipitation: implications for the linear emissions-deposition hypothesis, *EOS-Trans. AGU*, **65**, 177, 1984.

[1102] Bowyer, J. M., *Rocket Motor Exhaust Products Generated by the Space Shuttle Vehicle During Its Launch Phase (1976 Design Data)*, JPL Pub. 77-9, Jet Propulsion Lab., Pasadena, CA, 1977.

[1103] Wauters, E., Gaschromatografische analyse van de organische fractie van atmosferische aerosolen, *Chem. Mag.*, **4**(4), 24-27, 1978 (CA, **90**, 11475t, 1979).

[1104] Adams, D. F., Sulfur gas emissions from flue gas desulfurization sludge ponds, *J. Air Poll. Contr. Assoc.*, **29**, 963-968, 1979.

[1105] Tanaka, S., Y. Odagiri, T. Kato, and Y. Hashimoto, The loss of Cl as HCl from sea salt particles by air pollutants in the marine atmosphere, *E. Nippon Kagaku Kaishi*, **12**, 1946-1952, 1982 (CA, **98**, 77377, 1983).

[1106] Quann, R. J., M. Neville, M. Janghorbani, C. A. Mims, and A. F. Sarofim, Mineral matter and trace-element vaporization in a laboratory-pulverized coal combustion system, *Environ. Sci. Technol.*, **16**, 776-781, 1982.

[1107] Joseph, K. T., and R. F. Browner, Analysis of particulate combustion products of polyurethane foam by high performance liquid chromatography and gas chromatography-mass spectrometry, *Anal. Chem.*, **52**, 1083-1085, 1980.

[1108] Wallace, L., S. Bromberg, E. Pellizzari, T. Hartwell, H. Zelon, and L. Sheldon, Plan and preliminary results of the U.S. Environmental Protection Agency's indoor air monitoring program: 1982, *Proc. 3rd Int'l Conf. on Indoor Air Quality and Climate*, **1**, 173-178, 1984.

[1109] Weschler, C. J., and K. L. Fong, Characterization of organic species associated with indoor aerosol particles, *Proc. 3rd Int'l. Conf. on Indoor Air Quality and Climate*, **2**, 203-208, 1984.

[1110] DeBortoli, M., H. Knöppel, E. Pecchio, A. Peil, L. Rogora, H. Schauenberg, H. Schlitt, and H. Vissers, Integrating 'real life' measurements of organic pollution in indoor and outdoor air of homes in Northern Italy, *Proc. 3rd Int'l. Conf. on Indoor Air Quality and Climate*, **4**, 21-26, 1984.

[1111] Girman, J. R., A. T. Hodgson, A. S. Newton, and A. W. Winkes, Volatile organic emissions from adhesive with indoor applications, *Proc. 3rd Int'l. Conf. on Indoor Air Quality and Climate*, **4**, 271-276, 1984.

[1112] Monteith, D. K., T. H. Stock, and W. E. Seifert, Jr., Sources and characterization of organic species associated with indoor aerosol particles, *Proc. 3rd Int'l. Conf. on Indoor Air Quality and Climate*, **4**, 285-290, 1984.

[1113] Pellizzari, E. D., L. S. Sheldon, C. M. Sparacino, J. T. Bursey, L. Wallace, and S. Bromberg, Volatile organic levels in indoor air, *Proc. 3rd Int'l. Conf. on Indoor Air Quality and Climate*, **4**, 303-308, 1984.

[1114] Neftel, A., P. Jacob, and D. Klockow, Measurements of hydrogen peroxide in polar ice samples, *Nature*, **311**, 43-45, 1984.

[1115] Schöne, E.,
Berichte, **7**, 1693-1708, 1874.

[1116] Harrison, R. M., and W. T. Sturges, Physico-chemical speciation and transformation reactions of particulate atmospheric nitrogen and sulphur compounds, *Atmos. Environ.*, **18**, 1829-1833, 1984.

[1117] Voigt, R., K. H. Lieser, and B. Baumgartner, Identification of chemical species in air dust by powder diffractometry, *Naturwiss.*, **71**, 377-378, 1984.

[1118] Glotfelty, D. E., and J. H. Caro, Introduction, transport, and fate of persistent pesticides in the atmosphere, in *Removal of Trace Contaminants from the Air*, V. R. Deitz, Ed., ACS Symp Ser. 17, Amer. Chem. Soc., Washington, D.C., 1975.

[1119] Mackinnon, I. D. R., and F. J. M. Rietmeijer, Bismuth in interplanetary dust, *Nature*, **311**, 135-138, 1984.

[1120] Kawamura, K., and I. R. Kaplan, Capillary gas chromatography determination of volatile organic acids in rain and fog samples, *Anal. Chem.*, **56**, 1616-1620, 1984.

[1121] Bohren, C. F., and J. J. Olivero, Evidence for haematite particles at 60 km altitude, *Nature*, **310**, 216-218, 1984.

[1122] Eimutis, E. C., and R. P. Quill, *Source Assessment: Noncriteria Pollutant Emissions*, Rpt. EPA-600/2-77-107e, Environ. Prot. Agency, Research Triangle Park, NC, 1977 (NTIS Document PB 270550).

[1123] Wang, T. C., A study of bioeffluents in a college classroom, *ASHRAE Trans.*, **81** (Pt. 1), 32-44, 1975.

[1124] Schmidt, H. E., C. D. Hollowell, R. R. Miksch, and A. S. Newton, *Trace Organics in Offices*, LBL Report 11378, Lawrence Berkeley Lab., Berkeley, CA, 1980.

[1125] Giam, C. S., E. Atlas, and K. Sullivan, Widespread occurrence of polyhalogenated anisoles and related compounds in the marine atmosphere, *SEAREX Newsletter*, **7**(2), 10-12, 1984.

[1126] Christofferson, R., and P. R. Buseck, Epsilon carbide: a low-temperature component of interplanetary dust particles, *Science*, **222**, 1327-1329, 1983.

[1127] Bradley, J. P., D. E. Brownlee, and P. Fraundorf, Carbon compounds in interplanetary dust: evidence for formation by heterogeneous catalysis, *Science*, **223**, 56-58, 1984.

[1128] Fraundorf, P., Inter-planetary dust in the transmission electron microscope-diverse materials from the early solar system, *Geochim. Cosmochim. Acta*, **45**, 915, 1981.

[1129] Hesse, C. S., *Investigation of Rainwater for the Presence of Asbestos*, Report W79-00437, OWRT-R-071-ILL(4), Univ. Illinois Med. Ctr., Chicago, 1977 (NTIS Document PB 288084).

[1130] Biggins, P. D. E., and R. M. Harrison, Atmospheric chemistry of automotive lead, *Environ. Sci. Technol.*, **13**, 558-565, 1979.

[1131] Sundstroem, G., Toxaphene in the Swedish environment-delivery by atmospheric fallout, *Miljoevardsserien*, **1**, 331-336, 1981.

[1132] Ter Haar, G. L., and M. A. Bayard, Composition of airborne lead particles, *Nature*, **232**, 553-554, 1971.

[1133] Habibi, K., Characterization of particulate matter in vehicle exhaust, *Environ. Sci. Technol.*, **7**, 223-234, 1973.

[1134] Hirschler, D. A., L. F. Gilbert, F. W. Lamb, and L. M. Niebylski, Particulate lead compounds in automobile exhaust gas, *Ind. Eng. Chem.*, **49**, 1131-1142, 1957.

[1135] Dutta, P. K., D. C. Rigano, R. A. Hofstader, E. Denoyer, D. F. S. Natusch, and F. C. Adams, Laser microprobe mass analysis of refinery source emissions and ambient samples, *Anal. Chem.*, **56**, 302-304, 1984.

[1136] Tomkins, B. A., R. S. Brazell, M. E. Roth, and V. H. Ostrum, Isolation of mononitrated polycyclic aromatic hydrocarbons in particulate matter by liquid chromatography and determination by gas chromatography with the thermal energy analyzer, *Anal. Chem.*, **56**, 781-786, 1984.

[1137] Hard, T. M., R. J. O'Brien, C. Y. Chan, and A. A. Mehrabzadeh, Tropospheric free radical determination by FAGE, *Environ. Sci. Technol.*, **18**, 768-777, 1984.

[1138] Grochmalicka-Mikolajczyk, J., J. R. Ochocka, and J. Lulek, Determination of polynuclear aromatic hydrocarbons accumulated in snow cover, *Bromatol. Chem. Toksykol.*, **15**, 67-69, 1982 (CA, **97**, 188049c, 1982).

[1139] Thomas, W., Concentrations and inputs of PAH, chlorinated hydrocarbons and trace metals in precipitation - comparison of suburb and rural stations, *Dtsch. Gewaesserkd. Mitt.*, **25**, 120-129, 1981 (CA, **96**, 109911f, 1982).

[1140] Kobayashi, R., and Y. Hashimoto, A study on emission sources of selinium in the atmosphere, *Taiki Osen Gakkaishi*, **17**, 96-101, 1982 (CA, **97**, 97489, 1982).

[1141] Khatamov, S. H., R. Khamidova, and A. A. Kist, Evaluation of snow pollution around nonferrous metallurgy plants by neutron activation, *Zavod. Lab.*, **46**, 417-419, 1980 (CA, **93**, 224864, 1980).

[1142] Konig, J., E. Balfanz, W. Funcke, and T. Romanowski, Quinone- and ketone-derivatives of PAH in particulate matter from ambient air, in *Polynuclear Aromatic Hydrocarbons: Formation, Metabolism and Measurement*, M. Cooke and A. J. Dennis, Eds., Columbus, OH: Battelle Press, pp. 711-720, 1983.

[1143] Nielsen, T., B. Seitz, A. M. Hansen, K. Keiding, and B. Westerberg, The presence of nitro-PAH in samples of airborne particulate matter, in *Polynuclear Aromatic Hydrocarbons: Formation, Metabolism and Measurement*, M. Cooke and A. J. Dennis, Eds., Columbus, OH: Battelle Press, pp. 961-970, 1983.

[1144] Salmeen, I., R. A. Gorse, Jr., T. Riley, and D. Schuetzle, Identification and quantification of direct acting Ames assay mutagens in diesel particulate extracts, in *Polynuclear Aromatic Hydrocarbons: Formation, Metabolism, and Measurement*, M. Cooke and A. J. Dennis, Eds., Columbus, OH: Battelle Press, pp. 1057-1066, 1983.

[1145] Tomkins, B. A., R. R. Reagan, M. P. Maskarinec, S. H. Harmon, W. H. Griest, and J. E. Caton, Analytical chemistry of polycyclic aromatic hydrocarbons present in coal-fired power plant fly ash, in *Polynuclear Aromatic Hydrocarbons: Formation, Metabolism, and Measurement*, M. Cooke and A. J. Dennis, Eds., Columbus, OH: Battelle Press, pp. 1173-1187, 1983.

[1146] Satsumabayashi, H., K. Sasaki, and Y. Nakazawa, Measurement of lower fatty acids in air from a factory of fatty acids, *Nagano-Ken Eisei Kogai Kenkyusho Kenkyu Hokoku*, **4**, 11-14, 1981 (CA, **97**, 60107c, 1982).

[1147] Horiba, H. and N. Yamanaka, Relation between odorant strength and odorant concentration for complex odors in pig farms and chemical plants, *Akushu no Kenkyu*, **9**, 33-40, 1980 (CA, **94**, 144554x, 1981).

[1148] Serth, R. W., and T. W. Hughes, Polycyclic organic matter (POM) and trace element contents of carbon black vent gas, *Environ. Sci. Technol*, **14**, 298-301, 1980.

[1149] Newell, R. E., and A. Deepak, Eds., *Mount St. Helens Eruption of 1980*, NASA SP-458, Washington, D.C.: Nat. Aeronautics and Space Admin., 1982.

[1150] Zellner, R., The reactive removal of anthropogenic emissions from the atmosphere, in *Proc. Int'l. Workshop on Test Methods and Assessment Procedures for the Determination of the Photochemical Degradation Behavior of Chemical Substances*, W. Funcke, J. König, A. W. Klein, W. Stöber, and F. Schmidt-Bleck, Eds., Fraunhofer-Gesellschaft, Inst. für Toxikologie und Aerosolforschung, Münster, FRG, 1980.

[1151] Okuno, T., Control and chemical assessment of leather industry wastes. Air pollutants, *Akushu no Kenkyu*, **10**, 12-29, 1981 (CA, **95**, 208578p, 1981).

[1152] Sugita, M., H. Ando, S. Okada, H. Hattori, T. Okuno, M. Tuji, T. Yamasaki, and Y. Shintani, Studies on offensive odor in leather factory. 1. Offensive odor substances of tanning process, *Hikaku Kagaku*, **24**, 29-35, 1978 (CA, **90**, 141630c, 1979).

[1153] Eatough, D. J., *Determination of Sulfur Speciation in Industrial Aerosols in an Sulfur Dioxide Rich Environment*, Report DOE/EV/10405-10, Brigham Young Univ., Provo, UT, 1983.

[1154] Committee on Aldehydes, *Formaldehyde and Other Aldehydes*, Washington, D.C.: National Academy Press, 1981.

[1155] Boubel, R. W., E. F. Darley, and E. A. Schuck, Emissions from burning grass stubble and straw, *J. Air Poll. Contr. Assoc.*, **19**, 497-500, 1969.

[1156] Howie, S. J. and E. W. Koesters, *Ambient Acrylonitrile Levels Near Major Acrylonitrile Production and Use Facilities*, PEDCO Environmental, Inc., Cincinnati, OH, 1983 (NTIS Document PB 83-196154).

[1157] Feairheller, W. R., Measurement of the volatile organic compound emissions from process streams and fugative emission sources by on-site gas chromatography, Paper ENVT-16, 13th Middle Atlantic Reg. Mtg., Amer. Chem. Soc., Monmouth College, NJ, 1979.

[1158] Furutani, C., T. Hayata, and Y. Sadayoshi, Release of smelly substances from fermentation plants and their impacts on environments, *Akushu no Kenkyu*, **7**, 24-34, 1979 (CA, **91**, 215859b, 1979).

[1159] Griest, W. H., C. E. Higgins, R. W. Holmberg, J. H. Moneyhun, J. E. Caton, J. S. Wike, and R. R. Reagan, Characterization of vapor- and particulate-phase organics from ambient air sampling at the Kosovo gasifier, in *Energy and Environmental Chemistry*, L. H. Keith, Ed., Ann Arbor, MI: Ann Arbor Sci. Pub., pp. 395-410, 1982.

[1160] Furuya, C., T. Hayada, Y. Kitakawa, Y. Sadayoshi, and H. Tanabe, Environmental pollution by odor in petrochemical plants. 1. Formation of offensive odor and its environmental effect, *Akushu no Kenkyu*, **10**, 34-41, 1982 (CA, **97**, 132548x, 1982).

[1161] Olenik, T. J., Domestic sewage and refuse odor control, in *Industrial Odor Technology Assessment*, P. N. Cheremisinoff and R. A. Young, Eds., Ann Arbor, MI: Ann Arbor Sci. Pub., pp. 117-146, 1975.

[1162] Andersen, K. A., D. T. Bernstein, R. L. Caret, and L. J. Romanczyk, Chemical constituents of the defensive secretion of the striped skunk (*Mephitis mephitis*), *Tetrahedron*, **38**, 1965-1970, 1982.

[1163] Snow, R. H., Investigation of odor control in the rendering industry, in *Industrial Odor Technology Assessment*, P. N. Cheremisinoff and R. A. Young, Eds., Ann Arbor, MI: Ann Arbor Sci. Pub., pp. 147-174, 1975.

[1164] Habib, Y. H., Odor emission sources in the chemical and petroleum industries, in *Industrial Odor Technology Assessment*, P. N. Cheremisinoff and R. A. Young, Eds., Ann Arbor, MI: Ann Arbor Sci. Pub., pp. 189-201, 1975.

[1165] Clemo, G. R., Some phenolic constituents of coal soot, *Tetrahedron*, **26**, 5845-5846, 1970.

[1166] G. R. Clemo, Some aromatic basic constituents of coal soot, *Tetrahedron*, **29**, 3987-3990, 1973.

[1167] Hauck, G., and F. Arnold, Improved positive-ion composition measurements in the upper troposphere and lower stratosphere and the detection of acetone, *Nature*, **311**, 547-550, 1984.

[1168] Arijs, E., D. Nevejans, J. Ingels, and P. Frederick, Negative ion composition and sulfuric acid vapour in the upper stratosphere, *Planet. Space Sci.*, **31**, 1459-1464, 1983.

[1169] Seader, J. D., I. N. Einhorn, W. O. Drake, and C. M. Mihlfeith, Analysis of volatile combustion products and a study of their toxicological effects, *Polymer Eng. Sci.*, **12**, 125-133, 1972.

[1170] Westberg, H. H., Biogenic hydrocarbon measurements, in *Atmospheric Biogenic Hydrocarbons, V.2*, J. J. Bufalini and R. R. Arnts, Eds., Ann Arbor, MI: Ann Arbor Sci. Pubs., pp. 25-49, 1981.

[1171] Khalil, M. A. K., S. A. Edgerton, and R. A. Rasmussen, A gaseous tracer model for air pollution from residential wood burning, *Environ. Sci. Technol.*, **17**, 555-559, 1983.

[1172] Penkett, S. A., N. J. D. Prosser, R. A. Rasmussen, and M. A. K. Khalil, Atmospheric measurements of CF_4 and other fluorocarbons containing the CF_3 group, *J. Geophys. Res.*, **86**, 5172-5178, 1981.

[1173] Singh, H. B., L. J. Salas, and R. E. Stiles, Methyl halides in and over the Eastern Pacific (40°N-32°S), *J. Geophys. Res.*, **88**, 3684-3690, 1983.

[1174] Seiler, W., and H. Giehl, Influence of plants on the atmospheric carbon monoxide, *Geophys. Res. Lett.*, **4**, 329-332, 1977.

[1175] Conrad, R., W. Seiler, G. Bunse, and H. Giehl, Carbon monoxide in seawater (Atlantic Ocean), *J. Geophys. Res.*, **87**, 8839-8852, 1982.

[1176] Conrad, R., and W. Seiler, Arid soils as a source of atmospheric carbon monoxide, *Geophys. Res. Lett.*, **9**, 1353-1356, 1982.

[1177] Saint-Jalm, Y., and P. Moree-Testa, Study of nitrogen-containing compounds in cigarette smoke by gas chromatography-mass spectrometry, *J. Chromatogr.*, **198**, 188-192, 1980.

[1178] Friedli, H., E. Moor, H. Oeschger, U. Siegenthaler, and B. Stauffer, $^{13}C/^{12}C$ ratios in CO_2 extracted from Antarctic ice, *Geophys. Res. Lett.*, **11**, 1145-1148, 1984.

[1179] Chamberlain, W. J., M. E. Snook, J. L. Baker, and O. T. Chortyk, Gel permeation chromatography of oxygenated components of cigarette smoke condensate, *Anal. Chim. Acta*, **111**, 235-241, 1979.

[1180] Nielsen, T., B. Seitz, and T. Ramdahl, Occurrence of nitro-PAH in the atmosphere in a rural area, *Atmos. Environ.*, **18**, 2159-2165, 1984.

[1181] Grosjean, D., Distribution of atmospheric nitrogenous pollutants at a Los Angeles area smog receptor site, *Environ. Sci. Technol.*, **17** 13-19, 1983.

[1182] Noxon, J. F., Tropospheric NO_2, *J. Geophys. Res.*, **83**, 3051-3057, 1978.

[1183] Maroulis, P. J., A. L. Torres, A. B. Goldberg, and A. R. Bandy, Atmospheric SO_2 measurements on Project GAMETAG, *J. Geophys. Res.*, **85**, 7345-7349, 1980.

[1184] Weschler, C. J., Identification of selected organics in the Arctic aerosol, *Atmos. Environ.*, **15**, 1365-1369, 1981.
[1185] Bandy, A. R., Private communication, 1984.
[1186] Tanaka, S., M. Darzi, and J. W. Winchester, Sulfur and associated elements and acidity in continental and marine rain from North Florida, *J. Geophys. Res.*, **85**, 4519-4526, 1980.
[1187] Struempler, A. W., Trace metals in rain and snow during 1973 at Chadron, Nebraska, *Atmos. Environ.*, **10**, 33-37, 1976.
[1188] Arimoto, R., R. A. Duce, B. J. Ray, and C. K. Unni, Atmospheric trace elements at Enewetak Atoll: 2. Transport to the ocean by wet and dry deposition, *J. Geophys. Res.*, **90**, 2391-2408, 1985.
[1189] Sutton, D. C., R. S. Morse, P. A. Legotte, and W. C. Rosa, Determination of ten selected trace metals in precipitation samples using atomic absorption and direct current plasma emission spectrometry, EML-356, US Dept. of Energy, New York, NY, 1979.
[1190] Luten, J. B., The determination of some trace elements in rainwater by neutron activation analysis, *J. Radioanal. Chem.*, **37**, 897-904, 1977.
[1191] Chan, K. C., B. L. Cohen, J. O. Frohliger, and L. Shabason, Pittsburgh rainwater analysis by PIXE, *Tellus*, **28**, 24-30, 1976.
[1192] Hamilton, E. P., and A. Chatt, Determination of trace elements in atmospheric wet precipitation by instrumental neutron activation analysis, *J. Radioanal. Chem.*, **71**, 29-45, 1979.
[1193] Landsberger, S., R. E. Jervis, S. Aufreiter, and J. C. Van Loon, The determination of heavy metals (aluminum, manganese, iron, nickel, copper, zinc, cadmium and lead) in urban snow using an atomic absorption graphite furnace, *Chemosphere*, **11**, 237-247, 1982.
[1194] Li, Y. H., Geochemical cycles of elements and human perturbation, *Geochim, Cosmochim. Acta*, **45**, 2073-2084, 1981.
[1195] Liljestrand, H. M., Chemical composition of acid precipitation in Texas, Report TENRAC/EDF-100, Center for Energy Studies, Univ. of Texas, Austin, TX, 1983.
[1196] Krey, P. W., R. Leifer, W. K. Benson, L. A. Dietz, H. C. Hendrikson, and J. L. Coluzza, Atmospheric burnup of the Cosmos-954 reactor, *Science*, **205**, 583-585, 1979.
[1197] Middleton, A. P., Analysis of airborne dust samples for chrysotile by x-ray diffraction: interference by non-asbestiform serpentine, *Ann. Occup. Hyg.*, **25**, 443-447, 1982.
[1198] Norton, R. B., Private communication, 1985.
[1199] Grosjean, D., and J. D. Nies, Sampling and ion chromatographic analysis of pyruvic acid and methane sulfonic acid in air, *Anal. Lett.*, **17**, (A2), 89-96, 1984.
[1200] Ilnitsky, A. P., V. S. Mischenko, and L. M. Shabad, New data on volcanoes as natural sources of carcinogenic substances, *Cancer Lett.*, **3**, 227-230, 1977.
[1201] Dickinson, R., Water (hydrological cycle), in *Global Tropospheric Chemistry: A Plan for Action*, Global Tropospheric Chemistry Panel, National Research Council, Washington, D.C., pp. 106-108, 1984.
[1202] Bach, W., Global air pollution and climatic change, *Rev. Geophys. Space Phys.*, **14**, 429-474, 1976.
[1203] Pruppacher, H. R., and J. D. Klett, *Microphysics of Clouds and Precipitation*, Dordrecht: D. Reidel, 714 pp., 1978.
[1204] Peterson, J. T., *Calculated Actinic Fluxes (290-700 nm) for Air Pollution Photochemistry Applications*, Rep. EPA-600/4-76-025, Environ. Prot. Agency, Research Triangle Park, NC, 1976.

[1205] Leighton, P. A., *Photochemistry of Air Pollution*, New York: Academic Press, pp. 29ff, 1961.
[1206] Fabian, P., R. Borchers, G. Flentje, W. A. Matthews, W. Seiler, H. Giehl, K. Bunse, F. Müller, U. Schmidt, A. Volz, A. Khedim, and F. J. Johnen, The vertical distribution of stable trace gases at mid-latitudes, *J. Geophys. Res.*, **86**, 5179-5184, 1981.
[1207] Rudolph, J., D. H. Ehhalt, U. Schmidt, and A. Khedim, Vertical distributions of some C_2–C_5 hydrocarbons in the nonurban troposphere, *2nd Symp. on the Composition of the Nonurban Troposphere*, Amer. Meteorological Society, Boston, pp. 56-59, 1982.
[1208] Whitby, K. T., Aerosol formation in urban plumes, *Ann. NY Acad. Sci.*, **338**, 258-275, 1980.
[1209] Jaenicke, R., Natural aerosols, *Ann. NY Acad. Sci.*, **338**, 317-329, 1980.
[1210] Gill, P. S., T. E. Graedel, and C. J. Weschler, Organic films on atmospheric aerosol particles, fog droplets, cloud droplets, raindrops, and snowflakes, *Rev. Geophys. Space Phys.*, **21**, 903-920, 1983.
[1211] Cahn, R. S., and O. C. Dermer, *Introduction to Chemical Nomenclature*, Fifth Ed., London: Butterworths, 1979.
[1212] Willaert, G. A., P. J. Dirinck, H. L. DePooter, and N. M. Schamp, Objective measurement of aroma quality of Golden Delicious apples as a function of controlled-atmosphere storage time, *J. Agric. Food Chem.*, **31**, 809-813, 1983.
[1213] Calvert, J. G., and J. N. Pitts, Jr., *Photochemistry*, New York: John Wiley, 1966.
[1214] Benson, S. W., *The Foundations of Chemical Kinetics*, New York: McGraw-Hill, 1960.
[1215] Benson, S. W., *Thermochemical Kinetics*, 2nd. Ed., New York: John Wiley, 1976.
[1216] Heicklen, J., *Atmospheric Chemistry*, New York: Academic Press, 1976.
[1217] Seinfeld, J. H., *Air Pollution: Physical and Chemical Fundamentals*, New York: McGraw-Hill, 1975.
[1218] Péczely, G., Precipitation patterns of the earth, *Acta Climatologica*, **12**, 3-17, 1973.
[1219] Likens, G. E., R. F. Wright, J. N. Galloway, and T. J. Butler, Acid rain, *Scient. Amer.*, **241**(4), 43-51, 1979.
[1220] Wofsy, S. C., and M. B. McElroy, On vertical mixing in the upper stratosphere and lower mesosphere, *J. Geophys. Res.*, **78**, 2619-2624, 1973.
[1221] Atkinson, R., K. R. Darnall, A. C. Lloyd, A. M. Winer, and J. N. Pitts, Jr., Kinetics and mechanisms of the reaction of the hydroxyl radical with organic compounds in the gas phase, *Adv. Photochem.*, **11**, 375-488, 1979.
[1222] Perry, R. A., R. A. Atkinson, and J. N. Pitts, Jr., Kinetics and mechanism of the gas phase reaction of OH radicals with methoxybenzene and *o*-cresol over the temperature range 299-435K, *J. Phys. Chem.*, **81**, 1607-1611, 1977.
[1223] Atkinson, R., C. N. Plum, W. P. L. Carter, A. M. Winer, and J. N. Pitts, Jr., Rate constants for the gas-phase reactions of nitrate radicals with a series of organics in air at 298±1K, *J. Phys. Chem.*, **88**, 1210-1215, 1984.
[1224] Atkinson, R., and S. M. Aschmann, Rate constants for the reactions of O_3 and OH radicals with a series of alkynes, *Int. J. Chem. Kinet.*, **16**, 259-268, 1984.
[1225] Gorse, R. A., and D. H. Volman, Photochemistry of the gaseous hydrogen peroxide-carbon monoxide system. II. Rate constants for hydroxyl radical reactions with hydrocarbons and for hydrogen atom reactions with hydrogen peroxide, *J. Photochem.*, **3**, 115-122, 1974.
[1226] Atkinson, R., S. M. Aschmann, A. M. Winer, and J. N. Pitts, Jr., Rate constants for the reaction of OH radicals with a series of alkanes and alkenes at 299±2K, *Int. J. Chem. Kinet.*, **14**, 507-516, 1982.

[1227] Atkinson, R., S. M. Aschmann, and W. P. L. Carter, Effects of ring strain on gas-phase rate constants. 2. OH radical reactions with cycloalkenes, *Int. J. Chem. Kinet.*, **15**, 1161-1177, 1983.

[1228] Takagi, H., N. Washida, H. Bandow, H. Akimoto, and M. Okuda, Photooxidation of C_5—C_7 cycloalkanes in the NO-H_2O-air system, *J. Phys. Chem.*, **85**, 2701-2705, 1981.

[1229] Atkinson, R., S. M. Aschmann, A. M. Winer, and J. N. Pitts, Jr., Kinetics of the gas-phase reactions of NO_3 radicals with a series of dialkenes, cycloalkenes, and monoterpenes at 295 ± 1K, *Environ. Sci. Technol.*, **18**, 370-375, 1984.

[1230] Kleindienst, T. E., G. W. Harris, and J. N. Pitts, Jr., Rates and temperature dependences of the reaction of OH with isoprene, its oxidation products, and selected terpenes, *Environ. Sci. Technol.*, **16**, 844-846, 1982.

[1231] Grosjean, D., Atmospheric reactions of ortho cresol: gas phase and aerosol products, *Atmos. Environ.*, **18**, 1641-1652, 1984.

[1232] Atkinson, R., S. M. Aschmann, and W. P. L. Carter, Rate constants for the gas-phase reactions of OH radicals with a series of bi- and tricycloalkanes at 299 ± 2K: effects of ring strain, *Int. J. Chem. Kinet.*, **15**, 37-50, 1983.

[1233] Atkinson, R., S. M. Aschmann, and J. N. Pitts, Jr., Kinetics of the reactions of naphthalene and biphenyl with OH radicals and with O_3 at 294 ± 1K, *Environ. Sci. Technol.*, **18**, 110-113, 1984.

[1234] Carter, W. P. L., A. M. Winer, and J. N. Pitts, Jr., Major atmospheric sink for phenol and the cresols. Reaction with the nitrate radical, *Environ. Sci. Technol.*, **15**, 829-831, 1981.

[1235] Atkinson, R., S. M. Aschmann, W. P. L. Carter, and J. N. Pitts, Jr., Rate constants for the gas-phase reaction of OH radicals with a series of ketones at 299 ± 2K, *Int. J. Chem. Kinet.*, **14**, 839-847, 1982.

[1236] Atkinson, R., S. M. Aschmann, A. M. Winer, and J. N. Pitts, Jr., Rate constants for the gas-phase reactions of O_3 with a series of carbonyls at 296K, *Int. J. Chem. Kinet.*, **13**, 1133-1142, 1981.

[1237] Kerr, J. A., and D. W. Sheppard, Kinetics of the reactions of hydroxyl radicals with aldehydes studied under atmospheric conditions, *Environ. Sci. Technol.*, **15**, 960-963, 1981.

[1238] Plum, C. N., E. Sanhueza, R. Atkinson, W. P. L. Carter, and J. N. Pitts, Jr., OH radical rate constants and photolysis rates of α-dicarbonyls, *Environ. Sci. Technol.*, **17**, 479-484, 1983.

[1239] Atkinson, R., S. M. Aschmann, and J. N. Pitts, Jr., Kinetics of the gas-phase reactions of OH radicals with a series of α, β-unsaturated carbonyls at 299 ± 2K, *Int. J. Chem. Kinet.*, **15**, 75-81, 1983.

[1240] Wine, P. H., and R. J. Thompson, Kinetics of OH reactions with furan, thiophene, and tetrahydrothiophene, *Int. J. Chem. Kinet.*, **16**, 867-878, 1984.

[1241] Harris, G. W., T. E. Kleindienst, and J. N. Pitts, Jr., Rate constants for the reaction of OH radicals with CH_3CN, C_2H_5CN and CH_2=CH-CN in the temperature range 298-424K, *Chem. Phys. Lett.*, **80**, 479-483, 1981.

[1242] Harris, G. W., and J. N. Pitts, Jr., Rates of reaction of hydroxyl radicals with 2-(dimethylamino) ethanol and 2-amino-2-methyl-1-propanol in the gas phase at 300 ± 2K, *Environ. Sci. Technol.*, **17**, 50-51, 1983.

[1243] Dod, R. L., L. A. Gundel, W. H. Benner, and T. Novakov, Non-ammonium reduced nitrogen species in atmospheric aerosol particles, *Sci. Total Environ.*, **36**, 277-282, 1984.

[1244] Atkinson, R., R. A. Perry, and J. N. Pitts, Jr., Rate constants for the reaction of the OH radical with $(CH_3)_2NH$, $(CH_3)_3N$, and $C_2H_5NH_2$ over the temperature range 298-426°K, *J. Chem. Phys.*, **68**, 1850-1853, 1978.

[1245] Pitts, J. N., Jr., D. Grosjean, K. A. Van Cauwenberghe, and J. P. Schmid, Photooxidation of aliphatic amines under simulated atmospheric conditions: formation of nitrosamines, nitramines, amides, and photochemical oxidant, *Environ. Sci. Technol.*, **12**, 946-953, 1978.

[1246] Harris, G. W., R. Atkinson, and J. N. Pitts, Jr., Kinetics of the reactions of the OH radical with hydrazine and methylhydrazine, *J. Phys. Chem.*, **83**, 2557-2559, 1979.

[1247] Senum, G. I., Y.-N. Lee, and J. S. Gaffney, Ultraviolet absorption spectrum of peroxyacetyl/nitrate and peroxypropionyl nitrate, *J. Phys. Chem.*, **88**, 1269-1270, 1984.

[1248] Atkinson, R., S. M. Aschmann, W. P. L. Carter, A. M. Winer, and J. N. Pitts, Jr., Alkyl nitrate formation from the NO_x-air photooxidations of C_2-C_8 *n*-alkanes, *J. Phys. Chem.*, **86**, 4563-4569, 1982.

[1249] Atkinson, R., S. M. Aschmann, W. P. L. Carter, and A. M. Winer, Kinetics of the gas-phase reactions of OH radicals with alkyl nitrates at 299 ± 2K, *Int. J. Chem. Kinet.*, **14**, 919-926, 1982.

[1250] Taylor, W. D., T. D. Allston, M. J. Moscato, G. B. Fazekas, R. Kozlowski, and G. A. Takacs, Atmospheric photodissociation lifetimes for nitromethane, methyl nitrite, and methyl nitrate, *Int. J. Chem. Kinet.*, **12**, 231-240, 1980.

[1251] Atkinson, R., S. M. Aschmann, A. M. Winer, and W. P. L. Carter, Rate constants for the gas phase reactions of OH radicals and O_3 with pyrrole at 295 ± 1K and atmospheric pressure, *Atmos. Environ.*, **18**, 2105-2107, 1984.

[1252] Steenken, S., and P. O'Neill, Reaction of OH radicals with 2- and 4-pyridones in aqueous solution. An electron spin resonance and pulse radiolysis study, *J. Phys. Chem.*, **83**, 2407-2412, 1979.

[1253] Hatekayama, S., and H. Akimoto, Reactions of OH radicals with methanethiol, dimethyl sulfide, and dimethyl disulfide in air, *J. Phys. Chem.*, **87**, 2387-2395, 1983.

[1254] Grosjean, D., Photooxidation of methyl sulfide, ethyl sulfide, and methanethiol, *Environ. Sci. Technol.*, **18**, 460-468, 1984.

[1255] Atkinson, R., and J. N. Pitts, Jr., Absolute rate constants for the reactions of $O(^3P)$ atoms and OH radicals with SiH_4 over the temperature range of 297-438°K, *Int. J. Chem. Kinet.*, **10**, 1151-1160, 1978.

[1256] Patrick, R., and D. M. Golden, Kinetics of the reactions of NH_2 radicals with O_3 and O_2, *J. Phys. Chem.*, **88**, 491-495, 1984.

[1257] Graedel, T. E., Atmospheric photochemistry, in *Handbook of Environmental Chemistry*, vol. 2A, O. Hutzinger, Ed., Berlin: Springer-Verlag, pp. 107-143, 1980.

[1258] Snider, J. R., and G. A. Dawson, Tropospheric light alcohols, carbonyls, and acetonitrile: concentrations in the southwestern United States and Henry's law data, *J. Geophys. Res.*, **90**, 3797-3805, 1985.

[1259] Graedel, T. E., Terpenoids in the atmosphere, *Rev. Geophys. Space Phys.*, **17**, 937-947, 1979.

[1260] Brewer, D. A., M. A. Ogliaruso, T. R. Augustsson, and J. S. Levine, The oxidation of isoprene in the troposphere: mechanism and model calculations, *Atmos. Environ.*, **18**, 2723-2744, 1984.

[1261] Falk, H. L., I. Markul, and P. Kotin, Aromatic hydrocarbons. IV. Their fate following emission into the atmosphere and experimental exposure to washed air and synthetic smog, *Arch. Ind. Health*, **13**, 13-17, 1956.

[1262] Fatiadi, A. J., Effects of temperature and of ultraviolet radiation on pyrene adsorbed on garden soil, *Environ. Sci. Technol.*, **1**, 570-572, 1967.

[1263] Barofsky, D. F., and E. J. Baum, Field desorption mass analysis of the photooxidation products of adsorbed polycyclic aromatic hydrocarbons, Paper 76-20.5, 69th Annual Meeting, Air Pollut. Contr. Assoc., Portland, OR, 1976.

[1264] Ogren, J. A., and R. J. Charlson, Elemental carbon in the atmosphere: cycle and lifetime, *Tellus*, **35B**, 241-254, 1983.

[1265] Rosen, H., A. D. A. Hansen, R. L. Dod, and T. Novakov, Soot in urban atmospheres: determination by an optical absorption technique, *Science*, **208**, 741-744, 1980.

[1266] Brodzinsky, R., S. G. Chang, S. S. Markowitz, and T. Novakov, Kinetics and mechanism for the catalytic oxidation of sulfur dioxide on carbon in aqueous suspensions, *J. Phys. Chem.*, **84**, 3354-3358, 1980.

[1267] Calvert, J. G., B. G. Heikes, W. R. Stockwell, V. A. Mohnen, and J. A. Kerr, Some considerations of the important chemical processes in acid deposition, in Chemistry of Multiphase Atmospheric Systems, W. Jaeschke, Ed., Dordrecht: D. Reidel, 1984.

[1268] Tuazon, E. C., A. M. Winer, and J. N. Pitts, Jr., Trace pollutant concentrations in a multiday smog episode in the California South Coast Air Basin by long path length fourier transform infrared spectroscopy, *Environ. Sci. Technol.*, **15**, 1232-1237, 1981.

[1269] Appel, B. R., S. M. Wall, and R. L. Knights, Characterization of carbonaceous materials in atmospheric aerosols by high-resolution mass spectrometric thermal analysis, *Adv. Environ. Sci. Technol.*, **9**, 353-364, 1980.

[1270] Niki, H., P. D. Maker, C. M. Savage, and L. P. Breitenbach, Relative rate constants for the reaction of hydroxyl radical with aldehydes, *J. Phys. Chem.*, **82**, 132-134, 1978.

[1271] Graedel, T. E., *Chemical Compounds in the Atmosphere*, New York: Academic Press, 440 pp., 1978.

[1272] Crutzen, P. J., The influence of nitrogen oxides on the atmospheric ozone content, *Q. J. Roy. Met. Soc.*, **96**, 320-325, 1970.

[1273] Farlow, N. H., G. V. Ferry, H. Y. Lem, and D. M. Hayes, Latitudinal variations of stratospheric aerosols, *J. Geophys. Res.*, **84**, 733-743, 1979.

[1274] Martin, L. R., Kinetic studies of sulfite oxidation in aqueous solution, in *Acid Precipitiation,* SO_2, NO, *and* NO_2 *Oxidation Mechanisms: Atmospheric Considerations*, J. G. Calvert, Ed., Ann Arbor, MI: Ann Arbor Science Pubs., 1983.

[1275] Keyser, T. R., D. F. S. Natusch, C. A. Evans, Jr., and R. W. Linton, Characterizing the surfaces of environmental particles, *Environ. Sci. Technol.*, **12**, 768-773, 1978.

[1276] Lee, Y.-N., and S. E. Schwartz, Evaluation of rate of uptake of nitrogen dioxide by atmospheric and surface liquid water, *J. Geophys. Res.*, **86**, 11971-11983, 1981.

[1277] Chameides, W. L., and D. D. Davis, Chemistry in the troposphere, *Chem. and Eng. News*, **60**(40), 38-52, 1982.

[1278] Crutzen, P. J., Atmospheric interactions – homogeneous gas reactions of C, N, and S containing compounds, in *The Major Biogeochemical Cycles and Their Interactions*, Ed. B. Bolin and R. B. Cook, SCOPE Report No. 21, New York: John Wiley, 1983.

[1279] Graedel, T. E., Concentrations and metal interactions of atmospheric trace gases involved in corrosion, *Proc. 9th Intl. Congr. Metallic Corrosion*, V.1, pp. 396-401, National Research Council of Canada, Ottawa, 1984.

[1280] Leifer, R., Z. R. Juzdan, and R. Larsen, The high altitude sampling program: radioactivity in the stratosphere, Report EML-434, US Department of Energy, New York, NY, 1984.

[1281] Singh, H. B., and P. L. Hanst, Peroxyacetyl nitrate (PAN) in the unpolluted atmosphere: an important reservoir for nitrogen oxides, *Geophys. Res. Lett.*, **8**, 941-944, 1981.

[1282] Spence, J. W., and P. L. Hanst, Oxidation of chlorinated ethanes, *J. Air Poll. Contr. Assoc.*, **28**, 250-253, 1978.

[1283] Wahner, A., and C. Zetzsch, Rate constants for the addition of OH to aromatics (benzene, *p*-chloroaniline, and *o*-, *m*-, and *p*-dichlorobenzene) and the unimolecular decay of the adduct. Kinetics into a quasi-equilibrium. 1, *J. Phys. Chem.*, **87**, 4945-4951, 1983.

[1284] Harrison, R. M., and D. P. H. Laxen, Sink processes for tetraalkyllead compounds in the atmosphere, *Environ. Sci. Technol.*, **12**, 1384-1392, 1978.

[1285] Frye, K., *Modern Mineralogy*, Englewood Cliffs, NJ: Prentice-Hall, 1974.

[1286] Hellman, T. M., and F. H. Small, Characterization of the odor properties of 101 photochemicals using sensory methods, *J. Air. Poll. Contr. Assoc.*, **24**, 979-982, 1974.

[1287] Stahl, W. H., Ed., *Compilation of Odor and Taste Threshhold Values Data*, ASTM Pub. D5-48, Philadelphia, PA: Amer. Soc. for Testing and Materials, 1973.

[1288] Moschandreas, D. J., J. Zabransky, and D. J. Pelton, *Comparison of Indoor-Outdoor Concentrations of Atmospheric Pollutants*, GEOMET Report ES-823, Electric Power Research Inst., Palo Alto, CA, 1980.

[1289] Grot, R. A., and R. B. Clark, Air leakage characteristics and weatherization techniques for low-income housing, Paper presented at DOE/ASHRAE Conf. on Thermal Performance of Exterior Envelopes of Buildings, Orlando, FL, 1979.

[1290] Committee on Indoor Pollutants, *Indoor Pollutants*, Washington, D.C.: National Academy Press, 1981.

[1291] Camuffo, D., Indoor dynamic climatology: investigations on the interactions between walls and indoor environment, *Atmos. Environ.*, **17**, 1803-1809, 1983.

[1292] Wynder, E. L., and D. Hoffmann, Tobacco and health. A societal challenge, *New Engl. J. Med.*, **300**, 894-903, 1979.

[1293] Whitby, K. T., R. B. Husar, and B. Y. H. Liu, The aerosol size distribution of Los Angeles smog, *J,. Colloid Interface Sci.*, **39**, 177-204, 1972.

[1294] Friedlander, S. K., Chemical element balances and identification of air pollution sources, *Environ. Sci. Technol.*, **7**, 235-240, 1973.

[1295] Ho, W., G. M. Hidy, and R. M. Govan, Microwave measurements of the liquid water content of atmospheric aerosols, *J. Appl. Met.*, **13**, 871-879, 1974.

[1296] Grosjean, D., and S. K. Friedlander, Gas-particle distribution factors for organic and other pollutants in the Los Angeles atmosphere, *J. Air Poll. Contr. Assoc.*, **25**, 1038-1044, 1975.

[1297] Frevert, T., and O. Klemm, Wie ändern sich pH-Werte im Regen- und Nebelwasser beim Abtrocknen auf Pflanzenoberflächen?, *Arch. Met. Geoph. Biocl.*, **B34**, 75-81, 1984.

[1298] Steelink, C., What is humic acid?, *J. Chem. Ed.*, **40**, 379-384, 1963.

[1299] Stelson, A. W., and J. H. Seinfeld, Chemical mass accounting of urban aerosol, *Environ. Sci. Technol.*, **15**, 671-679, 1981.

[1300] Winkler, P., Observations on acidity in continental and in marine atmospheric aerosols and in precipitation, *J. Geophys. Res.*, **85**, 4481-4486, 1980.

[1301] Korfmacher, W. A., D. F. S. Natusch, D. R. Taylor, G. Mamantov, and E. L. Wehry, Oxidative transformations of polycyclic aromatic hydrocarbons adsorbed on coal fly ash, *Science*, **207**, 763-765, 1980.

[1302] Fox, M. A., and S. Olive, Photooxidation of anthracene on atmospheric particulate matter, *Science*, **205**, 582-583, 1979.

[1303] Pitts, J. N., Jr., Photochemical and biological implications of the atmospheric reactions of amines and benzo (a) pyrene, *Phil. Trans. Roy. Soc. London, Ser. A.*, **290**, 551-576, 1979.

[1304] Yocom, J. E., and N. S. Baer, Effects on materials, in *The Acidic Deposition Phenomenon and Its Effects*, EPA-600/8-83-016B, Environ. Prot. Agency, Washington, D.C., 1984.

[1305] Mackay, D., and P. J. Leinonen, Rate of evaporation of low-solubility contaminants from water bodies to atmosphere, *Environ. Sci. Technol.*, **9**, 1178-1180, 1975.

[1306] Zepp, R. G., N. L. Wolfe, L. V. Azarraga, and R. H. Cox, Photochemical transformation of the 1,1-diaryl-2,2-dichloroethylenes, DDE and DMDE, by sunlight, Paper ENVT-69, 172nd Meeting, Amer. Chem. Soc., San Francisco, CA, 1976.

[1307] Glotfelty, D. E., B. C. Turner, and A. W. Taylor, Transport and reactions of pesticides in the atmosphere, Paper PEST-3, 172nd Meeting, Amer. Chem. Soc., San Francisco, CA, 1976.

[1308] Kaufman, J. E., Ed., IES Lighting Handbook, 5th ed., New York: Illuminating Engineering Soc., 1972.

[1309] Chang, S. G., and T. Novakov, Formation of pollution particulate nitrogen compounds by NO-soot and NH_3-soot gas-particle surface reactions, *Atmos. Environ.*, **9**, 495-504, 1975.

[1310] Craig, N. L., A. B. Harker, and T. Novakov, Determination of the chemical states of sulfur in ambient pollution aerosols by x-ray photoelectron spectroscopy, *Atmos. Environ.*, **8**, 15-21, 1974.

[1311] Junge, C. E., Basic considerations about trace constituents in the atmosphere as related to the fate of global pollutants, in Fate of Pollutants in the Air and Water Environments, I. H. Suffet, Ed., Part I, pp. 7-25, New York: John Wiley, 1977.

[1312] Yamasaki, H., K. Kuwata, and H. Miyamoto, Effects of ambient temperature on aspects of airborne polycyclic aromatic hydrocarbons, *Environ. Sci. Technol.*, **16**, 189-194, 1982.

[1313] U. S. Air Force and U. S. Dept. of Commerce, *Global Atlas of Relative Cloud Cover, 1967-1970*, Washington, D.C., 1971.

[1314] Neville, R. C., *Solar Energy Conversion: The Solar Cell*, Amsterdam: Elsevier Pub. Co., 297 pp., 1978.

[1315] Gast, P. R., in *Handbook of Geophysics and Space Environments*, S. L. Valley, Ed., New York: McGraw-Hill, 1965.

[1316] Bacastow, R. B., and C. D. Keeling, Atmospheric carbon dioxide concentration and the observed airborne fraction, *SCOPE Rpt.*, **16**, 103-112, 1981.

[1317] Rasmussen, R. A., and M. A. K. Khalil, Atmospheric methane (CH_4): trends and seasonal cycles, *J. Geophys. Res.*, **86**, 9826-9832, 1981.

[1318] Cunnold, D. M., R. G. Prinn, R. A. Rasmussen, P. G. Simmonds, F. N. Alyea, C. A. Cardelino, A. J. Crawford, P. J. Frazer, and R. D. Rosen, The atmospheric lifetime experiment, 3, lifetime methodology and application to three years of $CFCl_3$ data, *J. Geophys. Res.*, **88**, 8379-8400, 1983.

[1319] Rudloff, W., *World Climates*, Stuttgart: Wiss. Verlag. mbH, 1981.

[1320] Altshuller, A. P., Review: Natural volatile organic substances and their effect on air quality in the United States, *Atmos. Environ*, **17**, 2131-2165, 1983.

[1321] Graedel, T. E., and N. Schwartz, Air quality reference data for corrosion assessment, *Materials Performance*, **16**(8), 17-25, 1977.

[1322] Ferguson, E. E., F. C. Fehsenfeld, and D. L. Albritton, Ion chemistry of the earth's atmosphere, in *Gas Phase Ion Chemistry*, **1**, M. T. Bowers, Ed., New York: Academic Press, pp. 45-82, 1979.

[1323] Graedel, T. E., and C. J. Weschler, Chemistry within aqueous atmospheric aerosols and raindrops, *Rev. Geophys. Space Phys.*, **19**, 505-539, 1981.

[1324] Atkinson, R., and W. P. L. Carter, Kinetics and mechanisms of the gas-phase reactions of ozone with organic compounds under atmospheric conditions, *Chem. Rev.*, **84**, 437-470, 1984.

[1325] Atkinson, R., Kinetics and mechanisms of the gas phase reactions of the hydroxyl radical with organic compounds under atmospheric conditions, *Chem. Rev.*, **85**, in press, 1985.

[1326] Graedel, T. E., C. J. Weschler, and M. L. Mandich, Kinetic studies of atmospheric droplet chemistry. 2. Homogeneous transition metal chemistry in raindrops, Submitted for publication, 1985.

[1327] Humphreys, W. J., *Physics of the Air*, 3rd ed., New York: Dover Publications, 1964.

[1328] Graedel, T. E., and K. I. Goldberg, Kinetic studies of raindrop chemistry. 1. Inorganic and organic processes, *J. Geophys. Res.*, **88**, 10865-10882, 1983.

[1329] Tuazon, E. C., R. Atkinson, and W. P. L. Carter, Atmospheric chemistry of *cis*- and *trans*-3-hexene-2,5-dione, *Environ. Sci. Technol.*, **19**, 265-269, 1985.

[1330] Hanst, P. L., J. W. Spence, and M. Miller, Atmospheric chemistry of N-nitroso dimethylamine, *Environ. Sci. Technol.*, **11**, 403-405, 1977.

[1331] Siple, G. W., C. K. Fitzsimmons, D. F. Zeller, and R. B. Evans, Long range airborne measurements of ozone off the coast of the Northeastern United States, in *Proc. Intl. Conf. on Photochemical Oxidant Pollution and Its Control*, EPA-600/3-777-001a, Environmental Protection Agency, Research Triangle Park, NC, 1977.

[1332] Committee on Medical and Biologic Effects of Environmental Pollutants, *Ozone and Other Photochemical Oxidants*, National Academy of Sciences, Washington, D.C., 1977.

[1333] Ashenden, T. W., and T. A. Mansfield, Extreme pollution sensitivity of grasses when SO_2 and NO_2 are present in the atmosphere together, *Nature*, **273**, 142-143, 1978.

[1334] Biermann, H. W., H. MacLeod, R. Atkinson, A. M. Winer, and J. N. Pitts, Jr., Kinetics of the gas-phase reactions of the hydroxyl radical with naphthalene, phenanthrene, and anthracene, *Environ. Sci. Technol.*, **19**, 244-248, 1985.

Bioassay References

B1. Brusick, D. J., V. F. Simmon, H. S. Rosenkranz, V. A. Ray, and R. S. Stafford, An evaluation of the *Escherichia coli* WP_2 and WP_2 *uvrA* revese mutation assay, *Mutation Res.*, **76**, 169-190, 1980.

B2. Generoso, W. M., J. B. Bishop, D. G. Gosslee, G. W. Newell, C. Sheu, and E. von Halle, Heritable translocation test in mice, *Mutation Res.*, **76**, 191-215, 1980.

B3. Hsie, A. W., D. A. Casciano, D. B. Couch, D. F. Krahn, J. P. O'Neill, and B. L. Whitfield, The use of Chinese hamster ovary cells to quantify specific locus mutation and to determine mutagenicity of chemicals, *Mutation Res.*, **86**, 193-214, 1981.

B4. Russell, L. B., P. B. Selby, E. von Halle, W. Sheridan, and L. Valcovic, The mouse specific-locus test with agents other than radiations: interpretation of data and recommendations for future work, *Mutation Res.*, **86**, 329-354, 1981.

B5. Russell, L. B., P. B. Selby, E. von Halle, W. Sheridan, and L. Valcovic, Use of the mouse spot test in chemical mutagenesis: interpretation of past data and recommendations for future work, *Mutation Res.*, **86**, 355-379, 1981.

B6. Latt, S. A., J. Allen, S. E. Bloom, A. Carrano, E. Falke, D. Kram, E. Schneider, R. Schreck, R. Tice, B. Whitfield, and S. Wolff, Sister-chromatid exchanges: A report of the Gene-Tox Program, *Mutation Res.*, **87**, 17-62, 1981.

B7. Bradley, M. O., B. Bhuyan, M. C. Francis, R. Langenbach, A. Peterson, and E. Huberman, Mutagenesis by chemical agents in V79 Chinese hamster cells: A review and analysis of the literature, *Mutation Res.*, **87**, 81-142, 1981.

B8. Preston, R. J., W. Au, M. A. Bender, J. G. Brewen, A. V. Carrano, J. A. Heddle, A. F. McFee, S. Wolff, and J. S. Wassom, Mammalian *in vivo* and *in vitro* cytogenetic assays, *Mutation Res.*, **87**, 143-188, 1981.

B9. Leifer, Z., T. Kada, M. Mandel, E. Zeiger, R. Stafford, and H. Rosenkranz, An evaluation of tests using DNA repair-deficient bacteria for predicting genotoxicity and carcinogenicity, *Mutation Res.*, **87**, 211-297, 1981.

B10. Kafer, E., B. R. Scott, G. L. Dorn, and R. Stafford, *Aspergillus nidulans:* Systems and results of tests for chemical induction of mitotic segregation and mutation, I. Diploid and duplication assay systems, *Mutation Res.*, **98**, 1-48, 1982.

B11. Scott, B. R., G. L. Dorn, E. Kafer, and R. Stafford, *Aspergillus nidulans:* Systems and results of tests for induction of mitotic segregation and mutation, II. Haploid assay systems and overall response of all systems, *Mutation Res.*, **98**, 49-94, 1982.

B12. Larsen, K. H., D. Brash, J. E. Cleaver, R. W. Hart, V. M. Maher, R. B. Painter, and G. A. Sega, DNA repair assays as tests for environmental mutagens, *Mutation Res.*, **98**, 287-318, 1982.

B13. Legator, M. S., E. Bueding, R. Batzinger, T. H. Connor, E. Eisenstadt, M. G. Farrow, G. Ficsor, A. Hsie, J. Seed, and R. S. Stafford, An evaluation of the host-mediated assay and body fluid analysis, *Mutation Res.*, **98**, 319-374, 1982.

B14. Constantin, M. J., and E. T. Owens, Introduction and perspective of plant genetic and cytogenetic assays, *Mutation Res.*, **99**, 1-12, 1982.

B15. Constantin, M. J., and R. A. Nilan, Chromosome aberration assays in barley (*Hordeum vulgare*), *Mutation Res.*, **99**, 13-36, 1982.

B16. Constantin, M. J., and R. A. Nilan, The chorophyll-deficient mutant assay in barley (*Hordeum vulgare*), *Mutation Res.*, **99**, 37-49, 1982.

B17. Redei, G. P., Mutagen assay with Arabidopsis, *Mutation Res.*, **99**, 243-255, 1982.

B18. Ma, T.,. Vicia cytogenetic tests for environmental mutagens, *Mutation Res.*, **99**, 257-271, 1982.

B19. Grant, W. F., Chromosome aberration assays in Allium, *Mutation Res.*, **99**, 273-291, 1982.

B20. Ma., T., Tradescantia cytogenetic tests (root-tip mitosis, pollen mitosis, pollen mother-cell meiosis), *Mutation Res.*, **99**, 293-302, 1982.

B21. Hof, J. V., and L. A. Schairer, Tradescantia assay system for gaseous mutagens, *Mutation Res.*, **99**, 303-315, 1982.

B22. Plewa, M. J., Specific-locus mutation assays in *Zea mays, Mutation Res.*, **99**, 317-337, 1982.

B23. Vig., B. K., Soybean (*Glycine max* [L.] merrill) as a short-term assay for study of environmental mutagens, *Mutation Res.*, **99**, 339-347, 1982.

B24. Heidelberger, C., A. E. Freeman, R. J. Pienta, A. Sivak, J. S. Bertram, B. C. Casto, V. C. Dunkel, M. W. Francis, T. Kakunaga, J. B. Little, and L. M. Schechtman, Cell transformation by chemical agents — a review and analysis of the literature, *Mutation Res.*, **114**, 283-385, 1983.

B25. Wyrobek, A. J., L. A. Gordon, J. G. Burkhart, M. W. Francis, R. W. Kapp, Jr., G. Letz, H. V. Malling, J. C. Topham, and M. D. Whorton, An evaluation of the mouse sperm morphology test and other sperm tests in nonhuman mammals, *Mutation Res.*, **115**, 1-72, 1983.

B26. Wyrobek, A. J., L. A. Gordon, J. G. Burkhart, M. W. Francis, R. W. Kapp, Jr., G. Letz, H. V. Malling, J. C. Topham, and M. D. Whorton, An evaluation of human sperm as indicators of chemically induced alterations of spermatogenic function, *Mutation Res.*, **115**, 73-148, 1983.

B27. Loprieno, N., R. Barale, E. S. von Halle, and R. C. von Borstel, Testing of chemicals for mutagenic acitivity with *Schizosaccharomyces pombe, Mutation Res.*, **115**, 215-223, 1983.

B28. Clive, D., R. McCuen, J. F. S. Spector, C. Piper, and K. H. Mavournin, Specific gene mutations in L5178Y cells in culture, *Mutation Res.*, **115**, 225-251, 1983.

B29. Heddle, J. A., M. Hite, B. Kirkhart, K. Mavournin, J. T. MacGregor, G. W. Newell, and M. F. Salamone, The induction of micronuclei as a measure of genotoxicity, *Mutation Res.*, **123**, 61-118, 1983.

B30. Lee, W. R., S. Abrahamson, R. Valencia, E. S. von Halle, F. E. Wurgler, and S. Zimmering, The sex-linked recessive lethal test for mutagenesis in *Drosophila melanogaster, Mutation Res.*, **123**, 183-279, 1983.

B31. Hayes, W. A., *Principles and Methods of Toxicology*, New York: Raven Press, 1982.

B32. Brockman, H. E., F. J. de Serres, T. Ong, D. M. DeMarini, A. J. Katz, A. J. F. Griffiths, and R. S. Stafford, Mutation tests in *Neurospora crassa, Mutation Res.*, **133**, 87-134, 1984.

B33. McKusik, V. A., *Mendelian Inheritance in Man*, 3rd edition. Baltimore: Johns Hopkins Press, 1971.

B34. Kennedy, W. P., Epidemiologic aspects of the problem of congenital malformations, *Birth Defects Original Article Series*, **3**, 2-22, 1967.

B35. Levy, H. L., and J. P. Kennedy, Jr., Genetic screening for inborn errors of metabolism, In: Harris, H., and K. Hirschborn (eds.), *Advances in Human Genetics*, New York: Plenum Press, 1973.

B36. Kier, L. D., D. J. Brusick, A. E. Auletta, E. S. von Halle, V. F. Simmon, M. M. Brown, V. C. Dunkel, J. McCann, K. Mortelmans, M. J. Prival, T. K. Rao, and V. A. Ray, The *Salmonella typhimurium*/mammalian microsome mutagenicity assay. [In final draft for *Mutation Res.*; data provided on computer tape by Dr. John Wassom, Environmental Mutagen Information Center, Oak Ridge, Tennessee, October, 1984].

B37. Nesnow, S., M. Argus, H. Bergman, K. Chu, C. Frith, T. Helmes, R. McGaughey, V. Ray, T. J. Slaga, R. Tennant, and E. Weisburger, Chemical carcinogens: A review and analysis of the literature of selected chemicals and the establishment of the Gene-Tox carcinogen data base, [In final draft for *Mutation Res.*; data provided by Dr. Steve Nesnow, U.S. Environmental Protection Agency, Research Triangle Park, North Carolina, November, 1984].

B38. Haworth, S., T. Lawlor, K. Mortelmans, W. Speck, and E. Zeiger, Salmonella mutagenicity test results for 250 chemicals, *Environmental Mutagenesis, Supplement 1*, 3-142, 1983.

B39. Carr, D. H., Population cytogenetics of human abortuses, In: Hook, E. B., and I. H. Porter (eds.) *Population Cytogenetics*, New York: Academic Press, 1975.

B40. Ratcliffe, S. G., A. L. Stewart, M. M. Melville, and P. A. Jacobs, Chromosome studies of 3500 new born male infants, *Lancet*, **7638**, 121-128, 1970.

B41. Jacobs, P. A., Chromosome abnormalities and fertility in man, In: Beatty, R. A., and J. Glaechsohn-Waelsh (eds.), *The Genetics of the Spermatozoon*, Copenhagen: Bogtujkkeriert Forum, 1972.

B42. Benirschke, K., G. Carpenter, C. Epstein, C. Fraser, L. Jackson, A. Motusky, and W. Nyhan, Genetic diseases, In: Brent, R. L., and M. I. Harris (eds.), *Prevention of Embryonic, Fetal, and Perinatal Disease*, Washington: DHEW Publication No. (NIH) 76-853, 1976.

B43. Fialkow, P. J., The origin and development of human tumors studied with cell markers, *New England J. Medicine*, **291**(1), 26-34, 1974.

B44. Rowley, J. D., Mapping of human chromosome regions related to neoplasia: Evidence from chromosome 1 and 17. *Pub. Natl. Acad. Sci.*, **74**(12), 5729-5733, 1977.

B45. Land, H., L. F. Parada, and R. A. Weinberg, Cellular oncogenes and multistep carcinogenesis, *Science*, **222**, 771-778, 1983.

B46. Knudson, A., Mutation and human cancer, *Advances in Cancer Research*, **17**, 317-352, 1973.

B47. Miller, R. W., Ethnic differences in cancer occurrence: Genetic and environmental influence with particular reference to neuroblastoma, In: Malvihill, J. J. (ed.), *Genetics of Human Cancer*, New York: Raven Press, 1977.

B48. Kobayoshi, N., Congenital anomalies in children with malignancy, *Paediatri Univ. Tokyo*, **16**, 31-37, 1968.

B49. McCann, J., and B. N. Ames, Detection of carcinogens and mutations in the Salmonella microsome test: Assay of 300 chemicals. *Pub. Natl. Acad. Sci.*, **72**(12), 5135-5139, 1975.

B50. Malling, H. V., and J. S. Wassom, Action of mutagenic agents, In: Wilson, J. G., and F. C. Fraser, *Handbook of Teratology, Volume 1*, New York: Plenum Press, 1977.

B51. Brusick, D., *Principles of Genetic Toxicology*, New York: Plenum Press, 1980.

B52. McElheny, V. K., and S. Abrahamson (eds.), *Banbury Report 1, Assessing Chemical Mutagens, The Risk to Humans*, Cold Spring Harbor: Cold Spring Harbor Laboratory, 1979.

B53. Hsie, A., J. P. O'Neill, and V. K. McElheny, *Banbury Report 2: Mammalian Cell Mutagenesis, The Maturation of Test Systems*, Cold Spring Harbor: Cold Spring Harbor Laboratory, 1979.

B54. Bridges, B. A., B. E. Butterworth, and I. B. Weinstein, *Banbury Report 13: Indicators of Genotoxic Exposure*, Cold Spring Harbor: Cold Spring Harbor Laboratory, 1982.

B55. de Serres, F. J., and M. Shelby, *Comparative Chemical Mutagenesis*, New York: Plenum Press, 1981.

B56. Frei, R. W., and U. A. T. Brinkman (eds.), *Mutagenicity Testing and Related Analytical Topics*, New York: Gordon and Breach Scientific Publishers, 1981.

B57. Heddle, J. A. (ed.), *Mutagenicity: New Horizons in Genetic Toxicology*, New York: Academic Press, 1982.

B58. Claxton, L. D., and P. Z. Barry, Chemical mutagenesis: An emerging issue for public health, *Amer. J. Pub. Health*, **67**, 1037-1042, 1977.

B59. Hollstein, M., J. McCann, F. A. Angelosanto, and W. W. Nichols, Short-term tests for carcinogens and mutagens, *Mutation Res.*, **65**, 133-226, 1979.

B60. Walters, D. B. (ed), *Safe Handling of Chemical Carcinogens, Mutagens, Teratogens, and Highly Toxic Substances, Volume 1*, Ann Arbor, MI: Ann Arbor Science Pub., 381 pp., 1980.

B61. Straus, D. S., Somatic mutation, cellular differentiation, and cancer causation, *J. National Cancer Inst.*, **67**, 233-241, 1981.

B62. Waters, M. D. and A. Auletta, The GENE-TOX Program: Genetic activity evaluation, *J. Chem. Inf. Comput. Sci.*, **21**, 35-40, 1981.

B63. Green, S. and A. Auletta, Editorial introduction to the reports of "The Gene-Tox Program." An evaluation of bioassays in genetic toxicology, *Mutation Res.*, **76**, 165-168, 1980.

B64. Bridges, B.A., The three-tier approach to mutagenicity screening and the concept of radiation-equivalent dose, *Mutation Res.*, **26**, 335-340, 1974.

B65. Flamm, W. G., A tier system approach to mutagen testing, *Mutation Res.*, **26**, 329-333, 1974.

B66. Claxton, L. D., and P. Z. Barry, Chemical mutagenesis: An emerging issue for public health, *Amer. J. Pub. Health*, **67**, 1037-1042, 1977.

B67. Stephens, E. R., E. F. Darley, O. C. Taylor, and W. E. Scott, Photochemical reaction products in air pollution, *Int. J. Air Water Pollut.*, **4**, 79-100, 1961.

B68. Claxton, L. D., and H. M. Barnes, The mutagenicity of diesel-exhaust particle extracts collected under smog-chamber conditions using the *Salmonella typhimurium* test system, *Mutation Res.*, **88**, 255-272, 1981.

B69. Shepson, P. B., T. E. Kleindienst, E. O. Edney, G. R. Namie, J. H. Pittman, L. T. Cupitt, and L. D. Claxton, The mutagenic activity of irradiated toluene/NO_x/H_2O/air mixtures, *Environ. Sci. Technol.*, **19**, 249-255, 1985.

B70. Sawicki, E., Atmospheric genotoxicants -- What numbers do we collect?, In: M. D. Waters, S. Nesnow, J. L. Huisingh, S. S. Sandhu, and L. Claxton (eds.) *Application of Short-Term Bioassays in the Fractionation and Analysis of Complex Environmental Mixtures*, New York: Plenum Press, 171-194, 1978.

INDEX 1

Authors

This is an alphabetical listing of all authors cited in this book, together with the item numbers of the references which they authored or co-authored. Reference numbers preceded by B indicate those in the bioassay reference list. In a few cases, varying representations of an author's name have been edited for consistency. The References, however, always reflect the names as they appear in the literature.

INDEX 2

Species Names

This is an alphabetical listing of the chemical species names included in this book, together with the species number assigned to each. The numbers to the left of the hyphen in the species number represent the chapter and table where data concerning that species will be found. The number to the right of the hyphen is the sequential number of the species within the table.

The name index is presented in the "inverted" format (e.g., hexane, 3-ethyl) familiar to chemists. The choice of a name for species which may be named in more than one way is discussed in Section 1.4.3. Synonyms for a number of the species are included here, with cross-referencing to the preferred name. The synonym list is in no way comprehensive, but consists of those names which were encountered in the preparation of this book and which are, therefore, contained in the atmospheric chemical literature. Because of the possibility of different names for the same compound, a compound whose name does not appear in this index should not be assumed to be absent from this book unless its Registry Number is absent from Index 3 as well.

Substance	Species Number
Butyric acid, 5-formyl	9.1-29
Butyric acid, 2-hydroxy-4-phenyl	9.4-30
Butyric acid, 2,3-dimethyl	9.1-19
Butyric acid, phenyl	9.4-29
Butyrolactone, γ-	10.1-2
Cadalene, see Naphalene, 1-isopropyl, 4,7-dimethyl	
Cadaverine	11.2-19
Cadinene	3.4-20
Cadinol	5.3-20
Cadmium (II) oxide	2.6-41
Caffeine	11.5-135
Calcite, see Calcium carbonate	
Calcium carbonate	2.6-58
Calcium fluoride	2.5-24
Calcium ion	2.1-13
Calcium oxide	2.6-20
Calcium sulfate	2.4-43
Campesterol	5.3-26
Camphene	3.4-13
Camphor	6.3-11
Capric acid, see Decanoic acid	
Caproic acid, see Hexanoic acid	
Caprylic acid, see Octanoic acid	
Carbaryl	11.2-69
Carbazole, 9H-	11.5-99
Carbazole, benzo[*a*]	11.5-103
Carbazole, benzo[*c*]	11.5-104
Carbazole, benzo[*def*]	11.5-105
Carbazole, dibenzo[*b,def*]	11.5-107
Carbazole, dibenzo[*c,g*]	11.5-108
Carbazole, chloro	13.3-87
Carbazole, chlorobenzo[*c*]	13.3-88
Carbazole, 9-methyl	11.5-100
Carbazole, dimethyl	11.5-101
Carbazole, trimethyl	11.5-102
Carbazole, methylbenzo[*def*]	11.5-106
Carbinol, methyl, ethyl, see Pentane-2,3-diol	
Carbofuran	11.2-68
Carbon (soot)	2.6-1
Carbon tetrachloride	13.1-5
Carbon tetrafluoride	13.1-1
Carbon monoxide	2.6-6
Carbon dioxide	2.6-7
Carbon disulfide	2.4-3
Carbonate ion	2.6-10
Carbonates	2.6
Carbonic acid	2.6-8
Carbonyl fluoride	13.1-60
Carbonyl sulfide	2.4-2
Carene, Δ^3-	3.4-11
Carene, Δ^4-	3.4-12

Substance	Species Number
Carvacrol, see Phenol, 3-isopropyl, 5-methyl	
Carvenol, *l*-	5.3-10
Carvone	6.3-16
Caryophyllene, α-	3.4-25
Caryophyllene, β-	3.4-26
Caryophyllene alcohol	5.3-23
Catechol, 4-ethyl	5.4-47
Catechol, 3-methyl	5.4-45
Catechol, 4-methyl	5.4-46
Catechol, 4-propyl	5.4-48
Cedrene	10.1-12
Cedrenol	5.3-22
Cedrol	5.3-21
Cerium oxide	2.6-48
Chamazulene	3.4-24
Chavicol, see Phenol, 4-allyl	
Chloral	13.1-62
Chlorate-nitric acid ion	2.5-44
Chlordane	13.4-30
Chloride ion	2.1-10
Chlorine atom	2.1-9
Chlorine molecule (Dichlorine)	2.5-2
Chlorine monoxide radical	2.5-1
Chlorite	2.7-5
Chlorocyanide	13.1-68
Chloroform	13.1-4
Chloroform, methyl, see Ethane, 1,1,1-trichloro	
Chloroprene, see 1,3-Butadiene, 2-chloro	
Chloropropham	13.4-14
Cholanthrene, 3-methyl	3.8-26
Cholest-5-en-3b-ol, 24-ethyl, see β-Sitosterol	
Cholest-5-en-3b-ol, 24-methyl, see Campesterol	
Cholesta-5,22-dien-3b-ol, 24-methyl, see Brassicasterol	
Cholesterol	5.3-24
Chromic acid	2.6-70
Chromium (III) dioxide	2.6-24
Chromium (II) sulfate	2.4-49
Chromone, 2-phenyl	10.3-27
Chrysen-4-one, 4H-cyclopenta[*def*]	6.6-26
Chrysen-4-one, 4H-benzo[*b*]cyclopenta[*mno*]	6.6-28
Chrysen-4-one, 4H-benzo[*c*]cyclopenta[*mno*]	6.6-29
Chrysen-4-one, 4H-benzo[*def*]cyclopenta[*mno*]	6.6-31
Chrysen-5-one, 5H-benzo[*b*]cyclopenta[*def*]	6.6-27

Substance	Species Number
Cyclohexane, phenyl	3.3-76
Cyclohexane, propyl	3.3-44
Cyclohexane, triacontyl	3.3-75
Cyclohexane, vinyl	3.3-43
Cyclohexanol	5.3-3
Cyclohexanol, methyl	5.3-4
Cyclohexanone	6.3-4
Cyclohexen-3-one	6.3-7
Cyclohexene	3.3-78
Cyclohexene, 4-isopropyl-1-methyl	3.3-83
Cyclohexene, 1-methyl	3.3-79
Cyclohexene, 4-methyl	3.3-80
Cyclohexene, 2,4-dimethyl-4-vinyl	3.3-82
Cyclohexene, phenyl	3.3-84
Cyclohexene, 4-vinyl	3.3-81
Cyclohexyl peroxide carbonate, di-	8.5-25
Cyclohexylamine	11.2-46
Cyclooctadiene, 1,5-	3.3-89
Cyclooctane	3.3-88
Cyclooctanone, 2-methyl	6.3-23
Cyclooctatetraene	3.3-90
Cyclopentadecalactone	10.1-10
Cyclopentadiene	3.3-26
Cyclopentane	3.3-9
Cyclopentane, ethyl	3.3-18
Cyclopentane, 1-ethyl-2-methyl	3.3-19
Cyclopentane, 1-ethyl-3-methyl	3.3-20
Cyclopentane, methyl	3.3-10
Cyclopentane, 1,1-dimethyl	3.3-11
Cyclopentane, *cis*-1,2-dimethyl	3.3-12
Cyclopentane, *trans*-1,2-dimethyl	3.3-13
Cyclopentane, *cis*-1,3-dimethyl	3.3-14
Cyclopentane, *trans*-1,3-dimethyl	3.3-15
Cyclopentane, 1,2,3-trimethyl	3.3-16
Cyclopentane, 1,2,4-trimethyl	3.3-17
Cyclopentane, propyl	3.3-21
Cyclopentanediol, 3-methyl-1,2-	5.3-2
Cyclopentanedione, 1,2-	6.3-3
Cyclopentanol	5.3-1
Cyclopentanone	6.3-2
Cyclopentanone, 2-ethoxy-5-(1-hydroxy-ethyl)-	6.3-25
Cyclopentasiloxane, decamethyl	14.2-9
Cyclopentene	3.3-22
Cyclopentene, 1-methyl	3.3-23
Cyclopentene, 3-methyl	3.3-24
Cyclopentene, 4-methyl	3.3-25
Cyclopropane	3.3-1
Cyclopropane, 1-ethyl-1-methyl	3.3-3
Cyclopropane, *trans*-1,2-dimethyl	3.3-2
Cyclopropane, isopropyl	3.3-4
Cyclotetrasiloxane, octamethyl	14.2-8
Cyclotrisiloxane, hexamethyl	14.2-7

Substance	Species Number
Cyclohexanedione, methyl	6.3-17
Cyclohexane	3.3-29
Cyclohexane, butyl	3.3-47
Cyclohexane, *sec*-butyl	3.3-49
Cyclohexane, *tert*-butyl	3.3-50
Cyclohexane, α-hexachloro	13.4-3
Cyclohexane, β-hexachloro	13.4-4
Cyclohexane, γ-hexachloro	13.4-5
Cyclohexane, tricosyl	3.3-68
Cyclohexane, tetracosyl	3.3-69
Cyclohexane, pentacosyl	3.3-70
Cyclohexane, hexacosyl	3.3-71
Cyclohexane, heptacosyl	3.3-72
Cyclohexane, octacosyl	3.3-73
Cyclohexane, nonacosyl	3.3-74
Cyclohexane, decyl	3.3-56
Cyclohexane, undecyl	3.3-57
Cyclohexane, dodecyl	3.3-58
Cyclohexane, tridecyl	3.3-59
Cyclohexane, tetradecyl	3.3-60
Cyclohexane, pentadecyl	3.3-61
Cyclohexane, hexadecyl	3.3-62
Cyclohexane, heptadecyl	3.3-63
Cyclohexane, nonadecyl	3.3-64
Cyclohexane, dichloro	13.3-3
Cyclohexane, diethyl	3.3-42
Cyclohexane, docosyl	3.3-67
Cyclohexane, eicosyl	3.3-65
Cyclohexane, ethyl	3.3-40
Cyclohexane, ethylmethyl	3.3-41
Cyclohexane, ethylphenyl	3.3-77
Cyclohexane, fluoro	13.3-2
Cyclohexane, heneicosyl	3.3-66
Cyclohexane, heptyl	3.3-53
Cyclohexane, hexyl	3.3-52
Cyclohexane, 1,2,4,5-tetrahydroxy	5.3-5
Cyclohexane, hexahydroxy	5.3-6
Cyclohexane, isobutyl	3.3-48
Cyclohexane, isopropyl	3.3-45
Cyclohexane, 1-isopropyl-4-methyl	3.3-46
Cyclohexane, methyl	3.3-30
Cyclohexane, 1,1-dimethyl	3.3-31
Cyclohexane, *cis*-1,2-dimethyl	3.3-32
Cyclohexane, *trans*-1,2-dimethyl	3.3-33
Cyclohexane, 1,3-dimethyl	3.3-34
Cyclohexane, *trans*-1,4-dimethyl	3.3-35
Cyclohexane, 1,1,3-trimethyl	3.3-36
Cyclohexane, 1,2,3-trimethyl	3.3-37
Cyclohexane, 1,2,4-trimethyl	3.3-38
Cyclohexane, 1,3,5-trimethyl	3.3-39
Cyclohexane, nonyl	3.3-55
Cyclohexane, octyl	3.3-54
Cyclohexane, pentyl	3.3-51

Substance	Species Number
Dodecene, 1-	3.2-111
Dodecene, 2-methyl-1-	3.2-112
Dodecenoic acid	9.2-7
Dolomite	2.7-8
Dotriacontane	3.1-127
Dotriacontane, 2-methyl	3.1-128
Dotriacontanoic acid	9.1-107
Dotriacontanol	5.1-79
Dotriacontene, 1-	3.2-138
Durane, see Benzene, 1,2,4,5-tetramethyl	
Dursban	14.3-14
Eicosane	3.1-109
Eicosanedioic acid	9.1-91
Eicosanoate, ethyl	8.2-133
Eicosanoate, methyl	8.2-132
Eicosanoic acid	9.1-89
Eicosanoic acid, 20-hydroxy	9.1-90
Eicosanol	5.1-65
Eicosene, 1-	3.2-126
Eicosenoic acid	9.2-17
Elements	2.1
Elemicin, see Benzene, 1,2,3-trimethoxy, 5-allyl	
Elliotinoic acid	9.3-4
Enanthic acid, see Heptanoic acid	
Endosulfan	13.4-37
Endrin	13.4-33
Epichlorhydrin, see Propane, 3-chloro, 1,2-epoxy	
EPTC	12.4-5
Esters	8.2-8.4
Esters, alkanic	8.2
Esters, aromatic	8.4
Esters, cyclic	8.4
Esters, olefinic	8.3
Estragol, see Benzene, 4-allyl, 1-methoxy	
Ethanal, hydroxy	7.1-4
Ethane	3.1-2
Ethane, 1,2-dibromo	13.1-33
Ethane, 1-bromo-2-chloro	13.1-40
Ethane, chloropentafluoro	13.1-35
Ethane, 1-chloro-1,1-difluoro	13.1-34
Ethane, 1,2-dichloro	13.1-27
Ethane, 1,2-dichlorotetrafluoro	13.1-36
Ethane, 1,1,1-trichloro	13.1-28
Ethane, 1,1,2-trichloro	13.1-29
Ethane, 1,1,1-trichloro-2,2,2-trifluoro	13.1-38
Ethane, 1,2,2-trichloro-1,1,2-trifluoro	13.1-37
Ethane, 1,1,2,2-tetrachloro	13.1-30
Ethane, 1,1,1,2-tetrachloro	13.1-31
Ethane, tetrachloro-1,2-difluoro	13.1-39
Ethane, epoxy	10.2-1
Ethane, diethoxy	4.1-7
Ethane, 1,1-difluoro	13.1-24
Ethane, perfluoro	13.1-25
Ethane, 1,2-dimethoxy-1-phenyl	4.3-19
Ethane, α-methoxy-2-phenyl	4.3-18
Ethane, nitro	11.3-4
Ethane, diphenyl	3.5-73
Ethane, phenylepoxy	10.3-1
Ethane dinitrile, see Cyanogen	
Ethane-1-thiol, 1-phenyl	12.1-20
Ethanedioic acid, see Oxalic acid	
Ethanenitrile	11.1-2
Ethanenitrile, phenyl	11.1-22
Ethanethiol	12.1-2
Ethanoic acid, 2-formyl, see Glyoxylic acid	
Ethanol	5.1-4
Ethanol, 2-butoxy	5.1-9
Ethanol, 2-ethoxy	5.1-5
Ethanol, 2-bis(ethoxy)	5.1-6
Ethanol, dodecoxy	5.1-10
Ethanol, 2-phenyl	5.4-66
Ethen-1-ol, 2-butoxy	5.2-3
Ethen-1-ol, 2-ethoxy	5.2-2
Ethene	3.2-1
Ether, allyl methyl	4.2-10
Ether, diallyl	4.2-11
Ether, dibutyl	4.1-9
Ether, bis(2-chloroethyl)	13.1-65
Ether, bis(chloromethyl)	13.1-64
Ether, *tert*-butyl methyl	4.1-4
Ether, butyl vinyl	4.2-5
Ether, chloromethyl methyl	13.1-63
Ether, ethyl methoxyvinyl	4.2-3
Ether, ethyl methyl	4.1-2
Ether, ethyl vinyl	4.2-2
Ether, diethyl	4.1-6
Ether, difurfuryl	10.2-36
Ether, hexyl vinyl	4.2-7
Ether, isobutyl vinyl	4.2-6
Ether, isopropyl vinyl	4.2-4
Ether, diisopropyl	4.1-8
Ether, methyl propyl	4.1-3
Ether, methyl vinyl	4.2-1
Ether, dimethyl	4.1-1
Ether, biphenyl	4.3-20
Ether, bitolyl	4.3-22
Ether, vinyl-1,2-dihydroxyvinyl	5.2-4
Ether, divinyl	4.2-9
Ether, octadecyl vinyl	4.2-8
Ethers	4.1-4.3
Ethers, alkanic	4.1

Substance	Species Number
Indanone, hydroxymethyl	6.5-7
Indanone, hydroxytrimethyl	6.5-8
Indanone, hydroxytetramethyl	6.5-9
Indanone, dimethyl-1-	6.5-2
Indanone, 3,3-dimethyl-5-*tert*-butyl	6.5-4
Indanone, 3,4,7-trimethyl-1-	6.5-3
Indanone, methoxy	6.5-5
Indene	3.6-10
Indene, ethyl	3.6-14
Indene, 3,3a,7,7a-tetrahydro-4,7-methano	3.4-18
Indene, 1-methyl	3.6-11
Indene, 3-methyl	3.6-12
Indene, dimethyl	3.6-13
Indenone	6.5-11
Indenone, hydroxy	6.5-14
Indenone, hydroxymethyl	6.5-15
Indenone, hydroxydimethyl	6.5-16
Indenone, dimethyl	6.5-12
Indenone, pentamethyl	6.5-13
Indole	11.5-64
Indole, 7-aza	11.5-75
Indole, ethyl	11.5-73
Indole, 2-methyl	11.5-65
Indole, 3-methyl	11.5-66
Indole, 5-methyl	11.5-67
Indole, 7-methyl	11.5-68
Indole, methylphenyl	11.5-74
Indole, 1,2-dimethyl	11.5-69
Indole, 2,3-dimethyl	11.5-70
Indole, 2,5-dimethyl	11.5-71
Indole, 1,7-methylene-2,3-dimethyl	11.5-72
Indolizine, methyl	11.5-63
Iodine molecule (Diiodine)	2.5-4
Ionone, α-	6.3-20
Ionone, β-	6.3-21
Iron (II) chloride	2.5-28
Iron (II) fluoride	2.5-27
Iron (II) oxide	2.6-28
Iron (II) sulfate	2.4-51
Iron (II) sulfate heptahydrate	2.4-52
Iron (III) hydroxide ion	2.6-30
Iron (III) ion	2.1-14
Iron (III) oxide	2.6-29
Iron (III) phosphate	2.6-68
Iron (III) sulfate-ammonium sulfate (1/3)	2.4-53
Iron (III) sulfite	2.4-17
Iron carbonyl	2.6-63
Iron tetroxide, tri-	2.6-31
Iron-nickel carbide	2.7-14
Irone, α-	6.3-22
Isobutanal	7.1-7

Substance	Species Number
Isobutane	3.1-5
Isobutanedioic acid	9.1-23
Isobutanol	5.1-19
Isobutene	3.2-16
Isobutylnitrile, azodi	11.1-17
Isobutyramide	11.2-41
Isobutyrate, butyl	8.2-79
Isobutyrate, geranyl	8.2-81
Isobutyrate, ethyl	8.2-78
Isobutyrate, linalyl	8.2-82
Isobutyrate, octyl	8.2-80
Isobutyric acid	9.1-22
Isocitric acid, see 1,2,3-Propanetricarboxylic acid, 1-hydroxy	
Isocyanate, methyl	11.1-16
Isoelemicin, see Benzene, 5-allyl, 1,2,3-trimethoxy	
Isoeugenol	5.4-37
Isoindole, 1,3-dioxo	11.5-134
Isomenthone	6.3-9
Isopentanal	7.1-11
Isopentanamide	11.2-42
Isopentane	3.1-11
Isopentanedioic acid	9.1-39
Isopentanoate, bornyl	8.2-89
Isopentanoate, ethyl	8.2-88
Isopentanol	5.1-27
Isophorone	6.3-14
Isophthalic acid	9.4-19
Isoprene	3.2-19
Isopulegol	5.3-11
Isoquinoline	11.5-76
Isoquinoline, benzo[*f*]	11.5-80
Isoquinoline, ethyl	11.5-79
Isoquinoline, 1-methyl	11.5-77
Isoquinoline, dimethyl	11.5-78
Isosqualene	3.4-60
Isothiocyanate, allyl	12.4-4
Isovaleric acid	9.1-38
Kamacite	2.7-15
Kaolinite	2.7-16
Ketene	6.2-1
Ketone, cyclopropyl methyl	6.3-1
Ketone, ethyl phenyl	6.4-21
Ketone, ethyl-2,4-dimethoxyphenyl	6.4-22
Ketone, (2,4,6-trihydroxy-3-methylphenyl) propyl	6.4-26
Ketone, phenyl vinyl	6.4-23
Ketone, 3-pyridyl propyl	11.5-131
Ketones	6.1-6.6
Ketones, alkanic	6.1
Ketones, aromatic	6.4-6.6

Substance	Species Number
Octadien-1-ol, 3,7-dimethyl-*cis*-2,6-	5.2-25
Octadien-1-ol, 3,7-dimethyl-*trans*-2,6-	5.2-26
Octadien-2-ol, *trans*-2-methyl-6-methylene-3,7-	5.2-27
Octadien-3-ol, *cis*-1,5-	5.2-22
Octadien-3-ol, *trans*-1,5-	5.2-24
Octadien-3-ol, 3,7-dimethyl-1,6-	5.2-23
Octadien-3-one, 1,5-	6.2-13
Octadienal	7.2-15
Octadienal, 3,7-dimethyl-2,6-	7.2-16
Octadiene	3.2-95
Octalactone, γ-	10.1-5
Octalactone, δ-	10.1-9
Octanal	7.1-26
Octane	3.1-66
Octane, 2-methyl	3.1-67
Octane, 3-methyl	3.1-68
Octane, 4-methyl	3.1-69
Octane, 2,3-dimethyl	3.1-70
Octane, 2,5-dimethyl	3.1-71
Octane, 2,6-dimethyl	3.1-72
Octane, 2,3,4-trimethyl	3.1-73
Octane, 3,3,5-trimethyl	3.1-74
Octanedioic acid	9.1-55
Octanedioic acid, 3-methyl	9.1-56
Octanethiol	12.1-16
Octanoic acid	9.1-54
Octanol, 1-	5.1-53
Octanol, 3-	5.1-54
Octanone, 2-	6.1-28
Octanone, 3-	6.1-29
Octasulfur, *cyclo*-	2.4-9
Octatriacontane	3.1-134
Octen-1-ol, 2-	5.2-18
Octen-1-ol, 3,7-dimethyl-6-	5.2-19
Octen-2-ol, 2-methyl-6-methylene-7-	5.2-20
Octen-3-ol, 1-	5.2-21
Octen-3-one, 1-	6.2-12
Octenal, 3,7-dimethyl-6-	7.2-13
Octenal, 4-hydroxy-3,7-dimethyl-6-	7.2-14
Octene, 1-	3.2-92
Octene, 2-methyl-1-	3.2-93
Octene, 2,6-dimethyl-1-	3.2-94
Octene, *cis*-2-	3.2-97
Octene, *trans*-2-	3.2-98
Octene, *trans*-4-	3.2-99
Octenoate, methyl-*cis*-4-	8.3-18
Octylnitrate, 2-oxo	11.3-17
Octyne	3.2-96
Olivine	2.7-24
Organometallics	14.1-14.2
Orthoclase	2.7-25
Oxalate ion	9.1-4
Oxalic acid	9.1-8
Oxides	2.6
Oxopropanal, 2-	7.1-39
Oxygen atom	2.1-3
Oxygen molecule (Dioxygen)	2.2-2
Oxyprolin, see Pyrrolidine, 5-oxo, 2-carboxylic acid	
Ozone	2.2-3
Palmitic acid, see Hexadecanoic acid	
Palustric acid	9.3-7
Palygorskite	2.7-26
Parathion	14.3-10
Parathion, methyl	14.3-11
Pelargonic acid, see Nonanoic acid	
Pentacene	3.8-87
Pentacosane	3.1-114
Pentacosanoic acid	9.1-100
Pentacosanol	5.1-70
Pentacosene, 1-	3.2-131
Pentadecadien-2-one, 6,9-	6.2-14
Pentadecan-2-one, 6,10,14-trimethyl	6.1-35
Pentadecanal	7.1-33
Pentadecane	3.1-98
Pentadecane, 2-methyl	3.1-99
Pentadecane, 2,6,10,14-tetramethyl, see Pristane	
Pentadecanoate, methyl	8.2-120
Pentadecanoate, methyl-14-methyl	8.2-121
Pentadecanoic acid	9.1-76
Pentadecanoic acid, methyl	9.1-77
Pentadecanol, 5-	5.1-60
Pentadecanone, 2-	6.1-34
Pentadecene	3.2-118
Pentadecenoate, methyl	8.3-19
Pentadecenoic acid	9.2-10
Pentadiene, 1,2-	3.2-37
Pentadiene, 1,3-	3.2-38
Pentadiene, 2-methyl-1,3-	3.2-39
Pentadiene, 2,4-dimethyl-1,3-	3.2-40
Pentadiene, 1,4-	3.2-41
Pentadiene, 2,4-dimethyl-2,3-	3.2-54
Pentanal	7.1-10
Pentanal, 5-hydroxy	7.1-14
Pentanal, 2-methyl	7.1-13
Pentanal, phenyl	7.3-33
Pentane	3.1-10
Pentane, 1,5-diamine, see Cadaverine	
Pentane, 3-ethyl	3.1-25
Pentane, 3-ethyl-2-methyl	3.1-26
Pentane, 3-ethyl-3-methyl	3.1-27
Pentane, 3-ethyl-2,3-dimethyl	3.1-28
Pentane, 3,3-diethyl	3.1-29
Pentane, 2,3-epoxy-4-methyl	10.2-6

Substance	Species Number
Phellandrene, α-	3.4-16
Phellandrene, β-	3.4-17
Phenalen-1-one, 1H-	6.5-45
Phenalen-1-one,2,3-dihydro, 1H-	6.5-44
Phenalene, cyclopenta[*cd*]	3.7-27
Phenanthren-4-one, 4H-cyclopenta[*def*]	6.5-48
Phenanthren-4-one, methyl-4H-cyclopenta[*def*]	6.5-49
Phenanthren-7-one, 7H-indeno[*2,1-a*]	6.5-52
Phenanthren-8-one, 8H-dibenzo[*b,mn*]	6.5-50
Phenanthren-8-one, 8H-indeno[*2,1-b*]	6.5-55
Phenanthren-9-one, 9H-dibenzo[*c,mn*]	6.5-51
Phenanthren-9-one, 9H-indeno[*2,1-c*]	6.5-56
Phenanthren-11-one, 11H-indeno[*2,1-a*]	6.5-53
Phenanthren-12-one, 12H-indeno[*1,2-b*]	6.5-54
Phenanthren-13-one, 13H-indeno[*1,2-c*]	6.5-57
Phenanthren-13-one, 13H-indeno[*1,2-l*]	6.5-58
Phenanthrene	3.7-40
Phenanthrene, 3,8-diaza	11.5-112
Phenanthrene, benzo[*ac*]	3.7-61
Phenanthrene, dibenzo[*b,k*], see Pentacene	
Phenanthrene, 4H-cyclopenta[*def*]	3.7-54
Phenanthrene, ethyl	3.7-51
Phenanthrene, ethyl-4H-cyclopenta[*def*]	3.7-58
Phenanthrene, ethyldihydromethylene-	3.7-60
Phenanthrene, 7-ethyl-1-methyl	3.7-52
Phenanthrene, ethylmethyl-4H-cyclopenta[*def*]	3.7-59
Phenanthrene, dihydro	3.7-39
Phenanthrene, tetrahydro	3.7-36
Phenanthrene, octahydro	3.7-30
Phenanthrene, perhydro	3.7-28
Phenanthrene, dihydrobenzo[*ac*]	3.7-62
Phenanthrene, 7-isopropyl-1-methyl	3.7-53
Phenanthrene, methoxy	4.3-24
Phenanthrene, 1-methyl	3.7-41
Phenanthrene, 2-methyl	3.7-42
Phenanthrene, 3-methyl	3.7-43
Phenanthrene, 4-methyl	3.7-44
Phenanthrene, 9-methyl	3.7-45
Phenanthrene, 1,7-dimethyl	3.7-46
Phenanthrene, 2,5-dimethyl	3.7-47
Phenanthrene, 3,6-dimethyl	3.7-48
Phenanthrene, methylbenzo[*ac*]	3.7-63
Phenanthrene, dimethylbenzo[*ac*]	3.7-64
Phenathrene, methyl-4-cyclopenta[*def*]	3.7-55
Phenanthrene, dimethyl-4-cyclopenta[*def*]	3.7-56
Phenanthrene, methylnitro	11.4-26
Phenanthrene, dimethylnitro	11.4-27
Phenanthrene, 4,5-dimethylperhydro	3.7-29
Phenanthrene, trimethyl	3.7-49
Phenanthrene, trimethyl-4H-cyclopenta[*def*]	3.7-57
Phenanthrene, tetramethyl	3.7-50
Phenanthrene, 4,5-methylene, see Phenanthrene, cyclopenta[*def*]	
Phenanthrene, 4,5-methylenenitro	11.4-28
Phenanthrene, 2-nitro	11.4-25
Phenanthrenecarbaldehyde, 2-	7.3-46
Phenanthrenecarbaldehyde, methyl	7.3-47
Phenanthrenecarbaldehyde dimethylnitro	11.4-92,
Phenanthrenecarbaldehyde, 1,2,3,4,4a,9,10,10a-octahydro-1,4a-, 1-	7.3-45
Phenanthrenecarboxylic acid	9.4-35
Phenanthrenemethanol-1,2,3,4,4a,9,10,10a-octahydro-1,4a-, 1-	5.4-89
Phenanthrenequinone, 9,10-	6.5-46
Phenanthrenequinone, methyl-9,10-	6.5-47
Phenanthrenol	5.4-86
Phenanthrenol, dimethyl	5.4-87
Phenanthrenol, trimethyl	5.4-88
Phenanthridine	11.5-111
Phenanthridine, benzo[*lmn*], see Pyrene, 4-aza	
Phenanthridine, 5,6-dihydro	11.5-110
Phenanthrone, nitro	11.4-86
Phenol	5.4-1
Phenol, 4-allyl	5.4-21
Phenol, bromodichloromethyl	13.3-52
Phenol, bromomethylenedi	13.3-54
Phenol, 2-*tert*-butyl	5.4-24
Phenol, di-*tert*-butyl	5.4-26
Phenol, 2,6-di-*tert*-butyl-4-ethyl	5.4-28
Phenol, 2-*tert*-butyl-4-methoxy	5.4-25
Phenol, 2,6-di-*tert*-butyl-4-methyl	5.4-27
Phenol, chloro	13.3-47
Phenol, 2,4-dichloro	13.3-48
Phenol, trichloro	13.3-49
Phenol, tetrachloro	13.3-50
Phenol, pentachloro	13.3-51
Phenol, chloromethylenedi	13.3-53
Phenol, (8-ethoxynonyl)	5.4-40
Phenol, *o*-ethyl	5.4-13
Phenol, *m*-ethyl	5.4-14
Phenol, *p*-ethyl	5.4-15
Phenol, 2-ethyl-6-methyl-4-nitro	11.4-71
Phenol, ethylmethyl	5.4-16
Phenol, (3-hydroxybenzyl)methylenedi	5.4-73
Phenol, isobutyl	5.4-23
Phenol, 4-isopropyl	5.4-18
Phenol, 2-isopropyl-5-methyl	5.4-19
Phenol, 3-isopropyl-5-methyl	5.4-20
Phenol, *o*-methoxy	5.4-33
Phenol, *p*-methoxy	5.4-34

Substance	Species Number
Pyridine, 3-propionyl	11.5-130
Pyridine, 3-propyl	11.5-34
Pyridine, 3-vinyl	11.5-40
Pyridine-3-carbaldehyde	11.5-136
Pyridinecarbonitrile, 2-	11.5-141
Pyridinecarboxylic acid, 3-	11.5-139
Pyrite	2.4-5
Pyrocatechol	5.4-44
Pyrocatechol, allyl	5.4-49
Pyrocatechol, phenyl	5.4-50
Pyrogallol, 4,6-dimethoxy	5.4-56
Pyrone, nitro	11.4-90
Pyroxene	2.7-28
Pyrrole	11.5-1
Pyrrole, methyl	11.5-2
Pyrrole, 2,4-diphenyl	11.5-3
Pyrrolidine, 1-acetyl	11.5-125
Pyrrolidine-2-carboxylic acid, 5-oxo	11.5-137
Pyrrolidone, 2-	11.5-126
Pyrrolidone, methyl	11.5-127
Pyrrolidone, N-methyl	11.5-128
Pyruvic acid	9.1-13
Quaterphenyl, *o*-	3.5-70
Quaterphenyl, *m*-	3.5-71
Quinoline	11.5-81
Quinoline, benzo[*f*]	11.5-91
Quinoline, benzo[*h*]	11.5-92
Quinoline, 2-ethyl	11.5-89
Quinoline, 4-ethyl	11.5-90
Quinoline, 11H-indeno[*1,2-6*]	11.5-96
Quinoline, 2-methyl	11.5-82
Quinoline, 4-methyl	11.5-83
Quinoline, 7-methyl	11.5-84
Quinoline, 2,4-dimethyl	11.5-85
Quinoline, 2,6-dimethyl	11.5-86
Quinoline, 2,8-dimethyl	11.5-87
Quinoline, trimethyl	11.5-88
Quinoline, 3-methylbenzo[*h*]	11.5-93
Quinoline, dimethylbenzo[*h*]	11.5-94
Quinoline, trimethylbenzo[*h*]	11.5-95
Quinoline, 5-nitro	11.4-98
Quinoline, 8-nitro	11.4-99
Radon atom	2.1-23
Reductic acid	9.1-14
Resorcinol	5.4-51
Resorcinol, 2-methyl	5.4-52
Resorcinol, 5-methyl	5.4-53
Retene, see Phenanthrene, 7-isopropyl, 1-methyl	
Ronnel	14.3-13
Rovral	13.4-36
Sabinene	3.4-15
Safranal	7.3-6
Safrole	10.3-2
Salicylaldehyde	7.3-18
Salicylate, benzyl	8.4-16
Salicylate, methyl	8.4-15
Sandaracopimaric acid	9.3-10
Santalol, α-	5.3-16
Scopoletin	10.3-26
Selenide, dimethyl	14.1-6
Selenium atom	2.1-17
Selenium dioxide	2.6-39
Selenomercaptan, ethyl	14.1-5
Selenone, dimethyl	14.1-8
Senecioic acid, see Butyric acid, 2,3-dimethyl	
Serine, see Propionic acid, 2-amino, 3-hydroxy	
Silane, methyl	14.2-1
Silane, tetramethyl	14.2-2
Silane, trimethyl(phenoxy)	14.2-4
Silanol, trimethyl	14.2-3
Silicon dihydrogen hexafluoride	2.5-11
Silicon dioxide	2.6-16
Silicon tetrafluoride	2.5-22
Siloxane, hexamethyldi	14.2-5
Simonellite	3.7-38
Sinapylaldehyde	7.3-32
Sitosterol, β-	5.3-30
Sitosterol, γ-	5.3-31
Skatole, see Indole, 3-methyl	
Sodium carbonate	2.6-56
Sodium chloride	2.5-15
Sodium fluoride	2.5-14
Sodium hydrogen sulfate	2.4-36
Sodium ion	2.1-6
Sodium nitrate	2.3-18
Sodium oxide	2.6-11
Sodium sulfate	2.4-37
Sodium sulfate decahydrate	2.4-38
Sodium sulfide	2.4-4
Solanesol	11.5-124
Sorbaldehyde, see 2,4-hexadienal	
Squalene	3.4-59
Stearic acid, see Octadecanoic acid	
Sterane	3.4-38
Sterane, ethyl	3.4-40
Sterane, methyl	3.4-39
Stigmasta-5,24(28)-dien-3b-ol, (24E), see Fucosterol	
Stigmasterol	5.3-25
Strontium carbonate	2.6-60
Strontium sulfate	2.4-61
Styrene	3.5-27
Styrene, ethyl	3.5-34

Substance	Species Number
Thiophene, methyldibenzo	12.3-25
Thiophene, ethyldibenzo	12.3-26
Thiophene, benzo[*b*]	12.3-21
Thiophene, benzo[*b*]naphtho[*1,2-d*]	12.3-27
Thiophene, benzo[*b*]naphtho[*2,1-d*]	12.3-28
Thiophene, benzo[*b*]naphtho[*2,3-d*]	12.3-30
Thiophene, 2-butyl	12.3-11
Thiophene, ethyl	12.3-7
Thiophene, 2-ethyl-5-isopentyl	12.3-13
Thiophene, ethylpropyl	12.3-10
Thiophene, 2-formyl-5-propynyl	12.3-17
Thiophene, 2-hexyl	12.3-14
Thiophene, 2,3,4,5-tetrahydro	12.3-1
Thiophene, dihydrobenzo[*b*]	12.3-20
Thiophene, methyl	12.3-4
Thiophene, dimethyl	12.3-5
Thiophene, trimethyl	12.3-6
Thiophene, methylbenzo[*b*]	12.3-22
Thiophene, dimethylbenzo[*b*]	12.3-23
Thiophene, methylpropyl	12.3-9
Thiophene, methyltetrahydro	12.3-2
Thiophene, nitrobenzo	11.4-103
Thiophene, (1-oxoethyl)	12.3-18
Thiophene, (1-oxopropyl)	12.3-19
Thiophene, 2-pentyl	12.3-12
Thiophene, phenyl	12.3-15
Thiophene, diphenyl	12.3-16
Thiophene, 2-propyl	12.3-8
Thiophenol, see Benzenethiol	
Thiotriazine, chloromethyl	13.3-89
Threonine, see Butyric acid, 2-amino, 3-hydroxy	
Thujate, methyl	8.4-1
Thujene, α-	3.4-14
Thujone	6.3-12
Thujopsene	3.4-21
Thymol, see Phenol, 2-isopropyl, 5-methyl	
Tiglaldehyde, see Crotonaldehyde, 2-methyl	
Tiglate, allyl	8.3-15
Tin (II) oxide	2.6-42
Tin (IV) oxide	2.6-43
Titanite	2.7-30
Titanium dioxide	2.6-21
Titanium tetrachloride	2.5-26
Toluene	3.5-2
Toluene, bromo	13.3-25
Toluene, *o*-chloro	13.3-20
Toluene, *p*-chloro	13.3-21
Toluene, dichloro	13.3-22
Toluene, trichloro	13.3-23
Toluene, tetrachloro	13.3-24
Toluene, *m*-ethyl	3.5-15
Toluene, *o*-ethyl	3.5-14
Toluene, *p*-ethyl	3.5-16
Toluene, perfluoro	13.3-19
Toluene, 2-isopropyl, see *o*-cymene	
Toluene, 3-isopropyl, see *m*-cymene	
Toluene, 4-isopropyl, see *p*-cymene	
Toluene, *m*-nitro	11.4-3
Toluene, *o*-nitro	11.4-2
Toluene, *p*-nitro	11.4-4
Toluene, 2-propyl	3.5-39
Toluene, 3-propyl	3.5-40
Toluene, 4-propyl	3.5-41
Toluidine, *p*-	11.2-48
Toluidine, N-phenyl-*p*-	11.2-53
Toluonitrile	11.1-21
Toxaphene	13.4-1
Triacontanal	7.1-38
Triacontane	3.1-123
Triacontane, 2-methyl	3.1-124
Triacontanoic acid	9.1-105
Triacontanol	5.1-77
Triacontene, 1-	3.2-136
Trichlorofon	14.3-5
Tricosane	3.1-112
Tricosanoic acid	9.1-96
Tricosanol	5.1-68
Tricosene, 1-	3.2-129
Tridecanal	7.1-31
Tridecane	3.1-92
Tridecane, 2-methyl	3.1-93
Tridecane, 7-methyl	3.1-94
Tridecanoic acid	9.1-70
Tridecanoic acid, methyl	9.1-71
Tridecene, 1-	3.2-114
Tridecene, 2-methyl-1-	3.2-115
Tridecenoic acid	9.2-8
Trifluralin	13.4-15
Triphenylen-4-one, 4H-benzo[*m*]cyclopenta[*cde*]	6.6-7
Triphenylen-4-one, 4H-benzo[*l*]cyclopenta[*cde*]	6.6-8
Triphenylen-4-one, 4H-cyclopenta[*def*]	6.6-5
Triphenylen-13-one, 13H-benzo[*b*]cyclopenta[*def*]	6.6-6
Triphenylen-13-one, 13H-dibenzo[*b,jkl*]cyclopenta[*def*]	6.6-9
Triphenylene, methylnitro	11.4-46
Triphenylene	3.8-27
Triphenylene, methyl	3.8-28
Triphenylene, benzo[*b*]	3.8-29
Trisiloxane, octamethyl	14.2-6
Trisulfide, allyl methyl	12.2-32

INDEX 3

Chemical Abstracts Service (CAS) Registry Numbers

In this index the CAS Registry Numbers for all species listed in this book are presented in numerical order, together with the species number assigned. The numbers to the left of the hyphen in the species number represent the chapter and table where data concerning that species will be found. The number to the right of the hyphen is the sequential number of the species within the table.

CAS Registry Number	Species Number	CAS Registry Number	Species Number	CAS Registry Number	Species Number
50-00-0	7.1-1	67-64-1	6.1-1	75-44-5	13.1-61
50-24-8	3.4-38	67-66-3	13.1-4	75-45-6	13.1-13
50-29-3	13.4-19	67-71-0	12.2-35	75-50-3	11.2-5
50-32-8	3.8-39	68-11-1	12.4-1	75-52-5	11.3-1
50-99-7	10.2-40	68-12-2	11.2-35	75-56-9	10.2-2
51-28-5	11.4-73	69-72-7	9.4-8	75-63-8	13.1-17
51-44-5	13.3-66	71-23-8	5.1-11	75-65-0	5.1-20
52-68-6	14.3-5	71-36-3	5.1-18	75-66-1	12.1-7
53-19-0	13.4-22	71-41-0	5.1-26	75-68-3	13.1-34
53-70-3	3.7-86	71-43-2	3.5-1	75-69-4	13.1-14
54-11-5	11.5-41	71-47-6	9.1-2	75-71-8	13.1-15
55-18-5	11.2-44	71-52-3	2.6-9	75-72-9	13.1-16
56-12-2	11.2-28	71-55-6	13.1-28	75-73-0	13.1-1
56-23-5	13.1-5	72-03-7	9.1-10	75-74-1	14.1-12
56-38-2	14.3-10	72-19-5	11.2-29	75-75-2	12.2-34
56-40-6	11.2-25	72-20-8	13.4-33	75-76-3	14.2-2
56-41-7	11.2-26	72-43-5	13.4-23	75-83-2	3.1-6
56-45-1	11.2-27	72-54-8	13.4-21	75-87-6	13.1-62
56-49-5	3.8-26	72-55-9	13.4-24	75-97-8	6.1-7
56-55-3	3.7-73	74-82-8	3.1-1	76-13-1	13.1-37
56-81-5	5.1-15	74-83-9	13.1-6	76-14-2	13.1-36
56-84-8	11.2-30	74-84-0	3.1-2	76-15-3	13.1-35
57-06-7	12.4-4	74-85-1	3.2-1	76-16-4	13.1-25
57-10-3	9.1-78	74-86-2	3.2-2	76-44-8	13.4-28
57-11-4	9.1-84	74-87-3	13.1-2	76-49-3	8.2-41
57-14-7	11.2-2	74-88-4	13.1-9	77-53-2	5.3-21
57-55-6	5.1-13	74-89-5	11.2-3	77-73-6	3.4-18
57-88-5	5.3-24	74-90-8	11.1-1	77-75-8	5.2-12
57-97-6	3.7-83	74-93-1	12.1-1	77-78-1	12.2-37
58-08-2	11.5-135	74-95-3	13.1-7	77-92-9	9.1-36
58-27-5	6.5-21	74-96-4	13.1-32	77-94-1	8.2-92
58-89-9	13.4-5	74-97-5	13.1-18	78-00-2	14.1-16
59-67-6	11.5-139	74-98-6	3.1-3	78-35-3	8.2-82
59-89-2	11.2-63	74-99-7	3.2-5	78-42-2	14.3-3
60-12-8	5.4-66	75-00-3	13.1-26	78-48-8	14.3-6
60-29-7	4.1-6	75-01-4	13.2-2	78-51-3	14.3-2
60-35-5	11.2-36	75-04-7	11.2-6	78-59-1	6.3-14
60-51-5	14.3-8	75-05-8	11.1-2	78-70-6	5.2-23
60-57-1	13.4-34	75-07-0	7.1-3	78-78-4	3.1-11
61-90-5	11.2-31	75-08-1	12.1-2	78-79-5	3.2-19
62-53-3	11.2-47	75-09-2	13.1-3	78-83-1	5.1-19
62-73-7	14.3-4	75-15-0	2.4-3	78-84-2	7.1-7
62-75-9	11.2-43	75-18-3	12.2-1	78-85-3	7.2-2
63-25-2	11.2-69	75-19-4	3.3-1	78-86-4	13.1-52
64-17-5	5.1-4	75-21-8	10.2-1	78-87-5	13.1-45
64-18-6	9.1-1	75-25-2	13.1-8	78-92-2	5.1-23
64-19-7	9.1-3	75-27-4	13.1-19	78-93-3	6.1-5
65-85-0	9.4-1	75-28-5	3.1-5	78-94-4	6.2-2
66-25-1	7.1-16	75-29-6	13.1-43	78-98-8	7.1-39
66-77-3	7.3-36	75-31-0	11.2-10	78-99-9	13.1-44
66-99-9	7.3-37	75-33-2	12.1-6	79-00-5	13.1-29
67-47-0	10.2-35	75-35-4	13.2-3	79-01-6	13.2-5
67-56-1	5.1-1	75-37-6	13.1-24	79-05-0	11.2-39
67-63-0	5.1-16	75-43-4	13.1-12	79-09-4	9.1-9

CAS Registry Number	Species Number	CAS Registry Number	Species Number	CAS Registry Number	Species Number
99-89-8	5.4-18	105-30-6	5.1-28	107-51-7	14.2-6
99-93-4	6.4-17	105-31-7	5.2-17	107-52-8	14.2-10
99-96-7	9.4-10	105-37-3	8.2-59	107-83-5	3.1-13
99-99-0	11.4-4	105-46-4	8.2-17	107-84-6	13.1-53
100-01-6	11.4-96	105-54-4	8.2-71	107-87-9	6.1-10
100-02-7	11.4-65	105-67-9	5.4-6	107-92-6	9.1-18
100-06-1	6.4-14	105-68-0	8.2-61	108-05-4	8.2-31
100-21-0	9.4-20	105-85-1	8.2-5	108-08-7	3.1-17
100-40-3	3.3-81	105-87-3	8.2-36	108-10-1	6.1-12
100-41-4	3.5-13	105-91-9	8.2-63	108-18-9	11.2-11
100-42-5	3.5-27	106-02-5	10.1-10	108-20-3	4.1-8
100-44-7	13.3-26	106-22-9	5.2-19	108-21-4	8.2-14
100-47-0	11.1-19	106-23-0	7.2-13	108-24-7	8.1-1
100-51-6	5.4-57	106-24-1	5.2-26	108-29-2	10.1-3
100-52-7	7.3-1	106-25-2	5.2-25	108-31-6	8.1-2
100-53-8	12.1-19	106-30-9	8.2-99	108-38-3	3.5-4
100-61-8	11.2-51	106-33-2	8.2-113	108-39-4	5.4-3
100-66-3	4.3-1	106-35-4	6.1-25	108-46-3	5.4-51
100-70-9	11.5-141	106-40-1	13.3-84	108-47-4	11.5-20
100-80-1	3.5-30	106-42-3	3.5-5	108-48-5	11.5-22
100-97-0	11.5-62	106-43-4	13.3-21	108-50-9	11.5-56
101-21-3	13.4-14	106-44-5	5.4-4	108-57-6	3.5-35
101-77-9	11.2-57	106-46-7	13.3-8	108-59-8	8.2-57
101-81-5	3.5-72	106-47-8	13.3-83	108-64-5	8.2-88
101-84-8	4.3-20	106-49-0	11.2-48	108-67-8	3.5-8
102-71-6	11.2-24	106-51-4	6.4-1	108-68-9	5.4-10
103-09-3	8.2-22	106-55-8	11.5-51	108-70-3	13.3-11
103-23-1	8.2-101	106-65-0	8.2-68	108-82-7	5.1-47
103-24-2	8.2-104	106-67-2	5.1-30	108-83-8	6.1-27
103-36-6	8.4-19	106-68-3	6.1-29	108-87-2	3.3-30
103-41-3	8.4-20	106-73-0	8.2-98	108-88-3	3.5-2
103-54-8	8.2-51	106-89-8	13.3-67	108-89-4	11.5-18
103-65-1	3.5-38	106-93-4	13.1-33	108-90-7	13.3-5
103-71-9	11.1-29	106-97-8	3.1-4	108-91-8	11.2-46
103-73-1	4.3-14	106-98-9	3.2-7	108-93-0	5.3-3
103-79-7	6.4-24	106-99-0	3.2-18	108-94-1	6.3-4
103-82-2	9.4-22	107-00-6	3.2-14	108-95-2	5.4-1
104-46-1	4.3-7	107-02-8	7.2-1	108-98-5	12.1-17
104-50-7	10.1-5	107-03-9	12.1-3	108-99-6	11.5-17
104-51-8	3.5-50	107-04-0	13.1-40	109-06-8	11.5-16
104-53-0	7.3-30	107-05-1	13.2-12	109-08-0	11.5-53
104-54-1	5.4-68	107-06-2	13.1-27	109-15-9	8.2-80
104-55-2	7.3-28	107-10-8	11.2-9	109-21-7	8.2-72
104-57-4	8.2-8	107-12-0	11.1-4	109-29-5	10.1-11
104-61-0	10.1-6	107-13-1	11.1-8	109-49-9	6.2-10
104-62-1	8.2-9	107-16-4	11.1-13	109-52-4	9.1-24
104-67-6	10.1-7	107-18-6	5.2-5	109-53-5	4.2-6
104-72-3	3.5-61	107-21-1	5.2-1	109-60-4	8.2-12
104-76-7	5.1-38	107-30-2	13.1-63	109-66-0	3.1-10
104-83-6	13.3-65	107-31-3	8.2-1	109-67-1	3.2-26
104-87-0	7.3-4	107-39-1	3.2-34	109-73-9	11.2-12
104-90-5	11.5-31	107-40-4	3.2-52	109-75-1	11.1-12
104-93-8	4.3-3	107-44-8	14.1-1	109-79-5	12.1-9
105-05-5	3.5-25	107-46-0	14.2-5	109-87-5	4.1-5

CAS Registry Number	Species Number	CAS Registry Number	Species Number	CAS Registry Number	Species Number
124-40-3	11.2-4	143-08-8	5.1-55	217-59-4	3.8-27
124-48-1	13.1-20	143-13-5	8.2-25	218-01-9	3.8-57
126-44-3	9.1-37	144-39-8	8.2-64	220-97-3	3.7-26
126-64-7	8.4-8	144-62-7	9.1-8	222-93-5	3.8-85
126-98-7	11.1-9	150-30-1	11.2-70	224-41-9	3.7-88
126-99-8	13.2-19	150-76-5	5.4-34	224-42-0	11.5-118
127-17-3	9.1-13	150-78-7	4.3-10	225-11-6	11.5-115
127-18-4	13.2-6	150-84-5	8.2-35	225-51-4	11.5-116
127-19-5	11.2-37	150-86-7	5.2-32	226-36-8	11.5-117
127-27-5	9.3-8	151-50-8	2.3-5	226-86-8	3.8-72
127-91-3	3.4-2	189-55-9	3.8-86	226-88-0	3.8-69
128-37-0	5.4-27	189-64-0	3.8-67	227-04-3	3.8-71
128-66-5	6.6-32	189-96-8	3.8-79	229-67-4	11.5-80
129-00-0	3.8-32	190-70-5	3.8-89	229-70-9	11.5-112
129-79-3	11.4-85	190-95-4	3.8-84	229-87-8	11.5-111
131-11-3	8.5-1	191-07-1	3.8-88	230-17-1	11.5-97
132-64-9	10.3-22	191-24-2	3.8-81	230-27-3	11.5-92
132-65-0	12.3-24	191-26-4	3.8-51	236-02-2	11.5-122
134-20-3	11.2-64	191-30-0	3.8-50	238-79-9	3.7-19
134-32-7	11.2-61	191-33-3	3.8-42	238-84-6	3.7-18
134-81-6	6.4-34	191-37-7	10.3-31	239-01-0	11.5-103
134-96-3	7.3-22	192-51-8	3.8-52	239-30-5	10.3-24
135-01-3	3.5-23	192-65-4	3.8-48	239-35-0	12.3-28
135-19-3	5.4-79	192-97-2	3.8-44	239-60-1	3.7-25
135-48-8	3.8-87	193-09-9	3.8-73	243-17-4	3.7-22
135-98-8	3.5-53	193-39-5	3.8-53	243-42-5	10.3-25
137-18-8	6.4-4	193-43-1	3.8-19	243-46-9	12.3-30
137-32-6	5.1-21	194-03-6	11.5-121	243-51-6	11.5-96
138-86-3	3.4-7	194-26-3	3.7-27	244-99-5	11.5-109
138-87-4	5.3-13	194-59-2	11.5-108	259-79-0	3.7-1
139-66-2	12.2-12	195-19-7	3.7-61	260-94-6	11.5-113
140-11-4	8.2-47	196-42-9	3.8-74	268-40-6	3.6-48
140-29-4	11.1-22	198-55-0	3.8-80	271-63-6	11.5-75
140-39-6	8.2-46	201-06-9	3.8-22	271-89-6	10.3-13
140-67-0	4.3-6	202-03-9	3.8-25	275-51-4	3.4-23
140-88-5	8.3-2	203-12-3	3.8-14	287-23-0	3.3-5
141-12-8	8.2-37	203-33-8	3.8-10	287-92-3	3.3-9
141-16-2	8.2-75	203-64-5	3.7-54	288-13-1	11.5-10
141-32-2	8.3-3	203-65-6	11.5-105	288-32-4	11.5-4
141-43-5	11.2-22	205-12-9	3.7-23	290-37-9	11.5-52
141-78-6	8.2-11	205-25-4	11.5-104	291-64-5	3.3-87
141-79-7	6.2-6	205-43-6	12.3-27	292-64-8	3.3-88
141-82-2	9.1-15	205-82-3	3.8-17	293-96-9	3.3-91
141-93-5	3.5-24	205-99-2	3.8-23	294-62-2	3.3-92
142-09-6	8.3-6	206-44-0	3.8-3	298-00-0	14.3-11
142-28-9	13.1-46	206-56-4	11.5-119	298-12-4	9.1-7
142-29-0	3.3-22	208-96-8	3.7-6	299-84-3	14.3-13
142-50-7	5.2-31	210-79-7	10.3-20	300-57-2	3.5-48
142-62-1	9.1-40	213-46-7	3.8-75	302-01-2	11.2-1
142-82-5	3.1-48	214-17-5	3.8-65	309-00-2	13.4-32
142-83-6	7.2-9	215-26-9	3.7-89	313-80-4	11.5-120
142-92-7	8.2-21	215-58-7	3.8-29	316-14-3	3.7-79
142-96-1	4.1-9	216-00-2	3.8-70	316-49-4	3.7-77
143-07-7	9.1-66	216-53-5	3.8-66	319-84-6	13.4-3

CAS Registry Number	Species Number	CAS Registry Number	Species Number	CAS Registry Number	Species Number
544-85-4	3.1-127	584-02-1	5.1-35	598-62-9	2.6-59
546-80-5	6.3-12	584-84-9	11.1-30	598-63-0	2.6-61
546-93-0	2.6-57	584-94-1	3.1-34	598-76-5	13.1-47
548-39-0	6.5-45	586-27-6	5.3-10	598-77-6	13.1-49
551-08-6	10.1-15	586-62-9	3.4-10	598-82-3	9.1-11
553-26-4	11.5-49	586-81-2	5.3-14	600-14-6	6.1-16
553-97-9	6.4-2	589-34-4	3.1-32	601-97-8	9.4-18
554-12-1	8.2-58	589-35-5	5.1-29	602-38-0	11.4-17
554-61-0	3.4-12	589-38-8	6.1-21	602-60-8	11.4-31
555-10-2	3.4-17	589-43-5	3.1-35	605-02-7	3.6-41
556-67-2	14.2-8	589-53-7	3.1-51	605-71-0	11.4-16
556-82-1	5.2-7	589-66-2	8.3-10	607-34-1	11.4-98
557-17-5	4.1-3	589-81-1	3.1-50	607-35-2	11.4-99
557-40-4	4.2-11	589-93-5	11.5-21	607-42-1	9.4-36
557-59-5	9.1-97	590-18-1	3.2-21	607-91-0	10.3-4
558-37-2	3.2-11	590-19-2	3.2-17	607-99-8	13.3-45
560-21-4	3.1-21	590-35-2	3.1-15	608-25-3	5.4-52
562-49-2	3.1-18	590-50-1	6.1-13	608-73-1	13.3-15
562-74-3	5.3-15	590-66-9	3.3-31	608-93-5	13.3-14
563-16-6	3.1-37	590-73-8	3.1-33	609-26-7	3.1-26
563-45-1	3.2-9	590-86-3	7.1-11	610-48-0	3.7-68
563-46-2	3.2-8	591-21-9	3.3-34	611-14-3	3.5-14
563-47-3	13.2-13	591-22-0	11.5-24	611-15-4	3.5-29
563-54-2	13.2-14	591-47-9	3.3-80	612-60-2	11.5-84
563-78-0	3.2-10	591-49-1	3.3-79	612-78-2	3.6-45
563-79-1	3.2-24	591-76-4	3.1-31	612-94-2	3.6-42
563-80-4	6.1-6	591-78-6	6.1-20	613-12-7	3.7-69
563-83-7	11.2-41	591-87-7	8.2-13	613-20-7	6.5-24
564-02-3	3.1-19	591-93-5	3.2-41	613-31-0	3.7-66
565-59-3	3.1-16	591-95-7	3.2-37	613-46-7	11.1-24
565-61-7	6.1-11	592-13-2	3.1-36	614-96-0	11.5-67
565-69-5	6.1-18	592-20-1	8.2-56	615-13-4	6.5-10
565-75-3	3.1-22	592-27-8	3.1-49	615-87-2	13.3-28
565-77-5	3.2-51	592-41-6	3.2-55	616-19-3	13.2-10
565-80-0	6.1-19	592-46-1	3.2-74	616-45-5	11.5-126
569-41-5	3.6-29	592-76-7	3.2-82	616-47-7	11.5-5
569-51-7	9.4-21	592-78-9	3.2-88	617-62-9	9.1-31
571-58-4	3.6-27	592-84-7	8.2-3	617-65-2	11.2-32
573-56-5	11.4-74	592-88-1	12.2-5	617-78-7	3.1-25
573-98-8	3.6-26	593-45-3	3.1-106	619-66-9	9.4-14
575-43-9	3.6-28	593-49-7	3.1-117	619-99-8	3.1-44
576-26-1	5.4-8	593-57-7	14.1-3	620-02-0	10.2-33
577-16-2	6.4-10	593-69-1	14.1-5	620-14-4	3.5-15
578-58-5	4.3-2	593-70-4	13.1-10	620-17-7	5.4-14
580-51-8	5.4-31	593-74-8	14.1-11	620-23-5	7.3-3
581-40-8	3.6-30	593-79-3	14.1-6	620-84-8	11.2-53
581-42-0	3.6-31	593-88-4	14.1-4	620-92-8	5.4-72
581-49-7	11.5-46	594-18-3	13.1-21	621-63-6	5.1-6
581-50-0	11.5-47	594-56-9	3.2-12	621-82-9	9.4-25
581-89-5	11.4-14	594-60-5	5.1-24	622-85-5	4.3-15
582-16-1	3.6-32	594-70-7	11.3-8	622-96-8	3.5-16
583-48-2	3.1-38	594-82-1	3.1-9	622-97-9	3.5-31
583-58-4	11.5-23	598-58-3	11.3-3	623-05-2	5.4-65
583-61-9	11.5-19	598-61-8	3.3-6	623-27-8	7.3-26

CAS Registry Number	Species Number	CAS Registry Number	Species Number	CAS Registry Number	Species Number
623-39-2	5.1-14	630-19-3	7.1-12	789-02-6	13.4-20
623-42-7	8.2-70	630-20-6	13.1-31	789-07-1	11.4-48
623-43-8	8.3-7	631-65-2	13.1-54	814-29-9	14.3-1
623-69-8	5.1-17	634-66-2	13.3-12	814-78-8	6.2-3
623-70-1	8.3-8	634-90-2	13.3-13	816-79-5	3.2-53
623-91-6	8.3-11	637-27-4	9.4-23	821-09-0	5.2-11
624-45-3	8.2-91	637-78-5	8.2-60	821-38-5	9.1-75
624-64-6	3.2-22	638-36-8	3.1-103	821-55-6	6.1-30
624-83-9	11.1-16	638-67-5	3.1-112	821-95-4	3.2-109
624-89-5	12.2-2	638-68-6	3.1-123	822-23-1	8.2-30
624-92-0	12.2-13	641-13-4	6.6-21	822-50-4	3.3-13
625-27-4	3.2-44	641-96-3	3.5-70	824-22-6	3.6-4
625-30-9	11.2-18	642-31-9	7.3-48	824-63-5	3.6-3
625-33-2	6.2-5	643-79-8	7.3-25	824-90-8	3.5-56
625-50-3	11.2-38	644-08-6	3.5-63	827-52-1	3.3-76
625-58-1	11.3-5	646-04-8	3.2-43	832-64-4	3.7-44
625-65-0	3.2-48	646-07-1	9.1-26	832-69-9	3.7-41
625-80-9	12.2-6	646-30-0	9.1-88	832-71-3	3.7-43
625-86-5	10.2-19	646-31-1	3.1-113	849-99-0	8.2-102
626-51-7	9.1-32	661-11-0	13.1-57	870-23-5	12.1-8
626-70-0	9.1-46	666-14-8	9.1-6	871-70-5	9.1-87
626-93-7	5.1-40	687-97-4	3.2-15	871-83-0	3.1-76
627-13-4	11.3-9	691-37-2	3.2-29	872-05-9	3.2-105
627-17-3	6.4-6	692-24-0	3.2-79	872-50-4	11.5-128
627-19-0	3.2-36	692-70-6	3.2-80	874-35-1	3.6-5
627-20-3	3.2-42	692-96-6	3.2-89	874-41-9	3.5-18
627-40-7	4.2-10	693-02-7	3.2-66	875-79-6	11.5-69
628-41-1	3.3-85	693-23-2	9.1-69	877-43-0	11.5-86
628-63-7	8.2-19	693-54-9	6.1-32	881-03-8	11.4-13
628-97-7	8.2-123	693-89-0	3.3-23	883-20-5	3.7-45
629-11-8	5.1-41	693-98-1	11.5-6	892-21-7	11.4-42
629-19-6	12.2-17	695-06-7	10.1-4	921-47-1	3.1-41
629-20-9	3.3-90	695-12-5	3.3-43	922-28-1	3.1-58
629-30-1	5.1-48	696-29-7	3.3-45	922-61-2	3.2-45
629-50-5	3.1-92	698-27-1	7.3-19	925-54-2	7.1-17
629-59-4	3.1-95	698-76-0	10.1-9	925-78-0	6.1-31
629-62-9	3.1-98	707-07-3	4.3-11	926-65-8	4.2-4
629-70-9	8.2-29	719-22-2	6.3-18	926-66-9	4.2-1
629-73-2	3.2-119	759-94-4	12.4-5	928-45-0	11.3-11
629-78-7	3.1-104	760-20-3	3.2-28	928-81-4	9.1-43
629-80-1	7.1-34	760-21-4	3.2-13	929-17-9	11.2-33
629-90-3	7.1-35	762-62-9	3.2-33	929-77-1	8.2-135
629-92-5	3.1-108	762-75-4	8.2-4	930-27-8	10.2-17
629-94-7	3.1-110	763-29-1	3.2-27	930-61-0	11.5-9
629-97-0	3.1-111	763-93-9	6.2-9	930-68-7	6.3-7
629-99-2	3.1-114	764-41-0	13.2-18	933-67-5	11.5-68
630-01-3	3.1-115	764-93-2	3.2-107	933-98-2	3.5-17
630-02-4	3.1-119	765-43-5	6.3-1	934-74-7	3.5-22
630-03-5	3.1-121	767-58-8	3.6-2	934-80-5	3.5-21
630-04-6	3.1-125	767-59-9	3.6-11	935-92-2	6.4-5
630-05-7	3.1-129	767-60-2	3.6-12	937-30-4	6.4-13
630-06-8	3.1-132	768-03-6	6.4-23	939-27-5	3.6-37
630-07-9	3.1-131	779-02-2	3.7-70	949-87-1	11.2-71
630-08-0	2.6-6	781-43-1	3.7-71	970-06-9	3.6-44

CAS Registry Number	Species Number
992-94-9	14.2-1
999-40-6	8.2-76
1000-86-8	3.2-40
1000-87-9	3.2-54
1002-43-3	3.1-89
1002-69-3	13.1-58
1002-84-2	9.1-76
1003-38-9	10.2-9
1008-88-4	11.5-38
1013-08-7	3.7-36
1022-22-6	13.4-26
1024-57-3	13.4-29
1066-40-6	14.2-3
1067-08-9	3.1-27
1067-20-5	3.1-29
1069-53-0	3.1-42
1070-66-2	7.2-7
1071-26-7	3.1-52
1071-81-4	3.1-43
1072-05-5	3.1-56
1072-21-5	7.1-22
1074-17-5	3.5-39
1074-43-7	3.5-40
1074-55-1	3.5-41
1075-38-3	3.5-55
1077-16-3	3.5-58
1079-71-6	3.7-65
1081-75-0	3.5-75
1081-77-2	3.5-60
1090-13-7	6.6-33
1118-58-7	3.2-39
1119-14-8	3.2-63
1119-40-0	8.2-84
1120-21-4	3.1-87
1120-28-1	8.2-132
1120-36-1	3.2-116
1120-62-3	3.3-24
1121-55-7	11.5-40
1124-39-6	5.4-47
1126-61-0	5.4-49
1127-76-0	3.6-36
1131-62-0	6.4-15
1135-24-6	9.4-28
1162-65-8	10.3-32
1166-18-3	3.5-71
1186-53-4	3.1-24
1188-02-9	9.1-50
1191-25-9	9.1-42
1191-99-7	10.2-13
1192-18-3	3.3-12
1192-31-0	10.2-6
1195-32-0	3.5-32
1195-79-5	6.3-10
1196-79-8	11.5-71
1198-37-4	11.5-85
1231-35-2	9.3-4
1242-76-8	3.8-78
1242-77-9	3.8-77
1289-45-8	9.2-7
1302-27-8	2.7-4
1303-18-0	2.4-8
1303-86-2	2.6-5
1304-29-6	2.6-46
1304-56-9	2.6-4
1304-76-3	2.6-54
1305-78-8	2.6-20
1306-05-4	2.5-25
1306-19-0	2.6-41
1306-41-8	2.7-21
1308-04-9	2.6-32
1308-14-1	2.6-70
1308-38-9	2.6-24
1309-36-0	2.4-5
1309-37-1	2.6-29
1309-48-4	2.6-12
1309-60-0	2.6-52
1309-64-4	2.6-44
1310-14-1	2.7-9
1313-13-9	2.6-26
1313-27-5	2.6-40
1313-59-3	2.6-11
1313-82-2	2.4-4
1313-99-1	2.6-33
1314-13-2	2.6-35
1314-41-6	2.6-53
1314-56-3	2.6-18
1314-62-1	2.6-23
1314-87-0	2.4-7
1314-98-3	2.4-6
1317-35-7	2.6-27
1317-36-8	2.6-51
1317-38-0	2.6-34
1317-61-9	2.6-31
1317-63-1	2.7-18
1317-71-1	2.7-24
1318-59-8	2.7-5
1318-74-7	2.7-16
1318-93-0	2.7-22
1318-94-1	2.7-23
1320-18-9	13.4-12
1321-64-8	13.3-38
1322-06-1	5.4-29
1327-41-9	2.5-18
1327-53-3	2.6-38
1331-43-7	3.3-42
1333-47-7	10.3-9
1333-52-4	6.5-20
1333-74-0	2.2-1
1334-76-5	8.4-3
1335-08-6	9.4-6
1335-87-1	13.3-39
1336-21-6	2.3-6
1341-24-8	13.3-55
1344-28-1	2.6-13
1344-43-0	2.6-25
1345-25-1	2.6-28
1430-97-3	3.7-11
1455-20-5	12.3-11
1463-17-8	11.5-87
1487-15-6	10.2-14
1487-18-9	10.2-26
1504-74-1	7.3-29
1529-17-5	14.2-4
1531-37-9	11.1-17
1551-27-5	12.3-8
1552-67-6	8.3-17
1559-81-5	3.6-18
1560-42-5	3.1-102
1560-72-1	3.1-124
1560-75-4	3.1-122
1560-88-9	3.1-107
1560-89-0	3.1-105
1560-93-6	3.1-99
1560-95-8	3.1-96
1560-96-9	3.1-93
1560-97-0	3.1-91
1560-98-1	3.1-120
1561-00-8	3.1-118
1561-02-0	3.1-116
1561-49-5	8.5-25
1563-66-2	11.2-68
1570-48-5	11.5-130
1576-67-6	3.7-48
1582-09-8	13.4-15
1589-49-7	5.1-12
1599-67-3	3.2-128
1599-68-4	3.2-127
1604-11-1	8.2-69
1606-30-0	11.3-14
1606-31-1	11.3-13
1611-21-8	3.3-82
1613-34-9	11.5-89
1629-58-9	6.2-8
1632-16-2	3.2-61
1632-73-1	5.3-8
1633-05-2	2.6-60
1633-89-2	12.1-13
1633-97-2	12.1-14
1634-04-4	4.1-4
1636-27-7	9.1-16
1638-26-2	3.3-11
1640-89-7	3.3-18

CAS Registry Number	Species Number	CAS Registry Number	Species Number	CAS Registry Number	Species Number
3131-63-3	10.3-18	3978-81-2	11.5-35	5315-79-7	5.4-91
3133-01-5	5.1-68	4013-37-0	4.3-19	5363-64-4	4.2-7
3155-42-8	9.1-86	4030-18-6	11.5-125	5379-20-4	3.5-33
3161-99-7	3.2-81	4032-86-4	3.1-57	5392-40-5	7.2-16
3170-83-0	2.2-7	4038-04-4	3.2-35	5405-53-8	11.4-24
3178-22-1	3.3-50	4050-45-7	3.2-69	5502-88-5	3.3-83
3178-29-8	3.1-65	4107-64-6	11.1-28	5522-43-0	11.4-47
3218-36-8	7.3-34	4130-42-1	5.4-28	5617-41-4	3.3-53
3221-61-2	3.1-67	4159-29-9	8.4-12	5622-73-1	12.2-9
3229-98-9	8.5-24	4161-60-8	6.1-14	5623-32-5	6.5-72
3268-87-9	13.3-82	4164-28-7	11.3-33	5656-82-6	10.3-21
3274-56-4	11.5-3	4170-30-3	7.2-3	5676-32-4	3.5-36
3351-28-8	3.8-58	4181-12-8	4.2-2	5737-13-3	6.5-48
3351-30-2	3.8-61	4181-95-7	3.1-136	5743-97-5	3.7-28
3351-31-3	3.8-60	4206-58-0	7.3-32	5746-02-1	9.1-29
3351-32-4	3.8-59	4218-48-8	3.5-47	5756-24-1	12.2-33
3352-57-6	2.2-6	4221-03-8	7.1-14	5796-89-4	11.3-10
3353-12-6	3.8-35	4221-98-1	3.4-16	5802-82-4	8.2-137
3358-28-9	10.2-10	4250-38-8	9.1-104	5910-85-0	7.2-12
3387-41-5	3.4-15	4265-25-2	10.3-14	5910-89-4	11.5-54
3391-86-4	5.2-21	4292-75-5	3.3-52	5911-04-6	3.1-77
3404-62-4	3.2-71	4292-92-6	3.3-51	5954-72-3	12.1-11
3404-63-5	3.2-20	4312-99-6	6.2-12	5989-27-5	3.4-6
3404-72-6	3.2-30	4316-65-8	3.2-60	6006-33-3	3.3-59
3404-78-2	3.2-73	4318-57-9	3.3-86	6006-95-7	3.3-61
3424-82-6	13.4-25	4359-57-3	7.1-26	6028-61-1	12.2-31
3442-78-2	3.8-34	4437-22-3	10.2-36	6032-29-7	5.1-33
3452-07-1	3.2-126	4443-55-4	3.3-65	6040-04-6	9.4-34
3452-09-3	3.2-102	4461-48-7	3.2-46	6051-98-5	6.5-36
3522-94-9	3.1-40	4489-61-6	8.5-7	6066-49-5	10.1-14
3524-73-0	3.2-58	4536-23-6	9.1-41	6068-69-5	11.2-52
3558-60-9	4.3-18	4558-16-1	6.6-16	6094-02-6	3.2-56
3586-69-4	11.4-30	4588-18-5	3.2-93	6117-91-5	5.2-6
3600-24-6	12.2-29	4599-92-2	6.5-53	6263-65-6	12.1-20
3625-52-3	9.1-107	4599-94-4	6.5-38	6443-92-1	3.2-86
3638-35-5	3.3-4	4602-84-0	5.2-30	6484-52-2	2.3-19
3648-21-3	8.5-11	4606-15-9	8.2-50	6531-35-7	6.5-63
3658-80-8	12.2-28	4673-31-8	11.5-34	6703-99-7	3.3-66
3674-66-6	3.7-47	4706-90-5	3.5-46	6728-26-3	7.2-8
3697-24-3	3.8-62	4712-34-9	9.2-13	6750-03-4	7.2-17
3710-42-7	9.1-51	4746-86-5	7.1-4	6753-98-6	3.4-25
3710-43-8	10.2-18	4754-26-1	6.4-7	6765-39-5	3.2-121
3724-65-0	9.2-2	4766-40-9	3.8-24	6812-38-0	3.3-62
3726-46-3	3.3-19	4786-20-3	11.1-10	6874-29-9	3.2-94
3726-47-4	3.3-20	4806-61-5	3.3-7	6876-23-9	3.3-33
3734-49-4	13.4-31	4812-22-0	11.3-16	6899-04-3	11.2-34
3748-83-2	11.5-27	4861-58-9	12.3-12	6915-15-7	9.1-21
3752-42-9	11.1-27	4904-61-4	3.3-93	6915-18-0	9.2-5
3769-23-1	3.2-57	4912-92-9	3.6-6	6917-35-7	5.3-6
3772-55-2	5.4-89	5084-46-8	7.3-38	6929-04-0	8.2-125
3777-69-3	10.2-25	5095-43-2	7.2-10	6962-60-3	6.4-32
3782-00-1	10.3-17	5129-60-2	8.2-121	6975-92-4	3.2-59
3812-32-6	2.6-10	5129-61-3	8.2-127	6975-98-0	3.1-83
3910-35-8	3.6-9	5264-33-5	11.1-14	6994-89-4	5.2-27

CAS Registry Number	Species Number	CAS Registry Number	Species Number	CAS Registry Number	Species Number
12648-30-5	2.6-50	14807-96-6	2.7-29	18281-05-5	8.2-133
12680-02-3	2.6-47	14808-79-8	2.4-23	18282-10-5	2.6-43
12768-75-1	2.1-2	14850-23-8	3.2-99	18294-85-4	9.1-48
12789-03-6	13.4-30	14871-68-2	2.4-34	18432-25-2	2.4-49
13019-22-2	5.2-29	14989-30-1	2.5-1	18435-22-8	3.1-97
13073-29-5	11.4-72	15110-74-4	11.4-23	18435-45-5	3.2-125
13129-35-6	12.1-21	15181-46-1	2.4-13	18435-53-5	3.2-136
13151-27-4	3.2-106	15281-98-8	2.6-63	18435-54-6	3.2-137
13151-34-3	3.1-84	15306-27-1	3.2-133	18435-55-7	3.2-138
13151-35-4	3.1-86	15502-74-6	2.6-36	18472-36-1	5.3-29
13153-14-5	7.2-5	15584-04-0	2.6-37	18516-37-5	3.2-110
13177-28-1	11.4-40	15594-90-8	5.1-66	18647-78-4	5.3-17
13177-29-2	11.4-41	15595-02-5	3.7-87	18667-07-7	9.1-45
13177-31-6	11.4-43	15668-97-0	2.4-33	18669-52-8	3.2-64
13177-32-7	11.4-44	15679-24-0	3.8-36	18787-63-8	6.1-36
13190-97-1	11.5-124	15840-60-5	3.2-77	18794-77-9	12.3-14
13201-46-2	9.2-3	15869-80-4	3.1-61	18804-49-4	3.2-91
13209-09-1	11.4-38	15869-85-9	3.1-79	18835-32-0	3.2-129
13269-52-8	3.2-78	15869-89-3	3.1-71	18835-33-1	3.2-132
13366-73-9	13.4-35	15870-10-7	3.2-83	18835-34-2	3.2-134
13389-42-9	3.2-98	15887-88-4	2.5-32	18835-35-3	3.2-135
13392-69-3	9.1-27	16056-34-1	14.1-9	18936-17-9	11.1-5
13397-24-5	2.4-44	16135-81-2	3.8-13	19020-26-9	11.5-90
13463-39-3	2.6-65	16389-88-1	2.7-8	19031-80-2	11.3-15
13463-67-7	2.6-21	16434-59-6	3.7-74	19060-13-0	11.1-15
13466-78-9	3.4-11	16435-49-7	3.2-112	19345-99-4	3.7-7
13475-78-0	3.1-63	16651-91-5	2.5-34	19353-76-5	3.7-52
13494-91-2	2.4-60	16747-26-5	3.1-39	19407-17-1	3.7-32
13597-44-9	2.4-19	16747-33-4	3.1-28	19407-18-2	3.7-33
13601-88-2	7.3-45	16887-00-6	2.1-10	19407-28-4	3.7-35
13717-92-5	11.5-61	16961-83-4	2.5-11	19407-37-5	9.3-6
13775-30-9	2.4-35	16980-85-1	3.2-131	19694-02-1	9.4-37
13775-53-6	2.5-19	16984-48-8	2.1-4	19700-21-1	5.3-19
13778-36-4	2.5-35	17024-18-9	11.4-25	19781-73-8	3.3-63
13814-87-4	2.4-59	17030-74-9	7.1-2	19876-64-3	5.1-31
13826-67-0	2.4-28	17059-52-8	10.3-15	20074-52-6	2.1-14
13849-96-2	3.4-45	17068-62-1	2.7-11	20184-91-2	3.2-104
13891-87-7	6.2-7	17179-91-8	9.1-33	20268-51-3	11.4-37
13925-00-3	11.5-58	17278-74-9	9.1-74	20296-29-1	5.1-54
13952-84-6	11.2-15	17301-94-9	3.1-78	20333-39-5	12.2-14
13983-17-0	2.7-31	17302-23-7	3.1-81	20395-23-7	13.2-22
14035-94-0	8.2-85	17302-27-1	3.1-80	20485-57-8	3.8-8
14127-61-8	2.1-13	17341-25-2	2.1-6	20490-42-0	6.5-22
14167-59-0	3.1-130	17605-67-3	5.3-28	20589-63-3	11.4-63
14168-73-1	2.4-41	17618-77-8	3.2-70	20589-85-9	13.2-16
14265-44-2	2.6-17	17619-36-2	12.2-30	20615-64-9	7.3-44
14280-50-3	2.1-22	17640-08-3	13.1-67	20633-03-8	6.1-22
14287-61-7	9.1-19	17656-09-6	11.1-11	21228-90-0	12.2-38
14667-55-1	11.5-57	17778-80-2	2.1-3	21444-04-2	9.2-10
14686-13-6	3.2-87	17836-08-7	5.1-2	21645-51-2	2.6-14
14701-22-5	2.1-15	18060-77-0	8.3-9	21651-19-4	2.6-42
14797-55-8	2.3-11	18094-01-4	3.2-115	22054-14-3	7.1-24
14797-65-0	2.3-9	18252-79-4	2.6-22	22089-69-6	14.1-8
14798-03-9	2.3-3	18281-04-4	8.2-131	22104-78-5	5.2-18

CAS Registry Number	Species Number
31551-45-8	11.4-84
32021-53-7	11.4-67
32021-54-8	11.4-71
32073-03-3	3.2-96
32074-25-2	11.1-20
32368-69-7	11.4-97
32391-38-1	5.4-35
33081-34-4	10.2-11
33081-35-5	10.2-12
33543-31-6	3.8-5
34006-76-3	8.5-2
34067-76-0	7.1-20
34135-85-8	12.2-32
34284-35-0	9.1-56
34322-84-0	6.5-3
35241-40-8	3.4-37
35788-00-2	2.4-18
35913-09-8	13.3-64
35923-65-0	9.1-52
35976-82-0	12.2-7
36541-21-6	3.7-3
36617-18-2	8.2-114
36643-74-0	5.4-75
36729-58-5	5.1-57
36734-19-7	13.4-36
36854-53-2	6.2-4
36884-28-3	7.3-33
36889-28-8	12.2-24
36925-31-2	11.4-29
37062-82-1	11.5-93
37480-21-0	5.4-76
38232-01-8	9.1-106
38232-03-0	9.1-108
38232-04-1	9.1-109
38743-20-3	7.2-15
38888-98-1	3.5-73
38998-75-3	13.3-74
39001-02-0	13.3-75
39255-32-8	8.2-90
39379-95-8	3.8-15
39407-42-6	6.6-1
39437-98-4	13.2-17
39461-53-5	6.6-11
39723-60-9	11.5-60
40356-65-8	11.5-8
40463-15-8	3.6-20
40529-66-6	3.5-65
40589-14-8	5.1-56
40868-73-3	5.1-39
41593-21-9	3.7-9
41593-22-0	3.8-1
41593-24-2	3.8-2
41593-27-5	3.7-24
41593-29-7	3.8-55

CAS Registry Number	Species Number
41593-31-1	3.8-56
41637-86-9	3.7-72
41637-89-2	3.8-28
41637-92-7	3.8-64
41637-94-9	3.8-12
41699-04-1	3.8-45
41699-09-6	3.8-82
42027-23-6	5.1-34
42135-22-8	11.4-82
42140-26-1	12.3-16
42231-82-3	8.2-115
42296-74-2	3.2-62
42397-64-8	11.4-54
42397-65-9	11.4-55
42441-75-8	3.2-84
42523-54-6	6.5-31
42764-74-9	3.2-122
43022-03-3	6.1-2
43047-99-0	13.3-69
43048-00-6	13.3-70
43145-54-6	7.3-9
43211-62-7	8.2-124
45043-50-3	11.1-18
49612-49-9	10.2-37
50306-14-4	5.2-24
50306-18-8	5.2-22
50764-83-5	11.4-102
50820-24-1	2.4-17
50984-52-6	7.3-16
51001-44-6	3.8-16
51013-18-4	11.5-127
51366-52-0	11.5-88
51855-29-9	3.6-19
52006-63-0	12.3-7
52006-64-1	12.3-31
52428-35-0	3.7-90
52554-38-3	3.2-117
52642-16-7	11.5-14
52645-53-1	13.4-27
52783-43-4	5.1-64
52783-44-5	5.1-62
52783-45-6	5.1-69
52896-91-0	3.1-62
52896-99-8	3.1-47
53123-74-3	11.5-26
53185-69-6	7.1-25
53197-58-3	11.4-83
53223-75-9	6.5-41
53452-81-6	13.3-42
53742-07-7	13.3-34
53778-43-1	3.3-3
53951-50-1	7.3-7
54105-66-7	3.3-57
54116-90-4	11.5-114

CAS Registry Number	Species Number
54598-10-6	8.4-1
54751-98-3	11.5-123
54811-53-9	3.7-20
54833-05-5	13.1-50
54886-36-1	13.3-87
55124-97-5	8.2-130
55320-57-5	7.1-18
55345-04-5	11.4-21
55599-95-6	10.3-10
55684-94-1	13.3-73
55775-16-1	3.8-30
55880-77-8	13.2-20
56908-81-7	3.6-35
57652-57-0	6.6-12
57706-44-2	3.7-31
57765-50-1	6.4-26
57835-92-4	11.4-50
58037-87-9	3.4-14
58207-38-8	2.6-64
58429-99-5	3.7-84
58548-39-3	3.7-55
58924-35-9	3.6-14
59173-59-0	10.2-34
59534-35-9	5.4-78
60584-82-9	3.6-8
60826-62-2	10.3-23
60826-74-6	3.8-9
60848-01-3	6.5-71
61128-87-8	4.3-24
61828-07-7	3.3-67
61828-08-8	3.3-68
61828-09-9	3.3-69
61828-10-2	3.3-70
61828-11-3	3.3-71
61828-12-4	3.3-72
61828-13-5	3.3-73
61828-14-6	3.3-74
61828-15-7	3.3-75
61868-19-7	3.2-120
61868-20-0	3.2-124
61878-57-7	13.3-23
62016-31-3	3.1-73
62016-41-5	3.1-74
62549-24-0	8.2-74
62643-46-3	9.1-90
62656-49-9	12.3-19
62708-42-3	7.3-11
62716-20-5	6.5-51
63041-47-4	6.5-37
63041-90-7	11.4-59
63311-56-8	2.4-46
63449-79-6	6.4-28
63597-41-1	3.2-95
64503-02-2	3.8-83

INDEX 4

Sources of Atmospheric Compounds

This index is derived from the tables in Chapters 2-14. It presents the source information contained therein in a format cross-indexed by source name instead of by compound. Each source name in this index is followed by one or more species numbers. For "air conditioning", for example, three species numbers are listed. Reference to Table 13.1 indicates that the compounds emitted into the atmosphere as a result of air conditioner operation and maintenance are dichlorofluoromethane (13.1-12), trichlorofluoromethane (13.1-14), and dichlorodifluoromethane (13.1-15). Similar analyses may be performed for sources of interest to a particular user.

As with other data in this book, that in this index should not be regarded as complete or perfectly accurate. It represents, rather, the information available when this compilation was performed. Sources that have been studied intensively for one reason or another will thus be more completely described than those which have received less attention. Nonetheless, considerable chemical emissions information is available from this listing. Some of the implications of these data from a broader perspective are discussed in Section 15.3.

Abbreviations used in this index are as follows: calc., calcium; comb., combustion; cond., conditioning; mfr., manufacturing; mtls., materials; proc., processing; stor., storage; xch., exchange.

INDEX 5

Subjects

A

B

C

D

E

F

G

H

I

K

L

M

N